Cambridge Studies in Advanced Mathematics

Commutative ring theory

Already published

1 W.M.L. Holcombe *Algebraic automata theory*
2 K. Petersen *Ergodic theory*
3 P.T. Johnstone *Stone spaces*
4 W.H. Schikhof *Ultrametric calculus*
5 J.-P. Kahane *Some random series of functions, 2nd edition*
6 H. Cohn *Introduction to the construction of class fields*
7 J. Lambek & P.J. Scott *Introduction to higher-order categorical logic*
8 H. Matsumura *Commutative ring theory*
9 C.B. Thomas *Characteristic classes and the cohomology of finite groups*
10 M. Aschbacher *Finite group theory*
11 J.L. Alperin *Local representation theory*
12 P. Koosis *The logarithmic integral I*
13 A. Pietsch *Eigenvalues and s-numbers*
14 S.J. Patterson *An introduction to the theory of the Riemann zeta-function*
15 H.J. Baues *Algebraic homotopy*
16 V.S. Varadarajan *Introduction to harmonic analysis on semisimple Lie groups*
17 W. Dicks & M. Dunwoody *Groups acting on graphs*
18 L.J. Corwin & F.P. Greenleaf *Representations of nilpotent Lie groups and their applications*
19 R. Fritsch & R. Piccinini *Cellular structures in topology*
20 H Klingen *Introductory lectures on Siegel modular forms*
22 M.J. Collins *Representations and characters of finite groups*
24 H. Kunita *Stochastic flows and stochastic differential equations*
25 P. Wojtaszczyk *Banach spaces for analysts*
26 J.E. Gilbert & M.A.M. Murray *Clifford algebras and Dirac operators in harmonic analysis*
27 A. Frohlich & M.J. Taylor *Algebraic number theory*
28 K. Goebel & W.A. Kirk *Topics in metric fixed point theory*
29 J.F. Humphreys *Reflection groups and Coxeter groups*
30 D.J. Benson *Representations and cohomology I*
31 D.J. Benson *Representations and cohomology II*
32 C. Allday & V. Puppe *Cohomological methods in transformation groups*
33 C. Soulé et al *Lectures on Arakelov geometry*
34 A. Ambrosetti & G. Prodi *A primer of nonlinear analysis*
35 J. Palis & F. Takens *Hyperbolicity, stability and chaos at homoclinic bifurcations*
37 Y. Meyer *Wavelets and operators I*
38 C. Weibel *An introduction to homological algebra*
39 W. Bruns & J. Herzog *Cohen-Macaulay rings*
40 V. Snaith *Explicit Brauer induction*
41 G. Laumon *Cohomology of Drinfield modular varieties I*
42 E.B. Davies *Spectral theory and differential operators*
43 J. Diestel, H. Jarchow & A. Tonge *Absolutely summing operators*
44 P. Mattila *Geometry of sets and measures in Euclidean spaces*
45 R. Pinsky *Positive harmonic functions and diffusion*
46 G. Tenenbaum *Introduction to analytic and probabilistic number theory*
47 C. Peskine *An algebraic introduction to complex projective geometry I*
50 I. Porteous *Clifford algebras and the classical groups*
51 M. Audin *Spinning Tops*
54 J. Le Potier *Lectures on Vector bundles*

Commutative ring theory

HIDEYUKI MATSUMURA
Department of Mathematics, Faculty of Sciences
Nagoya University, Nagoya, Japan

Translated by M. Reid

CAMBRIDGE
UNIVERSITY PRESS

CAMBRIDGE UNIVERSITY PRESS
Cambridge, New York, Melbourne, Madrid, Cape Town, Singapore, São Paulo, Delhi

Cambridge University Press
The Edinburgh Building, Cambridge CB2 8RU, UK

Published in the United States of America by Cambridge University Press, New York

www.cambridge.org
Information on this title: www.cambridge.org/9780521367646

Originally published in Japanese as *Kakan kan ron,*
Kyoritsu Shuppan Kabushiki Kaisha, Kyoritsu texts on Modern Mathematics, 4,
Tokyo, 1980 and © H. Matsumura, 1980.
English translation © Cambridge University Press 1986

First published in English by Cambridge University Press 1986 as
Commutative ring theory
First paperback edition (with corrections) 1989
Ninth printing 2006

A catalogue record for this publication is available from the British Library

Library of Congress Cataloguing in Publication data
Matsumura, Hideyuki, 1930–
Commutative ring theory.
Translation of: Kakan kan ron.
Includes index.
1. Cummutative rings. I. Title.
QA251.3.M37213 1986 512′.4 86-11691

ISBN 978-0-521-36764-6 paperback

Transferred to digital printing 2008

Contents

Preface

In publishing this English edition I have tried to make a rather extensive revision. Most of the mistakes and insufficiencies in the original edition have, I hope, been corrected, and some theorems have been improved. Some topics have been added in the form of Appendices to individual sections. Only Appendices A, B and C are from the original. The final section, §33, of the original edition was entitled 'Kunz' Theorems' and did not substantially differ from a section in the second edition of my previous book Commutative Algebra (Benjamin, 2nd edn 1980), so I have replaced it by the present §33. The bibliography at the end of the book has been considerably enlarged, although it is obviously impossible to do justice to all of the ever-increasing literature.

Dr Miles Reid has done excellent work of translation. He also pointed out some errors and proposed some improvements. Through his efforts this new edition has become, I believe, more readable than the original. To him, and to the staff of Cambridge University Press and Kyoritsu Shuppan Co., Tokyo, who cooperated to make the publication of this English edition possible, I express here my heartfelt gratitude.

Hideyuki Matsumura
Nagoya

Preface

In publishing this English edition I have tried to make a rather extensive revision. Most of the mistakes and insufficiencies in the original edition have, I hope, been corrected, and some theorems have been improved. Some topics have been added in the form of Appendices to individual sections. Only Appendices A, B and C are from the original Japanese edition. Section §27 of the original edition was entitled 'Kunz' Theorems' and did not substantially differ from a section in the second edition of my previous book *Commutative Algebra* (Benjamin/Cummings 1980), so I have replaced it by the present §27. The bibliography at the end of the book has been considerably enlarged, although it is obviously impossible to do justice to all of the ever-increasing literature.

Dr Miles Reid has done much more than just translation; he also corrected some errors and proposed some improvements. Through his efforts this new edition has become I am sure, more readable than the original. To him, and to the staff of Cambridge University Press, and Kyoritsu Shuppan Co., Tokyo, who cooperated to make the publication of this English edition possible, I express here my heartfelt gratitude.

Hideyuki Matsumura

Nagoya

Introduction

In addition to being a beautiful and deep theory in its own right, commutative ring theory is important as a foundation for algebraic geometry and complex analytic geometry. Let us start with a historical survey of its development.

The most basic commutative rings are the ring $\mathbb{Z}$ of rational integers, and the polynomial rings over a field. $\mathbb{Z}$ is a principal ideal ring, and so is too simple to be ring-theoretically very interesting, but it was in the course of studying its extensions, the rings of integers of algebraic number fields, that Dedekind first introduced the notion of an ideal in the 1870s. For it was realised that only when prime ideals are used in place of prime numbers do we obtain the natural generalisation of the number theory of $\mathbb{Z}$.

Meanwhile, in the second half of the 19th century, polynomial rings gradually came to be studied both from the point of view of algebraic geometry and of invariant theory. In his famous papers of the 1890s on invariants, Hilbert proved that ideals in polynomial rings are finitely generated, as well as other fundamental theorems. After the turn of the present century had seen the deep researches of Lasker and Macaulay on primary decomposition of polynomial ideals came the advent of the age of abstract algebra. A forerunner of the abstract treatment of commutative ring theory was the Japanese Shōzō Sono (On congruences, I–IV, *Mem. Coll. Sci. Kyoto*, **2** (1917), **3** (1918–19)); in particular he succeeded in giving an axiomatic characterisation of Dedekind rings. Shortly after this Emmy Noether discovered that primary decomposition of ideals is a consequence of the ascending chain condition (1921), and gave a different system of axioms for Dedekind rings (1927), in work which was to have a decisive influence on the direction of subsequent development of commutative ring theory. The central position occupied by Noetherian rings in commutative ring theory became evident from her work.

However, the credit for raising abstract commutative ring theory to a substantial branch of science belongs in the first instance to Krull (1899–1970). In the 1920s and 30s he established the dimension theory of Noetherian rings, introduced the methods of localisation and completion, and the notion of a regular local ring, and went beyond the framework of Noetherian rings to create the theory of general valuation rings and Krull rings. The contribution of Akizuki in the 1930s was also considerable; in particular, a counter-example, which he obtained after a year's hard struggle, of an integral domain whose integral closure is not finite as a module was to become the model for many subsequent counter-examples.

In the 1940s Krull's theory was applied to algebraic geometry by Chevalley and Zariski, with remarkable success. Zariski applied general valuation theory to the resolution of singularities and the theory of birational transformations, and used the notion of regular local ring to give an algebraic formulation of the theory of simple (non-singular) point of a variety. Chevalley initiated the theory of multiplicities of local rings, and applied it to the computation of intersection multiplicities of varieties. Meanwhile, Zariski's student I.S. Cohen proved the structure theorem for complete local rings [1], underlining the importance of completion.

The 1950s opened with the profound work of Zariski on the problem of whether the completion of a normal local ring remains normal (Sur la normalité analytique des variétés normales, *Ann. Inst. Fourier* **2** (1950)), taking Noetherian ring theory from general theory deeper into precise structure theorems. Multiplicity theory was given new foundations by Samuel and Nagata, and became one of the powerful tools in the theory of local rings. Nagata, who was the most outstanding research worker of the 1950s, also created the theory of Hensel rings, constructed examples of non-catenary Noetherian rings and counter-examples to Hilbert's 14th problem, and initiated the theory of Nagata rings (which he called pseudo-geometric rings). Y. Mori carried out a deep study of the integral closure of Noetherian integral domains.

However, in contrast to Nagata and Mori's work following the Krull tradition, there was at the same time a new and completely different movement, the introduction of homological algebra into commutative ring theory by Auslander and Buchsbaum in the USA, Northcott and Rees in Britain, and Serre in France, among others. In this direction, the theory of regular sequences and depth appeared, giving a new treatment of Cohen–Macaulay rings, and through the homological characterisation of regular local rings there was dramatic progress in the theory of regular local rings.

The early 1960s saw the publication of Bourbaki's *Algèbre commutative*, which emphasised flatness, and treated primary decomposition from a new angle. However, without doubt, the most characteristic aspect of this decade was the activity of Grothendieck. His scheme theory created a fusion of commutative ring theory and algebraic geometry, and opened up ways of applying geometric methods in ring theory. His local cohomology is an example of this kind of approach, and has become one of the indispensable methods of modern commutative ring theory. He also initiated the theory of Gorenstein rings. In addition, his systematic development, in Chapter IV of EGA, of the study of formal fibres, and the theory of excellent rings arising out of it, can be seen as a continuation and a final conclusion of the work of Zariski and Nagata in the 1950s.

In the 1960s commutative ring theory was to receive another two important gifts from algebraic geometry. Hironaka's great work on the resolution of singularities [1] contained an extremely original piece of work within the ideal theory of local rings, the ring-theoretical significance of which is gradually being understood. The theorem on resolution of singularities has itself recently been used by Rotthaus in the study of excellent rings. Secondly, in 1969 M. Artin proved his famous approximation theorem; roughly speaking, this states that if a system of simultaneous algebraic equations over a Hensel local ring A has a solution in the completion $\hat{A}$, then there exist arbitrarily close solutions in A itself. This theorem has a wide variety of applications both in algebraic geometry and in ring theory. A new homology theory of commutative rings constructed by M. André and Quillen is a further important achievement of the 1960s.

The 1970s was a period of vigorous research in homological directions by many workers. Buchsbaum, Eisenbud, Northcott and others made detailed studies of properties of complexes, while techniques discovered by Peskine and Szpiro [1] and Hochster [H] made ingenious use of the Frobenius map and the Artin approximation theorem. Cohen–Macaulay rings, Gorenstein rings, and most recently Buchsbaum rings have been studied in very concrete ways by Hochster, Stanley, Kei-ichi Watanabe and S. Goto among others. On the other hand, classical ideal theory has shown no sign of dying off, with Ratliff and Rotthaus obtaining extremely deep results.

To give the three top theorems of commutative ring theory in order of importance, I have not much doubt that Krull's dimension theorem (Theorem 13.5) has pride of place. Next perhaps is I.S. Cohen's structure theorem for complete local rings (Theorems 28.3, 29.3 and 29.4). The fact that a complete local ring can be expressed as a quotient of a well-understood ring, the formal power series ring over a field or a discrete valuation ring, is something to feel extremely grateful for. As a third, I would give Serre's characterisation of a regular local ring (Theorem 19.2); this grasps the essence of regular local rings, and is also an important meeting-point of ideal theory and homological algebra.

This book is written as a genuine textbook in commutative algebra, and is as self-contained as possible. It was also the intention to give some thought to the applications to algebraic geometry. However, both for reasons of space and limited ability on the part of the author, we are not able to touch on local cohomology, or on the many subsequent results of the cohomological work of the 1970s. There are readable accounts of these subjects in [G6] and [H], and it would be useful to read these after this book.

This book was originally to have been written by my distinguished friend Professor Masao Narita, but since his tragic early death through illness, I have taken over from him. Professor Narita was an exact contemporary of mine, and had been a close friend ever since we met at the age of 24. Well-respected and popular with all, he was a man of warm character, and it was a sad loss when he was prematurely called to a better place while still in his forties. Believing that, had he written the book, he would have included topics which were characteristic of him, UFDs, Picard groups, and so on, I have used part of his lectures in §20 as a memorial to him. I could wish for nothing better than to present this book to Professor Narita and to hear his criticism.

Hideyuki Matsumura
Nagoya

Conventions and terminology

(1) Some basic definitions are given in Appendixes A–C. The index contains references to all definitions, including those of the appendixes.

(2) In this book, by a *ring* we always understand a commutative ring with unit; ring homomorphisms $A \longrightarrow B$ are assumed to take the unit element of A into the unit element of B. When we say that A is a subring of B it is understood that the unit elements of A and B coincide.

(3) If $f: A \longrightarrow B$ is a ring homomorphism and J is an ideal of B, then $f^{-1}(J)$ is an ideal of B, and we denote this by $A \cap J$; if A is a subring of B and f is the inclusion map then this is the same as the usual set-theoretic notion of intersection. In general this is not true, but confusion does not arise.

Moreover, if I is an ideal of A, we will write IB for the ideal $f(I)B$ of B.

(4) If A is a ring and $a_1, \ldots, a_n$ elements of A, the ideal of A generated by these is written in any of the following ways: $a_1A + a_2A + \cdots + a_nA$, $\sum a_iA$, $(a_1, \ldots, a_n)$ or $(a_1, \ldots, a_n)A$.

(5) The sign $\subset$ is used for inclusion of a subset, including the possibility of equality; in [*M*] the sign $\subseteq$ was used for this purpose. However, when we say that '$M_1 \subset M_2 \subset \cdots$ is an ascending chain', $M_1 \subsetneqq M_2 \subsetneqq \cdots$ is intended.

(6) When we say that R is a ring of characteristic p, or write char $R = p$, we always mean that $p > 0$ is a prime number.

(7) In the exercises we generally omit the instruction 'prove that'. Solutions or hints are provided at the end of the book for most of the exercises. Many of the exercises are intended to supplement the material of the main text, so it is advisable at least to glance through them.

(8) The numbering Theorem 7.1 refers to Theorem 1 of §7; within one paragraph we usually just refer to Theorem 1, omitting the section number.

1

Commutative rings and modules

This chapter discusses the very basic definitions and results.

§1 centres around the question of the existence of prime ideals. In §2 we treat Nakayama's lemma, modules over local rings and modules of finite presentation; we give a complete proof, following Kaplansky, of the fact that a projective module over a local ring is free (Theorem 2.5), although, since we will not make any subsequent use of this in the infinitely generated case, the reader may pass over it. In §3 we give a detailed treatment of finiteness conditions in the form of Emmy Noether's chain condition, discussing among other things Akizuki's theorem, I.S. Cohen's theorem and Formanek's proof of the Eakin–Nagata theorem.

1 Ideals

If A is a ring and I an ideal of A, it is often important to consider the residue class ring A/I. Set $\bar{A}=A/I$, and write $f:A\longrightarrow\bar{A}$ for the natural map; then ideals $\bar{J}$ of $\bar{A}$ and ideals $J=f^{-1}(\bar{J})$ of A containing I are in one-to-one correspondence, with $\bar{J}=J/I$ and $A/J\simeq\bar{A}/\bar{J}$. Hence, when we just want to think about ideals of A containing I, it is convenient to shift attention to A/I. (If I' is any ideal of A then $f(I')$ is an ideal of $\bar{A}$, with $f^{-1}(f(I'))=I+I'$, and $f(I')=(I+I')/I$.)

A is itself an ideal of A, often written (1) since it is generated by the identity element 1. An ideal distinct from (1) is called a *proper ideal*. An element $a\in A$ which has an inverse in A (that is, for which there exists $a'\in A$ with $aa'=1$) is called a *unit* (or *invertible element*) of A; this holds if and only if the principal ideal (a) is equal to (1). If a is a unit and x is nilpotent then $a+x$ is again a unit: indeed, if $x^n=0$ then setting $y=-a^{-1}x$, we have $y^n=0$; now

$$(1-y)(1+y+\cdots+y^{n-1})=1-y^n=1,$$

so that $a+x=a(1-y)$ has an inverse.

In a ring A we are allowed to have $1=0$, but if this happens then it follows that $a=1\cdot a=0\cdot a=0$ for every $a\in A$, so that A has only one element 0; in this case we write $A=0$. In definitions and theorems about rings, it may sometimes happen that the condition $A \neq 0$ is omitted even when it is actually necessary. A ring A is an *integral domain* (or simply a *domain*) if $A \neq 0$, and if A has no zero-divisors other than 0. If A is an integral domain and every non-zero element of A is a unit then A is a *field*. A field is characterised by the fact that it is a ring having exactly two ideals (0) and (1).

An ideal which is maximal among all proper ideals is called a *maximal ideal*; an ideal $\mathfrak{m}$ of A is maximal if and only if $A/\mathfrak{m}$ is a field. Given a proper ideal I, let M be the set of ideals containing I and not containing 1, ordered by inclusion; then Zorn's lemma can be applied to M. Indeed, $I \in M$ so that M is non-empty, and if $L \subset M$ is a totally ordered subset then the union of all the ideals belonging to L is an ideal of A and obviously belongs to M, so is the least upper bound of L in M. Thus by Zorn's lemma M has got a maximal element. This proves the following theorem.

Theorem 1.1. If I is a proper ideal then there exists at least one maximal ideal containing I.

An ideal P of A for which A/P is an integral domain is called a *prime ideal*. In other words, P is prime if it satisfies

(i) $P \neq A$ and (ii) $x, y \notin P \Rightarrow xy \notin P$ for $x, y \in A$

A field is an integral domain, so that a maximal ideal is prime.

If I and J are ideals and P a prime ideal, then

$$I \not\subset P,\ J \not\subset P \Rightarrow IJ \not\subset P.$$

Indeed, taking $x \in I$ and $y \in J$ with $x, y \notin P$, we have $xy \in IJ$ but $xy \notin P$.

A subset S of A is *multiplicative* if it satisfies

(i) $x, y \in S \Rightarrow xy \in S$, and (ii) $1 \in S$;

(here condition (ii) is not crucial: given a subset S satisfying (i), there will usually not be any essential change on replacing S by $S \cup \{1\}$). If I is an ideal disjoint from S, then exactly as in the proof of Theorem 1 we see that the set of ideals containing I and disjoint from S has a maximal element. If P is an ideal which is maximal among ideals disjoint from S then P is prime. For if $x \notin P$, $y \notin P$, then since $P + xA$ and $P + yA$ both meet S, the product $(P + xA)(P + yA)$ also meets S. However,

$$(P + xA)(P + yA) \subset P + xyA,$$

so that we must have $xy \notin P$. We have thus obtained the following theorem.

Theorem 1.2. Let S be a multiplicative set and I an ideal disjoint from S; then there exists a prime ideal containing I and disjoint from S.

If I is an ideal of A then the set of elements of A, some power of which belongs to I, is an ideal of A (for $x^n \in I$ and $y^m \in I \Rightarrow (x + y)^{n+m-1} \in I$ and

$(ax)^n \in I$). This set is called the *radical* of I, and is sometimes written $\sqrt{I}$:

$$\sqrt{I} = \{a \in A \mid a^n \in I \text{ for some } n > 0\}.$$

If P is a prime ideal containing I then $x^n \in I \subset P$ implies that $x \in P$, and hence $\sqrt{I} \subset P$; conversely, if $x \notin \sqrt{I}$ then $S_x = \{1, x, x^2, \ldots\}$ is a multiplicative set disjoint from I, and by the previous theorem there exists a prime ideal containing I and not containing x. Thus, the radical of I is the intersection of all prime ideals containing I:

$$\sqrt{I} = \bigcap_{P \supset I} P.$$

In particular if we take $I = (0)$ then $\sqrt{(0)}$ is the set of all nilpotent elements of A, and is called the *nilradical* of A; we will write nil(A) for this. Then nil(A) is intersection of all the prime ideals of A. When nil(A) $= 0$ we say that A is *reduced*. For any ring A we write A_{red} for $A/\text{nil}(A)$; A_{red} is of course reduced.

The intersection of all maximal ideals of a ring $A(\neq 0)$ is called the *Jacobson radical*, or simply the *radical* of A, and written rad(A). If $x \in \text{rad}(A)$ then for any $a \in A$, $1 + ax$ is an element of A not contained in any maximal ideal, and is therefore a unit of A by Theorem 1. Conversely if $x \in A$ has the property that $1 + Ax$ consists entirely of units of A then $x \in \text{rad}(A)$ (prove this!).

A ring having just one maximal ideal is called a *local ring*, and a (non-zero) ring having only finitely many maximal ideals a *semilocal ring*. We often express the fact that A is a local ring with maximal ideal $\mathfrak{m}$ by saying that $(A, \mathfrak{m})$ is a local ring; if this happens then the field $k = A/\mathfrak{m}$ is called the *residue field* of A. We will say that $(A, \mathfrak{m}, k)$ is a local ring to mean that A is a local ring, $\mathfrak{m} = \text{rad}(A)$ and $k = A/\mathfrak{m}$. If $(A, \mathfrak{m})$ is a local ring then the elements of A not contained in $\mathfrak{m}$ are units; conversely a (non-zero) ring A whose non-units form an ideal is a local ring.

In general the product II' of two ideals I, I' is contained in $I \cap I'$, but does not necessarily coincide with it. However, if $I + I' = (1)$ (in which case we say that I and I' are *coprime*), then $II' = I \cap I'$; indeed, then $I \cap I' = (I \cap I')(I + I') \subset II' \subset I \cap I'$. Moreover, if I and I', as well as I and I'' are coprime, then I and $I'I''$ are coprime:

$$(1) = (I + I')(I + I'') \subset I + I'I'' \subset (1).$$

By induction we obtain the following theorem.

Theorem 1.3. If $I_1, I_2, \ldots, I_n$ are ideals which are coprime in pairs then

$$I_1 I_2 \ldots I_n = I_1 \cap I_2 \cap \cdots \cap I_n.$$

In particular if A is a semilocal ring and $\mathfrak{m}_1, \ldots \mathfrak{m}_n$ are all of its maximal ideals then

$$\text{rad}(A) = \mathfrak{m}_1 \cap \cdots \cap \mathfrak{m}_n = \mathfrak{m}_1 \ldots \mathfrak{m}_n.$$

Furthermore, if $I+I'=(1)$ then $A/II' \simeq A/I \times A/I'$. To see this it is enough to prove that the natural injective map from $A/II' = A/I\cap I'$ to $A/I \times A/I'$ is surjective; taking $e\in I$, $e'\in I'$ such that $e+e'=1$, we have $ae'+a'e \equiv a \pmod{I}$ $ae'+a'e\equiv a' \pmod{I'}$ for any a, $a'\in A$, giving the surjectivity. By induction we get the following theorem.

Theorem 1.4. If $I_1,\ldots,I_n$ are ideals which are coprime in pairs then

$$A/I_1\ldots I_n \simeq A/I_1 \times \cdots \times A/I_n.$$

Example 1. Let A be a ring, and consider the ring $A[\![X]\!]$ of formal power series over A. A power series $f=a_0+a_1X+a_2X^2+\cdots$ with $a_i\in A$ is a unit of $A[\![X]\!]$ if and only if a_0 is a unit of A. Indeed, if there exists an inverse $f^{-1}=b_0+b_1X+\cdots$ then $a_0b_0=1$; and conversely if $a_0^{-1}\in A$, then

$$\begin{aligned}1&=(a_0+a_1X+\cdots)(b_0+b_1X+\cdots)\\&=a_0b_0+(a_0b_1+a_1b_0)X+(a_0b_2+a_1b_1+a_2b_0)X^2+\cdots\end{aligned}$$

can be solved for b_0, $b_1,\ldots$: we just find b_0, $b_1,\ldots$ successively from $a_0b_0=1$, $a_0b_1+a_1b_0=0,\ldots$.

Since the formal power series ring in several variables $A[\![X_1,\ldots,X_n]\!]$ can be thought of as $(A[\![X_1,\ldots,X_{n-1}]\!])[\![X_n]\!]$, here also $f=a_0+\sum a_iX_i+\sum a_{ij}X_iX_j+\cdots$ is a unit if and only if the constant term a_0 is a unit of A; from this we see that if $g\in(X_1,\ldots,X_n)$ then $1+gh$ is a unit for any power series h, so that $g\in \operatorname{rad}(A[\![X_1,\ldots,X_n]\!])$, and hence

$$(X_1,\ldots,X_n)\subset \operatorname{rad}(A[\![X_1,\ldots,X_n]\!]).$$

If k is a field then $k[\![X_1,\ldots,X_n]\!]$ is a local ring with maximal ideal $(X_1,\ldots,X_n)$. If A is any ring and we set $B=A[\![X_1,\ldots,X_n]\!]$, then since any maximal ideal of B contains $(X_1,\ldots,X_n)$, it corresponds to a maximal ideal of $B/(X_1,\ldots,X_n)\simeq A$, and so is of the form $\mathfrak{m}B+(X_1,\ldots,X_n)$, where $\mathfrak{m}$ is a maximal ideal of A. If we write $\mathbf{m}$ for this then $\mathbf{m}\cap A=\mathfrak{m}$.

By contrast the case of polynomial rings is quite complicated; here it is just not true that a maximal ideal of $A[X]$ must contain X. For example, $X-1$ is a non-unit of $A[X]$, and so there exists a maximal ideal $\mathfrak{m}$ containing it, and $X\notin\mathfrak{m}$. Also, if $\mathfrak{m}$ is a maximal ideal of $A[X]$, it does not necessarily follow that $\mathfrak{m}\cap A$ is a maximal ideal of A.

If A is an integral domain then so are both $A[X]$ and $A[\![X]\!]$: if $f=a_rX^r+a_{r+1}X^{r+1}+\cdots$ and $g=b_sX^s+b_{s+1}X^{s+1}+\cdots$ with $a_r\neq 0$, $b_s\neq 0$ then $fg=a_rb_sX^{r+s}+\cdots\neq 0$. If I is an ideal of A we write $I[X]$ or $I[\![X]\!]$ for the set of polynomials or power series with coefficients in I; these are ideals of $A[X]$ or $A[\![X]\!]$, the kernels of the homomorphisms

$$A[X]\longrightarrow(A/I)[X] \text{ or } A[\![X]\!]\longrightarrow(A/I)[\![X]\!]$$

obtained by reducing coefficients modulo I. Hence

$$A[X]/I[X] \simeq (A/I)[X], \quad \text{and} \quad A[\![X]\!]/I[\![X]\!] \simeq (A/I)[\![x]\!];$$

in particular if P is a prime ideal then $P[X]$ and $P[\![X]\!]$ are prime ideals of $A[X]$ and $A[\![X]\!]$, respectively.

If I is finitely generated, that is $I = a_1A + \cdots + a_rA$, then $I[\![X]\!] = a_1A[\![X]\!] + \cdots + a_rA[\![X]\!] = I \cdot A[\![X]\!]$; however, if I is not finitely generated then $I[\![X]\!]$ is bigger than $I \cdot A[\![X]\!]$. In the polynomial ring this distinction does not arise, and we always have $I[X] = I \cdot A[X]$.

Example 2. For a ring A and $a, b \in A$, we have $aA \subset bA$ if and only if a is divisible by b, that is $a = bc$ for some $c \in A$. We assume that A is an integral domain in what follows. An element $a \in A$ is said to be *irreducible* if a is not a unit of A and satisfies the condition

$$a = bc \Rightarrow b \text{ or } c \text{ is a unit of } A.$$

This is equivalent to saying that aA is maximal among proper principal ideals. If aA is a prime ideal then a is said to be *prime*. As one sees easily, a prime element is irreducible, but the converse does not always hold.

Suppose that an element a has two expressions as products of prime elements:

$$a = p_1p_2 \ldots p_n = p'_1 \ldots p'_m, \text{ with } p_i \text{ and } p'_j \text{ prime.}$$

Then $n = m$, and after a suitable reordering of the p'_j we have $p_iA = p'_iA$; for $p'_1 \cdots p'_m$ is divisible by p_1, and so one of the factors, say p'_1, is divisible by p_1. Now since both p_1 and p'_1 are irreducible, $p_1A = p'_1A$ hence $p'_1 = up_1$, with u a unit, and $p_2 \cdots p_n = up'_2 \cdots p'_m$. We can replace p'_2 by up'_2, and induction on n completes the proof. In this sense, factorisation into prime elements (whenever possible) is unique.

An integral domain in which any element which is neither 0 nor a unit can be expressed as a product of prime elements is called a *unique factorisation domain* (abbreviated to UFD), or a factorial ring. It is well known that a principal ideal domain, that is an integral domain in which every ideal is principal, is a UFD (see Ex. 1.4). If A is a principal ideal domain then the prime ideals are of the form (0) or pA with p a prime element, and the latter are maximal ideals.

If k is a field then $k[X_1, \ldots, X_n]$ is a UFD, as is well-known (see Ex. 20.2). If $f(X_1, \ldots, X_n)$ is an irreducible polynomial then (f) is a prime ideal, but is not maximal if $n > 1$ (see §5).

$\mathbb{Z}[\sqrt{-5}]$ is not a UFD; indeed if $\alpha = n + m\sqrt{-5}$ with $n, m \in \mathbb{Z}$ then $\alpha\bar{\alpha} = n^2 + 5m^2$, and since $2 = n^2 + 5m^2$ has no integer solutions it follows that 2 is an irreducible element of $\mathbb{Z}[\sqrt{-5}]$, but we see from $2.3 = (1 + \sqrt{-5})(1 - \sqrt{-5})$ that 2 is not a prime element. We write

$A = \mathbb{Z}[\sqrt{-5}] = \mathbb{Z}[X]/(X^2+5)$; then setting $k = \mathbb{Z}/2\mathbb{Z}$ we have

$$A/2A = \mathbb{Z}[X]/(2, X^2+5) = k[X]/(X^2-1) = k[X]/(X-1)^2.$$

Then $P = (2, 1-\sqrt{-5})$ is a maximal ideal of A containing 2.

Exercises to §1. Prove the following propositions.

1.1. Let A be a ring, and $I \subset \operatorname{nil}(A)$ an ideal made up of nilpotent elements; if $a \in A$ maps to a unit of A/I then a is a unit of A.

1.2. Let $A_1, \ldots, A_n$ be rings; then the prime ideals of $A_1 \times \cdots \times A_n$ are of the form
$$A_1 \times \cdots \times A_{i-1} \times P_i \times A_{i+1} \times \cdots \times A_n,$$
where P_i is a prime ideal of A_i.

1.3. Let A and B be rings, and $f: A \longrightarrow B$ a surjective homomorphism.
(*a*) Prove that $f(\operatorname{rad} A) \subset \operatorname{rad} B$, and construct an example where the inclusion is strict.
(*b*) Prove that if A is a semilocal ring then $f(\operatorname{rad} A) = \operatorname{rad} B$.

1.4. Let A be an integral domain. Then A is a UFD if and only if every irreducible element is prime and the principal ideals of A satisfy the ascending chain condition. (Equivalently, every non-empty family of principal ideals has a maximal element.)

1.5. Let $\{P_\lambda\}_{\lambda \in \Lambda}$ be a non-empty family of prime ideals, and suppose that the P_λ are totally ordered by inclusion; then $\bigcap P_\lambda$ is a prime ideal. Also, if I is any proper ideal, the set of prime ideals containing I has a minimal element.

1.6. Let A be a ring, $I, P_1, \ldots, P_r$ ideals of A, and suppose that $P_3, \ldots, P_r$ are prime, and that I is not contained in any of the P_i; then there exists an element $x \in I$ not contained in any P_i.

2 Modules

Let A be a ring and M an A-module. Given submodules N, N' of M, the set $\{a \in A \mid aN' \subset N\}$ is an ideal of A, which we write $N:N'$ or $(N:N')_A$. Similarly, if $I \subset A$ is an ideal then $\{x \in M \mid Ix \subset N\}$ is a submodule of M, which we write $N:I$ or $(N:I)_M$. For $a \in A$ we define $N:a$ similarly. The ideal $0:M$ is called the *annihilator* of M, and written $\operatorname{ann}(M)$. We can consider M as a module over $A/\operatorname{ann}(M)$. If $\operatorname{ann}(M) = 0$ we say that M is a *faithful* A-module. For $x \in M$ we write $\operatorname{ann}(x) = \{a \in A \mid ax = 0\}$.

If M and M' are A-modules, the set of A-linear maps from M to M' is written $\operatorname{Hom}_A(M, M')$. This becomes an A-module if we define the sum $f+g$ and the scalar product af by

$$(f+g)(x) = f(x) + g(x), \quad (af)(x) = a \cdot f(x);$$

(the fact that af is A-linear depends on A being commutative).

To say that M is an A-module is to say that M is an Abelian group under addition, and that a scalar product ax is defined for $a \in A$ and $x \in M$ such that the following hold:

(*) $a(x+y) = ax + ay, \quad (ab)x = a(bx), \quad (a+b)x = ax + bx, \quad 1x = x;$

for fixed $a \in A$ the map $x \mapsto ax$ is an endomorphism of M as an additive group. Let E be the set of endomorphisms of the additive group M; defining the sum and product of $\lambda, \mu \in E$ by

$$(\lambda + \mu)(x) = \lambda(x) + \mu(x), \quad (\lambda\mu)(x) = \lambda(\mu(x))$$

makes E into a ring (in general non-commutative), and giving M an A-module structure is the same thing as giving a homomorphism $A \longrightarrow E$. Indeed, if we write a_L for the element of E defined by $x \mapsto ax$ then (*) become

$$(ab)_L = a_L b_L, \quad (a+b)_L = a_L + b_L, \quad (1_A)_L = 1_E.$$

We can express the fact that $\varphi: M \longrightarrow M$ is A-linear by saying that $\varphi \in E$ and that φ commutes with a_L for $a \in A$, that is $a_L \varphi = \varphi a_L$. Since A is commutative, a_L is itself an A-linear map of M for $a \in A$. We normally write simply $a: M \longrightarrow M$ for the map a_L.

If M is a B-module and $f: A \longrightarrow B$ a ring homomorphism, then we can make M into an A-module by defining $a \cdot x = f(a) \cdot x$ for $a \in A$ and $x \in M$. This is the A-module structure defined by the composite of $f: A \longrightarrow B$ with $B \longrightarrow E$, where E is the endomorphism ring of the additive group of M, and $B \longrightarrow E$ is the map defining the B-module structure of M.

If M is finitely generated as an A-module we say simply that M is a *finite* A-module, or is *finite over* A. A standard technique applicable to finite A-modules is the 'determinant trick', one form of which is as follows (taken from Atiyah and Macdonald [AM]).

Theorem 2.1. Suppose that M is an A-module generated by n elements, and that $\varphi \in \mathrm{Hom}_A(M, M)$; let I be an ideal of A such that $\varphi(M) \subset IM$. Then there is a relation of the form

(**) $\varphi^n + a_1 \varphi^{n-1} + \cdots + a_{n-1}\varphi + a_n = 0,$

with $a_i \in I^i$ for $1 \leqslant i \leqslant n$ (where both sides are considered as endomorphisms of M).

Proof. Let $M = A\omega_1 + \cdots + A\omega_n$; by the assumption $\varphi(M) \subset IM$ there exist $a_{ij} \in I$ such that $\varphi(\omega_i) = \sum_{j=1}^n a_{ij}\omega_j$. This can be rewritten

$$\sum_{j=1}^{n} (\varphi\delta_{ij} - a_{ij})\,\omega_j = 0 \quad (\text{for } 1 \leqslant i \leqslant n),$$

where δ_{ij} is the Kronecker symbol. The coefficients of this system of linear equations can be viewed as a square matrix $(\varphi\delta_{ij} - a_{ij})$ of elements of $A'[\varphi]$, the commutative subring of the endomorphism ring E of M generated by the image A' of A together with φ; let b_{ij} denote its (i,j)th cofactor, and

d its determinant. By multiplying the above equation through by b_{ik} and summing over i, we get $d\omega_k = 0$ for $1 \leqslant k \leqslant n$. Hence $d \cdot M = 0$, so that $d = 0$ as an element of E. Expanding the determinant d gives a relation of the form (**). ∎

Remark. As one sees from the proof, the left-hand side of (**) is the characteristic polynomial of (a_{ij}),

$$f(X) = \det(X\delta_{ij} - a_{ij})$$

with φ substituted for X. If M is the free A-module with basis $\omega_1, \ldots, \omega_n$ and $I = A$, the above result is nothing other than the classical Cayley–Hamilton theorem: let $f(X)$ be the characteristic polynomial of the square matrix $\varphi = (a_{ij})$; then $f(\varphi) = 0$.

Theorem 2.2 (NAK). Let M be a finite A-module and I an ideal of A. If $M = IM$ then there exists $a \in A$ such that $aM = 0$ and $a \equiv 1 \bmod I$. If in addition $I \subset \operatorname{rad}(A)$ then $M = 0$.

Proof. Setting $\varphi = 1_M$ in the previous theorem gives the relation $a = 1 + a_1 + \cdots + a_n = 0$ as endomorphisms of M, that is $aM = 0$, and $a \equiv 1 \bmod I$. If $I \subset \operatorname{rad}(A)$ then a is a unit of A, so that on multiplying both sides of $aM = 0$ by a^{-1} we get $M = 0$. ∎

Remark. This theorem is usually referred to as Nakayama's lemma, but the late Professor Nakayama maintained that it should be referred to as a theorem of Krull and Azumaya; it is in fact difficult to determine which of these three first had the result in the case of commutative rings, so we refer to it as NAK in this book. Of course, this result can easily be proved without using determinants, by induction on the number of generators of M.

Corollary. Let A be a ring and I an ideal contained in $\operatorname{rad}(A)$. Suppose that M is an A-module and $N \subset M$ a submodule such that M/N is finite over A. Then $M = N + IM$ implies $M = N$.

Proof. Setting $\bar{M} = M/N$ we have $\bar{M} = I\bar{M}$ so that, by the theorem, $\bar{M} = 0$. ∎

If W is a set of generators of an A-module M which is minimal, in the sense that any proper subset of W does not generate M, then W is said to be a *minimal basis* of M. Two minimal bases do not necessarily have the same number of elements; for example, when $M = A$, if x and y are non-units of A such that $x + y = 1$ then both $\{1\}$ and $\{x, y\}$ are minimal bases of A. However, if A is a local ring then the situation is clear:

Theorem 2.3. Let $(A, \mathfrak{m}, k)$ be a local ring and M a finite A-module; set $\bar{M} = M/\mathfrak{m}M$. Now $\bar{M}$ is a finite-dimensional vector space over k, and we write n for its dimension. Then:

(i) If we take a basis $\{\bar{u}_1,\ldots,\bar{u}_n\}$ for $\bar{M}$ over k, and choose an inverse image $u_i \in M$ of each $\bar{u}_i$, then $\{u_1,\ldots,u_n\}$ is a minimal basis of M;

(ii) conversely every minimal basis of M is obtained in this way, and so has n elements.

(iii) If $\{u_1,\ldots,u_n\}$ and $\{v_1,\ldots,v_n\}$ are both minimal bases of M, and $v_i = \sum a_{ij}u_j$ with $a_{ij} \in A$ then $\det(a_{ij})$ is a unit of A, so that (a_{ij}) is an invertible matrix.

Proof. (i) $M = \sum Au_i + \mathfrak{m}M$, and M is finitely generated (hence also $M/\sum Au_i$), so that by the above corollary $M = \sum Au_i$. If $\{u_1,\ldots,u_n\}$ is not minimal, so that, for example, $\{u_2,\ldots,u_n\}$ already generates M then $\{\bar{u}_2,\ldots,\bar{u}_n\}$ generates $\bar{M}$, which is a contradiction. Hence $\{u_1,\ldots,u_n\}$ is a minimal basis.

(ii) If $\{u_1,\ldots,u_m\}$ is a minimal basis of M and we set $\bar{u}_i$ for the image of u_i in $\bar{M}$, then $\bar{u}_1,\ldots,\bar{u}_m$ generate $\bar{M}$, and are linearly independent over k; indeed, otherwise some proper subset of $\{\bar{u}_1,\ldots,\bar{u}_m\}$ would be a basis of $\bar{M}$, and then by (i) a proper subset of $\{u_1,\ldots,u_m\}$ would generate M, which is a contradiction.

(iii) Write $\bar{a}_{ij}$ for the image in k of a_{ij}, so that $\bar{v}_i = \sum \bar{a}_{ij}\bar{u}_j$ holds in $\bar{M}$. Since $(\bar{a}_{ij})$ is the matrix transforming one basis of the vector space $\bar{M}$ into another, its determinant is non-zero. Since $\det(a_{ij}) \bmod \mathfrak{m} = \det(\bar{a}_{ij}) \neq 0$ it follows that $\det(a_{ij})$ is a unit of A. By Cramér's formula the inverse matrix of (a_{ij}) exists as a matrix with entries in A. ■

We give another interesting application of NAK, the proof of which is due to Vasconcelos [2].

Theorem 2.4. Let A be a ring and M a finite A-module. If $f: M \longrightarrow M$ is an A-linear map and f is surjective then f is also injective, and is thus an automorphism of M.

Proof. Since f commutes with scalar multiplication by elements of A, we can view M as an $A[X]$-module by setting $X \cdot m = f(m)$ for $m \in M$. Then by assumption $XM = M$, so that by NAK there exists $Y \in A[X]$ such that $(1 + XY)M = 0$. Now for $u \in \operatorname{Ker}(f)$ we have $0 = (1 + XY)(u) = u + Yf(u) = u$, so that f is injective. ■

Theorem 2.5. Let $(A, \mathfrak{m})$ be a local ring; then a projective module over A is free (for the definition of projective module, see Appendix B, p. 277).

Proof. This is easy when M is finite: choose a minimal basis $\omega_1,\ldots,\omega_n$ of M and define a surjective map $\varphi: F \longrightarrow M$ from the free module $F = Ae_1 \oplus \cdots \oplus Ae_n$ to M by $\varphi(\sum a_ie_i) = \sum a_i\omega_i$; if we set $K = \operatorname{Ker}(\varphi)$ then, from the minimal basis property,

$$\sum a_i\omega_i = 0 \Rightarrow a_i \in \mathfrak{m} \quad \text{for all } i.$$

Thus $K \subset \mathfrak{m}F$. Because M is projective, there exists $\psi: M \longrightarrow F$ such that $F = \psi(M) \oplus K$, and it follows that $K = \mathfrak{m}K$. On the other hand, K is a quotient of F, therefore finite over A, so that $K = 0$ by NAK and $F \simeq M$.

The result was proved by Kaplansky [2] without the assumption that M is finite. He proves first of all the following lemma, which holds for any ring (possibly non-commutative).

Lemma 1. Let R be any ring, and F an R-module which is a direct sum of countably generated submodules; if M is an arbitrary direct summand of F then M is also a direct sum of countably generated submodules.

Proof of Lemma 1. Suppose that $F = M \oplus N$, and that $F = \bigoplus_{\lambda \in \Lambda} E_\lambda$, where each E_λ is countably generated. By transfinite induction, we construct a well-ordered family $\{F_\alpha\}$ of submodules of F with the following properties:

(i) if $\alpha < \beta$ then $F_\alpha \subset F_\beta$,

(ii) $F = \bigcup_\alpha F_\alpha$,

(iii) if α is a limiting ordinal then $F_\alpha = \bigcup_{\beta < \alpha} F_\beta$,

(iv) $F_{\alpha+1}/F_\alpha$ is countably generated,

(v) $F_\alpha = M_\alpha \oplus N_\alpha$, where $M_\alpha = M \cap F_\alpha$, $N_\alpha = N \cap F_\alpha$,

(vi) each F_α is a direct sum of E_λ taken over a suitable subset of Λ.

We now construct such a family $\{F_\alpha\}$. Firstly, set $F_0 = (0)$. For an ordinal α, assume that F_β has been defined for all ordinals $\beta < \alpha$. If α is a limiting ordinal, set $F_\alpha = \bigcup_{\beta < \alpha} F_\beta$. If α is of the form $\alpha = \beta + 1$, let Q_1 be any one of the E_λ not contained in F_β (if $F_\beta = F$ then the construction stops at F_β). Take a set $x_{11}, x_{12}, \ldots$ of generators of Q_1, and decompose x_{11} into its M- and N-components; now let Q_2 be the direct sum of the finitely many E_λ which are necessary to write each of these two components in the decomposition $F = \oplus E_\lambda$, and let $x_{21}, x_{22}, \ldots$ be generators of Q_2. Next decompose x_{12} into its M- and N-components, let Q_3 be the direct sum of the finitely many E_λ needed to write these components, and let $x_{31}, x_{32}, \ldots$ be generators of Q_3. Then carry out the same procedure with x_{21}, getting $x_{41}, x_{42}, \ldots$, then do the same for x_{13}. Carrying out the same procedure for each of the x_{ij} in the order $x_{11}, x_{12}, x_{21}, x_{13}, x_{22}, x_{31}, \ldots$ we get countably many elements x_{ij}. We let F_α be the submodule of F generated by F_β and the x_{ij}, and this satisfies all our requirements. This gives the family $\{F_\alpha\}$.

Now $M = \bigcup M_\alpha$, with each M_α a direct summand of F, and $M_{\alpha+1} \supset M_\alpha$, so that M_α is also a direct summand of $M_{\alpha+1}$. Moreover,

$$F_{\alpha+1}/F_\alpha = (M_{\alpha+1}/M_\alpha) \oplus (N_{\alpha+1}/N_\alpha),$$

and hence $M_{\alpha+1}/M_\alpha$ is countably generated. Thus we can write

$$M_{\alpha+1} = M_\alpha \oplus M'_{\alpha+1}, \quad \text{with } M'_{\alpha+1} \text{ countably generated.}$$

When α is a limit ordinal, since $M_\alpha = \bigcup_{\beta<\alpha} M_\beta$, we set $M'_\alpha = 0$. Then finally we can write

$$M = \bigoplus_\alpha M'_\alpha \quad \text{with} \quad M'_\alpha \text{ (at most) countably generated.} \quad ■$$

Of course a free module satisfies the assumption of Lemma 1, so that, in particular, we see that any projective module is a direct sum of countably generated projective modules. Thus in the proof of Theorem 2.5 we can assume that M is countably generated.

Lemma 2. Let M be a projective module over a local ring A, and $x \in M$. Then there exists a direct summand of M containing x which is a free module.

Proof of Lemma 2. We write M as a direct summand of a free module $F = M \oplus N$. Choose a basis $B = \{u_i\}_{i \in I}$ of F such that the given element x has the minimum possible number of non-zero coordinates when expressed in this basis. Then if $x = u_1 a_1 + \cdots + u_n a_n$ with $0 \neq a_i \in A$, we have

$$a_i \notin \sum_{j \neq i} A a_j \quad \text{for} \quad i = 1, 2, \ldots, n;$$

indeed, if, say, $a_n = \sum_1^{n-1} b_i a_i$ then $x = \sum_1^{n-1} (u_i + u_n b_i) a_i$, which contradicts the choice of B. Now set $u_i = y_i + z_i$ with $y_i \in M$ and $z_i \in N$; then

$$x = \sum a_i u_i = \sum a_i y_i.$$

If we write $y_i = \sum_{j=1}^n c_{ij} u_j + t_i$, with t_i linear combinations of elements of B other than $u_1, \ldots, u_n$, we get relations $a_i = \sum_{j=1}^n a_j c_{ji}$, and, hence, in view of what we have seen above, we must have

$$1 - c_{ii} \in \mathfrak{m} \quad \text{and} \quad c_{ij} \in \mathfrak{m} \quad \text{for} \quad i \neq j.$$

It follows that the matrix (c_{ij}) has an inverse (this can be seen from the fact that the determinant is $\equiv 1 \bmod \mathfrak{m}$, or by elimination). Thus replacing $u_1, \ldots, u_n$ by $y_1, \ldots, y_n$ in B, we still have a basis of F. Hence, $F_1 = \sum y_i A$ is a direct summand of F, and hence also of M, and satisfies all the requirements of Lemma 2. ■

To prove the theorem, let M be a countably generated projective module, $M = \omega_1 A + \omega_2 A + \cdots$. By Lemma 2, there exists a free module F_1 such that $\omega_1 \in F_1$, and $M = F_1 \oplus M_1$, where M_1 is a projective module. Let ω'_2 be the M_1-component of ω_2 in the decomposition $M = F_1 \oplus M_1$, and take a free module F_2 such that $\omega'_2 \in F_2$ and $M_1 = F_2 \oplus M_2$, where M_2 is a projective module. Let ω'_3 be the M_2-component of ω_3 in $M = F_1 \oplus F_2 \oplus M_2$; proceeding in the same way, we get

$$M = F_1 \oplus F_2 \oplus \ldots,$$

so that M is a free module. ■

We say that an A-module $M \neq 0$ is a *simple* module if it has no submodules other than 0 and M itself. For any $0 \neq \omega \in M$, we then have $M = A\omega$. Now $A\omega \simeq A/\mathrm{ann}(\omega)$, but in order for this to be simple, $\mathrm{ann}(\omega)$ must be a maximal ideal of A. Hence, any simple A-module is isomorphic to $A/\mathfrak{m}$ with $\mathfrak{m}$ a maximal ideal, and conversely an A-module of this form is simple. If M is an A-module, a chain

$$M = M_0 \supset M_1 \supset \cdots \supset M_r = 0$$

of submodules of M is called a *composition series* of M if every M_i/M_{i+1} is simple; r is called the *length* of the composition series. If a composition series of M exists, its length is an invariant of M independent of the choice of composition series. More precisely, if M has a composition series of length r, and if $M \supset N_1 \supset \cdots \supset N_s$ is a strictly descending chain of submodules, then we have $s \leqslant r$. This invariance corresponds to part of the basic Jordan–Hölder theorem in group theory, but it can easily be proved on its own by induction, and the reader might like to do this as an exercise. The length of a composition series of M is called the *length* of M, and written $l(M)$; if M does not have a composition series we set $l(M) = \infty$. A necessary and sufficient condition for the existence of a composition series of M is that the submodules of M should satisfy both the ascending and descending chain conditions (for which see §3). In general, if $N \subset M$ is a submodule, we have

$$l(M) = l(N) + l(M/N).$$

If $0 \to M_1 \longrightarrow M_2 \longrightarrow \cdots \longrightarrow M_n \to 0$ is an exact sequence of A-modules and each M_i has finite length then

$$\sum_{i=1}^{n} (-1)^i l(M_i) = 0.$$

If $\mathfrak{m}$ is a maximal ideal of A and is finitely generated over A then $l(A/\mathfrak{m}^\nu) < \infty$. In fact,

$$l(A/\mathfrak{m}^\nu) = l(A/\mathfrak{m}) + l(\mathfrak{m}/\mathfrak{m}^2) + \cdots + l(\mathfrak{m}^{\nu-1}/\mathfrak{m}^\nu);$$

now each $\mathfrak{m}^i/\mathfrak{m}^{i+1}$ is a finite-dimensional vector space over the field $k = A/\mathfrak{m}$, and since its A-submodules are the same thing as its vector subspaces, $l(\mathfrak{m}^i/\mathfrak{m}^{i+1})$ is equal to the dimension of $\mathfrak{m}^i/\mathfrak{m}^{i+1}$ as k-vector space. (This shows that $A/\mathfrak{m}^\nu$ is an Artinian ring, see §3.)

Considering $l(A/\mathfrak{m}^\nu)$ for all ν, we get a function of ν which is intimately related to the ring structure of A, and which also plays a role in the resolution of singularities in algebraic and complex analytic geometry; this is studied in Chapter 5.

We say that an A-module M is *of finite presentation* if there exists an exact sequence of the form

$$A^p \longrightarrow A^q \longrightarrow M \to 0.$$

This means that M can be generated by q elements $\omega_1, \dots, \omega_q$ in such a way that the module $R = \{(a_1, \dots, a_q) \in A^q | \sum a_i\omega_i = 0\}$ of linear relations holding between the ω_i can be generated by p elements.

Theorem 2.6. Let A be a ring, and suppose that M is an A-module of finite presentation. If

$$0 \to K \longrightarrow N \longrightarrow M \to 0$$

is an exact sequence and N is finitely generated then so is K.

Proof. By assumption there exists an exact sequence of the form $L_2 \xrightarrow{g} L_1 \xrightarrow{f} M \to 0$, where L_1 and L_2 are free modules of finite rank. From this we get the following commutative diagram (see Appendix B):

$$\begin{array}{ccccccc} L_2 & \xrightarrow{g} & L_1 & \xrightarrow{f} & M & \to & 0 \\ \downarrow{\scriptstyle\beta} & & \downarrow{\scriptstyle\alpha} & & \| & & \\ 0 \to K & \xrightarrow{\psi} & N & \xrightarrow{\varphi} & M & \to & 0. \end{array}$$

If we write $N = A\xi_1 + \cdots + A\xi_m$, then there exist $v_i \in L_1$ such that $\varphi(\xi_i) = f(v_i)$. Set $\xi_i' = \xi_i - \alpha(v_i)$; then $\varphi(\xi_i') = 0$, so that we can write $\xi_i' = \psi(\eta_i)$ with $\eta_i \in K$. Let us now prove that

$$K = \beta(L_2) + A\eta_1 + \cdots + A\eta_m.$$

For any $\eta \in K$, set $\psi(\eta) = \sum a_i\xi_i$; then

$$\psi(\eta - \sum a_i\eta_i) = \sum a_i(\xi_i - \xi_i') = \alpha(\sum a_iv_i),$$

and since $0 = \varphi\alpha(\sum a_iv_i) = f(\sum a_iv_i)$, we can write $\sum a_iv_i = g(u)$ with $u \in L_2$. Now

$$\psi\beta(u) = \alpha g(u) = \alpha(\sum a_iv_i) = \psi(\eta - \sum a_i\eta_i),$$

so that $\eta = \beta(u) + \sum a_i\eta_i$, and this proves our assertion. ■

Exercises to §2. Prove the following propositions.

2.1. Let A be a ring and I a finitely generated ideal satisfying $I = I^2$; then I is generated by an idempotent e (an element e satisfying $e^2 = e$).

2.2. Let A be a ring, I an ideal of A and M a finite A-module; then $\sqrt{\operatorname{ann}(M/IM)} = \sqrt{(\operatorname{ann}(M) + I)}$.

2.3. Let M and N be submodules of an A-module L. If $M + N$ and $M \cap N$ are finitely generated then so are M and N.

2.4. Let A be a (commutative) ring, $A \neq 0$. An A-module is said to be free of rank n if it is isomorphic to A^n.

(*a*) If $A^n \simeq A^m$ then $n = m$; prove this by reducing to the case of a field. (Note that there are counter-examples to this for non-commutative rings.)

(*b*) Let $C = (c_{ij})$ be an $n \times m$ matrix over A, and suppose that C has a non-zero $r \times r$ minor, but that all the $(r+1) \times (r+1)$ minors are 0. Show then that if $r < m$, the m column vectors of C are linearly dependent. (Hint: you can assume that $m = r + 1$.) Deduce from this an alternative proof of (*a*).

(*c*) If A is a local ring, any minimal basis of the free module A^n is a basis (that is, a linearly independent set of generators).

2.5. Let A be a ring, and $0 \to L \longrightarrow M \longrightarrow N \to 0$ an exact sequence of A-modules.

(*a*) If L and N are both of finite presentation then so is M.

(*b*) If L is finitely generated and M is of finite presentation then N is of finite presentation.

3 Chain conditions

The following two conditions on a partially ordered set Γ are equivalent:

(*) any non-empty subset of Γ has a maximal element;

(**) any ascending chain $\gamma_1 < \gamma_2 < \cdots$ of elements of Γ must stop after a finite number of steps.

The implication (*)⇒(**) is obvious. We prove (**)⇒(*). Let Γ' be a non-empty subset of Γ. If Γ' does not have a maximal element, then by the axiom of choice, for each $\gamma \in \Gamma'$ we can choose a bigger element of Γ', say $\varphi(\gamma)$. Now if we choose any $\gamma_1 \in \Gamma'$ and set $\gamma_2 = \varphi(\gamma_1), \gamma_3 = \varphi(\gamma_2), \ldots$ then we get an infinite ascending chain $\gamma_1 < \gamma_2 < \cdots$, contradicting (**).

When these conditions are satisfied we say that Γ has the *ascending chain condition* (a.c.c.), or the *maximal condition*. Reversing the order we can define the *descending chain condition* (d.c.c.), or *minimal condition* in the same way.

If the set of ideals of a ring A has the a.c.c., we say that A is a *Noetherian ring*, and if it has the d.c.c., that A is an *Artinian ring*. If A is Noetherian (or Artinian) and B is a quotient of A then B has the same property; this is obvious, since the set of ideals of B is order-isomorphic to a subset of that of A.

The a.c.c. and d.c.c. were first used in a paper of Emmy Noether (1882–1935), Idealtheorie in Ringbereichen, *Math. Ann.*, 83 (1921). Emil Artin (1898–1962) was, together with Emmy Noether, one of the founders of modern abstract algebra. As well as studying non-commutative rings whose one-sided ideals satisfy the d.c.c., he also discovered the Artin–Rees lemma, which will turn up in §8.

In the same way, we say that a module is Noetherian or Artinian if its set of submodules satisfies the a.c.c. or the d.c.c. If M has either of these properties, then so do both its quotient modules and its submodules. (A subring of a Noetherian or Artinian ring does not necessarily have the same property: why not?)

A ring A is Noetherian if and only if every ideal of A is finitely generated. (Proof, 'only if': given an ideal I, consider a maximal element of the set of finitely generated ideals contained in I; this must coincide with I. 'If': given an ascending chain $I_1 \subset I_2 \subset \cdots$ of ideals, $\bigcup I_n$ is also an ideal, so that by assumption it can be generated by finitely many elements $a_1, \ldots, a_r$. There is some I_n which contains all the a_i, and the chain must stop there.)

In exactly the same way, an A-module M is Noetherian if and only if every submodule of M is finitely generated. In particular M itself must be finitely generated, and if A is Noetherian then this is also sufficient. Thus we have the well-known fact that finite modules over a Noetherian ring are Noetherian; we now give a proof of this in a more general form.

Theorem 3.1. Let A be a ring and M an A-module.

(i) Let $M' \subset M$ be a submodule and $\varphi: M \longrightarrow M/M'$ the natural map. If N_1 and N_2 are submodules of M such that $N_1 \subset N_2$, $N_1 \cap M' = N_2 \cap M'$ and $\varphi(N_1) = \varphi(N_2)$ then $N_1 = N_2$.

(ii) Let $0 \rightarrow M' \longrightarrow M \longrightarrow M'' \rightarrow 0$ be an exact sequence of A-modules; if M' and M'' are both Noetherian (or both Artinian), then so is M.

(iii) Let M be a finite A-module; then if A is Noetherian (or Artinian), so is M.

Proof. (i) is easy, and we leave it to the reader.

(ii) is obtained by applying (i) to an ascending (respectively descending) chain of submodules of M.

(iii) If M is generated by n elements then it is a quotient of the free module A^n, so that it is enough to show that A^n is Noetherian (respectively Artinian). However, this is clear from (ii) by induction on n. ∎

For a module M, it is equivalent to say that M has both the a.c.c. and the d.c.c., or that M has finite length. Indeed, if $l(M) < \infty$ then $l(M_1) < l(M_2)$ for any two distinct submodules $M_1 \subset M_2 \subset M$, so that the two chain conditions are clear. Conversely, if M has the d.c.c. then we let M_1 be a minimal non-zero submodule of M, let M_2 be a minimal element among all submodules of M strictly containing M_1, and proceed in the same way to obtain an ascending chain $0 = M_0 \subset M_1 \subset M_2 \subset \cdots$; if M also has the a.c.c. then this chain must stop by arriving at M, so that M has a composition series.

Every submodule of the $\mathbb{Z}$-module $\mathbb{Z}$ is of the form $n\mathbb{Z}$, so that $\mathbb{Z}$ is Noetherian, but not Artinian. Let p be a prime, and write W for the $\mathbb{Z}$-module of rational numbers whose denominator is a power of p; then the $\mathbb{Z}$-module $W/\mathbb{Z}$ is not Noetherian, but it is Artinian, since every proper submodule of $W/\mathbb{Z}$ is either 0 or is generated by p^{-n} for $n = 1, 2, \ldots$. This shows that the a.c.c. and d.c.c. for modules are independent conditions, but this is not the case for rings, as shown by the following result.

Theorem 3.2 (Y. Akizuki). An Artinian ring is Noetherian.

Proof. Let A be an Artinian ring. It is sufficient to prove that A has finite length as an A-module. First of all, A has only finitely many maximal ideals. Indeed, if $\mathfrak{p}_1, \mathfrak{p}_2, \ldots$ is an infinite set of distinct maximal ideals then it is easy to see that $\mathfrak{p}_1 \supset \mathfrak{p}_1\mathfrak{p}_2 \supset \mathfrak{p}_1\mathfrak{p}_2\mathfrak{p}_3 \cdots$ is an infinite descending chain of ideals, which contradicts the assumption. Thus, we let $\mathfrak{p}_1, \mathfrak{p}_2, \ldots, \mathfrak{p}_r$ be all the maximal ideals of A and set $I = \mathfrak{p}_1\mathfrak{p}_2 \ldots \mathfrak{p}_r = \operatorname{rad}(A)$. The descending chain $I \supset I^2 \supset \cdots$ stops after finitely many steps, so that there is an s such that $I^s = I^{s+1}$. If we set $(0:I^s) = J$ then

$$(J:I) = ((0:I^s):I) = (0:I^{s+1}) = J;$$

let's prove that $J = A$. By contradiction, suppose that $J \neq A$; then there exists an ideal J' which is minimal among all ideals strictly bigger than J. For any $x \in J' - J$ we have $J' = Ax + J$. Now $I = \operatorname{rad}(A)$ and $J \neq J'$, so that by NAK $J' \neq Ix + J$, and hence by minimality of J' we have $Ix + J = J$, and this gives $Ix \subset J$. Thus $x \in (J:I) = J$, which is a contradiction. Therefore $J = A$, so that $I^s = 0$. Now consider the chain of ideals

$$A \supset \mathfrak{p}_1 \supset \mathfrak{p}_1\mathfrak{p}_2 \supset \cdots \supset \mathfrak{p}_1 \cdots \mathfrak{p}_{r-1} \supset I \supset I\mathfrak{p}_1 \supset I\mathfrak{p}_1\mathfrak{p}_2$$
$$\supset \cdots \supset I^2 \supset I^2\mathfrak{p}_1 \supset \cdots \supset I^s = 0.$$

Let M and $M\mathfrak{p}_i$ be any two consecutive terms in this chain; then $M/M\mathfrak{p}_i$ is a vector space over the field $A/\mathfrak{p}_i$, and since it is Artinian, it must be finite-dimensional. Hence, $l(M/M\mathfrak{p}_i) < \infty$, and therefore the sum $l(A)$ of these terms is also finite. ■

Remark. This theorem is sometimes referred to as Hopkins' theorem, but it was proved in the above form by Akizuki [2] in 1935. It was rediscovered four years later by Hopkins [1], and he proved it for non-commutative rings (a left-Artinian ring with unit is also left-Noetherian).

Theorem 3.3. If A is Noetherian then so are $A[X]$ and $A[\![X]\!]$.

Proof. The statement for $A[X]$ is the well-known Hilbert basis theorem (see, for example Lang, Algebra, or [AM], p. 81), and we omit the proof. We now briefly run through the proof for $A[\![X]\!]$. Set $B = A[\![X]\!]$, and let I be an ideal of B; we will prove that I is finitely generated. Write $I(r)$ for the ideal of A formed by the leading coefficients a_r of $f = a_rX^r + a_{r+1}X^{r+1} + \cdots$ as f runs through $I \cap X^rB$; then we have

$$I(0) \subset I(1) \subset I(2) \subset \cdots.$$

Since A is Noetherian, there is an s such that $I(s) = I(s+1) = \cdots$; moreover, each $I(i)$ is finitely generated. For each i with $0 \leqslant i \leqslant s$ we take finitely many elements $a_{i\nu} \in A$ generating $I(i)$, and choose $g_{i\nu} \in I \cap X^iB$ having $a_{i\nu}$ as the coefficient of X^i. These $g_{i\nu}$ now generate I. Indeed, for $f \in I$ we can take a linear combination g_0 of the $g_{0\nu}$ with coefficients in A such that

$f - g_0 \in I \cap XB$, then take a linear combination g_1 of the $g_{1\nu}$ with coefficients in A such that $f - g_0 - g_1 \in I \cap X^2 B$, and proceeding in the same way we get

$$f - g_0 - g_1 - \cdots - g_s \in I \cap X^{s+1} B.$$

Now $I(s+1) = I(s)$, so we can take a linear combination g_{s+1} of the $Xg_{s\nu}$ with coefficients in A such that

$$f - g_0 - g_1 - \cdots - g_{s+1} \in I \cap X^{s+2} B.$$

We now proceed in the same way to get $g_{s+2}, \ldots$. For $i \leqslant s$, each g_i is a linear combination of the $g_{i\nu}$ with coefficients in A, and, for $i > s$, a combination of the elements $X^{i-s} g_{s\nu}$. For each $i \geqslant s$ we write $g_i = \sum_\nu a_{i\nu} X^{i-s} g_{s\nu}$, and then for each ν we set $h_\nu = \sum_{i=s}^{\infty} a_{i\nu} X^{i-s}$; h_ν is an element of B, and

$$f = g_0 + \cdots + g_{s-1} + \sum_\nu h_\nu g_{s\nu}. \quad \blacksquare$$

A ring $A[b_1, \ldots, b_n]$ which is finitely generated as a ring over a Noetherian ring A is a quotient of a polynomial ring $A[X_1, \ldots, X_n]$, and so by the Hilbert basis theorem is again Noetherian. We now give some other criteria for a ring to be Noetherian.

Theorem 3.4 (I. S. Cohen). If all the prime ideals of a ring A are finitely generated then A is Noetherian.

Proof. Write Γ for the set of ideals of A which are not finitely generated. If $\Gamma \neq \varnothing$ then by Zorn's lemma Γ contains a maximal element I. Then I is not a prime ideal, so that there are elements $x, y \in A$ with $x \notin I$, $y \notin I$ but $xy \in I$. Now $I + Ay$ is bigger than I, and hence is finitely generated, so that we can choose $u_1, \ldots, u_n \in I$ such that

$$I + Ay = (u_1, \ldots, u_n, y).$$

Moreover, $I{:}y = \{a \in A \mid ay \in I\}$ contains x, and is thus bigger than I, so it has a finite system of generators $\{v_1, \ldots, v_m\}$. Finally, it is easy to check that $I = (u_1, \ldots, u_n, v_1 y, \ldots, v_m y)$; hence, $I \notin \Gamma$, which is a contradiction. Therefore $\Gamma = \varnothing$. $\blacksquare$

Theorem 3.5. Let A be a ring and M an A-module. Then if M is a Noetherian module, $A/\mathrm{ann}(M)$ is a Noetherian ring.

Proof. If we set $\bar{A} = A/\mathrm{ann}(M)$ and view M as an $\bar{A}$-module, then the submodules of M as an A-module or $\bar{A}$-module coincide, so that M is also Noetherian as an $\bar{A}$-module. We can thus replace A by $\bar{A}$, and then $\mathrm{ann}(M) = (0)$. Now letting $M = A\omega_1 + \cdots + A\omega_n$, we can embed A in M^n by means of the map $a \mapsto (a\omega_1, \ldots, a\omega_n)$. By Theorem 1, M^n is a Noetherian module, so that its submodule A is also Noetherian. (This theorem can be expressed by saying that a ring having a faithful Noetherian module is Noetherian.) $\blacksquare$

Theorem 3.6 (E. Formanek [1]). Let A be a ring, and B an A-module which is finitely generated and faithful over A. Assume that the set of submodules of B of the form IB with I an ideal of A satisfies the a.c.c.; then A is Noetherian.

Proof. It will be enough to show that B is a Noetherian A-module. By contradiction, suppose that it is not; then the set

$$\left\{ IB \,\middle|\, \begin{array}{l} I \text{ is an ideal of } A \text{ and } B/IB \text{ is} \\ \text{non-Noetherian as } A\text{-module} \end{array} \right\}$$

contains $\{0\}$ and so is non-empty, so that by assumption it contains a maximal element. Let IB be one such maximal element; then replacing B by B/IB and A by $A/\text{ann}(B/IB)$ we see that we can assume that B is a non-Noetherian A-module, but for any non-zero ideal I of A the quotient B/IB is Noetherian.

Next we set

$$\Gamma = \{N \mid N \text{ is a submodule of } B \text{ and } B/N \text{ is a faithful } A\text{-module}\}.$$

If $B = Ab_1 + \cdots + Ab_n$ then for a submodule N of B,

$$N \in \Gamma \Leftrightarrow \forall\, a \in A - 0, \quad \{ab_1, \ldots, ab_n\} \not\subset N.$$

From this, one sees at once that Zorn's lemma applies to Γ; hence there exists a maximal element N_0 of Γ. If B/N_0 is Noetherian then A is Noetherian ring, and thus B is Noetherian, which contradicts our hypothesis. It follows that on replacing B by B/N_0 we arrive at a module B with the following properties:

(1) B is non-Noetherian as an A-module;

(2) for any ideal $I \neq (0)$ of A, B/IB is Noetherian;

(3) for any submodule $N \neq (0)$ of B, B/N is not faithful as an A-module.

Now let N be any non-zero submodule of B. By (3) there is an element $a \in A$ with $a \neq 0$ such that $a(B/N) = 0$, that is such that $aB \subset N$. By (2) B/aB is a Noetherian module, so that N/aB is finitely generated; but since B is finitely generated so is aB, and hence N itself is finitely generated. Thus, B is a Noetherian module, which contradicts (1). ∎

As a corollary of this theorem we get the following result.

Theorem 3.7.

(i) (Eakin–Nagata theorem). Let B be a Noetherian ring, and A a subring of B such that B is finite over A; then A is also a Noetherian ring.

(ii) Let B be a non-commutative ring whose right ideals have the a.c.c., and let A be a commutative subring of B. If B is finitely generated as a left A-module then A is a Noetherian ring.

(iii) Let B be a non-commutative ring whose two-sided ideals have the a.c.c., and let A be a subring contained in the centre of B; if B is finitely generated as an A-module then A is a Noetherian ring.

Proof. B has a unit, so is faithful as an A-module. Hence it is enough to apply the previous theorem. ■

Remark. Part (i) of Theorem 7 was proved in Eakin's thesis [1] in 1968, and the same result was obtained independently by Nagata [9] a little later. Subsequently many alternative proofs and extensions to the non-commutative case were published; the most transparent of these seems to be Formanek's result [1], which we have given above in the form of Theorem 6. However, this also goes back to the idea of the proofs of Eakin and Nagata.

Exercises to §3. Prove the following propositions.

3.1. Let $I_1, \ldots, I_n$ be ideals of a ring A such that $I_1 \cap \cdots \cap I_n = (0)$; if each A/I_i is a Noetherian ring then so is A.

3.2. Let A and B be Noetherian rings, and $f: A \longrightarrow C$ and $g: B \longrightarrow C$ ring homomorphisms. If both f and g are surjective then the fibre product $A \times_C B$ (that is, the subring of the direct product $A \times B$ given by $\{(a, b) \in A \times B \mid f(a) = g(b)\}$ is a Noetherian ring.

3.3. Let A be a local ring such that the maximal ideal $\mathfrak{m}$ is principal and $\bigcap_{n>0} \mathfrak{m}^n = (0)$. Then A is Noetherian, and every non-zero ideal of A is a power of $\mathfrak{m}$.

3.4. Let A be an integral domain with field of fractions K. A *fractional ideal* I of A is an A-submodule I of K such that $I \neq 0$ and $\alpha I \subset A$ for some $0 \neq \alpha \in K$. The product of two fractional ideals is defined in the same way as the product of two ideals. If I is a fractional ideal of A we set $I^{-1} = \{\alpha \in K \mid \alpha I \subset A\}$; this is also a fractional ideal, and $II^{-1} \subset A$. In the particular case that $II^{-1} = A$ we say that I is *invertible*. An invertible fractional ideal of A is finitely generated as an A-module.

3.5. If A is a UFD, the only ideals of A which are invertible as fractional ideals are the principal ideals.

3.6. Let A be a Noetherian ring, and $\varphi: A \longrightarrow A$ a homomorphism of rings. Then if φ is surjective it is also injective, and hence an automorphism of A.

3.7. If A is a Noetherian ring then any finite A-module is of finite presentation, but if A is non-Noetherian then A must have finite A-modules which are not of finite presentation.

2

Prime ideals

The notion of prime ideal is central to commutative ring theory. The set Spec$A$ of prime ideals of a ring A is a topological space, and the 'localisation' of rings and modules with respect to this topology is an important technique for studying them. These notions are discussed in §4. Starting with the topology of Spec A, we can define the dimension of A and the height of a prime ideal as notions with natural geometrical content. In §5 we treat elementary dimension theory using only field theory, developing especially the dimension theory of ideals in polynomial rings, including the Hilbert Nullstellensatz. We also discuss, as example of an application of the notion of dimension, the theory of Forster and Swan on estimates for the number of generators of a module. (Dimension theory will be the subject of a detailed study in Chapter 5 using methods of ring theory). In §6 we discuss the classical theory of primary decomposition as modernised by Bourbaki.

4 Localisation and Spec of a ring

Let A be a ring and $S \subset A$ a multiplicative set; that is (as in §1), suppose that

(i) $x, y \in S \Rightarrow xy \in S$, and (ii) $1 \in S$.

Definition. Suppose that $f: A \longrightarrow B$ is a ring homomorphism satisfying the two conditions

(1) $f(x)$ is a unit of B for all $x \in S$;

(2) if $g: A \longrightarrow C$ is a homomorphism of rings taking every element of S to a unit of C then there exists a unique homomorphism

$$h: B \longrightarrow C \quad \text{such that} \quad g = hf;$$

then B is uniquely determined up to isomorphism, and is called the *localisation* or the *ring of fractions* of A with respect to S. We write $B = S^{-1}A$ or A_S, and call $f: A \longrightarrow A_S$ the canonical map.

We prove the existence of B as follows: define the relation $\sim$ on the set $A \times S$ by

$$(a, s) \sim (b, s') \Leftrightarrow \exists t \in S \quad \text{such that} \quad t(s'a - sb) = 0;$$

it is easy to check that this is an equivalence relation (if we just had $s'a = sb$ in the definition, the transitive law would fail when S has zero-divisors). Write a/s for the equivalence class of (a, s) under $\sim$, and let B be the set of these; sums and products are defined in B by the usual rules for calculating with fractions:

$$a/s + b/s' = (as' + bs)/ss', \quad (a/s)\cdot(b/s') = ab/ss'.$$

This makes B into a ring, and defining $f: A \longrightarrow B$ by $f(a) = a/1$ we see that f is a homomorphism of rings satisfying (1) and (2) above. Indeed, if $s \in S$ then $f(s) = s/1$ has the inverse $1/s$; and if $g: A \longrightarrow C$ is as in (2) then we just have to set $h(a/s) = g(a)g(s)^{-1}$ (the reader should check that $a/s = b/s'$ implies $g(a)g(s)^{-1} = g(b)g(s')^{-1}$). From this construction we see that the kernel of the canonical map $f: A \longrightarrow A_S$ is given by

$$\operatorname{Ker} f = \{a \in A \mid sa = 0 \quad \text{for some} \quad s \in S\}.$$

Hence f is injective if and only if S does not contain any zero-divisors of A. In particular, the set of all non-zero-divisors of A is a multiplicative set; the ring of fractions with respect to S is called the *total ring of fractions* of A. If A is an integral domain then its total ring of fractions is the same thing as its field of fractions.

In general, let $f: A \longrightarrow B$ be any ring homomorphism, I an ideal of A and J an ideal of B. According to the conventions at the beginning of the book, we write IB for the ideal $f(I)B$ of B. This is called the extension of I to B, or the *extended ideal*, and is sometimes also written I^e. Moreover, we write $J \cap A$ for the ideal $f^{-1}(I)$ of A. This is called the *contracted ideal* of J, and is sometimes also written J^c. In this notation, the inclusions

$$I^{ec} \supset I \quad \text{and} \quad J^{ce} \subset J$$

follow immediately from the definitions; from the first inclusion we get $I^{ece} \supset I^e$, but substituting $J = I^e$ in the second gives $I^{ece} \subset I^e$, and hence

$$(*) \quad I^{ece} = I^e, \quad \text{and similarly} \quad J^{cec} = J^c.$$

This shows that there is a canonical bijection between the sets $\{IB \mid I$ is an ideal of $A\}$ and $\{J \cap A \mid J$ is an ideal of $B\}$.

If P is a prime ideal of B then B/P is an integral domain, and since A/P^c can be viewed as a subring of B/P it is also an integral domain, so that P^c is a prime ideal of A. (The extended ideal of a prime ideal does not have to be prime.)

An ideal J of B is said to be *primary* if it satisfies the two conditions: (1) $1 \notin J$, and (2) for $x, y \in B$, if $xy \in J$ and $x \notin J$ then $y^n \in J$ for some $n > 0$; in other words, all zero-divisors of B/J are nilpotent. The property that all zero-divisors are nilpotent passes to subrings, so that just as for prime ideals we see that the contraction of a primary ideal remains primary. If J is primary then $\sqrt{J}$ is a prime ideal (see Ex. 4.1).

The importance of rings of fractions for ring theory stems mainly from the following theorem.

Theorem 4.1.

(i) All the ideals of A_S are of the form IA_S, with I an ideal of A.

(ii) Every prime ideal of A_S is of the form $\mathfrak{p}A_S$ with $\mathfrak{p}$ a prime ideal of A disjoint from S, and conversely, $\mathfrak{p}A_S$ is prime in A_S for every such $\mathfrak{p}$; exactly the same holds for primary ideals.

Proof. (i) If J is an ideal of A_S, set $I = J \cap A$. If $x = a/s \in J$ then $x \cdot f(s) = f(a) \in J$, so that $a \in I$, and then $x = (1/s) \cdot f(a) \in IA_S$. The converse inclusion $IA_S \subset J$ is obvious, so that $J = IA_S$.

(ii) If P is a prime ideal of A_S and we set $\mathfrak{p} = P \cap A$, then $\mathfrak{p}$ is a prime ideal of A, and from the above proof $P = \mathfrak{p}A_S$. Moreover, since P does not contain units of A_S, we have $\mathfrak{p} \cap S = \emptyset$. Conversely, if $\mathfrak{p}$ is a prime ideal of A disjoint from S then

$$\frac{a}{s} \cdot \frac{b}{t} \in \mathfrak{p}A_S \quad \text{with} \quad s, t \in S \Rightarrow rab \in \mathfrak{p} \quad \text{for some} \quad r \in S,$$

and since $r \notin \mathfrak{p}$ we must have $a \in \mathfrak{p}$ or $b \in \mathfrak{p}$, so that a/s or $b/t \in \mathfrak{p}A_S$. One also sees easily that $1 \notin \mathfrak{p}A_S$, so that $\mathfrak{p}A_S$ is a prime ideal of A_S.

For primary ideals the argument is exactly the same: if $\mathfrak{p}$ is a primary ideal of A disjoint from S and if $rab \in \mathfrak{p}$ with $r \in S$, then since no power of r is in $\mathfrak{p}$ we have $ab \in \mathfrak{p}$. From this we get either $a/s \in \mathfrak{p}A_S$ or $(b/t)^n \in \mathfrak{p}A_S$ for some n. ■

Corollary. If A is Noetherian (or Artinian) then so is A_S.

Proof. This follows from (i) of the theorem. ■

We now give examples of rings of fractions A_S for various multiplicative sets S.

Example 1. Let $a \in A$ be an element which is not nilpotent, and set $S = \{1, a, a^2, \dots\}$. In this case we sometimes write A_a for A_S. (The reason for not allowing a to be nilpotent is so that $0 \notin S$. In general if $0 \in S$ then from the construction of A_S it is clear that $A_S = 0$, which is not very interesting.) The prime ideals of A_a correspond bijectively with the prime ideals of A not containing a.

Example 2. Let $\mathfrak{p}$ be a prime ideal of A, and set $S = A - \mathfrak{p}$. In this case we usually write $A_\mathfrak{p}$ for A_S. (Writing $A_\mathfrak{p}$ and $A_{(A-\mathfrak{p})}$ to denote the same thing is totally illogical notation, and the Bourbaki school avoids A_S, writing $S^{-1}A$ instead; however, the notation A_S does not lead to any confusion.) The localisation $A_\mathfrak{p}$ is a local ring with maximal ideal $\mathfrak{p}A_\mathfrak{p}$. Indeed, as we saw in Theorem 1, $\mathfrak{p}A_\mathfrak{p}$ is a prime ideal of $A_\mathfrak{p}$, and furthermore, if $J \subset A_\mathfrak{p}$ is any proper ideal then $I = J \cap A$ is an ideal of A disjoint from $A - \mathfrak{p}$, and so $I \subset \mathfrak{p}$, giving $J = IA_\mathfrak{p} \subset \mathfrak{p}A_\mathfrak{p}$. The prime ideals of $A_\mathfrak{p}$ correspond bijectively with the prime ideals of A contained in $\mathfrak{p}$.

Example 3. Let I be a proper ideal of A and set $S = 1 + I = \{1 + x \mid x \in I\}$. Then S is a multiplicative set, and the prime ideals of A_S correspond bijectively with the prime ideals $\mathfrak{p}$ of A such that $I + \mathfrak{p} \neq A$.

Example 4. Let S be a multiplicative set, and set $\tilde{S} = \{a \in A \mid ab \in S$ for some $b \in A\}$. Then $\tilde{S}$ is also a multiplicative set, called the *saturation* of S. Since quite generally a divisor of a unit is again a unit, we see from the definition of the ring of fractions that $A_S = A_{\tilde{S}}$, and $\tilde{S}$ is maximal among multiplicative sets T such that $A_S = A_T$. Indeed, one sees easily that $\tilde{S} = \{a \in A \mid a/1$ is a unit in $A_S\}$. The multiplicative set $S = A - \mathfrak{p}$ of Example 2 is already saturated.

Theorem 4.2. Localisation commutes with passing to quotients by ideals. More precisely, let A be a ring, $S \subset A$ a multiplicative set, I an ideal of A and $\bar{S}$ the image of S in A/I; then

$$A_S/IA_S \simeq (A/I)_{\bar{S}}.$$

Proof. Both sides have the universal property for ring homomorphisms $g: A \longrightarrow C$ such that

(1) every element of S maps to a unit of C,

and (2) every element of I maps to 0;

the isomorphism follows by the uniqueness of the solution to a universal mapping problem. In concrete terms the isomorphism is given by

$$a/s \bmod IA_S \leftrightarrow \bar{a}/\bar{s}, \quad \text{where} \quad \bar{a} = a + I, \quad \bar{s} = s + I. \quad \blacksquare$$

In particular, if $\mathfrak{p}$ is a prime ideal of A then

$$A_\mathfrak{p}/\mathfrak{p}A_\mathfrak{p} \simeq (A/\mathfrak{p})_{\overline{A-\mathfrak{p}}}.$$

The left-hand side is the residue field of the local ring $A_\mathfrak{p}$, whereas the right-hand side is the field of fractions of the integral domain $A/\mathfrak{p}$. This field is written $\kappa(\mathfrak{p})$ and called the *residue field of* $\mathfrak{p}$.

Theorem 4.3. Let A be a ring, $S \subset A$ a multiplicative set, and $f: A \longrightarrow A_S$ the canonical map. If B is a ring, with ring homomorphisms $g: A \longrightarrow B$ and $h: B \longrightarrow A_S$ satisfying

(1) $f = hg$,

and (2) for every $b \in B$ there exists $s \in S$ such that $g(s) \cdot b \in g(A)$, then A_S can also be regarded as a ring of fractions of B. More precisely,

$$A_S = B_{g(S)} = B_T, \quad \text{where} \quad T = \{t \in B \mid h(t) \text{ is a unit of } A_S\}.$$

Proof. We can factorise h as $B \longrightarrow B_T \longrightarrow A_S$; write $\alpha: B_T \longrightarrow A_S$ for the second of these maps. Now $g(S) \subset T$, so that the composite $A \longrightarrow B \longrightarrow B_T$ factorises as $A \longrightarrow A_S \longrightarrow B_T$; write $\beta: A_S \longrightarrow B_T$ for the second of these maps. Then

$$\alpha(\beta(a/s)) = \alpha(g(a)/g(s)) = hg(a)/hg(s) = f(a)/f(s) = a/s,$$

so that $\alpha\beta = 1$, the identity map of A_S. Moreover by assumption, for $b \in B$ there exist $a \in A$ and $s \in S$ such that $bg(s) = g(a)$. Hence, $\beta(a/s) = g(a)/g(s) = b/1$. In particular for $t \in T$, if we take $u \in A_S$ such that $t/1 = \beta(u)$ then $u = \alpha\beta(u) = \alpha(t/1) = h(t)$, so that u is a unit of A_S. Hence, $b/t = \beta(a/s)\beta(u^{-1})$, and β is surjective. Thus α and β are mutually inverse, giving an isomorphism $A_S \simeq B_T$. The fact that $A_S \simeq B_{g(S)}$ can be proved similarly. (Alternatively, this follows since T is the saturation of the multiplicative set $g(S)$. The reader should check this for himself.) ∎

Corollary 1. If $\mathfrak{p}$ is a prime ideal of $A, S = A - \mathfrak{p}$ and B satisfies the conditions of the theorem, then setting $P = \mathfrak{p}A_{\mathfrak{p}} \cap B$ we have $A_{\mathfrak{p}} = B_P$.
Proof. Under these circumstances the T in the theorem is exactly $B - P$.

Corollary 2. Let $S \subset A$ be a multiplicative set not containing any zero-divisors; then A can be viewed as a subring of A_S, and for any intermediate ring $A \subset B \subset A_S$, the ring A_S is a ring of fractions of B.

Corollary 3. If S and T are two multiplicative subsets of A with $S \subset T$, then writing T' for the image of T in A_S, we have $(A_S)_{T'} = A_T$.

Corollary 4. If $S \subset A$ is a multiplicative set and P is a prime ideal of A disjoint from S then $(A_S)_{PA_S} = A_P$. In particular if $P \subset Q$ are prime ideals of A, then

$$(A_Q)_{PA_Q} = A_P.$$

Definition. The set of all prime ideals of a ring A is called the *spectrum* of A, and written Spec A; the set of maximal ideals of A is called the maximal spectrum of A, and written m-Spec A.

By Theorem 1.1 we have

$$A \neq 0 \Leftrightarrow \text{m-Spec}\, A \neq \varnothing \Leftrightarrow \text{Spec}\, A \neq \varnothing.$$

If I is any ideal of A, we set

$$V(I) = \{\mathfrak{p} \in \text{Spec}\, A \mid \mathfrak{p} \supset I\}.$$

Then

$$V(I) \cup V(I') = V(I \cap I') = V(II'),$$

and for any family $\{I_\lambda\}_{\lambda \in \Lambda}$ of ideals we have

$$\bigcap_\lambda V(I_\lambda) = V\left(\sum_\lambda I_\lambda\right).$$

From this it follows that $\mathscr{F} = \{V(I) \mid I \text{ is an ideal of } A\}$ is closed under finite unions and arbitrary intersections, so that there is a topology on Spec A for which $\mathscr{F}$ is the set of closed sets. This is called the *Zariski topology*. From now on we will usually consider the spectrum of a ring together with its Zariski topology. m-Spec A will be considered with the subspace topology, which we will also call the Zariski topology.

For $a \in A$ we set $D(a) = \{\mathfrak{p} \in \operatorname{Spec} A \mid a \notin \mathfrak{p}\}$; this is the complement of $V(aA)$, and so is an open set. Conversely, any open set of Spec A can be written as the union of open sets of the form $D(a)$. Indeed, if $U = \operatorname{Spec} A - V(I)$ then $U = \bigcup_{a \in I} D(a)$. Hence, the open sets of the form $D(a)$ form a basis for the topology of Spec A.

Let $f: A \longrightarrow B$ be a ring homomorphism. For $P \in \operatorname{Spec} B$, the ideal $P \cap A = f^{-1}P$ is a point of Spec A. The map $\operatorname{Spec} B \longrightarrow \operatorname{Spec} A$ defined by taking P into $P \cap A$ is written ${}^a f$. As one sees easily, $({}^a f)^{-1}(V(I)) = V(IB)$, so that ${}^a f$ is continuous. If $g: B \longrightarrow C$ is another ring homomorphism then obviously ${}^a(gf) = {}^a f\, {}^a g$. Hence, the correspondence $A \longmapsto \operatorname{Spec} A$ and $f \longmapsto {}^a f$ defines a contravariant functor from the category of rings to the category of topological spaces. If ${}^a f(P) = \mathfrak{p}$, that is if $P \cap A = \mathfrak{p}$, we say that P *lies over* $\mathfrak{p}$.

Remark. For P a maximal ideal of B it does not necessarily follow that $P \cap A$ is a maximal ideal of A; for an example we need only consider the natural inclusion $A \longrightarrow B$ of an integral domain A in its field of fractions B. Thus the correspondence $A \mapsto$ m-Spec A is not functorial. This is one reason for thinking of Spec A as more important than m-Spec A. On the other hand, one could say that Spec A contains too many points; for example, the set $\{\mathfrak{p}\}$ consisting of a single point is closed in Spec A if and only if $\mathfrak{p}$ is a maximal ideal (in general the closure of $\{\mathfrak{p}\}$ coincides with $V(\mathfrak{p})$), so that Spec A almost never satisfies the separation axiom T_1.

Let M be an A-module and $S \subset A$ a multiplicative set; we define the localisation M_S of M in the same way as A_S. That is,

$$M_S = \left\{ \frac{m}{s} \mid m \in M,\ s \in S \right\},$$

and

$$\frac{m}{s} = \frac{m'}{s'} \Leftrightarrow t(s'm - sm') = 0 \quad \text{for some } t \in S.$$

If we define addition in M_S and scalar multiplication by elements of A_S by

$$m/s + m'/s' = (s'm + sm')/ss' \quad \text{and} \quad (a/s)\cdot(m/s') = am/ss'$$

then M_S becomes an A_S-module, and a canonical A-linear map $M \longrightarrow M_S$ is given by $m \mapsto m/1$; the kernel is $\{m \in M \mid sm = 0 \text{ for some } s \in S\}$. If $S = A - \mathfrak{p}$ is the complement of a prime ideal $\mathfrak{p}$ of A we write $M_\mathfrak{p}$ for M_S. The set $\{\mathfrak{p} \in \operatorname{Spec} A \mid M_\mathfrak{p} \neq 0\}$ is called the *support* of M, and written Supp (M). If M is finitely generated, and we let $M = A\omega_i + \cdots + A\omega_m$, then

$$\mathfrak{p} \in \operatorname{Supp}(M) \Leftrightarrow M_\mathfrak{p} \neq 0 \Leftrightarrow \exists i \text{ such that } \omega_i \neq 0 \text{ in } M_\mathfrak{p}$$
$$\Leftrightarrow \exists i \text{ such that } \operatorname{ann}(\omega_i) \subset \mathfrak{p} \Leftrightarrow \operatorname{ann}(M) = \bigcap_{i=1}^{n} \operatorname{ann}(\omega_i) \subset \mathfrak{p},$$

so that $\operatorname{Supp}(M)$ coincides with the closed subset $V(\operatorname{ann}(M))$ of $\operatorname{Spec} A$.

Theorem 4.4. $M_S \simeq M \otimes_A A_S$.

Proof. The map $M \times A_S \longrightarrow M_S$ defined by $(m, a/s) \mapsto am/s$ is A-bilinear, so that there exists a linear map $\alpha: M \otimes A_S \longrightarrow M_S$ such that $\alpha(m \otimes a/s) = am/s$. Conversely we can define $\beta: M_S \longrightarrow M \otimes A_S$ by $\beta(m/s) = m \otimes (1/s)$; indeed, if $m/s = m'/s'$ then $ts'm = tsm'$ for some $t \in S$, and so

$$m \otimes (1/s) = m \otimes (ts'/tss') = ts'm \otimes (1/tss') = tsm' \otimes (1/tss')$$
$$= m' \otimes (1/s').$$

Now it is easy to check that α and β are mutually inverse A_S-linear maps. Hence, M_S and $M \otimes_A A_S$ are isomorphic as A_S-modules.

Theorem 4.5. $M \mapsto M_S$ is an exact (covariant) functor from the category of A-modules to the category of A_S-modules. That is, for a morphism of A-modules $f: M \longrightarrow N$ there is a morphism $f_S: M_S \longrightarrow N_S$ of A_S-modules such that

$(\mathrm{id})_S = \mathrm{id}$ (where id is the identity map of M or M_S),
$(gf)_S = g_S f_S$;

and such that an exact sequence $0 \to M' \longrightarrow M \longrightarrow M'' \to 0$ goes into an exact sequence $0 \to M'_S \longrightarrow M_S \longrightarrow M''_S \to 0$.

Proof. To prove the exactness of $0 \to M'_S \longrightarrow M_S$ on the last line, view M' as a submodule of M; then for $x \in M'$ and $s \in S$,

$$x/s = 0 \text{ in } M_S \Leftrightarrow tx = 0 \text{ for some } t \in S$$
$$\Leftrightarrow x/s = 0 \text{ in } M'_S,$$

as required. The remaining assertions follow from the properties of the tensor product (see Appendix A) and from the previous theorem. (Of course they can easily be proved directly.) ■

It follows from this that localisation commutes with $\otimes$ and with Tor, and we will treat all this together in the section on flatness in §7.

Let A be a ring, M an A-module and $\mathfrak{p} \in \operatorname{Spec} A$. There are at least two interpretations of what it should mean that some property of A or M holds 'locally at $\mathfrak{p}$'. Namely, this could mean that $A_\mathfrak{p}$ (or $M_\mathfrak{p}$) has the property, or it could mean that $A_\mathfrak{q}$ (or $M_\mathfrak{q}$) has the property for all $\mathfrak{q}$ in some neighbourhood U of $\mathfrak{p}$ in $\operatorname{Spec} A$. The first of these is more commonly used, but there are cases when the two interpretations coincide. In any case, we now prove a number of theorems which assert that a local property implies a global one.

Theorem 4.6. Let A be a ring, M an A-module and $x \in M$. If $x = 0$ in $M_\mathfrak{p}$ for every maximal ideal $\mathfrak{p}$ of A, then $x = 0$.
Proof.

$$x = 0 \text{ in } M_\mathfrak{p} \Leftrightarrow sx = 0 \text{ for some } s \in A - \mathfrak{p} \Leftrightarrow \operatorname{ann}(x) \not\subset \mathfrak{p}.$$

However, if $1 \notin \operatorname{ann}(x)$ then by Theorem 1.1, there must exist a maximal ideal containing $\operatorname{ann}(x)$. Therefore $1 \in \operatorname{ann}(x)$, that is $x = 0$. ∎

Theorem 4.7. Let A be an integral domain with field of fractions K; set $X = \text{m-Spec}\, A$. We consider any ring of fractions of A as a subring of K. Then in this sense we have

$$A = \bigcap_{\mathfrak{m} \in X} A_\mathfrak{m}.$$

Proof. For $x \in K$ the set $I = \{a \in A \mid ax \in A\}$ is an ideal of A. Now $x \in A_\mathfrak{p}$ is equivalent to $I \not\subset \mathfrak{p}$, so that if $x \in A_\mathfrak{m}$ for every maximal ideal $\mathfrak{m}$ then $1 \in I$, that is $x \in A$. ∎

Remark. The above I is the ideal consisting of all possible denominators of x when written as a fraction of elements of A, together with 0, and this can be called the ideal of denominators of x; similarly Ix can be called the ideal of numerators of x.

Theorem 4.8. Let A be a ring and M a finite A-module. If $M \otimes_A \kappa(\mathfrak{m}) = 0$ for every maximal ideal $\mathfrak{m}$ then $M = 0$.
Proof. $\kappa(\mathfrak{m}) = A_\mathfrak{m}/\mathfrak{m}A_\mathfrak{m}$, so that $M \otimes \kappa(\mathfrak{m}) = M_\mathfrak{m}/\mathfrak{m}M_\mathfrak{m}$, and by NAK (Theorem 2.2), $M \otimes \kappa(\mathfrak{m}) = 0 \Leftrightarrow M_\mathfrak{m} = 0$. Thus the assertion follows from Theorem 4.6. ∎

The theorem just proved is easy, but we can weaken the assumption that M is finite over A; we have the following result.

Theorem 4.9. Let $f: A \longrightarrow B$ be a homomorphism of rings, and M a finite B-modules; if $M \otimes_A \kappa(\mathfrak{p}) = 0$ for every $\mathfrak{p} \in \operatorname{Spec} A$, then $M = 0$.
Proof. If $M \neq 0$ then by Theorem 6 there is a maximal ideal P of B such that $M_P \neq 0$, so that by NAK, $M_P/PM_P \neq 0$. If we now set $\mathfrak{p} = P \cap A$ then, since $\mathfrak{p}M_P \subset PM_P$, we have $M_P/\mathfrak{p}M_P \neq 0$. Set $T = B - P$ and $S = A - \mathfrak{p}$; then the localisation $M_S = M_\mathfrak{p}$ of M as an A-module and the localisation $M_{f(S)}$ of M as a B-module coincide (both of them are $\{m/s \mid m \neq M, s \in S\}$). We have $f(S) \subset T$, so that

$$M_P = M_T = (M_{f(S)})_T = (M_\mathfrak{p})_T,$$

and hence

$$M_P/\mathfrak{p}M_P = (M_\mathfrak{p}/\mathfrak{p}M_\mathfrak{p})_T = (M \otimes_A \kappa(\mathfrak{p}))_T;$$

it follows that $M \otimes_A \kappa(\mathfrak{p}) \neq 0$. ∎

Remark. In Theorem 4.9 we cannot restrict $\mathfrak{p}$ to be a maximal ideal of A. As one sees from the proof we have just given, $M=0$ provided that $M\otimes\kappa(\mathfrak{p})=0$ for every ideal $\mathfrak{p}$ which is the restriction of a maximal ideal of B. However, if for example $(A,\mathfrak{m})$ is a local integral domain with field of fractions B, and $M=B$, then $M\otimes\kappa(\mathfrak{m})=B/\mathfrak{m}B=0$, but $M\neq 0$.

Theorem 4.10. Let A be a ring and M a finite A-module.

(i) For any non-negative integer r set

$$U_r=\{\mathfrak{p}\in\operatorname{Spec}A\,|\,M_\mathfrak{p}\text{ can be generated over }A_\mathfrak{p}\text{ by }r\text{ elements}\};$$

then U_r is an open subset of $\operatorname{Spec}A$.

(ii) If M is a module of finite presentation then the set

$$U_F=\{\mathfrak{p}\in\operatorname{Spec}A\,|\,M_\mathfrak{p}\text{ is a free }A_\mathfrak{p}\text{-module}\}$$

is open in $\operatorname{Spec}A$.

Proof. (i) Suppose that $A_\mathfrak{p}=M_\mathfrak{p}\omega_1+\cdots+A_\mathfrak{p}\omega_r$. Each ω_i is of the form $\omega_i=m_i/s_i$ with $s_i\in A-\mathfrak{p}$ and $m_i\in M$, but since s_i is a unit of $A_\mathfrak{p}$ we can replace ω_i by m_i, and so assume that ω_i is (the image in $M_\mathfrak{p}$ of) an element of M. Define a linear map $\varphi\colon A^r\longrightarrow M$ from the direct sum of r copies of A to M by $(a_1,\ldots,a_r)\mapsto\sum a_i\omega_i$, and write C for the cokernel of φ. Localising the exact sequence $A^r\longrightarrow M\longrightarrow C\to 0$ at a prime ideal $\mathfrak{q}$, we get an exact sequence

$$A_\mathfrak{q}^r\longrightarrow M_\mathfrak{q}\longrightarrow C_\mathfrak{q}\to 0,$$

and when $\mathfrak{q}=\mathfrak{p}$ we get $C_\mathfrak{q}=0$. C is a quotient of M, so is finitely generated, so that the support $\operatorname{Supp}(C)$ is a closed set, and hence there is an open neighbourhood V of $\mathfrak{p}$ such that $C_\mathfrak{q}=0$ for $\mathfrak{q}\in V$. This means that $V\subset U_r$. (In short, if $\omega_i,\ldots,\omega_r\in M$ generate $M_\mathfrak{p}$ at $\mathfrak{p}$ then they generate $M_\mathfrak{q}$ for all $\mathfrak{q}$ in a neighbourhood of $\mathfrak{p}$.)

(ii) Suppose that $M_\mathfrak{p}$ is a free $A_\mathfrak{p}$-module, and let $\omega_1,\ldots,\omega_r$ be a basis. As in (i) there is no loss of generality in assuming that $\omega_i\in M$. Moreover, if we choose a suitable $D(a)$ as a neighbourhood of $\mathfrak{p}$ in $\operatorname{Spec}A$, $\omega_1,\ldots,\omega_r$ generate $M_\mathfrak{q}$ for every $\mathfrak{q}\in D(a)$. Thus, replacing A by A_a and M by M_a we can assume that the elements $\omega_1,\ldots,\omega_r$ satisfy $M_\mathfrak{q}=\sum A_\mathfrak{q}\omega_i$ for every prime ideal $\mathfrak{q}$ of A. Then by Theorem 6,

$$M/\textstyle\sum A\omega_i=0,\quad\text{that is } M=A\omega_1+\cdots+A\omega_r.$$

(We think of replacing A by A_a as shrinking $\operatorname{Spec}A$ down to the neighbourhood $D(a)$ of $\mathfrak{p}$.) Now, defining $\varphi\colon A^r\longrightarrow M$ as above, and letting K be its kernel, we get the exact sequence

$$0\to K\longrightarrow A^r\longrightarrow M\to 0;$$

moreover, $K_\mathfrak{p}=0$. By Theorem 2.6, K is finitely generated, so that applying

(i) with $r=0$, we have that $K_{\mathfrak{q}}=0$ for every $\mathfrak{q}$ in a neighbourhood V of $\mathfrak{p}$; this gives $(A_{\mathfrak{q}})^r \simeq M_{\mathfrak{q}}$, so that $V \subset U_F$. ∎

Exercise to §4. Prove the following propositions.

4.1. The radical of a primary ideal is prime; also, if I is a proper ideal containing a power $\mathfrak{m}^\nu$ of a maximal ideal $\mathfrak{m}$ then I is primary and $\sqrt{I}=\mathfrak{m}$.

4.2. If P is a prime ideal of a ring A then the *symbolic nth power* of P is the ideal $P^{(n)}$ given by

$$P^{(n)} = P^n A_P \cap A.$$

This is a primary ideal with radical P.

4.3. If S is a multiplicative set of a ring A then $\operatorname{Spec}(A_S)$ is homeomorphic to the subspace $\{\mathfrak{p} \mid \mathfrak{p} \cap S = \varnothing\} \subset \operatorname{Spec} A$; this is in general neither open nor closed in $\operatorname{Spec} A$.

4.4. If I is an ideal of A then $\operatorname{Spec}(A/I)$ is homeomorphic to the closed subset $V(I)$ of $\operatorname{Spec} A$.

4.5. The spectrum of a ring $\operatorname{Spec} A$ is quasi-compact, that is, given an open covering $\{U_\lambda\}_{\lambda\in\Lambda}$ of $X=\operatorname{Spec} A$ (with $X=\bigcup_\lambda U_\lambda$), a finite number of the U_λ already cover X.

4.6. If $\operatorname{Spec} A$ is disconnected then A contains an idempotent e (an element e satisfying $e^2=e$) distinct from 0 and 1.

4.7. If A and B are rings then $\operatorname{Spec}(A\times B)$ can be identified with the disjoint union $\operatorname{Spec} A \amalg \operatorname{Spec} B$, with both of these open and closed in $\operatorname{Spec}(A\times B)$.

4.8. If M is an A-module, N and N' submodules of M, and $S\subset A$ a multiplicative set, then $N_S \cap N'_S = (N\cap N')_S$, where both sides are considered as subsets of M_S.

4.9. A topological space is said to be *Noetherian* if the closed sets satisfy the descending chain condition. If A is a Noetherian ring then $\operatorname{Spec} A$ is a Noetherian topological space. (Note that the converse is not true in general.)

4.10. We say that a non-empty closed set V in a topological space is *reducible* if it can be expressed as a union $V=V_1\cup V_2$ of two strictly smaller closed sets V_1 and V_2, and *irreducible* if it does not have any such expression. If $\mathfrak{p}\in\operatorname{Spec} A$ then $V(\mathfrak{p})$ is an irreducible closed set, and conversely every irreducible closed set of $\operatorname{Spec} A$ can be written as $V(\mathfrak{p})$ for some $\mathfrak{p}\in\operatorname{Spec} A$.

4.11. Any closed subset of a Noetherian topological space can be written as a union of finitely many irreducible closed sets.

4.12. Use the results of the previous two exercises to prove the following: for I a proper ideal of a Noetherian ring, the set of prime ideals containing I has only finitely many minimal elements.

5 The Hilbert Nullstellensatz and first steps in dimension theory

Let X be a topological space; we consider strictly decreasing (or strictly increasing) chains $Z_0, Z_1, \ldots, Z_r$ of length r of irreducible closed subsets of X. The supremum of the lengths, taken over all such chains, is called the *combinatorial dimension* of X and denoted $\dim X$. If X is a Noetherian space then there are no infinite strictly decreasing chains, but it can nevertheless happen that $\dim X = \infty$.

Let Y be a subspace of X. If $S \subset Y$ is an irreducible closed subset of Y then its closure in X is an irreducible closed subset $\bar{S} \subset X$ such that $\bar{S} \cap Y = S$. Indeed, if $\bar{S} = V \cup W$ with V and W closed in X then $S = (V \cap Y) \cup (W \cap Y)$, so that say $S = V \cap Y$, but then $V = \bar{S}$. It follows easily from this that $\dim Y \leqslant \dim X$.

Let A be a ring. The supremum of the lengths r, taken over all strictly decreasing chains $\mathfrak{p}_0 \supset \mathfrak{p}_1 \supset \cdots \supset \mathfrak{p}_r$ of prime ideals of A, is called the *Krull dimension*, or simply the dimension of A, and denoted $\dim A$. As one sees easily from Ex. 4.10, the Krull dimension of A is the same thing as the combinatorial dimension of $\operatorname{Spec} A$. For a prime ideal $\mathfrak{p}$ of A, the supremum of the lengths, taken over all strictly decreasing chains of prime ideals $\mathfrak{p} = \mathfrak{p}_0 \supset \mathfrak{p}_1 \supset \cdots \supset \mathfrak{p}_r$ starting from $\mathfrak{p}$, is called the *height* of $\mathfrak{p}$, and denoted $\operatorname{ht} \mathfrak{p}$; (if A is Noetherian it will be proved in Theorem 13.5 that $\operatorname{ht} \mathfrak{p} < \infty$). Moreover, the supremum of the lengths, taken over all strictly increasing chain of prime ideals $\mathfrak{p} = \mathfrak{p}_0 \subset \mathfrak{p}_1 \subset \cdots \subset \mathfrak{p}_r$ starting from $\mathfrak{p}$, is called the *coheight* of $\mathfrak{p}$, and written $\operatorname{coht} \mathfrak{p}$. It follows from the definitions that

$$\operatorname{ht} \mathfrak{p} = \dim A_{\mathfrak{p}}, \quad \operatorname{coht} \mathfrak{p} = \dim A/\mathfrak{p} \quad \text{and} \quad \operatorname{ht} \mathfrak{p} + \operatorname{coht} \mathfrak{p} \leqslant \dim A.$$

Remark. In more old-fashioned terminology $\operatorname{ht} \mathfrak{p}$ was usually called the rank of $\mathfrak{p}$, and $\operatorname{coht} \mathfrak{p}$ the dimension of $\mathfrak{p}$; in addition, Nagata [N1] calls $\dim A$ the altitude of A.

Example 1. The prime ideals in the ring $\mathbb{Z}$ of rational integers are the ideals $p\mathbb{Z}$ generated by the primes $p = 2, 3, 5, \ldots$, together with (0). Hence, every $p\mathbb{Z}$ is a maximal ideal, and $\dim \mathbb{Z} = 1$. More generally, any principal ideal domain which is not a field is one-dimensional.

Example 2. An Artinian ring is zero-dimensional; indeed, we have seen in the proof of Theorem 3.2 that there are only a finite number of maximal ideals $\mathfrak{p}_1, \ldots, \mathfrak{p}_r$, and that the product of all of these is nilpotent. If then $\mathfrak{p}$ is a prime ideal, $\mathfrak{p} \supset (0) = (\mathfrak{p}_1 \ldots \mathfrak{p}_r)^\nu$ so that $\mathfrak{p} \supset \mathfrak{p}_i$ for some i; hence, $\mathfrak{p} = \mathfrak{p}_i$, so that every prime ideal is maximal.

Example 3. A zero-dimensional integral domain is just a field.

Example 4. The polynomial ring $k[X_1,\ldots,X_n]$ over a field k is an integral domain, and since

$$k[X_1,\ldots,X_n]/(X_1,\ldots,X_i)\simeq k[X_{i+1},\ldots,X_n],$$

$(X_1,\ldots,X_i)$ is a prime ideal of $k[X_1,\ldots,X_n]$. Thus

$$(0)\subset(X_1)\subset(X_1,X_2)\subset\cdots\subset(X_1,\ldots,X_n)$$

is a chain of prime ideals of length n, and $\dim k[X_1,\ldots,X_n]\geqslant n$. In fact we will shortly be proving that equality holds.

For an ideal I of a ring A we define the height of I to be the infimum of the heights of prime ideals containing I:

$$\operatorname{ht} I=\inf\{\operatorname{ht}\mathfrak{p}\mid I\subset\mathfrak{p}\in\operatorname{Spec}A\}.$$

Here also we have the inequality

$$\operatorname{ht} I+\dim A/I\leqslant\dim A.$$

If M is an A-module we define the dimension of M by

$$\dim M=\dim(A/\operatorname{ann}(M)).$$

If M is finitely generated then $\dim M$ is the combinatorial dimension of the closed subspace $\operatorname{Supp}(M)=V(\operatorname{ann}(M))$ of $\operatorname{Spec}A$.

A strictly increasing (or decreasing) chain $\mathfrak{p}_0,\mathfrak{p}_1,\ldots$ of prime ideals is said to be *saturated* if there do not exist prime ideals strictly contained between any two consecutive terms. We say that A is a *catenary ring* if the following condition is satisfied; for any prime ideals $\mathfrak{p}$ and $\mathfrak{p}'$ of A with $\mathfrak{p}\subset\mathfrak{p}'$, there exists a saturated chain of prime ideals starting from $\mathfrak{p}$ and ending at $\mathfrak{p}'$, and all such chains have the same (finite) length.

If a local domain $(A,\mathfrak{m})$ is catenary then for any prime ideal $\mathfrak{p}$ we have $\operatorname{ht}\mathfrak{p}+\operatorname{coht}\mathfrak{p}=\dim A$. Conversely, if A is a Noetherian local domain and this equality holds for all $\mathfrak{p}$ then A is catenary (Ratliff [3], 1972); the proof of this is difficult, and we postpone it to Theorem 31.4. Practically all the important Noetherian rings arising in applications are known to be catenary; the first example of a non-catenary Noetherian ring was discovered in 1956 by Nagata [5].

We now spend some time discussing the elementary theory of dimensions of rings which are finitely generated over a field k.

Theorem 5.1. Let k be a field, L an algebraic extension of k and $\alpha_1,\ldots,\alpha_n\in L$; then

(i) $k[\alpha_1,\ldots,\alpha_n]=k(\alpha_1,\ldots,\alpha_n)$.

(ii) Write $\varphi\colon k[X_1,\ldots,X_n]\longrightarrow k(\alpha_1,\ldots,\alpha_n)$ for the homomorphism over k which maps X_i to α_i; then $\operatorname{Ker}\varphi$ is the maximal ideal generated by n elements of the form $f_1(X_1)$, $f_2(X_1,X_2),\ldots,f_n(X_1,\ldots,X_n)$, where each f_i can be taken to be monic in X_i.

Proof. Let $f_1(X)$ be the minimal polynomial of α_1 over k; then $(f_1(X_1))$ is a maximal ideal of $k[X_1]$, so that $k[\alpha_1] \simeq k[X_1]/(f_1(X_1))$ is a field, and hence $k[\alpha_1] = k(\alpha_1)$. Now let $\varphi_2(X)$ be the minimal polynomial of α_2 over $k(\alpha_1)$; then since $k(\alpha_1) = k[\alpha_1]$, the coefficients of φ_2 can be expressed as polynomials in α_1, and there is a polynomial $f_2 \in k[X_1, X_2]$, monic in X_2, such that $\varphi_2(X_2) = f_2(\alpha_1, X_2)$. Thus

$$k[\alpha_1, \alpha_2] = k(\alpha_1)[\alpha_2] = k(\alpha_1, \alpha_2) \simeq k(\alpha_1)[X_2]/(f_2(\alpha_1, X_2)).$$

Proceeding in the same way, for $1 \leqslant i \leqslant n$ there is an $f_i(X_1, \ldots, X_i) \in k[X_1, \ldots, X_i]$, monic in X_i, such that

$$k[\alpha_1, \ldots, \alpha_i] = k(\alpha_1, \ldots, \alpha_i)$$
$$\simeq k(\alpha_1, \ldots, \alpha_{i-1})[X_i]/(f_i(\alpha_1, \ldots, \alpha_{i-1}, X_i)).$$

Now if $P(X) \in k[X_1, \ldots, X_n]$ is in the kernel of φ, we have $\varphi(P) = P(\alpha_1, \ldots, \alpha_n) = 0$, so that $P(\alpha_1, \ldots, \alpha_{n-1}, X_n)$ is divisible by $f_n(\alpha_1, \ldots, \alpha_{n-1}, X_n)$; dividing $P(X_1, \ldots, X_n)$ as a polynomial in X_n by the monic polynomial $f_n(X_1, \ldots, X_n)$ and letting $R_n(X_1, \ldots, X_n)$ be the remainder, we can write $P = Q_n f_n + R_n$, with $R_n(\alpha_1, \ldots, \alpha_{n-1}, X_n) = 0$. Similarly, dividing $R_n(X_1, \ldots, X_n)$ as a polynomial in X_{n-1} by $f_{n-1}(X_1, \ldots, X_{n-1})$ and letting $R_{n-1}(X_1, \ldots, X_n)$ be the remainder, we get

$$R_n = Q_{n-1} f_{n-1} + R_{n-1},$$

with

$$R_{n-1}(\alpha_1, \ldots, \alpha_{n-2}, X_{n-1}, X_n) = 0.$$

Proceeding in the same way we get $P = \sum Q_i f_i + R$, with $R(X_1, \ldots, X_n) = 0$; that is $R = 0$ and $P = \sum Q_i f_i$, so that $\operatorname{Ker} \varphi = (f_1, f_2, \ldots, f_n)$. ∎

The following theorem can be regarded as a converse of Theorem 1, (i).

Theorem 5.2. Let k be a field and $A = k[\alpha_1, \ldots, \alpha_n]$ an integral domain, and write $r = \operatorname{tr.deg}_k A$ for the transcendence degree of A (that is, of its field of fractions) over k. Then if $r > 0$, A is not a field.

Proof. Suppose that $\alpha_1, \ldots, \alpha_r$ is a transcendence basis for A over k, and set $K = k(\alpha_1, \ldots, \alpha_r)$. Then since $\alpha_{r+1}, \ldots, \alpha_n$ are algebraic over K, there exist polynomials $f_i(X_{r+1}, \ldots, X_i) \in K[X_{r+1}, \ldots, X_i]$, monic of degree d_i in X_i, such that

$$K[\alpha_{r+1}, \ldots, \alpha_n] \simeq K[X_{r+1}, \ldots, X_n]/(f_{r+1}, \ldots, f_n)$$

and

$$d_i = [K(\alpha_{r+1}, \ldots, \alpha_i) : K(\alpha_{r+1}, \ldots, \alpha_{i-1})].$$

The coefficients of f_i are in K, so that for suitable $0 \neq g \in k[\alpha_1, \ldots, \alpha_r]$ we have $g f_i \in k[\alpha_1, \ldots, \alpha_r][X_{r+1}, \ldots, X_n]$. In other words, if we set $B = k[\alpha_1, \ldots, \alpha_r, g^{-1}]$ then the f_i are polynomials with coefficients in B. We are now going to show that $A[g^{-1}] = B[\alpha_{r+1}, \ldots, \alpha_n]$ is a free module over B with basis $\{\prod_{i=r+1}^n \alpha_i^{e_i} | 0 \leqslant e_i < d_i\}$. Every element of $B[\alpha_{r+1}, \ldots, \alpha_n]$ can be written as $P(\alpha_{r+1}, \ldots, \alpha_n)$ for some $P \in B[X_{r+1}, \ldots, X_n]$; dividing P as a polynomial in X_n by f_n, and replacing P by the remainder, we can assume that P has degree at most $d_n - 1$ in X_n; then dividing P as a polynomial in X_{n-1} by f_{n-1}, and replacing by the remainder, we can assume that P has degree at most $d_{n-1} - 1$ in X_{n-1}. Proceeding in the same way, we can assume that P has degree at most $d_i - 1$ in X_i for each i; in addition, the elements $\{\alpha_i^e | 0 \leqslant e < d_i\}$ are linearly independent over $K(\alpha_{r+1}, \ldots, \alpha_{i-1})$. Hence $A[g^{-1}]$ is a free B-module. However, B is not a field; for $k[\alpha_1, \ldots, \alpha_r]$ is a polynomial ring in r variables over k, and hence it contains infinitely many irreducible polynomials (the proof of this is exactly the same as Euclid's proof that there exist infinitely many primes). Hence, there is an irreducible polynomial $h \in k[\alpha_1, \ldots, \alpha_r]$ which does not divide g, and then obviously $h^{-1} \notin k[\alpha_1, \ldots, \alpha_r, g^{-1}]$. Therefore B contains an ideal I with $I \neq 0, B$, and since $A[g^{-1}]$ is a free module over B, $IA[g^{-1}]$ is a proper ideal of $A[g^{-1}]$. Thus $A[g^{-1}]$ is not a field. But if A were a field then we would have $A[g^{-1}] = A$, and hence A is not a field. ∎

Theorem 5.3. Let k be a field, and let $\mathfrak{m}$ be any maximal ideal of the polynomial ring $k[X_1, \ldots, X_n]$; then the residue class field $k[X_1, \ldots, X_n]/\mathfrak{m}$ is algebraic over k. Hence $\mathfrak{m}$ can be generated by n elements, and in particular if k is algebraically closed then $\mathfrak{m}$ is of the form $\mathfrak{m} = (X_1 - \alpha_1, \ldots, X_n - \alpha_n)$ for $\alpha_i \in k$.

Proof. Set $k[X_1, \ldots, X_n]/\mathfrak{m} = K$, and write α_i for the image of X_i in K; then $K = k[\alpha_1, \ldots, \alpha_n]$. By the previous theorem, since K is a field it is algebraic over k, and then by Theorem 1, (ii), $\mathfrak{m}$ is generated by n elements. If k is algebraically closed then $k = K$, so that each X_i is congruent modulo $\mathfrak{m}$ to some $\alpha_i \in k$; then $(X_1 - \alpha_1, \ldots, X_n - \alpha_n) \subset \mathfrak{m}$. On the other hand $(X_1 - \alpha_1, \ldots, X_n - \alpha_n)$ is obviously a maximal ideal, so that equality must hold. ∎

Let k be a field and $\bar{k}$ its algebraic closure. Suppose that $\Phi \subset k[X_1, \ldots, X_n]$ is a subset. An n-tuple $\alpha = (\alpha_1, \ldots, \alpha_n)$ of elements $\alpha_i \in \bar{k}$ is an *algebraic zero* of Φ if it satisfies $f(\alpha) = 0$ for every $f(X) \in \Phi$.

Theorem 5.4 (The Hilbert Nullstellensatz).

(i) If Φ is a subset of $k[X_1, \ldots, X_n]$ which does not have any algebraic zeros then the ideal generated by Φ contains 1.

(ii) Given a subset Φ of $k[X_1, \ldots, X_n]$ and an element $f \in k[X_1, \ldots, X_n]$, suppose that f vanishes at every algebraic zero of Φ. Then some power of f belongs to the ideal generated by Φ, that is there exist $\nu > 0$, $g_i \in k[X_1, \ldots, X_n]$ and $h_i \in \Phi$ such that $f^\nu = \sum g_i h_i$.

Proof. (i) Let I be the ideal generated by Φ; if $1 \notin I$ then there exists a maximal ideal $\mathfrak{m}$ containing I. By the previous theorem, $k[X_1, \ldots, X_n]/\mathfrak{m}$ is algebraic over k, so that it has a k-linear isomorphic embedding θ into $\bar{k}$. If we set $\theta(X_i \bmod \mathfrak{m}) = \alpha_i$ then for all $g(X) \in \mathfrak{m}$ we have $0 = \theta(g(X)) =$

$g(\alpha_1,\ldots,\alpha_n)$, and therefore $\alpha=(\alpha_1,\ldots,\alpha_n)$ is an algebraic zero of m, and hence also of Φ. This contradicts the hypothesis; hence, $1\in I$.

(ii) Inside $k[X_1,\ldots,X_n,Y]$ we consider the set $\Phi\cup\{1-Yf(X)\}$; then this set has no algebraic zeros, so that by (i) it generates the ideal (1). Therefore there exists a relation of the form

$$1=\sum P_i(X,Y)h_i(X)+Q(X,Y)(1-Yf(X)),$$

with $h_i(X)\in\Phi$. This is an identity in $X_1,\ldots,X_n$ and Y, so that it still holds if we substitute $Y=f(X)^{-1}$. Hence we have

$$1=\sum P_i(X,f^{-1})h_i(X),$$

so that multiplying through by a suitable power of f and clearing denominators gives $f^\nu=\sum g_i(X)h_i(X)$, with $g_i\in k[X_1,\ldots,X_n]$ and $h_i\in\Phi$. ■

Remark. The above proof of (ii) is a classical idea due to Rabinowitch [1]. In a modern form it can be given as follows: let $I\subset k[X_1,\ldots,X_n]=A$ be the ideal generated by Φ; then in the localisation A_f with respect of f (see §4, Example 1), we have $IA_f=A_f$, so that a power of f is in I.

Theorem 5.5. Let k be a field, A a ring which is finitely generated over k, and I a proper ideal of A; then the radical of I is the intersection of all maximal ideals containing I, that is $\sqrt{I}=\bigcap_{I\subset \mathfrak{m}}\mathfrak{m}$.

Proof. Let $A=k[a_1,\ldots,a_n]$, so that A is a quotient of $k[X_1,\ldots,X_n]$. Considering the inverse image of I in $k[X]$ reduces to the case $A=k[X]$, and the assertion follows from Theorem 4, (ii). ■

Compared to the result $\sqrt{I}=\bigcap_{I\subset P}P$ proved in §1, the conclusion of Theorem 5 is much stronger. It is equivalent to the condition on a ring that every prime ideal P should be expressible as an intersection of maximal ideals. Rings for which this holds are called Hilbert rings or *Jacobson rings*, and they have been studied independently by O. Goldmann [1] and W. Krull [7]. See also Kaplansky [K] and Bourbaki [B5].

Theorem 5.6. Let k be a field and A an integral domain which is finitely generated over k; then

$$\dim A=\operatorname{tr.deg}_k A.$$

Proof. Let $A=k[X_1,\ldots,X_n]/P$, and set $r=\operatorname{tr.deg}_k A$. To prove that $r\geqslant\dim A$ it is enough to show that if P and Q are prime ideals of $k[X]=k[X_1,\ldots,X_n]$ with $Q\supset P$ and $Q\neq P$ then

$$\operatorname{tr.deg}_k k[X]/Q<\operatorname{tr.deg}_k k[X]/P.$$

The k-algebra homomorphism $k[X]/P\longrightarrow k[X]/Q$ is onto, so that $\operatorname{tr.deg}_k$ $k(X)/Q\leqslant\operatorname{tr.deg}_k k[X]/P$ is obvious. Suppose that equality holds. Let $k[X]/P=k[\alpha_1,\ldots,\alpha_n]$ and $k[X]/Q=k[\beta_1,\ldots,\beta_n]$; we can assume that $\beta_1,\ldots,\beta_r$ is a transcendence basis for $k(\beta)$ over k. Then $\alpha_1,\ldots,\alpha_r$ are also algebraically independent over k, so that they form a transcendence basis for $k(\alpha)$ over k. Now set $S = k[X_1,\ldots,X_r] - \{0\}$; S is a multiplicative set in $k[X]$ with $P \cap S = \emptyset$ and $Q \cap S = \emptyset$. Setting $R = k[X_1,\ldots,X_n]$ and $K = k(X_1,\ldots,X_r)$, we have $R_S = K[X_{r+1},\ldots,X_n]$, and

$$R_S/PR_S \simeq k(\alpha_1,\ldots,\alpha_r)[\alpha_{r+1},\ldots,\alpha_n],$$

so that R_S/PR_S is algebraic over $K = k(X_1,\ldots,X_r) \simeq k(\alpha_1,\ldots,\alpha_r)$, and therefore by Theorem 1, PR_S is a maximal ideal of R_S; but this contradicts the assumptions $P \subset Q$ with $P \neq Q$ and $Q \cap S = \emptyset$.

Now let us prove that $r \leqslant \dim A$ by induction on r. If $r = 0$ then, by Theorem 1, A is a field, so $\dim A = 0$ and the assertion holds. Now let $r > 0$, and suppose that $A = k[\alpha_1,\ldots,\alpha_n]$ with α_1 transcendental over k; setting $S = k[X_1] - \{0\}$ and $R = k[X_1,\ldots,X_n]$ we get

$$R_S = k(X_1)[X_2,\ldots,X_n] \quad \text{and} \quad R_S/PR_S \simeq k(\alpha_1)[\alpha_2,\ldots,\alpha_n].$$

Hence R_S/PR_S has transcendence degree $r-1$ over $k(X_1)$, so that by induction $\dim R_S/PR_S \geqslant r-1$. Thus there exists a strictly increasing chain $PR_S = Q_0 \subset Q_1 \subset \cdots \subset Q_{r-1}$ of prime ideals of R_S. If we set $P_i = Q_i \cap R$ then P_i is a prime ideal of R disjoint from S; in particular, the residue class of X_1 in R/P_{r-1} is not algebraic over k, and so $\operatorname{tr.deg}_k R/P_{r-1} > 0$. Then P_{r-1} is not a maximal ideal of R by Theorem 3, and therefore R has a maximal ideal P_r strictly bigger than P_{r-1}. Hence $\dim A = \operatorname{coht} P \geqslant r$. ∎

Corollary. If k is a field then $\dim k[X_1,\ldots,X_n] = n$.

We now turn to a different topic, the theorem of Forster and Swan on the number of generators of a module. Let A be a ring and M a finite A-module; for $\mathfrak{p} \in \operatorname{Spec} A$, write $\kappa(\mathfrak{p})$ for the residue field of $A_\mathfrak{p}$, and let $\mu(\mathfrak{p}, M)$ denote the dimension over $\kappa(\mathfrak{p})$ of the vector space $M \otimes \kappa(\mathfrak{p}) = M_\mathfrak{p}/\mathfrak{p}M_\mathfrak{p}$ (in the usual sense of linear algebra). This is the cardinality of a minimal basis of the $A_\mathfrak{p}$-module $M_\mathfrak{p}$. Hence, if $\mathfrak{p} \supset \mathfrak{p}'$ then $\mu(\mathfrak{p}, M) \geqslant \mu(\mathfrak{p}', M)$.

In 1964 the young function-theorist O. Forster surprised the experts in algebra by proving the following theorem [1].

Theorem 5.7. Let A be a Noetherian ring and M a finite A-module. Set

$$b(M) = \sup\{\mu(\mathfrak{p}, M) + \operatorname{coht}\mathfrak{p} \mid \mathfrak{p} \in \operatorname{Supp} M\};$$

then M can be generated by at most $b(M)$ elements.

This theorem is a very important link between the number of local and global generators. However, there was room for improvement in the bound for the number of generators, and in no time R. Swan obtained a better bound. We will prove Swan's bound. For this we need the concept of j-Spec A introduced by Swan. This is a space having the same irreducible closed subsets as m-Spec A, but has the advantage of having a 'generic point', not present in m-Spec A, for every irreducible closed subset.

A prime ideal which can be expressed as an intersection of maximal ideals is called a j-prime ideal, and we write j-Spec A for the set of all j-prime ideals. We consider j-Spec A also with its topology as a subspace of Spec A. Set $\mathbf{M}=\text{m-Spec}\,A$ and $\mathbf{J}=\text{j-Spec}\,A$. If F is a closed subset of $\mathbf{J}$ then there is an ideal I of A such that $F=V(I)\cap\mathbf{J}$. One sees easily that a prime ideal P belongs to F if and only if P can be expressed as an intersection of elements of $F\cap\mathbf{M}=V(I)\cap\mathbf{M}$. Hence F is determined by $F\cap\mathbf{M}$, so that there is a natural one-to-one correspondence between closed subsets of $\mathbf{J}$ and of $\mathbf{M}$. It follows that if $\mathbf{M}$ is Noetherian so is $\mathbf{J}$, and they both have the same combinatorial dimension. Now let B be an irreducible closed subset of $\mathbf{J}$, and let P be the intersection of all the elements of B. If $B=V(I)\cap\mathbf{J}$ then $I\subset P$ and we can also write $B=V(P)\cap\mathbf{J}$. If P is not a prime ideal then there exist $f, g\in A$ such that $f\notin P$, $g\notin P$ and $fg\in P$; but then

$$B=(V(P+fA)\cap\mathbf{J})\cup(V(P+gA)\cap\mathbf{J}),$$

and by definition of P there is a $Q\in B$ not containing f and a $Q'\in B$ not containing g, which implies that B is reducible, a contradiction. Therefore P is a prime ideal. Hence $P\in B$ and $B=V(P)\cap\mathbf{J}$. This P is called the *generic point* of B. Conversely if P is any element of $\mathbf{J}$ then $V(P)\cap\mathbf{J}$ is an irreducible closed subset of $\mathbf{J}$, and is the closure in $\mathbf{J}$ of $\{P\}$. We will write j-dim P for the combinatorial dimension of $V(P)\cap\mathbf{J}$.

For a finite A-module M and $\mathfrak{p}\in\mathbf{J}$ we set

$$b(\mathfrak{p}, M)=\begin{cases}0 & \text{if } M_\mathfrak{p}=0\\ \text{j-dim}\,\mathfrak{p}+\mu(\mathfrak{p}, M) & \text{if } M_\mathfrak{p}\neq 0.\end{cases}$$

Theorem 5.8 (Swan [1]). Let A be a ring, and suppose that m-Spec A is a Noetherian space. Let M be a finite A-module. If

$$\sup\{b(\mathfrak{p}, M)\mid \mathfrak{p}\in\text{j-Spec}\,A\}=r<\infty$$

then M is generated by at most r elements.

Proof.

Step 1. For $\mathfrak{p}\in\text{Spec}\,A$ and $x\in M$, we will say that x is *basic at* $\mathfrak{p}$ if x has non-zero image in $M\otimes\kappa(\mathfrak{p})$. It is easy to see that this condition is equivalent to $\mu(\mathfrak{p}, M/Ax)=\mu(\mathfrak{p}, M)-1$.

Lemma. Let M be a finite A-module and $\mathfrak{p}_1,\ldots,\mathfrak{p}_n\in\text{Supp}(M)$. Then there exists $x\in M$ which is basic at each of $\mathfrak{p}_1,\ldots,\mathfrak{p}_n$.

Proof. By reordering $\mathfrak{p}_1,\ldots,\mathfrak{p}_n$ we assume that $\mathfrak{p}_i$ is maximal among $\{\mathfrak{p}_i, \mathfrak{p}_{i+1},\ldots,\mathfrak{p}_n\}$ for each i. By induction on n suppose that $x'\in M$ is basic at $\mathfrak{p}_1,\ldots,\mathfrak{p}_{n-1}$. If x' is basic at $\mathfrak{p}_n$ then we can take $x=x'$. Suppose then that x' is not basic at $\mathfrak{p}_n$. By assumption $M_{\mathfrak{p}_n}\neq 0$ so that we can choose some $y\in M$ which is basic at $\mathfrak{p}_n$. We have $\mathfrak{p}_1\ldots\mathfrak{p}_{n-1}\not\subset\mathfrak{p}_n$, so that if we take an element

$a \in \mathfrak{p}_1 \ldots \mathfrak{p}_{n-1}$ not belonging to $\mathfrak{p}_n$ and set $x = x' + ay$, this x satisfies our requirements. This proves the lemma.

Step 2. Setting $\sup\{b(\mathfrak{p}, M) | \mathfrak{p} \in \text{j-Spec}\, A\} = r$, we now show that there are just a finite number of primes $\mathfrak{p}$ such that $b(\mathfrak{p}, M) = r$. Indeed, for $n = 1, 2, \ldots$, the subset $X_n = \{\mathfrak{p} \in \text{j-Spec}\, A \,|\, \mu(\mathfrak{p}, M) \geqslant n\}$ is closed in j-Spec A by Theorem 4.10; it has a finite number of irreducible components (by Ex. 4.11), and we let $\mathfrak{p}_{ni}$ (for $1 \leqslant i \leqslant \nu_n$) be their generic points. If M is generated by s elements then $X_n = \varnothing$ for $n > s$, so that the set $\{\mathfrak{p}_{nj}\}_{n,j}$ is finite. Let us prove that if $b(\mathfrak{p}, M) = r$ then $\mathfrak{p} \in \{\mathfrak{p}_{nj}\}_{n,j}$. Suppose $\mu(\mathfrak{p}, M) = n$; then $\mathfrak{p} \in X_n$, so that by construction $\mathfrak{p} \supset \mathfrak{p}_{ni}$ for some i. But if $\mathfrak{p} \neq \mathfrak{p}_{ni}$ then j-dim $\mathfrak{p} <$ j-dim $\mathfrak{p}_{ni}$, and since $\mu(\mathfrak{p}, M) = n = \mu(\mathfrak{p}_{ni}, M)$ we have $b(\mathfrak{p}, M) < b(\mathfrak{p}_{ni}, M)$, which is a contradiction. Hence $\mathfrak{p} = \mathfrak{p}_{ni}$.

Step 3. Let us choose an element $x \in M$ which is basic for each of the finitely many primes $\mathfrak{p}$ with $b(\mathfrak{p}, M) = r$, and set $\bar{M} = M/Ax$; then clearly $b(\mathfrak{p}, \bar{M}) \leqslant r - 1$ for every $\mathfrak{p} \in \text{j-Spec}\, A$. Hence by induction $\bar{M}$ is generated by $r - 1$ elements, and therefore M by r elements. ■

Swan's paper contains a proof of the following generalisation to non-commutative rings: let A be a commutative ring, Λ a possibly non-commutative A-algebra and M a finite left Λ-module. Suppose that m-Spec A is Noetherian, and that for every maximal ideal $\mathfrak{p}$ of A the $\Lambda_\mathfrak{p}$-module $M_\mathfrak{p}$ is generated by at most r elements; then M is generated as a Λ-module by at most $r + d$ elements, where d is the combinatorial dimension of m-Spec A.

The Forster–Swan theorem is a statement that local properties imply global ones; remarkable results in this direction have been obtained by Mohan Kumar [2] (see also Cowsik-Nori [1] and Eisenbud-Evans [1], [2]). The number of generators of ideals in local rings is the subject of a nice book by J. Sally [Sa].

Exercises to §5. Prove the following propositions.

5.1. Let k be a field, $R = k[X_1, \ldots, X_n]$ and let $P \in \operatorname{Spec} R$; then $\operatorname{ht} P + \operatorname{coht} P = n$.

5.2. A zero-dimensional Noetherian ring is Artinian (the converse to Example 2 above).

6 Associated primes and primary decomposition

Most readers will presumably have come across primary decomposition of ideals in Noetherian rings. This was the first big theorem obtained by Emmy Noether in her abstract treatment of commutative rings. Nowadays, as exemplified by Bourbaki [B4], the notion of associated prime is considered more important than primary decomposition itself.

Let A be a ring and M an A-module. A prime ideal P of A is called an *associated prime ideal* of M if P is the annihilator $\operatorname{ann}(x)$ of some $x \in M$. The set of associated primes of M is written $\operatorname{Ass}(M)$ or $\operatorname{Ass}_A(M)$. For I an ideal of A, the associated primes of the A-module A/I are referred to as the *prime divisors* of I. We say that $a \in A$ is a *zero-divisor* for M if there is a non-zero $x \in M$ such that $ax = 0$, and otherwise that a is *M-regular*.

Theorem 6.1. Let A be a Noetherian ring and M a non-zero A-module.

(i) Every maximal element of the family of ideals $F = \{\operatorname{ann}(x) | 0 \neq x \in M\}$ is an associated prime of M, and in particular $\operatorname{Ass}(M) \neq \varnothing$.

(ii) The set of zero-divisors for M is the union of all the associated primes of M.

Proof. (i) We have to prove that if $\operatorname{ann}(x)$ is a maximal element of F then it is prime: if $a, b \in A$ are such that $abx = 0$ but $bx \neq 0$ then by maximality $\operatorname{ann}(bx) = \operatorname{ann}(x)$; hence, $ax = 0$.

(ii) If $ax = 0$ for some $x \neq 0$ then $a \in \operatorname{ann}(x) \in F$, and by (i) there is an associated prime of M containing $\operatorname{ann}(x)$. ■

Theorem 6.2. Let $S \subset A$ be a multiplicative set, and N an A_S-module. Viewing $\operatorname{Spec}(A_S)$ as a subset of $\operatorname{Spec} A$, we have $\operatorname{Ass}_A(N) = \operatorname{Ass}_{A_S}(N)$. If A is Noetherian then for an A-module M we have $\operatorname{Ass}(M_S) = \operatorname{Ass}(M) \cap \operatorname{Spec}(A_S)$.

Proof. For $x \in N$ we have $\operatorname{ann}_A(x) = \operatorname{ann}_{A_S}(x) \cap A$, so that if $P \in \operatorname{Ass}_{A_S}(N)$ then $P \cap A \in \operatorname{Ass}_A(N)$. Conversely if $\mathfrak{p} \in \operatorname{Ass}_A(N)$ and we choose $x \in N$ such that $\mathfrak{p} = \operatorname{ann}_A(x)$ then $x \neq 0$, and hence, $\mathfrak{p} \cap S = \varnothing$ and $\mathfrak{p}A_S$ is a prime ideal of A_S with $\mathfrak{p}A_S = \operatorname{ann}_{A_S}(x)$. For the second part, if $\mathfrak{p} \in \operatorname{Ass}(M) \cap \operatorname{Spec}(A_S)$ then $\mathfrak{p} \cap S = \varnothing$, and $\mathfrak{p} = \operatorname{ann}_A(x)$ for some $x \in M$; if $(a/s)x = 0$ in M_S then there is a $t \in S$ such that $tax = 0$ in M, and $t \notin \mathfrak{p}$, $ta \in \mathfrak{p}$ gives $a \in \mathfrak{p}$, so that $\operatorname{ann}_{A_S}(x) = \mathfrak{p}A_S$ and $\mathfrak{p}A_S \in \operatorname{Ass}(M_S)$. Conversely, if $P \in \operatorname{Ass}(M_S)$ then without loss of generality we have $P = \operatorname{ann}_{A_S}(x)$ with $x \in M$. Setting $\mathfrak{p} = P \cap A$ we have $P = \mathfrak{p}A_S$. Now $\mathfrak{p}$ is finitely generated since A is Noetherian, and it follows that there exists some $t \in S$ such that $\mathfrak{p} = \operatorname{ann}_A(tx)$. Therefore $\mathfrak{p} \in \operatorname{Ass}_A(M)$. ■

Corollary. For a Noetherian ring A, an A-module M and a prime ideal P of A we have

$$P \in \operatorname{Ass}_A(M) \Leftrightarrow PA_P \in \operatorname{Ass}_{A_P}(M_P).$$

Theorem 6.3. Let A be a ring and $0 \to M' \longrightarrow M \longrightarrow M'' \to 0$ an exact sequence of A-modules; then

$$\operatorname{Ass}(M) \subset \operatorname{Ass}(M') \cup \operatorname{Ass}(M'').$$

Proof. If $P \in \operatorname{Ass}(M)$ then M contains a submodule N isomorphic to A/P. Since P is prime, for any non-zero element x of N we have $\operatorname{ann}(x) = P$.

Therefore if $N \cap M' \neq 0$ we have $P \in \mathrm{Ass}(M')$. If $N \cap M' = 0$ then the image of N in M'' is also isomorphic to A/P, so that $P \in \mathrm{Ass}(M'')$. ■

Theorem 6.4. Let A be a Noetherian ring and $M \neq 0$ a finite A-module. Then there exists a chain $0 = M_0 \subset M_1 \subset \cdots \subset M_n = M$ of submodules of M such that for each i we have $M_i/M_{i-1} \simeq A/P_i$ with $P_i \in \mathrm{Spec}\, A$.
Proof. Choose any $P_1 \in \mathrm{Ass}(M)$; then there exists a submodule M_1 of M with $M_1 \simeq A/P_1$. If $M_1 \neq M$ and we choose any $P_2 \in \mathrm{Ass}(M/M_1)$ then there exists $M_2 \subset M$ such that $M_2/M_1 \simeq A/P_2$. Continuing in the same way and using the ascending chain condition, we eventually arrive at $M_n = M$. ■

Theorem 6.5. Let A be a Noetherian ring and M a finite A-module.
(i) $\mathrm{Ass}(M)$ is a finite set.
(ii) $\mathrm{Ass}(M) \subset \mathrm{Supp}(M)$.
(iii) The set of minimal elements of $\mathrm{Ass}(M)$ and of $\mathrm{Supp}(M)$ coincide.
Proof. (i) follows from the previous two theorems; we need only note that $\mathrm{Ass}(A/P) = \{P\}$. For (ii), if $0 \to A/P \longrightarrow M$ is exact then so is $0 \to A_P/PA_P \longrightarrow M_P$, and therefore $M_P \neq 0$. For (iii) it is enough to show that if P is a minimal element of $\mathrm{Supp}(M)$ then $P \in \mathrm{Ass}(M)$. We have $M_P \neq 0$ so that by Theorem 2 and (ii),

$$\begin{aligned} \varnothing \neq \mathrm{Ass}(M_P) = \mathrm{Ass}(M) \cap \mathrm{Spec}(A_P) &\subset \mathrm{Supp}(M) \cap \mathrm{Spec}(A_P) \\ &= \{P\}. \end{aligned}$$

Therefore we must have $P \in \mathrm{Ass}(M)$. ■

Let A be a Noetherian ring and M a finite A-module. Let $P_1, \ldots, P_r$ be the minimal elements of $\mathrm{Supp}(M)$; then $\mathrm{Supp}(M) = V(P_1) \cup \cdots \cup V(P_r)$, and the $V(P_i)$ are the irreducible components of the closed set $\mathrm{Supp}(M)$ (see Ex. 4.11). The prime ideals $P_1, \ldots, P_r$ are called the *isolated associated primes* of M, and the remaining associated primes of M are called *embedded primes.* If I is an ideal of A then $\mathrm{Supp}_A(A/I)$ is the set of prime ideals containing I, and the minimal prime divisors of I (that is the minimal associated primes of the A-module A/I) are precisely the minimal prime ideals containing I. We have seen in Ex. 4.12 that there are only a finite number of such primes, and Theorem 5 now gives a new proof of this. (For examples of embedded primes see Ex. 6.6 and Ex. 8.9.)

Definition. Let A be a ring, M an A-module and $N \subset M$ a submodule. We say that N is a *primary submodule* of M if the following condition holds for all $a \in A$ and $x \in M$:

$$x \notin N \text{ and } ax \in N \Rightarrow a^\nu M \subset N \text{ for some } \nu.$$

This definition in fact only depends on the quotient module M/N. It can be restated as

$$\text{if } a \in A \text{ is a zero-divisor for } M/N \text{ then } a \in \sqrt{(\mathrm{ann}(M/N))}.$$

A primary ideal is just a primary submodule of the A-module A. One might wonder about trying to set up a notion of prime submodule generalising prime ideal, but this does not turn out to be useful.

Theorem 6.6. Let A be a Noetherian ring and M a finite A-module. Then a submodule $N \subset M$ is primary if and only if $\operatorname{Ass}(M/N)$ consists of one element only. In this case, if $\operatorname{Ass}(M/N) = \{P\}$ and $\operatorname{ann}(M/N) = I$ then I is primary and $\sqrt{I} = P$.

Proof. If $\operatorname{Ass}(M/N) = \{P\}$ then by the previous theorem $\operatorname{Supp}(M/N) = V(P)$, so that $P = \sqrt{(\operatorname{ann}(M/N))}$. Now if $a \in A$ is a zero-divisor for M/N it follows from Theorem 1 that $a \in P$, so that $a \in \sqrt{(\operatorname{ann}(M/N))}$; hence, N is a primary submodule of M. Conversely, if N is a primary submodule and $P \in \operatorname{Ass}(M/N)$ then every $a \in P$ is a zero-divisor for M/N, so that by assumption $a \in \sqrt{I}$, where $I = \operatorname{ann}(M/N)$. Hence $P \subset \sqrt{I}$, but from the definition of associated prime we obviously have $I \subset P$, and hence $\sqrt{I} \subset P$, so that $P = \sqrt{I}$. Thus $\operatorname{Ass}(M/N)$ has just one element $\sqrt{I}$. We prove that in this case I is a primary ideal: let $a, b \in A$ with $b \notin I$; if $ab \in I$ then $ab(M/N) = 0$, but $b(M/N) \neq 0$, so that a is a zero-divisor for M/N, and therefore $a \in P = \sqrt{I}$. ∎

Definition. If $\operatorname{Ass}(M/N) = \{P\}$ we say that $N \subset M$ is a *P-primary* submodule, or a *primary* submodule *belonging to P*.

Theorem 6.7. If N and N' are P-primary submodules of M then so is $N \cap N'$.

Proof. We can embed $M/(N \cap N')$ as a submodule of $(M/N) \oplus (M/N')$, so that

$$\operatorname{Ass}(M/(N \cap N')) \subset \operatorname{Ass}(M/N) \cup \operatorname{Ass}(M/N') = \{P\}. \quad ∎$$

If $N \subset M$ is a submodule, we say that N is *reducible* if it can be written as an intersection $N = N_1 \cap N_2$ of two submodules N_1, N_2 with $N_i \neq N$, and otherwise that N is *irreducible*; note that this has nothing to do with the notion of irreducible modules in representation theory (= no submodules other than 0 and M), which is a condition on M only.

If M is a Noetherian module then any submodule N of M can be written as a finite intersection of irreducible submodules. Proof: let $\mathscr{F}$ be the set of submodules $N \subset M$ having no such expression. If $\mathscr{F} \neq \emptyset$ then it has a maximal element N_0. Then N_0 is reducible, so that $N_0 = N_1 \cap N_2$, and $N_i \notin \mathscr{F}$. Now each of the N_i is an intersection of a finite number of irreducible submodules, and hence so is N_0. This is a contradiction.

Remark. The representation as an intersection of irreducible submodules is in general not unique. For example, if A is a field and M an n-dimensional vector space over A then the irreducible submodules of M are just its $(n-1)$-dimensional subspaces. An $(n-2)$-dimensional subspace can be written in lots of ways as an intersection of $(n-1)$-dimensional subspaces.

In general we say that an expression of a set N as an intersection $N = N_1 \cap \cdots \cap N_r$ is *irredundant* if we cannot omit any N_i, that is if $N \neq N_1 \cap \cdots \cap N_{i-1} \cap N_{i+1} \cap \cdots \cap N_r$. If M is an A-module, we call an expression $N = N_1 \cap \cdots \cap N_r$ of a submodule N as an intersection of a finite number of submodules $N_i \subset M$ a *decomposition* of N; if each of the N_i is irreducible we speak of an *irreducible decomposition*, if primary of a *primary decomposition*. Let $N = N_1 \cap \cdots \cap N_r$ be an irredundant primary decomposition with $\mathrm{Ass}(M/N_i) = \{P_i\}$; if $P_i = P_j$ then $N_i \cap N_j$ is again primary, so that grouping together all of the N_i belonging to the same prime ideal we get a primary decomposition such that $P_i \neq P_j$ for $i \neq j$. A decomposition with this property will be called a *shortest primary decomposition*, and the N_i appearing in it the *primary components* of N; if N_i belongs to a prime P we sometimes say that N_i is the P-primary component of N.

Theorem 6.8. Let A be a Noetherian ring and M a finite A-module.

(i) An irreducible submodule of M is a primary submodule.

(ii) If

$$N = N_1 \cap \cdots \cap N_r \quad \text{with} \quad \mathrm{Ass}(M/N_i) = \{P_i\}$$

is an irredundant primary decomposition of a proper submodule $N \subset M$ then $\mathrm{Ass}(M/N) = \{P_1, \ldots, P_r\}$.

(iii) Every proper submodule N of M has a primary decomposition. If N is a proper submodule of M and P is a minimal associated prime of M/N then the P-primary component of N is $\varphi_P^{-1}(N_P)$, where $\varphi_P: M \longrightarrow M_P$ is the canonical map, and therefore it is uniquely determined by M, N and P.

Proof. (i) It is enough to prove that a submodule $N \subset M$ which is not primary is reducible: replacing M by M/N we can assume that $N = 0$. By Theorem 6, $\mathrm{Ass}(M)$ has at least two elements P_1 and P_2. Then M contains submodules K_i isomorphic to A/P_i for $i = 1, 2$. Now since $\mathrm{ann}(x) = P_i$ for any non-zero $x \in K_i$, we must have $K_1 \cap K_2 = 0$, and hence 0 is reducible.

(ii) We can again assume that $N = 0$. If $0 = N_1 \cap \cdots \cap N_r$ then M is isomorphic to a submodule of $M/N_1 \oplus \cdots \oplus M/N_r$, so that

$$\mathrm{Ass}(M) \subset \mathrm{Ass}\left(\bigoplus_{i=1}^{r} M/N_i\right) = \bigcup_{i=1}^{r} \mathrm{Ass}(M/N_i) = \{P_1, \ldots, P_r\}.$$

On the other hand $N_2 \cap \cdots \cap N_r \neq 0$, and taking $0 \neq x \in N_2 \cap \cdots \cap N_r$ we have $\mathrm{ann}(x) = 0{:}x = N_1{:}x$. But $N_1{:}M$ is a primary ideal belonging to P_1, so that $P_1^{\nu} M \subset N_1$ for some $\nu > 0$. Therefore $P_1^{\nu} x = 0$; hence there exists $i \geqslant 0$ such that $P_1^i x \neq 0$ but $P_1^{i+1} x = 0$, and choosing $0 \neq y \in P_1^i x$ we have $P_1 y = 0$. However, since $y \in N_2 \cap \cdots \cap N_r$ it follows that $y \notin N_1$, and by the definition of primary submodule $\mathrm{ann}(y) \subset P_1$, so that $P_1 = \mathrm{ann}(y)$ and

$P_1 \in \mathrm{Ass}(M)$. The same works for the other P_i, and this proves that $\{P_1,\ldots,P_r\} \subset \mathrm{Ass}(M)$.

(iii) We have already seen that a proper submodule has an irreducible decomposition, so that by (i) it has a primary decomposition. Suppose that $N = N_1 \cap \cdots \cap N_r$ is a shortest primary decomposition, and that N_1 is the P-primary component with $P = P_1$. By Ex. 4.8 we know that $N_P = (N_1)_P \cap \cdots \cap (N_r)_P$, and for $i > 1$ a power of P_i is contained in $\mathrm{ann}(M/N_i)$; then since $P_i \not\subset P_1$ we have $(M/N_i)_P = 0$, and therefore $(N_i)_P = M_P$. Thus $N_P = (N_1)_P$, and hence $\varphi_P^{-1}(N_P) = \varphi_P^{-1}((N_1)_P)$; it is easy to check that the right-hand side is N_1. ■

Remark. The uniqueness of the P-primary component N, proved in (iii) for minimal primes P, does not hold in general; see Ex. 6.6.

Exercises to §6.

6.1. Find $\mathrm{Ass}(M)$ for the $\mathbb{Z}$-module $M = \mathbb{Z} \oplus (\mathbb{Z}/3\mathbb{Z})$.

6.2. If M is a finite module over a Noetherian ring A, and M_1, M_2 are submodules of M with $M = M_1 + M_2$ then can we say that $\mathrm{Ass}(M) = \mathrm{Ass}(M_1) \cup \mathrm{Ass}(M_2)$?

6.3. Let A be a Noetherian ring and let $x \in A$ be an element which is neither a unit nor a zero-divisor; prove that the ideals xA and $x^n A$ for $n = 1, 2\ldots$ have the same prime divisors:

$$\mathrm{Ass}_A(A/xA) = \mathrm{Ass}_A(A/x^n A).$$

6.4. Let I and J be ideals of a Noetherian ring A. Prove that if $JA_P \subset IA_P$ for every $P \in \mathrm{Ass}_A(A/I)$ then $J \subset I$.

6.5. Prove that the total ring of fractions of a reduced Noetherian ring A is a direct product of fields.

6.6. (Taken from [Nor 1], p. 30.) Let k be a field. Show that in $k[X, Y]$ we have $(X^2, XY) = (X) \cap (X^2, Y) = (X) \cap (X^2, XY, Y^2)$.

6.7. Let $f: A \longrightarrow B$ be a homomorphism of Noetherian rings, and M a finite B-module. Write ${}^a f: \mathrm{Spec}\, B \longrightarrow \mathrm{Spec}\, A$ as in §4. Prove that ${}^a f(\mathrm{Ass}_B(M)) = \mathrm{Ass}_A(M)$. (Consequently, $\mathrm{Ass}_A(M)$ is a finite set for such M.)

Appendix to §6. Secondary representations of a module

I.G. Macdonald [1] has developed the theory of attached prime ideals and secondary representations of a module, which is in a certain sense dual to the theory of associated prime ideals and primary decompositions. This theory was successfully applied to the theory of local cohomology by him and R.Y. Sharp (Macdonald & Sharp [1], Sharp [7]).

Let A be a commutative ring. An A-module M is said to be *secondary* if $M \neq 0$ and, for each $a \in A$, the endomorphism $\varphi_a: M \longrightarrow M$ defined by $\varphi_a(m) = am$ (for $m \in M$) is either surjective or nilpotent. If M is secondary, then $P = \sqrt{(\mathrm{ann}\, M)}$ is a prime ideal, and M is said to be P-secondary. Any non-zero quotient of a P-secondary module is P-secondary.

Example 1. If A is an integral domain, its quotient field K is a (0)-secondary A-module.

Example 2. Let $W = \mathbb{Z}[p^{-1}]$, where p is a prime number, and consider the Artinian $\mathbb{Z}$-module $W/\mathbb{Z}$ (see §3). This is also a (0)-secondary $\mathbb{Z}$-module.

Example 3. If A is a local ring with maximal ideal P and if every element of P is nilpotent, then A itself is a P-secondary A-module.

Example 4. If P is a maximal ideal of A, then A/P^n is a P-secondary A-module for every $n > 0$.

A *secondary representation* of an A-module M is an expression of M as a finite sum of secondary submodules:

(*) $M = N_1 + \cdots + N_n$.

The representation is *minimal* if (1) the prime ideals $P_i := \sqrt{(\mathrm{ann}\, N_i)}$ are all distinct, and (2) none of the N_i is redundant. It is easy to see that the sum of two P-secondary submodules is again P-secondary, hence if M has a secondary representation then it has a minimal one.

A prime ideal P is called an *attached prime ideal* of M if M has a P-secondary quotient. The set of the attached prime ideals of M is denoted by $\mathrm{Att}(M)$.

Theorem 6.9. If (*) is a minimal secondary representation of M and $P_i = \sqrt{(\mathrm{ann}\, N_i)}$, then $\mathrm{Att}(M) = \{P_1, \ldots, P_n\}$.

Proof. Since $M/(N_1 + \cdots + N_{i-1} + N_{i+1} + \cdots + N_n)$ is a non-zero quotient of N_i, it is a P_i-secondary module. Thus $\{P_1, \ldots, P_n\} \subset \mathrm{Att}(M)$. Conversely, let $P \in \mathrm{Att}(M)$ and let W be a P-secondary quotient of M. Then $W = \bar{N}_1 + \cdots + \bar{N}_n$, where $\bar{N}_i$ is the image of N_i in W. From this we obtain a minimal secondary representation $W = \bar{N}_{i_1} + \cdots + \bar{N}_{i_s}$, and then $\mathrm{Att}(W) \supset \{P_{i_1}, \ldots, P_{i_s}\}$. On the other hand $\mathrm{Att}(W) = \{P\}$ since W is P-secondary. Therefore $P = P_i$ for some i. ■

Theorem 6.10. If $0 \to M' \longrightarrow M \longrightarrow M'' \to 0$ is an exact sequence of A-modules, then $\mathrm{Att}(M'') \subset \mathrm{Att}(M) \subset \mathrm{Att}(M') \cup \mathrm{Att}(M'')$.

Proof. The first inclusion is trivial from the definition. For the second, let $P \in \mathrm{Att}(M)$ and let N be a submodule such that M/N is P-secondary. If $M' + N = M$ then M/N is a non-trivial quotient of M', hence $P \in \mathrm{Att}(M')$. If $M' + N \neq M$ then $M/(M' + N)$ is a non-trivial quotient of M'' as well as of M/N, hence M'' has a P-secondary quotient and $P \in \mathrm{Att}(M'')$. ■

An A-module M is said to be *sum-irreducible* if it is neither zero nor the sum of two proper submodules.

Lemma. If M is Artinian and sum-irreducible, then it is secondary.

Proof. Suppose M is not secondary. Then there is $a \in A$ such that $M \neq aM$ and $a^n M \neq 0$ for all $n > 0$. Since M is Artinian, we have $a^n M = a^{n+1} M$ for some n. Set $K = \{x \in M \mid a^n x = 0\}$. Then it is immediate that $M = K + aM$, and so M is not sum-irreducible. ■

Theorem 6.11. If M is Artinian, then it has a secondary representation.

Proof. Similar to the proof of Theorem 6.8, (iii). ■

The class of modules which have secondary representations is larger than that of Artinian modules. Sharp [8] proved that an injective module over a Noetherian ring has a secondary representation.

Exercises to Appendix to §6.

6.8. An A-module M is *coprimary* if $\mathrm{Ass}(M)$ has just one element. Show that a finite module $M \neq 0$ over a Noetherian ring A is coprimary if and only if the following condition is satisfied: for every $a \in A$, the endomorphism $a: M \longrightarrow M$ is either injective or nilpotent. In this case $\mathrm{Ass}\, M = \{P\}$, where $P = \sqrt{(\mathrm{ann}\, M)}$.

6.9. Show that if M is an A-module of finite length then M is coprimary if and only if it is secondary. Show also that such a module M is a direct sum of secondary modules belonging to maximal ideals, and $\mathrm{Ass}(M) = \mathrm{Att}(M)$.

3

Properties of extension rings

Flatness was formulated by Serre in the 1950s and quickly grew into one of the basic tools of both algebraic geometry and commutative algebra. This is an algebraic notion which is hard to grasp geometrically. Flatness is defined quite generally for modules, but is particularly important for extensions of rings. The model case is that of completion. Complete local rings have a number of wonderful properties, and passing to the completion of a local ring is an effective technique in many cases; this is analogous to studying an algebraic variety as an analytic space. The theory of integral extension of rings had been studied by Krull, and he discovered the so-called going-up and going-down theorems. We show that the going-down theorem also holds for flat extensions, and gather together flatness, completion and integral extensions in this chapter. We will use more sophisticated arguments to study flatness over Noetherian rings in Chapter 8, and completion in Chapter 10.

7 Flatness

Let A be a ring and M an A-module. Writing $\mathscr{S}$ to stand for a sequence $\cdots \longrightarrow N' \longrightarrow N \longrightarrow N'' \longrightarrow \cdots$ of A-modules and linear maps, we let $\mathscr{S} \otimes_A M$, or simply $\mathscr{S} \otimes M$ stand for the induced sequence $\cdots \longrightarrow N' \otimes_A M \longrightarrow N \otimes_A M \longrightarrow N'' \otimes_A M \longrightarrow \cdots$.

Definition. M is *flat* over A if for every exact sequence $\mathscr{S}$ the sequence $\mathscr{S} \otimes_A M$ is again exact. We sometimes shorten this to A-flat.

M is *faithfully flat* if for every sequence $\mathscr{S}$,

$$\mathscr{S} \text{ is exact} \Leftrightarrow \mathscr{S} \otimes_A M \text{ is exact.}$$

Any exact sequence $\mathscr{S}$ can be broken up into short exact sequences of the form $0 \to N_1 \longrightarrow N_2 \longrightarrow N_3 \to 0$, so that in the definition of flatness we need only consider short exact sequences $\mathscr{S}$. Moreover, in view of the right-exactness of tensor product (see Appendix A, Formula 8), we can restrict attention to exact sequences $\mathscr{S}$ of the form $0 \to N_1 \longrightarrow N$, and need only check the exactness of $\mathscr{S} \otimes M$: $0 \to N_1 \otimes M \longrightarrow N \otimes M$.

If $f: A \longrightarrow B$ is a homomorphism of rings and B is flat as an A-module,

we say that f is a *flat homomorphism*, or that B is a *flat* A-algebra. For example, the localisation A_S of A is a flat A-algebra (Theorems 4.4 and 4.5).
Transitivity. Let B be an A-algebra and M a B-module. Then the following hold;

(1) B is flat over A and M is flat over $B \Rightarrow M$ is flat over A;

(2) B is faithfully flat over A and M is faithfully flat over $B \Rightarrow M$ is faithfully flat over A;

(3) M is faithfully flat over B and flat over $A \Rightarrow B$ is flat over A;

(4) M is faithfully flat over both A and $B \Rightarrow B$ is faithfully flat over A.

Each of these follows easily from the fact that $(\mathscr{S} \otimes_A B) \otimes_B M = \mathscr{S} \otimes_A B$ for any sequence of A-modules $\mathscr{S}$.
Change of coefficient ring. Let B be an A-algebra and M an A-module. Then the following hold:

(1) M is flat over $A \Rightarrow M \otimes_A B$ is flat over B;

(2) M is faithfully flat over $A \Rightarrow M \otimes_A B$ is faithfully flat over B.

These follow from that fact that $\mathscr{S} \otimes_B (B \otimes_A M) = \mathscr{S} \otimes_A M$ for any sequence of B-modules $\mathscr{S}$.

Theorem 7.1. Let $A \longrightarrow B$ be a homomorphism of rings and M a B-module. A necessary and sufficient condition for M to be flat over A is that for every prime ideal P of B, the localisation M_P is flat over $A_\mathfrak{p}$ where $\mathfrak{p} = P \cap A$ (or the same condition for every maximal ideal P of B).
Proof. First of all we make the following observation: if $S \subset A$ is a multiplicative set and M, N are A_s-modules, then $M \otimes_{A_s} N = M \otimes_A N$. This follows from the fact that in $N \otimes_A M$ we have

$$\frac{a}{s} x \otimes y = \frac{ax}{s} \otimes \frac{sy}{s} = \frac{sx}{s} \otimes \frac{ay}{s} = x \otimes \frac{a}{s} y,$$

for $x \in M$, $y \in N$, $a \in A$ and $s \in S$. (In general, if B is an A-algebra and M and N are B-modules, it can be seen from the construction of the tensor product that $M \otimes_B N$ is the quotient of $M \otimes_A N$ by the submodule generated by $\{bx \otimes y - x \otimes by \mid x \in M,\ y \in N \text{ and } b \in B\}$.)

Assume now that M is A-flat. The map $A \longrightarrow B$ induces $A_\mathfrak{p} \longrightarrow B_P$, and M_P is a B_P-module, therefore an $A_\mathfrak{p}$-module. Let $\mathscr{S}$ be an exact sequence of $A_\mathfrak{p}$-modules; then, by the above observation,

$$\mathscr{S} \otimes_{A_\mathfrak{p}} M_P = \mathscr{S} \otimes_A M_P = (\mathscr{S} \otimes_A M) \otimes_B B_P,$$

and the right-hand side is an exact sequence, so that M_P is $A_\mathfrak{p}$-flat.

Next, suppose that M_P is $A_\mathfrak{p}$-flat for every maximal ideal P of B. Let $0 \rightarrow N' \longrightarrow N$ be an exact sequence of A-modules, and write K for the kernel of the B-linear map $N' \otimes_A M \longrightarrow N \otimes_A M$, so that $0 \rightarrow K \longrightarrow N' \otimes M \longrightarrow N \otimes M$ is an exact sequence of B-modules. For any $P \in \text{m-Spec}\, B$ the localisation

$$0 \to K_P \longrightarrow N' \otimes_A M_P \longrightarrow N \otimes_A M_P$$

is exact, and since $N' \otimes_A M_P = N' \otimes_A (A_\mathfrak{p} \otimes_{A_\mathfrak{p}} M_P) = N'_\mathfrak{p} \otimes_{A_\mathfrak{p}} M_P$, and similarly $N \otimes_A M_P = N_\mathfrak{p} \otimes_{A_\mathfrak{p}} M_P$, we have $K_P = 0$ by hypothesis. Therefore by Theorem 4.6 we have $K = 0$, and this is what we have to prove.

Theorem 7.2. Let A be a ring and M an A-module. Then the following conditions are equivalent:

(1) M is faithfully flat over A;

(2) M is A-flat, and $N \otimes_A M \neq 0$ for any non-zero A-module N;

(3) M is A-flat, and $\mathfrak{m}M \neq M$ for every maximal ideal $\mathfrak{m}$ of A.

Proof. (1)⇒(2). Let $\mathscr{S}$ be the sequence $0 \to N \to 0$. If $N \otimes M = 0$ then $\mathscr{S} \otimes M$ is exact, so $\mathscr{S}$ is exact, and therefore $N = 0$.

(2)⇒(3). This is clear from $M/\mathfrak{m}M = (A/\mathfrak{m}) \otimes_A M$.

(3)⇒(2). If $N \neq 0$ and $0 \neq x \in N$ then $Ax \simeq A/\mathrm{ann}(x)$, so that taking a maximal ideal $\mathfrak{m}$ containing $\mathrm{ann}(x)$, we have $M \neq \mathfrak{m}M \supset \mathrm{ann}(x)\cdot M$; hence, $Ax \otimes M \neq 0$. By the flatness assumption, $Ax \otimes M \longrightarrow N \otimes M$ is injective, so that $N \otimes M \neq 0$.

(2)⇒(1). Consider a sequence of A-modules

$$\mathscr{S}: N' \xrightarrow{f} N \xrightarrow{g} N''.$$

If

$$\mathscr{S} \otimes M: N' \otimes M \xrightarrow{f_M} N \otimes M \xrightarrow{g_M} N'' \otimes M$$

is exact then $g_M \circ f_M = (g \circ f)_M = 0$, so that by flatness, $\mathrm{Im}(g \circ f) \otimes M = \mathrm{Im}(g_M \circ f_M) = 0$. By assumption we then have $\mathrm{Im}(g \circ f) = 0$, that is $g \circ f = 0$; hence $\mathrm{Ker}\, g \supset \mathrm{Im}\, f$. If we set $H = \mathrm{Ker}\, g/\mathrm{Im}\, f$ then by flatness,

$$H \otimes M = \mathrm{Ker}\,(g_M)/\mathrm{Im}\,(f_M) = 0,$$

so that the assumption gives $H = 0$. Therefore $\mathscr{S}$ is exact. ■

A ring homomorphism $f: A \longrightarrow B$ induces a map ${}^a f: \mathrm{Spec}\, B \longrightarrow \mathrm{Spec}\, A$, under which a point $\mathfrak{p} \in \mathrm{Spec}\, A$ has an inverse image ${}^a f^{-1}(\mathfrak{p}) = \{P \in \mathrm{Spec}\, B \mid P \cap A = \mathfrak{p}\}$ which is homeomorphic to $\mathrm{Spec}(B \otimes_A \kappa(\mathfrak{p}))$. Indeed, setting $C = B \otimes_A \kappa(\mathfrak{p})$ and $S = A - \mathfrak{p}$, and defining $g: B \longrightarrow C$ by $g(b) = b \otimes 1$, then since $\kappa(\mathfrak{p}) = (A/\mathfrak{p}) \otimes A_S$, we have

$$C = B \otimes_A (A/\mathfrak{p}) \otimes_A A_S = (B/\mathfrak{p}B)_S = (B/\mathfrak{p}B)_{f(S)}.$$

Thus ${}^a g: \mathrm{Spec}\, C \longrightarrow \mathrm{Spec}\, B$ has the image

$$\{P \in \mathrm{Spec}\, B \mid P \supset \mathfrak{p}B \text{ and } P \cap f(S) = \varnothing\}$$
$$= \{P \in \mathrm{Spec}\, B \mid P \cap A = \mathfrak{p}\},$$

which is ${}^a f^{-1}(\mathfrak{p})$, and ${}^a g$ induces a homomorphism of $\mathrm{Spec}\, C$ with ${}^a f^{-1}(\mathfrak{p})$. For this reason we call $\mathrm{Spec}\, C = \mathrm{Spec}(B \otimes \kappa(\mathfrak{p}))$ the *fibre over* $\mathfrak{p}$. The inverse map ${}^a f^{-1}(\mathfrak{p}) \longrightarrow \mathrm{Spec}\, C$ takes $P \in {}^a f^{-1}(\mathfrak{p})$ into $PC = PB_S/\mathfrak{p}B_S$.

For $P^*\in\operatorname{Spec} C$ we set $P=P^*\cap B$; then by Theorems 4.2 and 4.3, we have

$$P^*=PC \text{ and } C_{P^*}=(B_S/\mathfrak{p}B_S)_{PC}=B_P/\mathfrak{p}B_P=B_P\otimes_A\kappa(\mathfrak{p}).$$

Theorem 7.3. Let $f:A\longrightarrow B$ be a ring homomorphism and M a B-module. Then

(i) M is faithfully flat over $A\Rightarrow {}^af(\operatorname{Supp}(M))=\operatorname{Spec} A$.

(ii) If M is a finite B-module then

M is A-flat and ${}^af(\operatorname{Supp}(M))\supset \text{m-Spec} A\Leftrightarrow M$ is faithfully flat over A.

Proof. (i) For $\mathfrak{p}\in\operatorname{Spec} A$, by faithful flatness we have $M\otimes_A\kappa(\mathfrak{p})\neq 0$. Hence, if we set $C=B\otimes_A\kappa(\mathfrak{p})$ and $M'=M\otimes_A\kappa(\mathfrak{p})=M\otimes_B C$, the C-module $M'\neq 0$, so that there is a $P^*\in\operatorname{Spec} C$ such that $M'_{P^*}\neq 0$. Now set $P=P^*\cap B$; then

$$M'_{P^*}=M\otimes_B C_{P^*}=M\otimes_B(B_P\otimes_{B_P}C_{P^*})=M_P\otimes_{B_P}C_{P^*}$$

so that $M_P\neq 0$, that is $P\in\operatorname{Supp}(M)$. But $P^*\in\operatorname{Spec}(B\otimes\kappa(\mathfrak{p}))$, so that as we have seen $P\cap A=\mathfrak{p}$. Therefore $\mathfrak{p}\in {}^af(\operatorname{Supp}(M))$.

(ii) It is enough to show that $M/\mathfrak{m}M\neq 0$ for any maximal ideal $\mathfrak{m}$ of A. By assumption there is a prime ideal P of B such that $P\cap A=\mathfrak{m}$ and $M_P\neq 0$. By NAK, since M_P is finite over B_P we have $M_P/PM_P\neq 0$, and *a fortiori* $M_P/\mathfrak{m}M_P=(M/\mathfrak{m}M)_P\neq 0$, so that $M/\mathfrak{m}M\neq 0$. ∎

Let $(A,\mathfrak{m})$ and $(B,\mathfrak{n})$ be local rings, and $f:A\longrightarrow B$ a ring homomorphism; f is said to be a *local homomorphism* if $f(\mathfrak{m})\subset\mathfrak{n}$. If this happens then by Theorem 2, or by Theorem 3, (ii), we see that it is equivalent to say that f is flat or faithfully flat.

Let S be a multiplicative set of A. Then it is easy to see that $\operatorname{Spec}(A_S)\longrightarrow\operatorname{Spec} A$ is surjective only if S consists of units, that is $A=A_S$. Thus from the above theorem, if $A\neq A_S$ then A_S is flat but not faithfully flat over A.

Theorem 7.4.

(i) Let A be a ring, M a flat A-module, and N_1, N_2 two submodules of an A-module N. Then as submodules of $N\otimes M$ we have

$$(N_1\cap N_2)\otimes M=(N_1\otimes M)\cap(N_2\otimes M).$$

(ii) Let $A\longrightarrow B$ be a flat ring homomorphism, and let I_1 and I_2 be ideals of A. Then

$$(I_1\cap I_2)B=I_1B\cap I_2B.$$

(iii) If in addition I_2 is finitely generated then

$$(I_1:I_2)B=I_1B:I_2B.$$

Proof. (i) Define $\varphi:N\longrightarrow N/N_1\oplus N/N_2$ by $\varphi(x)=(x+N_1, x+N_2)$; then $0\to N_1\cap N_2\longrightarrow N\longrightarrow N/N_1\oplus N/N_2$ is exact, and hence so is

$$0\to(N_1\cap N_2)\otimes M\longrightarrow N\otimes M\longrightarrow$$
$$(N\otimes M)/(N_1\otimes M)\oplus(N\otimes M)/(N_2\otimes M).$$

This is the assertion in (i).

(ii) This is a particular case of (i) with $N = A, M = B$, in view of the fact that for an ideal I of A the subset $I \otimes_A B$ of $A \otimes_A B = B$ coincides with IB.

(iii) If $I_2 = Aa_1 + \cdots + Aa_n$ then since $(I_1 : I_2) = \bigcap_i (I_1 : a_i)$, we can use (ii) to reduce to the case that I_2 is principal. For $a \in A$ we have the exact sequence

$$0 \to (I_1 : Aa) \longrightarrow A \xrightarrow{a} A/I_1,$$

and tensoring this with B gives the assertion. ■

Example. Let k be a field, and consider the subring $A = k[x^2, x^3]$ of the polynomial ring $B = k[x]$ in an indeterminate x. Then $x^2 A \cap x^3 A$ is the set of polynomials made up of terms of degree $\geqslant 5$ in x, so that $(x^2 A \cap x^3 A)B = x^5 B$, but on the other hand $x^2 B \cap x^3 B = x^3 B$. Therefore by the above theorem, B is not flat over A.

Theorem 7.5. Let $f : A \longrightarrow B$ be a faithfully flat ring homomorphism.

(i) For any A-module M, the map $M \longrightarrow M \otimes_A B$ defined by $m \mapsto m \otimes 1$ is injective; in particular $f : A \longrightarrow B$ is itself injective.

(ii) If I is an ideal of A then $IB \cap A = I$.

Proof. (i) Let $0 \neq m \in M$. Then $(Am) \otimes B$ is a B-submodule of $M \otimes B$ which can be identified with $(m \otimes 1)B$. But by Theorem 2, $(Am) \otimes B \neq 0$, so that $m \otimes 1 \neq 0$.

(ii) follows by applying (i) to $M = A/I$, using $(A/I) \otimes B = B/IB$.

Theorem 7.6. Let A be a ring and M a flat A-module. If $a_{ij} \in A$ and $x_j \in M$ (for $1 \leqslant i \leqslant r$ and $1 \leqslant j \leqslant n$) satisfy

$$\sum_j a_{ij} x_j = 0 \quad \text{for all} \quad i,$$

then there exists an integer s and $b_{jk} \in A$, $y_k \in M$ (for $1 \leqslant j \leqslant n$ and $1 \leqslant k \leqslant s$) such that

$$\sum_j a_{ij} b_{jk} = 0 \quad \text{for all} \quad i, k, \quad \text{and} \quad x_j = \sum_j b_{jk} y_k \quad \text{for all} \quad j.$$

Thus the solutions in a flat module M of a system of simultaneous linear equations with coefficients in A can be expressed as a linear combination of solutions in A. Conversely, if the above conclusion holds for the case of a single equation (that is for $r = 1$), then M is flat.

Proof. Set $\varphi : A^n \longrightarrow A^r$ for the linear map defined by the matrix (a_{ij}), and let $\varphi_M : M^n \longrightarrow M^r$ be the same thing for M; then $\varphi_M = \varphi \otimes 1$, where 1 is the identity map of M. Setting $K = \operatorname{Ker} \varphi$ and tensoring the exact sequence $K \xrightarrow{i} A^n \xrightarrow{\varphi} A^r$ with M, we get the exact sequence

$$K \otimes M \xrightarrow{i \otimes 1} M^n \xrightarrow{\varphi_M} M^r.$$

By assumption $\varphi_M(x_1,\dots,x_n)=0$, so that we can write

$$(x_1,\dots,x_n)=(i\otimes 1)\left(\sum_{k=1}^{s}\beta_k\otimes y_k\right)\quad\text{with } \beta_k\in K \quad\text{and}\quad y_k\in M.$$

If we write out β_k as an element of A^n in the form $\beta_k=(b_{1k},\dots,b_{nk})$ with $b_{ik}\in A$ then the conclusion follows. The converse will be proved after the next theorem. ■

Theorem 7.7. Let A be a ring and M an A-module. Then M is flat over A if and only if for every finitely generated ideal I of A the canonical map $I\otimes_A M\longrightarrow A\otimes_A M$ is injective, and therefore $I\otimes M\simeq IM$.

Proof. The 'only if' is obvious, and we prove the 'if'. Firstly, every ideal of A is the direct limit of the finitely generated ideals contained in it, so that by Theorems A1 and A2 of Appendix A, $I\otimes M\longrightarrow M$ is injective for every ideal I. Moreover, if N is an A-module and $N'\subset N$ a submodule, then since N is the direct limit of modules of the form $N'+F$, with F finitely generated, to prove that $N'\otimes M\longrightarrow N\otimes M$ is injective we can assume that $N=N'+A\omega_1+\dots+A\omega_n$. Then setting $N_i=N'+A\omega_1+\dots+A\omega_i$ (for $1\leqslant i\leqslant n$), we need only show that each step in the chain

$$N'\otimes M\longrightarrow N_1\otimes M\longrightarrow N_2\otimes M\longrightarrow\cdots\longrightarrow N\otimes M$$

is injective, and finally that if $N=N'+A\omega$ then $N'\otimes M\longrightarrow N\otimes M$ is injective. Now we set $I=\{a\in A\,|\,a\omega\in N'\}$, and get the exact sequence

$$0\to N'\longrightarrow N\longrightarrow A/I\to 0.$$

This induces a long exact sequence (see Appendix B, p. 279)

$$\cdots\longrightarrow \mathrm{Tor}_1^A(M,A/I)\longrightarrow N'\otimes M\longrightarrow N\otimes M\longrightarrow (A/I)\otimes M\to 0;$$

hence it is enough to prove that

(*) $\mathrm{Tor}_1^A(M,A/I)=0.$

For this consider the short exact sequence

$$0\to I\longrightarrow A\longrightarrow A/I\to 0$$

and the induced long exact sequence

$$\mathrm{Tor}_1^A(M,A)=0\longrightarrow \mathrm{Tor}_1^A(M,A/I)\longrightarrow I\otimes M\longrightarrow M\longrightarrow\cdots;$$

since $I\otimes M\longrightarrow M$ is injective, (*) must hold. ■

From this theorem we can prove the converse of Theorem 6. Indeed, if $I=Aa_1+\dots+Aa_n$ is a finitely generated ideal of A then an element ξ of $I\otimes M$ can be written as $\xi=\sum_1^n a_i\otimes m_i$ with $m_i\in M$. Suppose that ξ is 0 in M, that is that $\sum a_i m_i=0$. Now if the conclusion of Theorem 6 holds for M, there exist $b_{ij}\in A$ and $y_j\in M$ such that

$$\sum_i a_i b_{ij}=0 \quad\text{for all } j,\quad\text{and}\quad m_i=\sum b_{ij}y_j\quad\text{for all}\quad i.$$

Then $\xi=\sum a_i\otimes m_i=\sum_i\sum_j a_i b_{ij}\otimes y_j=0$, so that $I\otimes M\longrightarrow M$ is injective, and therefore M is flat.

Theorem 7.8. Let A be a ring and M an A-module. The following conditions are equivalent:

(1) M is flat;

(2) for every A-module N we have $\mathrm{Tor}_1^A(M,N)=0$;

(3) $\mathrm{Tor}_1^A(M,A/I)=0$ for every finitely generated ideal I.

Proof. (1)⇒(2) If we let $\cdots\longrightarrow L_i\longrightarrow L_{i-1}\longrightarrow\cdots\longrightarrow L_0\longrightarrow N\to 0$ be a projective resolution of N then

$$\cdots\longrightarrow L_i\otimes M\longrightarrow L_{i-1}\otimes M\longrightarrow\cdots\longrightarrow L_0\otimes M$$

is exact, so that $\mathrm{Tor}_i^A(M,N)=0$ for all $i>0$.

(2)⇒(3) is obvious.

(3)⇒(1) The short exact sequence $0\to I\longrightarrow A\longrightarrow A/I\to 0$ induces a long exact sequence

$$\mathrm{Tor}_1^A(M,A/I)=0\longrightarrow I\otimes M\longrightarrow M\longrightarrow M\otimes A/I\to 0,$$

and hence $I\otimes M\longrightarrow M$ is injective; therefore by the previous theorem M is flat. ■

Theorem 7.9. Let $0\to M'\longrightarrow M\longrightarrow M''\to 0$ be an exact sequence of A-modules; then if M' and M'' are both flat, so is M.

Proof. For any A-module N the sequence $\mathrm{Tor}_1(M',N)\longrightarrow\mathrm{Tor}_1(M,N)\longrightarrow\mathrm{Tor}_1(M'',N)$ is exact, and since the first and third groups are zero, also $\mathrm{Tor}_1(M,N)=0$. Therefore by the previous theorem M is flat. ■

A free module is obvious faithfully flat (if F is free and $\mathscr{S}$ is a sequence of A-modules then $\mathscr{S}\otimes F$ is just a sum of copies of $\mathscr{S}$ in number equal to the cardinality of a basis of F). Conversely, over a local ring the following theorem holds, so that for finite modules flat, faithfully flat and free are equivalent conditions.

Theorem 7.10. Let $(A,\mathfrak{m})$ be a local ring and M a flat A-module. If $x_1,\ldots,x_n\in M$ are such that their images $\bar{x}_1,\ldots,\bar{x}_n$ in $\bar{M}=M/\mathfrak{m}M$ are linearly independent over the field $A/\mathfrak{m}$ then $x_1,\ldots,x_n$ are linearly independent over A. Hence if M is finite, or if $\mathfrak{m}$ is nilpotent, then any minimal basis of M (see §2) is a basis of M, and M is a free module.

Proof. By induction on n. If $n=1$, and $a\in A$ is such that $ax_1=0$ then by Theorem 6 there are $b_1,\ldots,b_s\in A$ such that $ab_i=0$ and $x\in\sum b_iM$; by assumption $x_1\notin\mathfrak{m}M$, so that among the b_i there must be one not contained in $\mathfrak{m}$. This b_i is then a unit, so that we must have $a=0$.

For $n>1$, let $\sum a_ix_i=0$; then there are $b_{ij}\in A$ and $y_j\in M$ (for $1\leqslant j\leqslant s$) such that $\sum a_ib_{ij}=0$ and $x_i=\sum b_{ij}y_j$. Now $x_n\notin\mathfrak{m}M$, so that among the b_{nj} at least one is a unit. Hence a_n is a linear combination of $a_1,\ldots,a_{n-1}$, that is $a_n = \sum_{i=1}^{n-1} a_i c_i$ for some $c_i \in A$. Therefore we have

$$a_1(x_1 + c_1 x_n) + \cdots + a_{n-1}(x_{n-1} + c_{n-1} x_n) = 0;$$

however, the $(n-1)$ elements $\bar{x}_1 + \bar{c}_1 x_n, \ldots, \bar{x}_{n-1} + \bar{c}_{n-1}\bar{x}_n$ of $\bar{M}$ are linearly independent over $A/\mathfrak{m}$, so that by induction, $a_1 = \cdots = a_{n-1} = 0$. Hence also $a_n = 0$. ■

Theorem 7.11. Let A be a ring, M and N two A-modules, and B a flat A-algebra. If M is of finite presentation then we have

$$\mathrm{Hom}_A(M, N) \otimes_A B = \mathrm{Hom}_B(M \otimes_A B, N \otimes_A B).$$

Proof. Fixing N and B, we define contravariant functors F and G of an A-module M by

$$F(M) = \mathrm{Hom}_A(M, N) \otimes_A B$$

and

$$G(M) = \mathrm{Hom}_B(M \otimes_A B, N \otimes_A B);$$

then we can define a morphism of functors $\lambda: F \longrightarrow G$ by

$$\lambda(f \otimes b) = b \cdot (f \otimes 1_B) \quad \text{for} \quad f \in \mathrm{Hom}_A(M, N) \quad \text{and} \quad b \in B.$$

Both F and G are left-exact functors.

Now if M is of finite presentation there is an exact sequence of the form $A^p \longrightarrow A^q \longrightarrow M \to 0$, and from this we get a commutative diagram

$$\begin{array}{ccccc} 0 \to F(M) & \longrightarrow & F(A^q) & \longrightarrow & F(A^p) \\ \lambda\downarrow & & \lambda\downarrow & & \lambda\downarrow \\ 0 \to G(M) & \longrightarrow & G(A^q) & \longrightarrow & G(A^p) \end{array}$$

having two exact rows. Now $F(A^p) = N^p \otimes B$ and $G(A^p) = (N \otimes B)^p$, so that the right-hand λ is an isomorphism, and similarly the middle λ is an isomorphism. Thus, as one sees easily, the left-hand λ is also an isomorphism. ■

Corollary. Let A, M and N be as in the theorem, and let $\mathfrak{p}$ be a prime ideal of A. Then

$$\mathrm{Hom}_A(M, N) \otimes_A A_\mathfrak{p} = \mathrm{Hom}_{A_\mathfrak{p}}(M_\mathfrak{p}, N_\mathfrak{p}).$$

Theorem 7.12. Let A be a ring and M an A-module of finite presentation. Then M is a projective A-module if and only if $M_\mathfrak{m}$ is a free $A_\mathfrak{m}$-module for every maximal ideal $\mathfrak{m}$ of A.

Proof of 'only if'. If M is projective it is a direct summand of a free module, and this property is preserved by localisation, so that $M_\mathfrak{m}$ is projective over $A_\mathfrak{m}$, and is therefore free by Theorem 2.5.

Proof of 'if'. Let $N_1 \longrightarrow N_2 \to 0$ be an exact sequence of A-modules. Write C for the cokernel of

$$\mathrm{Hom}_A(M, N_1) \longrightarrow \mathrm{Hom}_A(M, N_2);$$

then for any maximal ideal $\mathfrak{m}$ of A we have

$$C_\mathfrak{m} = \operatorname{Coker}\{\operatorname{Hom}_{A_\mathfrak{m}}(M_\mathfrak{m}, (N_1)_\mathfrak{m}) \longrightarrow \operatorname{Hom}_{A_\mathfrak{m}}(M_\mathfrak{m}, (N_2)_\mathfrak{m})\} = 0.$$

Hence $C = 0$ by Theorem 4.6, and this is what we had to prove . ■

Corollary. If A is a ring and M is an A-module of finite presentation, then M is flat if and only if it is projective.

Proof. This follows from Theorems 1, 12 and 10

Exercises to §7. Prove the following propositions.

7.1. If B is a faithfully flat A-algebra then for an A-module M we have
$B \otimes_A M$ is B-flat $\Leftrightarrow M$ is A-flat,
and similarly for faithfully flat.

7.2. Let A and B be integral domains with $A \subset B$, and suppose that A and B have the same field of fractions; if B is faithfully flat over A then $A = B$.

7.3. Let B be a faithfully flat A-algebra; for an A-module M we can view M as a submodule of $B \otimes_A M$ (by Theorem 7.5). Then if $\{m_\lambda\}$ is a subset of M which generates $B \otimes M$ over B, it also generates M over A.

7.4. Let A be a Noetherian ring and $\{M_\lambda\}_{\lambda\in\Lambda}$ a family of flat A-modules; then the direct product module $\prod_{\lambda\in\Lambda} M$ is also flat. In particular the formal power series $A[\![X_1, \ldots, X_n]\!]$ is a flat A-algebra (Chase [1]).

7.5. Let A be a ring and N a flat A-module; if $a \in A$ is A-regular, it is also N-regular.

7.6. Let A be a ring, and $C_.$ a complex of A-modules; for an A-module N we write $C_. \otimes N$ for the complex $\cdots \longrightarrow C_{i+1} \otimes N \longrightarrow C_i \otimes N \longrightarrow \cdots$. If N is flat over A then $H_i(C_.) \otimes N = H_i(C_. \otimes N)$ for all i.

7.7. Let A be a ring and B a flat A-algebra; then if M and N are A-modules,
$$\operatorname{Tor}_i^A(M, N) \otimes_A B = \operatorname{Tor}_i^B(M \otimes B, N \otimes B) \quad \text{for all } i.$$
If in addition M is finitely generated and A is Noetherian then
$$\operatorname{Ext}_A^i(M, N) \otimes_A B = \operatorname{Ext}_B^i(M \otimes_A B, N \otimes_A B) \quad \text{for all } i.$$

7.8. Theorem 7.4, (i) does not hold for the intersection of infinitely many submodules; explain why, and construct a counter-example.

7.9. If B is a faithfully flat A-algebra and B is Noetherian then A is Noetherian.

***Appendix to* §7.** Pure submodules

Let A be a ring and M an A-module. A submodule N of M is said to be *pure* if the sequence $0 \to N \otimes E \longrightarrow M \otimes E$ is exact for every A-module E. Since tensor product and exactness commute with inductive limits, we need only consider A-modules E of finite presentation.

Example 1. If M/N is a flat A-module, then N is a pure submodule of M. This follows from the exact sequence $\mathrm{Tor}_1^A(M/N, E) \longrightarrow N \otimes E \longrightarrow M \otimes E$.

Example 2. Any direct summand of M is a pure submodule.

Example 3. If $A = \mathbb{Z}$, a submodule N of M is pure if and only if $N \cap mM = mN$ for all $m > 0$. In fact the condition is equivalent to the exactness of $0 \to N \otimes \mathbb{Z}/m\mathbb{Z} \longrightarrow M \otimes \mathbb{Z}/m\mathbb{Z}$, and every finitely generated $\mathbb{Z}$-module is a direct sum of cyclic modules.

Theorem 7.13. A submodule N of M is pure if and only if the following condition holds: if $x_i = \sum_{j=1}^{s} a_{ij} m_j$ (for $1 \leqslant i \leqslant r$), with $m_j \in M$, $x_i \in N$ and $a_{ij} \in A$, then there exist $y_j \in N$ (for $1 \leqslant j \leqslant s$) such that $x_i = \sum_{j=1}^{s} a_{ij} y_j$ (for $1 \leqslant i \leqslant r$).

Proof. Suppose N is pure in M. Consider the free module A^r with basis $e_1, \ldots, e_r$ and let D be the submodule of A^r generated by $\sum_i a_{ij} e_i$, $1 \leqslant j \leqslant s$. Set $E = A^r/D$, and let $\bar{e}_i$ denote the image of e_i in E. Then in $M \otimes E$ we have

$$\sum_i x_i \otimes \bar{e}_i = \sum_i \sum_j a_{ij} m_j \otimes \bar{e}_i = \sum_j m_j \otimes \sum_i a_{ij} \bar{e}_i = 0,$$

hence $\sum x_i \otimes \bar{e}_i = 0$ in $N \otimes E$ by purity. But this means that, in $N \otimes A^r$, the element $\sum_i x_i \otimes e_i$ is of the form $\sum_j y_j \otimes \sum_i a_{ij} e_i$ for some $y_j \in N$.

Conversely, suppose the condition is satisfied. Let E be an A-module of finite presentation. Then we can write $E = A^r/D$ with D generated by a finite number of elements of A^r, say $\sum_{i=1}^{r} a_{ij} e_i$, $1 \leqslant j \leqslant s$. Then reversing the preceding argument we can see that $N \otimes E \longrightarrow M \otimes E$ is injective. ■

Theorem 7.14. If N is a pure submodule and M/N is of finite presentation, then N is a direct summand of M.

Proof. We will prove that $0 \to N \xrightarrow{i} M \xrightarrow{p} M/N \to 0$ splits, where i and p are the natural maps. For this we need only construct a linear map $f: M/N \longrightarrow M$ such that pf is the identity map of M/N. Let $\{t_1, \ldots, t_r\}$ be a set of generators of M/N, so that $M/N \simeq A^r/R$, where R is the submodule of relations among the t_j; let $\{(a_{i1}, \ldots, a_{ir}) | 1 \leqslant i \leqslant s\}$ be a set of generators of R. Choose a pre-image ξ_j of t_j in M for each j. Then set $\eta_i = \sum a_{ij} \xi_j \in N$ (for $1 \leqslant i \leqslant s$). By the preceding theorem there exist $\xi'_j \in N$ such that $\eta_i = \sum a_{ij} \xi'_j$ (for $1 \leqslant i \leqslant s$). Then $\sum a_{ij}(\xi_j - \xi'_j) = 0$ (for $1 \leqslant i \leqslant s$), and setting $f(t_j) = \xi_j - \xi'_j$, we obtain a linear map $f: M/N \longrightarrow M$ which satisfies the requirement. ■

8 Completion and the Artin–Rees lemma

Let A be a ring and M an A-module; for a directed set Λ, suppose that $\mathscr{F} = \{M_\lambda\}_{\lambda\in\Lambda}$ is a family of submodules of M indexed by Λ and such that $\lambda < \mu \Rightarrow M_\lambda \supset M_\mu$. Then taking $\mathscr{F}$ as a system of neighbourhoods of 0 makes M into a topological group under addition. In this topology, for any $x \in M$ a system of neighbourhoods of x is given by $\{x + M_\lambda\}_{\lambda\in\Lambda}$. In M addition and subtraction are continuous, as is scalar multiplication $x \mapsto ax$ for any $a \in A$. When $M = A$ each M_λ is an ideal, so that multiplication is also continuous:

$$(a + M_\lambda)(b + M_\lambda) \subset ab + M_\lambda.$$

This type of topology is called a *linear topology* on M; it is separated (that is, Hausdorff) if and only if $\bigcap_\lambda M_\lambda = 0$. Each $M_\lambda \subset M$ is an open set, each coset $x + M_\lambda$ is again open, and the complement $M - M_\lambda$ of M_λ is a union of cosets, so is also open. Hence M_λ is an open and closed subset; the quotient module M/M_λ is then discrete in the quotient topology.

$M/\bigcap_\lambda M_\lambda$ is called the *separated module associated* with M. Moreover, since for $\lambda < \mu$ there is a natural linear map $\varphi_{\lambda\mu}: M/M_\mu \longrightarrow M/M_\lambda$, we can construct the inverse system $\{M/M_\lambda;\ \varphi_{\lambda\mu}\}$ of A-modules; its inverse limit $\varprojlim M/M_\lambda$ is called the *completion* of M, and is written $\hat{M}$. We give each M/M_λ the discrete topology, the direct product $\prod_\lambda M/M_\lambda$ the product topology, and $\hat{M}$ the subspace topology in $\prod_\lambda M/M_\lambda$. Let $\psi: M \longrightarrow \hat{M}$ be the natural A-linear map; then ψ is continuous, and $\psi(M)$ is dense in $\hat{M}$. Write $p_\lambda: \hat{M} \longrightarrow M/M_\lambda$ for the projection, and set $\operatorname{Ker} p_\lambda = M_\lambda^*$; it is easy to see that the topology of $\hat{M}$ coincides with the linear topology defined by $\mathscr{F} = \{M_\lambda^*\}_{\lambda\in\Lambda}$. The map p_λ is surjective (in fact $p_\lambda(\psi(M)) = M/M_\lambda$), so that $\hat{M}/M_\lambda^* \simeq M/M_\lambda$, and the completion of $\hat{M}$ coincides with $\hat{M}$ itself. If $\psi: M \longrightarrow \hat{M}$ is an isomorphism, we say that M is *complete*. (Caution: in Bourbaki terminology this is 'complete and separated'; we shorten this to 'complete' throughout.)

If $\mathscr{F}' = \{M'_\gamma\}_{\gamma\in\Gamma}$ is another family of submodules of M indexed by a directed set Γ, then $\mathscr{F}$ and $\mathscr{F}'$ give the same topology on M if and only if for each M_λ there is a $\gamma \in \Gamma$ such that $M'_\gamma \subset M_\lambda$, and for every M'_γ there is a $\mu \in \Lambda$ such that $M_\mu \subset M'_\gamma$. It is then easy to see that there is an isomorphism of topological modules $\varprojlim M/M_\lambda \simeq \varprojlim M/M'_\gamma$. Thus $\hat{M}$ depends only on the topology of M, as does the question of whether M is complete.

When $M = A$, $\{M/M_\lambda;\ \varphi_{\lambda\mu}\}$ becomes an inverse system of rings, $\hat{M} = \hat{A}$ is a ring, and $\psi: A \longrightarrow \hat{A}$ a ring homomorphism. $M_\lambda^* \subset \hat{A}$ is not just an A-submodule, but an ideal of $\hat{A}$; this is clear from the fact that $p_\lambda: \hat{A} \longrightarrow A/M_\lambda$ is a ring homomorphism.

If $N \subset M$ is a submodule, then the closure $\bar{N}$ of N in M is given by the following formula:

$$\bar{N} = \bigcap_{\lambda}(N + M_\lambda).$$

Indeed,

$$x \in \bar{N} \Leftrightarrow (x + M_\lambda) \cap N \neq \varnothing \quad \text{for all } \lambda.$$
$$\Leftrightarrow x \in N + M_\lambda \quad \text{for all } \lambda.$$

If we write M'_λ for the image of M_λ in the quotient module M/N, the quotient topology of M/N is just the linear topology defined by $\{M'_\lambda\}_{\lambda \in \Lambda}$. In fact, let $G \subset M$ be the inverse image of $G' \subset M/N$; then

G' is open in the quotient topology of M/N

$\Leftrightarrow G$ is open in M

$\Leftrightarrow$ for every $x \in G$ there is an M_λ such that $x + M_\lambda \subset G$

$\Leftrightarrow$ for every $x' \in G'$ there is an M'_λ such that $x' + M'_\lambda \subset G'$.

Hence the condition for M/N to be separated is that $\bigcap_\lambda M'_\lambda = 0$, that is $\bigcap(N + M_\lambda) = N$, or in other words, that N is closed in M. Moreover, the subspace topology of N is clearly the same thing as the linear topology defined by $\{N \cap M_\lambda\}_{\lambda \in \Lambda}$. Set $M/N = M'$; then

$$0 \to N/(N \cap M_\lambda) \longrightarrow M/M_\lambda \longrightarrow M'/M'_\lambda = M/(N + M_\lambda) \to 0$$

is an exact sequence, so that taking the inverse limit, we see that

$$0 \to \hat{N} \longrightarrow \hat{M} \longrightarrow (M/N)\hat{}$$

is exact. If we view $\hat{N}$ as a submodule of $\hat{M}$, the condition that $\xi = (\xi_\lambda)_{\lambda \in \Lambda} \in \hat{M}$ belongs to $\hat{N}$ is that each ξ_λ can be represented by an element of N, or in other words that $\xi \in \psi(N) + M^*_\lambda$ for each λ. Hence $\hat{N}$ is the same thing as the closure of $\psi(N)$ in $\hat{M}$. In general it is not clear whether $\hat{M} \longrightarrow (M/N)\hat{}$ is surjective, but this holds in the case $\Lambda = \{1, 2, \dots\}$. In fact then

$$(M/N)\hat{} = \varprojlim_n M/(N + M_n);$$

given an element $\xi' = (\xi'_1, \xi'_2, \dots) \in (M/N)\hat{}$, with $\xi'_n \in M/(N + M_n)$, let $x_1 \in M$ be an inverse image of ξ'_1, and $y_2 \in M$ an inverse image of ξ'_2; then $y_2 - x_1 \in N + M_1$, so that we can write

$$y_2 - x_1 = t + m_1 \quad \text{with} \quad t \in N \quad \text{and} \quad m_1 \in M_1.$$

If we set $x_2 = y_2 - t$ then $x_2 \in M$ is also an inverse image of ξ'_2, and satisfies $x_2 - x_1 \in M_1$. Similarly we can successively choose inverse images $x_n \in M$ of the ξ'_n in such a way that for $n = 1, 2, \dots$, we have $x_{n+1} - x_n \in M_n$. If we set $\xi_n \in M/M_n$ for the image of x_n, then by construction $\xi = (\xi_1, \xi_2, \dots)$ is an element of $\varprojlim M/M_n = \hat{M}$ which maps to ξ' in $(M/N)\hat{}$. This proves the following theorem.

Theorem 8.1. Let A be a ring, M an A-module with a linear topology, and $N \subset M$ a submodule. We give N the subspace topology, and M/N the quotient topology. Then these are both linear topologies, and we have:

(i) $0 \to \hat{N} \longrightarrow \hat{M} \longrightarrow (M/N)\hat{}$ is an exact sequence, and $\hat{N}$ is the closure of $\psi(N)$ in $\hat{M}$, where $\psi: M \longrightarrow \hat{M}$ is the natural map.

(ii) If moreover the topology of M is defined by a decreasing chain of submodules $M_1 \supset M_2 \supset \cdots$, then

$$0 \to \hat{N} \longrightarrow \hat{M} \longrightarrow (M/N)\hat{} \to 0$$

is exact. In other words, $(M/N)\hat{} \simeq \hat{M}/\hat{N}$. ∎

Now suppose that M and N are two A-modules with linear topologies, and let $f: M \longrightarrow N$ be a continuous linear map. If the topologies of M and N are given by $\{M_\lambda\}_{\lambda\in\Lambda}$ and $\{N_\gamma\}_{\gamma\in\Gamma}$, then for any $\gamma\in\Gamma$ there exists $\lambda\in\Lambda$ such that $M_\lambda \subset f^{-1}(N_\gamma)$. Define $\varphi_\gamma: \hat{M} \longrightarrow N/N_\gamma$ as the composite $\hat{M} \longrightarrow \hat{M}/M_\lambda^* \longrightarrow N/N_\gamma$, where the first arrow is the natural map, and the second is induced by f; one sees at once that φ_γ does not depend on the choice of λ for which $M_\lambda \subset f^{-1}(N_\gamma)$. Also, for $\gamma < \gamma'$ if we let $\psi_{\gamma\gamma'}$ denote the natural map $N/N_{\gamma'} \longrightarrow N/N_\gamma$, it is easy to see that $\varphi_\gamma = \psi_{\gamma\gamma'} \circ \varphi_{\gamma'}$; hence there is a continuous linear map $\hat{f}: \hat{M} \longrightarrow \hat{N}$ defined by the $(\varphi_\gamma)_{\gamma\in\Gamma}$, and the following diagram is commutative (the vertical arrows are the natural maps):

$$\begin{array}{ccc} M & \xrightarrow{f} & N \\ \downarrow & & \downarrow \\ \hat{M} & \xrightarrow{\hat{f}} & \hat{N}. \end{array}$$

Moreover, $\hat{f}$ is determined uniquely by this diagram and by continuity. Similarly, if A and B are rings with linear topologies, and $f: A \longrightarrow B$ is a continuous ring homomorphism, then f induces a continuous ring homomorphism $\hat{f}: \hat{A} \longrightarrow \hat{B}$.

Among the linear topologies, those defined by ideals are of particular importance. Let I be an ideal of A and M an A-module; the topology on M defined by $\{I^nM\}_{n=1,2,\ldots}$ is called the *I-adic topology*. If we also give A the I-adic topology, the completions $\hat{A}$ and $\hat{M}$ of A and M are called *I-adic completions*; it is easy to see that $\hat{M}$ is an $\hat{A}$-module: for $\alpha = (a_1, a_2, \ldots) \in \hat{A}$ with $a_n \in A/I^n$ and $\xi = (x_1, x_2, \ldots) \in \hat{M}$ with $x_n \in M/I^nM$ (for all n), we can just set

$$a\xi = (a_1x_1, a_2x_2, \ldots) \in \hat{M}.$$

As one can easily check, to say that M is complete for the I-adic topology is equivalent to saying that for every sequence $x_1, x_2, \ldots$ of elements of M satisfying $x_i - x_{i+1} \in I^iM$ for all i, there exists a unique $x \in M$ such that $x - x_i \in I^iM$ for all i. We can define a Cauchy sequence in M in the usual way ($\{x_i\}$ is Cauchy if and only if for every positive integer r there is an n_0 such that $x_{n+1} - x_n \in I^rM$ for $n > n_0$), and completeness can then be expressed as saying that a Cauchy sequence has a unique limit.

Theorem 8.2. Let A be a ring, I an ideal, and M an A-module.

(i) If A is I-adically complete then $I \subset \operatorname{rad}(A)$;

(ii) If M is I-adically complete and $a \in I$, then multiplication by $1+a$ is an automorphism of M.

Proof. (i) For $a \in I$, $1 - a + a^2 - a^3 + \cdots$ converges in A, and provides an inverse of $1+a$; hence $1+a$ is a unit of A. This means (see §1) that $I \subset \mathrm{rad}(A)$.

(ii) M is also an $\hat{A}$-module, and $1+a$ (or rather, its image in $\hat{A}$) is a unit in $\hat{A}$, so that this is clear. ■

The following two results show the usefulness of completeness.

Theorem 8.3 (Hensel's lemma). Let $(A, \mathfrak{m}, k)$ be a local ring, and suppose that A is $\mathfrak{m}$-adically complete. Let $F(X) \in A[X]$ be a monic polynomial, and let $\bar{F} \in k[X]$ be the polynomial obtained by reducing the coefficients of F modulo $\mathfrak{m}$. If there are monic polynomials g, $h \in k[X]$ with $(g, h) = 1$ and such that $\bar{F} = gh$, then there exist monic polynomials G, H with coefficients in A such that $F = GH$, $\bar{G} = g$ and $\bar{H} = h$.

Proof. If we take polynomials $G_1, H_1 \in A[X]$ such that $g = \bar{G}_1$, $h = \bar{H}_1$ then $F \equiv G_1 H_1 \bmod \mathfrak{m}[X]$. Suppose by induction that monic polynomials G_n, H_n have been constructed such that $F \equiv G_n H_n \bmod \mathfrak{m}^n[X]$, and $\bar{G}_n = g$, $\bar{H}_n = h$; then we can write

$$F - G_n H_n = \sum \omega_i U_i(X), \quad \text{with} \quad \omega_i \in \mathfrak{m}^n \quad \text{and} \quad \deg U_i < \deg F.$$

Since $(g, h) = 1$ we can find v_i, $w_i \in k[X]$ such that $\bar{U}_i = g v_i + h w_i$. Replacing v_i by its remainder modulo h, and making the corresponding correction to w_i we can assume $\deg v_i < \deg h$. Then

$$\deg h w_i = \deg(\bar{U}_i - g v_i) < \deg F, \quad \text{hence} \deg w_i < \deg g.$$

Choosing V_i, $W_i \in A[X]$ such that $\bar{V}_i = v_i$, $\deg V_i = \deg v_i$, $\bar{W}_i = w_i$, $\deg W_i = \deg w_i$, and setting $G_{n+1} = G_n + \sum \omega_i W_i$, $H_{n+1} = H_n + \sum \omega_i V_i$, we get

$$F \equiv G_{n+1} H_{n+1} \bmod \mathfrak{m}^{n+1}[X].$$

We construct in this way sequences of polynomials G_n, H_n for $n = 1, 2, \ldots$; then $\lim G_n = G$ and $\lim H_n = H$ clearly exist and satisfy $F = GH$. Obviously, $\bar{G} = \bar{G}_1 = g$, $\bar{H} = \bar{H}_1 = h$. ■

Theorem 8.4. Let A be a ring, I an ideal, and M and A-module. Suppose that A is I-adically complete, and M is separated for the I-adic topology. If M/IM is generated over A/I by $\bar{\omega}_1, \ldots, \bar{\omega}_n$, and $\omega_i \in M$ is an arbitrary inverse image of $\bar{\omega}_i$ in M, then M is generated over A by $\omega_1, \ldots, \omega_n$.

Proof. By assumption $M = \sum A\omega_i + IM$, so that $M = \sum A\omega_i + I(\sum A\omega_i + IM) = \sum A\omega_i + I^2 M$, and similarly, $M = \sum A\omega_i + I^\nu M$ for all $\nu > 0$. For any $\xi \in M$, write $\xi = \sum a_i \omega_i + \xi_1$ with $\xi_1 \in IM$, then $\xi_1 = \sum a_{i,1} \omega_i + \xi_2$ with $a_{i,1} \in I$ and $\xi_2 \in I^2 M$, and choose successively $a_{i,\nu} \in I^\nu$ and $\xi_\nu \in I^\nu M$ to satisfy

$$\xi_\nu = \sum a_{i,\nu} \omega_i + \xi_{\nu+1} \quad \text{for} \quad \nu = 1, 2, \ldots.$$

Then $a_i + a_{i,1} + a_{i,2} + \cdots$ converges in A. If we set b_i for this sum then

$$\xi - \sum_1^n b_i\omega_i \in \bigcap_{v>0} I^v M = (0). \quad \blacksquare$$

This theorem is extremely handy for proving the finiteness of M. For a Noetherian ring A, the I-adic topology has several more important properties, which are based on the following theorem, proved independently by E. Artin and D. Rees.

Theorem 8.5 (the Artin–Rees lemma). Let A be a Noetherian ring, M a finite A-module, $N \subset M$ a submodule, and I an ideal of A. Then there exists a positive integer c such that for every $n > c$, we have

$$I^n M \cap N = I^{n-c}(I^c M \cap N).$$

Proof. The inclusion $\supset$ is obvious, so that we only have to prove $\subset$. Suppose that I is generated by r elements $a_1, \ldots, a_r$, and M by s elements $\omega_1, \ldots, \omega_s$. An element of $I^n M$ can be written as $\sum_1^s f_i(a)\omega_i$, where $f_i(X) = f_i(X_1, \ldots, X_r)$ is a homogeneous polynomial of degree n with coefficients in A. Now set $A[X_1, \ldots, X_r] = B$, and for each $n > 0$ set

$$J_n = \left\{ (f_1, \ldots, f_s) \in B^s \,\middle|\, \begin{array}{l} f_i \text{ are homogeneous of degree } n \\ \text{and } \sum_1^s f_i(a)\omega_i \in N \end{array} \right\};$$

let $C \subset B^s$ be the B-submodule generated by $\bigcup_{n>0} J_n$. Since B is Noetherian, C is a finite B-module, so that $C = \sum_{j=1}^t Bu_j$, where each u_j is a linear combination of elements of $\bigcup J_n$; therefore C is generated by finitely many elements of $\bigcup J_n$. Suppose

$$C = Bu_1 + \cdots + Bu_t, \quad \text{where} \quad u_j = (u_{j1}, \ldots, u_{js}) \in J_{d_j} \quad \text{for } 1 \leqslant j \leqslant t.$$

Set $c = \max\{d_1, \ldots, d_t\}$. Now if $\eta \in I^n M \cap N$, we can write $\eta = \sum f_i(a)\omega_i$ with $(f_1, \ldots, f_s) \in J_n$, and hence

$$(f_1, \ldots, f_s) = \sum p_j(X) u_j, \quad \text{with} \quad p_j \in B = A[X_1, \ldots, X_r].$$

The left-hand side is a vector made up of homogeneous polynomials of degree n only, so that the terms of degree other than n on the right-hand side must cancel out to give 0. Hence we can suppose that the $p_j(X)$ are homogeneous of degree $n - d_j$. Then $\eta = \sum f_i(a)\omega_i = \sum_j p_j(a) \sum_i u_{ji}(a)\omega_i$, and $\sum_i u_{ji}(a)\omega_i \in I^{d_j} M \cap N$, so that if $n > c$, $p_j(a) \in I^{n-c} I^{c-d_j}$, giving

$$\eta \in I^{n-c}(I^c M \cap N) \quad \text{for any} \quad n > c. \quad \blacksquare$$

Theorem 8.6. In the notation of the above theorem, the I-adic topology of N coincides with the topology induced by the I-adic topology of M on the subspace $N \subset M$.

Proof. By the previous theorem, for $n > c$, we have $I^n N \subset I^n M \cap N \subset I^{n-c} N$. The topology of N as a subspace of M is the linear topology defined by $\{I^n M \cap N\}_{n=1,2,\dots}$, and the above formula says that this defines the same topology as $\{I^n N\}_{n=1,2,\dots}$. ■

Theorem 8.7. Let A be a Noetherian ring, I and ideal, and M a finite A-module. Writing $\hat{M}, \hat{A}$ for the I-adic completions of M and A we have

$$M \otimes_A \hat{A} \simeq \hat{M}.$$

Hence if A is I-adically complete, so is M.

Proof. By Theorems 1 and 6, the I-adic completion of an exact sequence of finite A-modules is again exact. Now given M, let $A^p \longrightarrow A^q \longrightarrow M \to 0$ be an exact sequence; the commutative diagram

$$\begin{array}{ccccccc} \hat{A}^p & \longrightarrow & \hat{A}^q & \longrightarrow & \hat{M} & \to 0 \\ \uparrow & & \uparrow & & \uparrow & \\ A^p \otimes \hat{A} & \longrightarrow & A^q \otimes \hat{A} & \longrightarrow & M \otimes \hat{A} & \to 0 \end{array}$$

has exact rows. Here the vertical arrows are the natural maps; since completion commutes with direct sums, the two left-hand arrows are obviously isomorphisms, and hence the right-hand arrow is an isomorphism, as required. ■

Theorem 8.8. Let A be a Noetherian ring, I an ideal, and $\hat{A}$ the I-adic completion of A; then $\hat{A}$ is flat over A.

Proof. By Theorem 7.7 it is enough to show that $\mathfrak{a} \otimes \hat{A} \longrightarrow \hat{A}$ is injective for every ideal $\mathfrak{a} \subset A$; but $\mathfrak{a} \otimes \hat{A} = \hat{\mathfrak{a}}$, and by Theorems 1 and 6, $\hat{\mathfrak{a}} \longrightarrow \hat{A}$ is injective. ■

Theorem 8.9 (Krull). Let A be a Noetherian ring, I an ideal, and M a finite A-module; set $\bigcap_{n>0} I^n M = N$. Then there exists $a \in A$ such that $a \equiv 1 \bmod I$ and $aN = 0$.

Proof. By NAK, it is enough to show that $N = IN$. By the Artin–Rees lemma, $I^n M \cap N \subset IN$ for sufficiently large n; now by definition of N, the left-hand side coincides with N.

Theorem 8.10 (the Krull intersection theorem).

(i) Let A be a Noetherian ring and I an ideal of A with $I \subset \operatorname{rad} A$; then for any finite A-module the I-adic topology is separated, and any submodule is a closed set.

(ii) If A is a Noetherian integral domain and $I \subset A$ a proper ideal, then

$$\bigcap_{n>0} I^n = (0).$$

Proof. (i) In this case the a of the previous theorem is a unit of A, so that $N = 0$, and M is separated. If $M' \subset M$ is a submodule then M/M' is also I-adically separated, which is the same as saying that M' is closed in M.

(ii) Setting $M = A$ in the previous theorem, from $1 \notin I$ we get that $a \neq 0$, and since a is not a zero-divisor, $N = 0$. ■

Theorem 8.11. Let A be a Noetherian ring, I and J ideals of A, and M a finite A-module; write $\hat{}$ for the completion of an A-module in the I-adic topology, and $\psi: M \longrightarrow \hat{M}$ for the natural map. Then

$$(JM)\hat{} = J\hat{M} = \text{the closure of } \psi(JM) \text{ in } \hat{M},$$

and

$$(M/JM)\hat{} = \hat{M}/J\hat{M}.$$

Proof. By Theorems 1 and 6, $(JM)\hat{}$ is the kernel of $\hat{M} \longrightarrow (M/JM)\hat{}$, and this is equal to the closure of $\psi(JM)$ in $\hat{M}$ by Theorem 1. Now suppose $J = \sum_1^r a_i A$ and define $\varphi: M^r \longrightarrow M$ by $(\xi_1, \ldots, \xi_r) \mapsto \sum a_i \xi_i$. Then the sequence

$$M^r \xrightarrow{\varphi} M \xrightarrow{\mu} M/JM \to 0,$$

where μ is the natural map, is exact. The I-adic completion,

$$\hat{M}^r \xrightarrow{\hat{\varphi}} \hat{M} \xrightarrow{\hat{\mu}} (M/JM)\hat{} \to 0,$$

is again exact. On the other hand $\hat{\varphi}$ is given by the same formula $(\xi_1, \ldots, \xi_r) \mapsto \sum a_i \xi_i$ as φ, hence $(JM)\hat{} = \operatorname{Ker}(\hat{\mu}) = \operatorname{Im}(\hat{\varphi}) = \sum a_i \hat{M} = J\hat{M}$. ■

As is easily seen, the $(X_1, \ldots, X_n)$-adic completion of the polynomial ring $A[X_1, \ldots, X_n]$ over A can be identified with the formal power series ring $A[\![X_1, \ldots, X_n]\!]$. Using this we get the following theorem.

Theorem 8.12. Let A be a Noetherian ring, and $I = (a_1, \ldots, a_n)$ an ideal of A. Then the I-adic completion $\hat{A}$ of A is isomorphic to $A[\![X_1, \ldots, X_n]\!]/(X_1 - a_1, \ldots, X_n - a_n)$. Hence $\hat{A}$ is a Noetherian ring.

Proof. Let $B = A[X_1, \ldots, X_n]$, and set $I' = \sum X_i B$, $J = \sum (X_i - a_i)B$; then $B/J \simeq A$, and the I'-adic topology on A considered as the B-module B/J coincides with the I-adic topology of A. Now writing $\hat{}$ for the I'-adic completion of B-modules, we have

$$\hat{A} = \hat{B}/\hat{J} = \hat{B}/J\hat{B} = A[\![X_1, \ldots, X_n]\!]/(X_1 - a_1, \ldots, X_n - a_n). \quad ■$$

Theorem 8.13. Let A be a Noetherian ring, I an ideal, M a finite A-module, and $\hat{M}$ the I-adic completion of M; then the topology of $\hat{M}$ is the I-adic topology of $\hat{M}$ as an A-module, and is the $I\hat{A}$-adic topology of $\hat{M}$ as an $\hat{A}$-module.

Proof. If we let M_n^* be the kernel of the map from $\hat{M} = \varprojlim (M/I^nM)$ to M/I^nM, the topology of $\hat{M}$ is that defined by $\{M_n^*\}$. Thus it is enough to prove that $M_n^* = I^n\hat{M}$. Since M/I^nM is discrete in the I-adic topology, we have $(M/I^nM)\hat{} = M/I^nM$ and the kernel of $\hat{M} \longrightarrow (M/I^nM)\hat{}$ is $I^n\hat{M}$ by Theorem 11. Therefore $M_n^* = I^n\hat{M}$. Moreover, $I^n\hat{M}$ can also be written $(I^n\hat{A})\hat{M}$, and $I^n\hat{A} = (I\hat{A})^n$, so that the topology of $\hat{M}$ is also the $I\hat{A}$-adic topology.

Theorem 8.14. Let A be a Noetherian ring and I an ideal. If we consider A with the I-adic topology, the following conditions are equivalent:

(1) $I \subset \operatorname{rad}(A)$;

(2) every ideal of A is a closed set;

(3) the I-adic completion $\hat{A}$ of A is faithfully flat over A.

Proof. We have already seen (1)$\Rightarrow$(2).

(2)$\Rightarrow$(3) Since $\hat{A}$ is flat over A, we need only prove that $\mathfrak{m}\hat{A} \neq \hat{A}$ for every maximal ideal $\mathfrak{m}$ of A. By assumption, $\{0\}$ is closed in A, so that we can assume that $A \subset \hat{A}$, and by Theorem 11, $\mathfrak{m}\hat{A}$ is the closure of $\mathfrak{m}$ in $\hat{A}$. However, $\mathfrak{m}$ is closed in A, so that $\mathfrak{m}\hat{A} \cap A = \mathfrak{m}$, and so $\mathfrak{m}\hat{A} \neq \hat{A}$.

(3) $\Rightarrow$(1) By Theorem 7.5, $\mathfrak{m}\hat{A} \cap A = \mathfrak{m}$ for every maximal ideal $\mathfrak{m}$ of A. Now $\mathfrak{m}\hat{A} \subset \hat{A}$ is a closed set by Theorems 2, (i) and 10, (i), and since the natural map $A \longrightarrow \hat{A}$ is continuous, $\mathfrak{m} = \mathfrak{m}\hat{A} \cap A$ is closed in A. If $I \not\subset \mathfrak{m}$. then $I^n + \mathfrak{m} = A$ for every $n > 0$, so that $\mathfrak{m}$ is not closed. Thus $I \subset \mathfrak{m}$. ■

If the conditions of the above theorem are satisfied, the topological ring A is said to be a *Zariski ring*, and I an *ideal of definition* of A. An ideal of definition is not uniquely determined; any ideal defining the same topology will do. The most important example of a Zariski ring is a Noetherian local ring $(A, \mathfrak{m})$ with the $\mathfrak{m}$-adic topology. When discussing the completion of a local ring, we will mean the $\mathfrak{m}$-adic completion unless otherwise specified.

Theorem 8.15. Let A be a semilocal ring with maximal ideals $\mathfrak{m}_1, \ldots, \mathfrak{m}_r$, and set $I = \operatorname{rad}(A) = \mathfrak{m}_1\mathfrak{m}_2 \ldots \mathfrak{m}_r$. Then the I-adic completion $\hat{A}$ of A decomposes as a direct product.

$$\hat{A} = \hat{A}_1 \times \cdots \times \hat{A}_r,$$

where $A_i = A_{\mathfrak{m}_i}$, and $\hat{A}_i$ is the completion of the local ring A_i.

Proof. Since for $i \neq j$ and any $n > 0$ we have $\mathfrak{m}_i^n + \mathfrak{m}_j^n = A$, Theorem 1.4 gives

$$A/I^n = A/\mathfrak{m}_1^n \times \cdots \times A/\mathfrak{m}_r^n \quad \text{for} \quad n > 0.$$

Hence taking the limit we get

$$\hat{A} = \varprojlim A/I^n = (\varprojlim A/\mathfrak{m}_1^n) \times \cdots \times (\varprojlim A/\mathfrak{m}_r^n).$$

If we set A_i for the localisation of A at $\mathfrak{m}_i$, then, since $A/\mathfrak{m}_i^n$ is already local,

$$A/\mathfrak{m}_i^n = (A/\mathfrak{m}_i^n)_{\mathfrak{m}_i} = A_i/(\mathfrak{m}_i A_i)^n,$$

and so $\varprojlim A/\mathfrak{m}_i^n$ can be identified with $\hat{A}_i$. ■

We now summarise the main points proved in this section for a local Noetherian ring. Let $(A, \mathfrak{m})$ be a local Noetherian ring; then we have:

(1) $\bigcap_{n>0} \mathfrak{m}^n = (0)$.

(2) For M a finite A-module and $N \subset M$ a submodule,

$$\bigcap_{n>0} (N + \mathfrak{m}^n M) = N.$$

(3) The completion $\hat{A}$ of A is faithfully flat over A; hence $A \subset \hat{A}$, and $I\hat{A} \cap A = I$ for any ideal I of A.

(4) $\hat{A}$ is again a Noetherian local ring, with maximal ideal $\mathfrak{m}\hat{A}$, and it has the same residue class field as A; moreover, $\hat{A}/\mathfrak{m}^n\hat{A} = A/\mathfrak{m}^n$ for all $n > 0$.

(5) If A is a complete local ring, then for any ideal $I \neq A$, A/I is again a complete local ring.

Remark 1. Even if A is complete, the localisation $A_\mathfrak{p}$ of A at a prime $\mathfrak{p}$ may not be.

Remark 2. An Artinian local ring $(A, \mathfrak{m})$ is complete; in fact, it is clear from the proof of Theorem 3.2 that there exists a ν such that $\mathfrak{m}^\nu = 0$, so that $\hat{A} = \varprojlim A/\mathfrak{m}^n = A$.

Exercises to §8. Prove the following propositions.

8.1. If A is a Noetherian ring, I and J are ideals of A, and A is complete both for the I-adic and J-adic topologies, then A is also complete for the $(I + J)$-adic topology.

8.2. Let A be a Noetherian ring, and $I \supset J$ ideals of A; if A is I-adically complete, it is also J-adically complete.

8.3. Let A be a Zariski ring and $\hat{A}$ its completion. If $\mathfrak{a} \subset A$ is an ideal such that $\mathfrak{a}\hat{A}$ is principal, then $\mathfrak{a}$ is principal.

8.4. According to Theorem 8.12, if $y \in \bigcap_\nu I^\nu$ then

$$y \in \sum_{i=1}^{n} (X_i - a_i)A[\![X_1, \ldots, X_n]\!].$$

Verify this directly in the special case $I = eA$, where $e^2 = e$.

8.5. Let A be a Noetherian ring and I a proper ideal of A; consider the multiplicative set $S = 1 + I$ as in §4, Example 3. Then A_S is a Zariski ring with ideal of definition IA_S, and its completion coincides with the I-adic completion of A.

8.6. If A is I-adically complete then $B = A[\![X]\!]$ is $(IB + XB)$-adically complete.

8.7. Let $(A, \mathfrak{m})$ be a complete Noetherian local ring, and $\mathfrak{a}_1 \supset \mathfrak{a}_2 \supset \cdots$ a chain of ideals of A for which $\bigcap_\nu \mathfrak{a}_\nu = (0)$; then for each n there exists $\nu(n)$ for which $\mathfrak{a}_{\nu(n)} \subset \mathfrak{m}^n$. In other words, the linear topology defined by $\{\mathfrak{a}_\nu\}_{\nu=1,2,\ldots}$ is stronger than or equal to the $\mathfrak{m}$-adic topology (Chevalley's theorem).

8.8. Let A be a Noetherian ring, $\mathfrak{a}_1, \ldots, \mathfrak{a}_r$ ideals of A; if M is a finite A-module and $N \subset M$ a submodule, then there exists $c > 0$ such that

$$n_1 \geqslant c, \ldots, n_r \geqslant c \Rightarrow \mathfrak{a}_1^{n_1} \ldots \mathfrak{a}_r^{n_r} M \cap N = \mathfrak{a}_1^{n_1 - c} \ldots \mathfrak{a}_r^{n_r - c}(\mathfrak{a}_1^c \ldots \mathfrak{a}_r^c M \cap N).$$

8.9. Let A be a Noetherian ring and $P \in \mathrm{Ass}(A)$. Then there is an integer $c > 0$ such that $P \in \mathrm{Ass}(A/I)$ for every ideal $I \subset P^c$ (hint: localise at P).

8.10. Show by example that the conclusion of Ex. 8.7. does not necessarily hold if A is not complete.

9 Integral extensions

If A is a subring of a ring B we say that B is an extension ring of A. In this case, an element $b \in B$ is said to be *integral* over A if b is a root of a monic polynomial with coefficients in A, that is if there is a relation of the form $b^n + a_1 b^{n-1} + \cdots + a_n = 0$ with $a_i \in A$. If every element of B is integral over A we say that B is *integral* over A, or that B is an *integral extension* of A.

Theorem 9.1. Let A be a ring and B an extension of A.

(i) An element $b \in B$ is integral over A if and only if there exists a ring C with $A \subset C \subset B$ and $b \in C$ such that C is finitely generated as an A-module.

(ii) Let $\bar{A} \subset B$ be the set of elements of B integral over A; then $\bar{A}$ is a subring of B.

Proof. (i) If b is a root of $f(X) = X^n + a_1 X^{n-1} + \cdots + a_n$, for any $P(X) \in A[X]$ let $r(X)$ be the remainder of P on dividing by f; then $P(b) = r(b)$ and $\deg r < n$. Hence

$$A[b] = A + Ab + \cdots + Ab^{n-1},$$

so that we can take C to be $A[b]$. Conversely if an extension ring C of A is a finite A-module then every element of C is integral over A: for if $C = A\omega_1 + \cdots + A\omega_n$ and $b \in C$ then

$$b\omega_i = \sum_j a_{ij}\omega_j \quad \text{with} \quad a_{ij} \in A,$$

so that by Theorem 2.1 we get a relation $b^n + a_1 b^{n-1} + \cdots + a_n = 0$. (The left-hand side is obtained by expanding out $\det(b\delta_{ij} - a_{ij})$.)

(ii) If $b, b' \in \bar{A}$ then we see easily that $A[b, b']$ is finitely generated as an A-module, so that its elements bb' and $b \pm b'$ are integral over A. ■

The $\bar{A}$ appearing in (ii) above is called the *integral closure* of A in B; if $A = \bar{A}$ we say that A is *integrally closed in B*. In particular, if A is an integral domain, and is integrally closed in its field of fractions, we say that A is an *integrally closed domain.* If for every prime ideal $\mathfrak{p}$ of A the localisation $A_\mathfrak{p}$ is an integrally closed domain we say that A is a *normal ring.*

Remark. 'Normal ring' is often used to mean 'integrally closed domain'; in this book we follow the usage of Serre and Grothendieck. If A is a Noetherian ring which is normal in our sense, and $\mathfrak{p}_1, \ldots, \mathfrak{p}_r$ are all the minimal prime ideals of A then it can be seen (see Ex. 9.11) that $A \simeq A/\mathfrak{p}_1 \times \cdots \times A/\mathfrak{p}_r$, and then each $A/\mathfrak{p}_i$ is an integrally closed domain (see Theorem 4.7). Conversely, the direct product of a finite number of integrally closed domains is normal (see Example 3 below).

Let $A \subset C \subset B$ be a chain of ring extensions; if an element $b \in B$ is integral over C and C is integral over A then b is integral over A. Indeed, if $b^n + c_1 b^{n-1} + \cdots + c_n = 0$ with $c_i \in C$ then

$$A[c_1, \ldots, c_n, b] = \sum_{\nu=0}^{n-1} A[c_1, \ldots, c_n]b^\nu,$$

and since $A[c_1,\dots,c_n]$ is a finite A-module, so is $A[c_1,\dots,c_n,b]$. In particular, if we take C to be the integral closure $\tilde{A}$ of A in B we see that $\tilde{A}$ is integrally closed in B.

Example 1. A UFD is an integrally closed domain – the proof is easy.

Example 2. Let k be a field and t an indeterminate over k; set $A = k[t^2,t^3] \subset B = k[t]$. Then A and B both have the same field of fractions $K = k(t)$. Since B is a UFD, it is integrally closed; but t is integral over A, so that B is the integral closure of A in K.

Note that in this example $A \simeq k[X,Y]/(Y^2 - X^3)$. Thus A is the coordinate ring of the plane curve $Y^2 = X^3$, which has a singularity at the origin. The fact that A is not integrally closed is related to the existence of this singularity.

Example 3. If B is an extension ring of A, $S \subset A$ is a multiplicative set, and $\tilde{A}$ is the integral closure of A in B, then the integral closure of A_S in B_S is $\tilde{A}_S$. The proof is again easy. It follows from this that if A is an integrally closed domain, so is A_S.

Theorem 9.2. Let A be an integrally closed domain, K the field of fractions of A, and L an algebraic extension of K. Then an element $\alpha \in L$ is integral over A if and only if its minimal polynomial over K has all its coefficients in A.
Proof. Let $f(X) = X^n + a_1X^{n-1} + \cdots + a_n$ be the minimal polynomial of α over K. We have $f(\alpha) = 0$, so that if all the a_i are in A then α is integral over A. Conversely, if α is integral over A, then letting $\bar{L}$ be an algebraic closure of L we have a splitting $f(X) = (X - \alpha_1)\dots(X - \alpha_n)$ of $f(X)$ in $\bar{L}[X]$ into linear factors; each of the α_i is conjugate to α over K, so that there is an isomorphism $K[\alpha] \simeq K[\alpha_i]$ fixing the elements of K and taking α into α_i, and therefore the α_i are also integral over A. Then $a_1,\dots,a_n \in A[\alpha_1,\dots,\alpha_n]$, and hence they are integral over A; but $a_i \in K$ and A is integrally closed, so that finally $a_i \in A$.

Example 4. Let A be a UFD in which 2 is a unit. Let $f \in A$ be square-free, (that is, not divisible by the square of any prime of A). Then $A[\sqrt{f}]$ is an integrally closed domain.
Proof. Let α be a square root of f. Let K be the field of fractions of A; then A is integrally closed in K by Example 1, so that if $\alpha \in K$ we have $\alpha \in A$ and $A[\alpha] = A$, and the assertion is trivial. If $\alpha \notin K$ then the field of fractions of $A[\alpha]$ is $K(\alpha) = K + K\alpha$, and every element $\xi \in K(\alpha)$ can be written in a unique way as $\xi = x + y\alpha$ with $x, y \in K$. The minimal polynomial of ξ over K is $X^2 - 2xX + (x^2 - y^2f)$, so that using the previous theorem, if ξ is integral over A we get $2x \in A$ and $x^2 - y^2f \in A$. By assumption, $2x \in A$ implies $x \in A$. Hence $y^2f \in A$. From this, if some prime p of A divides the denominator of y

we get $p^2|f$, which contradicts the square-free hypothesis. Thus $y\in A$, and $\xi\in A+A\alpha=A[\alpha]$, so that $A[\alpha]$ is integrally closed in $K(\alpha)$. ■

Lemma 1. Let B be an integral domain and $A\subset B$ a subring such that B is integral over A. Then

$$A \text{ is a field} \Leftrightarrow B \text{ is a field.}$$

Proof. ($\Rightarrow$) If $0\neq b\in B$ then there is a relation of the form $b^n+a_1b^{n-1}+\cdots+a_n=0$ with $a_i\in A$, and since B is an integral domain we can assume $a_n\neq 0$. Then

$$b^{-1}=-a_n^{-1}(b^{n-1}+a_1b^{n-2}+\cdots+a_{n-1})\in B.$$

($\Leftarrow$) If $0\neq a\in A$ then $a^{-1}\in B$, so that there is a relation $a^{-n}+c_1a^{-n+1}+\cdots+c_n=0$ with $c_i\in A$. Then

$$a^{-1}=-(c_1+c_2a+\cdots+c_na^{n-1})\in A. \quad ■$$

Lemma 2. Let A be a ring, and B an extension ring which is integral over A. If P is a maximal ideal of B then $P\cap A$ is a maximal ideal of A. Conversely if $\mathfrak{p}$ is a maximal ideal of A then there exists a prime ideal P of B lying over $\mathfrak{p}$, and any such P is a maximal ideal of B.

Proof. For $P\in\operatorname{Spec}B$ let $P\cap A=\mathfrak{p}$; then the extension $A/\mathfrak{p}\subset B/P$ is integral. Thus by Lemma 1 above, P is maximal if and only if $\mathfrak{p}$ is maximal. Next, to prove that there exists P lying over a given maximal ideal $\mathfrak{p}$ of A, it is enough to prove that $\mathfrak{p}B\neq B$. For then any maximal ideal P of B containing $\mathfrak{p}B$ will satisfy $P\cap A\supset\mathfrak{p}$ and $1\notin P\cap A$, so that $P\cap A=\mathfrak{p}$. By contradiction, assume that $\mathfrak{p}B=B$; then there is an expression $1=\sum_1^n\pi_ib_i$ with $b_i\in B$ and $\pi_i\in\mathfrak{p}$. If we set $C=A[b_1,\ldots,b_n]$ then C is finite over A and $\mathfrak{p}C=C$. Letting $C=Au_1+\cdots+Au_r$ we get $u_i=\sum\pi_{ij}u_j$ for some $\pi_{ij}\in\mathfrak{p}$, so that $\Delta=\det(\delta_{ij}-\pi_{ij})$ satisfies $\Delta u_j=0$ for each j, and hence $\Delta C=0$. But $1\in C$, so that $\Delta=0$, and on the other hand $\Delta\equiv 1 \bmod \mathfrak{p}$; therefore $1\in\mathfrak{p}$, which is a contradiction. ■

Theorem 9.3. Let A be a ring, B an extension ring which is integral over A and $\mathfrak{p}$ a prime ideal of A.

(i) There exists a prime ideal of B lying over $\mathfrak{p}$.

(ii) There are no inclusions between prime ideals of B lying over $\mathfrak{p}$.

(iii) Let A be an integrally closed domain, K its field of fractions, and L a normal field extension of K in the sense of Galois theory (that is $K\subset L$ is algebraic, and for any $\alpha\in L$, all the conjugates of α over K are in L); if B is the integral closure of A in L then all the prime ideals of B lying over $\mathfrak{p}$ are conjugate over K.

Proof. Localising the exact sequence $0\to A\longrightarrow B$ at $\mathfrak{p}$ gives an exact sequence $0\to A_\mathfrak{p}\longrightarrow B_\mathfrak{p}=B\otimes_A A_\mathfrak{p}$ in which $B_\mathfrak{p}$ is an extension ring integral over $A_\mathfrak{p}$. From the commutative diagram

$$\begin{array}{ccc} A_\mathfrak{p} & \longrightarrow & B_\mathfrak{p} \\ \uparrow & & \uparrow \\ A & \longrightarrow & B \end{array}$$

we see that the prime ideals of B lying over $\mathfrak{p}$ correspond bijectively with the maximal ideals of $B_\mathfrak{p}$ lying over the maximal ideal $\mathfrak{p}A_\mathfrak{p}$ of $A_\mathfrak{p}$. Hence, to prove (i) and (ii) it is enough to consider the case that $\mathfrak{p}$ is a maximal ideal, which has already been done in Lemma 2.

Now for (iii). Let P_1 and P_2 be prime ideals of B lying over $\mathfrak{p}$. First of all we consider the case $[L:K] < \infty$; let $G = \{\sigma_1, \ldots, \sigma_r\}$ be the group of K-automorphisms of L. If $P_2 \neq \sigma_j^{-1}(P_1)$ for any j then by (ii) we have $P_2 \not\subset \sigma_j^{-1}(P_1)$, so that there is an element $x \in P_2$ not contained in any $\sigma_j^{-1}(P_1)$ for $1 \leqslant j \leqslant r$ (see Ex. 1.6). Set $y = (\prod_j \sigma_j(x))^q$, where $q = 1$ if char $K = 0$, and $q = p^\nu$ for a sufficiently large integer ν if char $K = p > 0$. Then $y \in K$, and is integral over A, so that $y \in A$. However, the identity map of L is contained among the σ_j, so that $y \in P_2$, and hence $y \in P_2 \cap A = \mathfrak{p} \subset P_1$. This contradicts $\sigma_j(x) \notin P_1$ for all j. Therefore $P_2 = \sigma_j^{-1}(P_1)$ for some j.

If $[L:K] = \infty$ we need Galois theory for infinite extensions. Let $K' \subset L$ be the fixed subfield of $G = \operatorname{Aut}(L/K)$; then L is Galois over K' and $K \subset K'$ is a purely inseparable extension. If $K' \neq K$ we must have char $K = p > 0$, and setting A' for the integral closure of A in K' we see easily that

$$\mathfrak{p}' = \{x \in A' \mid x^q \in \mathfrak{p} \text{ for some } q = p^\nu\}$$

is the unique prime ideal of A' lying over $\mathfrak{p}$. Thus replacing K by K' we can assume that L is a Galois extension of K. For any finite Galois extension $K \subset L'$ contained in L we now set

$$F(L') = \{\sigma \in G \mid \sigma(P_1 \cap L') = P_2 \cap L'\};$$

then by the case of finite extensions we have just proved, $F(L') \neq \varnothing$. Moreover, $F(L') \subset G$ is closed in the Krull topology. (Recall that the Krull topology of G is the topology induced by the inclusion of G into the direct product of finite groups $\prod_{L'} \operatorname{Aut}(L'/K)$; with respect to this topology, G is compact. For details see textbooks on field theory.) If L'_i for $1 \leqslant i \leqslant n$ are finite Galois extensions of K contained in L then their composite L'' is also a finite Galois extension of K, and $\bigcap F(L'_i) \supset F(L'') \neq \varnothing$, so that the family $\{F(L') \mid L' \subset L$ is a finite Galois extension of $K\}$ of closed subsets of G has the finite intersection property; since G is compact, $\bigcap F(L') \neq \varnothing$. Taking $\sigma \in \bigcap_{L'} F(L')$ we obviously have $\sigma(P_1) = P_2$. ∎

For a ring A and an A-algebra B, the following statement is called the *going-up theorem*: given two prime ideals $\mathfrak{p} \subset \mathfrak{p}'$ of A and a prime ideal P of B lying over $\mathfrak{p}$, there exists $P' \in \operatorname{Spec} B$ such that $P \subset P'$ and $P' \cap A = \mathfrak{p}'$. Similarly, the *going-down theorem* is the following statement: given $\mathfrak{p} \subset \mathfrak{p}'$

and $P' \in \operatorname{Spec} B$ lying over $\mathfrak{p}'$, there exists $P \in \operatorname{Spec} B$ such that $P \subset P'$ and $P \cap A = \mathfrak{p}$.

Theorem 9.4. (i) If $B \supset A$ is an extension ring which is integral over A then the going-up theorem holds.

(ii) If in addition B is an integral domain and A is integrally closed, the going-down theorem also holds.

Proof. (i) Suppose $\mathfrak{p} \subset \mathfrak{p}'$ and P are given as above. Since $P \cap A = \mathfrak{p}$, we can think of B/P as an extension ring of $A/\mathfrak{p}$, and it is integral over $A/\mathfrak{p}$ because the condition that an element of B is integral over A is preserved by the homomorphism $B \longrightarrow B/P$. By (i) of the previous theorem there is a prime ideal of B/P lying over $\mathfrak{p}'/\mathfrak{p}$, and writing P' for its inverse image in B we have $P' \in \operatorname{Spec} B$ and $P' \cap A = \mathfrak{p}'$.

(ii) Let K be the field of fractions of A, and let L be a normal extension field of K containing B; set C for the integral closure of A in L. Suppose given prime ideals $\mathfrak{p} \subset \mathfrak{p}'$ of A and $P' \in \operatorname{Spec} B$ such that $P' \cap A = \mathfrak{p}'$, and choose $Q' \in \operatorname{Spec} C$ such that $Q' \cap B = P'$. Choose also a prime ideal Q of C over $\mathfrak{p}$, so that using the going-up theorem we can find a prime ideal Q_1 of C containing Q and lying over $\mathfrak{p}'$. Both Q_1 and Q' lie over $\mathfrak{p}'$, so that by (iii) of the previous theorem there is an automorphism $\sigma \in \operatorname{Aut}(L/K)$ such that $\sigma(Q_1) = Q'$. Setting $\sigma(Q) = Q_2$ we have $Q_2 \subset Q'$, and $Q_2 \cap A = Q \cap A = \mathfrak{p}$, so that setting $P = Q_2 \cap B$ we get $P \cap A = \mathfrak{p}$, $P \subset Q' \cap B = P'$. (For a different proof of (ii) which does not use Theorem 3, (iii), see [AM], (5.16), or [Kunz].) ■

We now treat another important case in which the going-down theorem holds.

Theorem 9.5. Let A be a ring and B a flat A-algebra; then the going-down theorem holds between A and B.

Proof. Let $\mathfrak{p} \subset \mathfrak{p}'$ be prime ideals of A, and let P' be a prime ideal of B lying over $\mathfrak{p}'$; then $B_{P'}$ is faithfully flat over $A_{\mathfrak{p}'}$, so that by Theorem 7.3 $\operatorname{Spec}(B_{P'}) \longrightarrow \operatorname{Spec}(A_{\mathfrak{p}'})$ is surjective. Hence there is a prime ideal $\mathfrak{P}$ of $B_{P'}$ lying over $\mathfrak{p}A_{\mathfrak{p}'}$, and setting $\mathfrak{P} \cap B = P$ we obviously have $P \subset P'$ and $P \cap A = \mathfrak{p}$. ■

Theorem 9.6. Let $A \subset B$ be integral domains such that A is integrally closed and B is integral over A; then the canonical map $f : \operatorname{Spec} B \longrightarrow \operatorname{Spec} A$ is open. More precisely, for $t \in B$, let $X^n + a_1 X^{n-1} + \cdots + a_n$ be a monic polynomial with coefficients in A having t as a root and of minimal degree; then

$$f(D(t)) = \bigcup_{i=1}^{n} D(a_i),$$

where the notation $D(\ )$ is as in §4.

Proof (H. Seydi [4]). By Theorem 2, $F(X) = X^n + a_1 X^{n-1} + \cdots + a_n$ is irreducible over the field of fractions of A; if we set $C=A[t]$ then $C \simeq A[X]/(F(X))$, so C is a free A-module with basis $1, t, t^2, \ldots, t^{n-1}$, and is hence faithfully flat over A. Suppose that $P \in D(t)$, so that $P \in \operatorname{Spec} B$ with $t \notin P$, and set $\mathfrak{p} = P \cap A$; then $\mathfrak{p} \in \bigcup_i D(a_i)$, since otherwise $a_i \in \mathfrak{p}$ for all i, and so $t^n \in P$, hence $t \in P$, which is a contradiction.

Conversely, given $\mathfrak{p} \in \bigcup_i D(a_i)$, suppose that $t \in \sqrt{(\mathfrak{p}C)}$; then for sufficiently large m we have $t^m = \sum_{i=1}^n b_i t^{n-i}$ with $b_i \in \mathfrak{p}$. We can take $m \geqslant n$. Then $X^m - \sum_1^n b_i X^{n-i}$ is divisible by $F(X)$ in $A[X]$, which implies that X^m is divisible by $\bar{F}(X) = X^n + \sum \bar{a}_i X^{n-i}$ in $(A/\mathfrak{p})[X]$; since at least one of the $\bar{a}_i$ is non-zero, this is a contradiction. Thus $t \notin \sqrt{(\mathfrak{p}C)}$, so that there exists $Q \in \operatorname{Spec} C$ with $t \notin Q$ and $\mathfrak{p}C \subset Q$. Setting $Q \cap A = \mathfrak{q}$ we have $\mathfrak{p} \subset \mathfrak{q}$, so that by the previous theorem there exists $P_1 \in \operatorname{Spec} C$ satisfying $P_1 \cap A = \mathfrak{p}$ and $P_1 \subset Q$. Since B is integral over C there exists $P \in \operatorname{Spec} B$ lying over P_1. We have $P \in D(t)$, since otherwise $t \in P \cap C = P_1 \subset Q$, which contradicts $t \notin Q$. This proves that

$$f(D(t)) = \bigcup D(a_i).$$

Any open set of $\operatorname{Spec} B$ is a union of open sets of the form $D(t)$, and hence $f\colon \operatorname{Spec} B \longrightarrow \operatorname{Spec} A$ is open.

Exercises to §9. Prove the following propositions.

9.1. Let A be a ring, $A \subset B$ an integral extension, and $\mathfrak{p}$ a prime ideal of A. Suppose that B has just one prime ideal P lying over $\mathfrak{p}$; then $B_P = B_{\mathfrak{p}}$.

9.2. Let A be a ring and $A \subset B$ an integral extension ring. Then $\dim A = \dim B$.

9.3. Let A be a ring, $A \subset B$ a finitely generated integral extension of A, and $\mathfrak{p}$ a prime ideal of A. Then B has only a finite number of prime ideals lying over $\mathfrak{p}$.

9.4. Let A be an integral domain and K its field of fractions. We say that $x \in K$ is *almost integral over* A if there exists $0 \neq a \in A$ such that $ax^n \in A$ for all $n > 0$. If x is integral over A it is almost integral, and if A is Noetherian the converse holds.

9.5. Let $A \subset K$ be as in the previous question. Say that A is *completely integrally closed* if every $x \in K$ which is almost integral over A belongs to A. If A is completely integrally closed, so is $A[\![X]\!]$.

9.6. Let A be an integrally closed domain, K its field of fractions, and let $f(X) \in A[X]$ be a monic polynomial. Then if $f(X)$ is reducible in $K[X]$ it is also reducible in $A[X]$.

9.7. Let $m \in \mathbb{Z}$ be square-free, and write A for the integral closure of $\mathbb{Z}$ in $\mathbb{Q}[\sqrt{m}]$. Then $A = \mathbb{Z}[(1+\sqrt{m})/2]$ if $m \equiv 1 \bmod 4$, and $A = \mathbb{Z}[\sqrt{m}]$ otherwise.

9.8. Let A be a ring and $A \subset B$ an integral extension. If P is a prime ideal of B with $\mathfrak{p} = P \cap A$ then $\operatorname{ht} P \leqslant \operatorname{ht} \mathfrak{p}$.

9.9. Let A be a ring and B an A-algebra, and suppose that the going-down theorem holds between A and B. If P is a prime ideal of B with $\mathfrak{p} = P \cap A$ then $\operatorname{ht} P \geqslant \operatorname{ht} \mathfrak{p}$.

9.10. Let K be a field and L an extension field of K. If P is a prime ideal of $L[X_1,\ldots,X_n]$ and $\mathfrak{p} = P \cap K[X_1,\ldots,X_n]$ then $\operatorname{ht} P \geqslant \operatorname{ht} \mathfrak{p}$, and equality holds if L is an algebraic extension of K. Moreover, if two polynomials $f(x), g(x) \in K[X_1,\ldots,X_n]$ have no common factors in $K[X_1,\ldots,X_n]$, they have none in $L[X_1,\ldots,X_n]$.

9.11. Let A be a Noetherian ring and $\mathfrak{p}_1,\ldots,\mathfrak{p}_r$ all the minimal prime ideals of A. Suppose that $A_\mathfrak{p}$ is an integral domain for all $\mathfrak{p} \in \operatorname{Spec} A$. Then (i) $\operatorname{Ass} A = \{\mathfrak{p}_1,\ldots,\mathfrak{p}_r\}$; (ii) $\mathfrak{p}_1 \cap \cdots \cap \mathfrak{p}_r = \operatorname{nil}(A) = 0$; (iii) $\mathfrak{p}_i + \bigcap_{j \neq i} \mathfrak{p}_j = A$ for all i. It follows that $A \simeq A/\mathfrak{p}_1 \times \cdots \times A/\mathfrak{p}_r$.

4

Valuation rings

From Hensel's theory of p-adic numbers onwards, valuation theory has been an important tool of number theory and the theory of function fields in one variable; the main object of study was however the multiplicative valuations which generalise the usual notion of absolute value of a number. In contrast, Krull defined and studied valuation rings from a more ring-theoretic point of view ([3], 1931). His theory was immediately applied to algebraic geometry by Zariski. In §10 we treat the elementary parts of their theory. Discrete valuation rings (DVRs) and Dedekind rings, the classical objects of study, are treated in the following §11, which also includes the Krull–Akizuki theorem, so that this section contains the theory of one-dimensional Noetherian rings. In §12 we treat Krull rings, which should be thought of as a natural extension of Dedekind rings; we go as far as a recent theorem of J. Nishimura.

This book is mainly concerned with Noetherian rings, and general valuation rings and Krull rings are the most important rings outside this category. The present chapter is intended as complementary to the theory of Noetherian rings, and we have left out quite a lot on valuation theory. The reader should consult [B6, 7], [ZS] or other textbooks for more information.

10 General valuations

An integral domain R is a *valuation ring* if every element x of its field of fractions K satisfies

$$x \notin R \Rightarrow x^{-1} \in R;$$

(if we write R^{-1} for the set of inverses of non-zero elements of R then this condition can be expressed as $R \cup R^{-1} = K$). We also say that R is a valuation ring of K. The case $R = K$ is the trivial valuation ring.

If R is a valuation ring then for any two ideals I, J of R either $I \subset J$ or $J \subset I$ must hold; indeed, if $x \in I$ and $x \notin J$ then for any $0 \neq y \in J$ we have $x/y \notin R$, so that $y/x \in R$ and $y = x(y/x) \in I$, therefore $J \subset I$. Thus the ideals of R form a totally ordered set. In particular, since R has only one maximal ideal, R is a local ring. We write $\mathfrak{m}$ for the maximal ideal of R. Then as one sees easily, $K - R = \{x \in K^* | x^{-1} \in \mathfrak{m}\}$, where we write K^* for the multiplicative group $K - \{0\}$. Thus R is determined by K and $\mathfrak{m}$.

If R is a valuation ring of a field K then any ring R' with $R \subset R' \subset K$ is obviously also a valuation ring, and in fact we have the following stronger statement.

Theorem 10.1. Let $R \subset R' \subset K$ be as above, let $\mathfrak{m}$ be the maximal ideal of R and $\mathfrak{p}$ that of R', and suppose that $R \neq R'$. Then

(i) $\mathfrak{p} \subset \mathfrak{m} \subset R \subset R'$ and $\mathfrak{p} \neq \mathfrak{m}$.

(ii) $\mathfrak{p}$ is a prime ideal of R and $R' = R_\mathfrak{p}$.

(iii) $R/\mathfrak{p}$ is a valuation ring of the field $R'/\mathfrak{p}$.

(iv) Given any valuation ring $\bar{S}$ of the field $R'/\mathfrak{p}$, let S be its inverse image in R'. Then S is a valuation ring having the same field of fractions K as R'.

Proof. (i) If $x \in \mathfrak{p}$ then $x^{-1} \notin R'$, so $x^{-1} \notin R$ and hence $x \in R$; x is not a unit of R, so that $x \in \mathfrak{m}$. Also, since $R \neq R'$ we have $\mathfrak{p} \neq \mathfrak{m}$.

(ii) We know that $\mathfrak{p} \subset R$, so that $\mathfrak{p} = \mathfrak{p} \cap R$, and this is a prime ideal of R. Since $R - \mathfrak{p} \subset R' - \mathfrak{p} = \{\text{units of } R'\}$ we have $R_\mathfrak{p} \subset R'$, and moreover by construction, the maximal ideal of $R_\mathfrak{p}$ is contained in the maximal ideal $\mathfrak{p}$ of R'. Thus by (i), $R_\mathfrak{p} = R'$.

(iii) Write $\varphi: R' \longrightarrow R'/\mathfrak{p}$ for the natural map; then for $x \in R' - \mathfrak{p}$, if $x \in R$ we have $\varphi(x) \in R/\mathfrak{p}$, and if $x \notin R$ we have $\varphi(x)^{-1} = \varphi(x^{-1}) \in R/\mathfrak{p}$, and therefore $R/\mathfrak{p}$ is a valuation ring of $R'/\mathfrak{p}$.

(iv) Note that $\mathfrak{p} \subset S$ and $S/\mathfrak{p} = \bar{S}$, so that if $x \in R'$ and $x \notin S$ then x is a unit of R', and $\varphi(x) \notin \bar{S}$. Thus $\varphi(x^{-1}) = \varphi(x)^{-1} \in \bar{S}$, and hence $x^{-1} \in S$. If on the other hand $x \in K - R'$ then $x^{-1} \in \mathfrak{p} \subset S$, so that we have proved that $S \cup S^{-1} = K$. ■

The valuation ring S in (iv) is called the *composite* of R' and $\bar{S}$. According to (iii), every valuation ring of K contained in R' is obtained as the composite of R' and a valuation ring of $R'/\mathfrak{p}$.

Quite generally, we write $\mathfrak{m}_R$ for the maximal ideal of a local ring R. If R and S are local rings with $R \supset S$ and $\mathfrak{m}_R \cap S = \mathfrak{m}_S$ we say that R *dominates* S, and write $R \geqslant S$. If $R \geqslant S$ and $R \neq S$, we write $R > S$.

Theorem 10.2. Let K be a field, $A \subset K$ a subring, and $\mathfrak{p}$ a prime ideal of A. Then there exists a valuation ring R of K satisfying

$$R \supset A \quad \text{and} \quad \mathfrak{m}_R \cap A = \mathfrak{p}.$$

Proof. Replacing A by $A_\mathfrak{p}$ we can assume that A is a local ring with $\mathfrak{p} = \mathfrak{m}_A$. Now write $\mathscr{F}$ for the set of all subrings B of K containing A and such that $1 \notin \mathfrak{p}B$. Now $A \in \mathscr{F}$, and if $\mathscr{L} \subset \mathscr{F}$ is a subset totally ordered by inclusion then the union of all the elements of $\mathscr{L}$ is again an element of $\mathscr{F}$, so that, by

Zorn's lemma, $\mathscr{F}$ has an element R which is maximal for inclusion. Since $\mathfrak{p}R \neq R$ there is a maximal ideal $\mathfrak{m}$ of R containing $\mathfrak{p}R$. Then $R \subset R_\mathfrak{m} \in \mathscr{F}$, so that $R = R_\mathfrak{m}$, and R is local. Also $\mathfrak{p} \subset \mathfrak{m}$ and $\mathfrak{p}$ is a maximal ideal of A, so that $\mathfrak{m} \cap A = \mathfrak{p}$. Thus it only remains to prove that R is a valuation ring of K. If $x \in K$ and $x \notin R$ then since $R[x] \notin \mathscr{F}$ we have $1 \in \mathfrak{p}R[x]$, and there exists a relation of the form

$$1 = a_0 + a_1 x + \cdots + a_n x^n \quad \text{with} \quad a_i \in \mathfrak{p}R.$$

Since $1 - a_0$ is unit of R we can modify this to get a relation

$$(*) \quad 1 = b_1 x + \cdots + b_n x^n \quad \text{with} \quad b_i \in \mathfrak{m}.$$

Among all such relations, choose one for which n is as small as possible. Similarly, if $x^{-1} \notin R$ we can find a relation

$$(**) \quad 1 = c_1 x^{-1} + \cdots + c_m x^{-m} \quad \text{with} \quad c_i \in \mathfrak{m},$$

and choose one for which m is as small as possible. If $n \geqslant m$ we multiply (**) by $b_n x^n$ and subtract from (*), and obtain a relation of the form (*) but with a strictly smaller degree n, which is a contradiction; if $n < m$ then we get the same contradiction on interchanging the roles of x and x^{-1}. Thus if $x \notin R$ we must have $x^{-1} \in R$. ■

Theorem 10.3. A valuation ring is integrally closed.

Proof. Let R be a valuation ring of a field K, and let $x \in K$ be integral over R, so that $x^n + a_1 x^{n-1} + \cdots + a_n = 0$ with $a_i \in R$. If $x \notin R$ then $x^{-1} \in \mathfrak{m}_R$, but then

$$1 + a_1 x^{-1} + \cdots + a_n x^{-n} = 0,$$

and we get $1 \in \mathfrak{m}_R$, which is a contradiction. Hence $x \in R$. ■

Theorem 10.4. Let K be a field, $A \subset K$ a subring, and let B be the integral closure of A in K. Then B is the intersection of all the valuation rings of K containing A.

Proof. Write B' for the intersection of all valuation rings of K containing A, so that by the previous theorem we have $B' \supset B$. To prove the opposite inclusion it is enough to show that for any element $x \in K$ which is not integral over A there is a valuation ring of K containing A but not x. Set $x^{-1} = y$. The ideal $yA[y]$ of $A[y]$ does not contain 1: for if $1 = a_1 y + a_2 y^2 + \cdots + a_n y^n$ with $a_i \in A$ then x would be integral over A, contradicting the assumption. Therefore there is a maximal ideal $\mathfrak{p}$ of $A[y]$ containing $yA[y]$, and by Theorem 2 there exists a valuation ring R of K such that $R \supset A[y]$ and $\mathfrak{m}_R \cap A[y] = \mathfrak{p}$. Now $y = x^{-1} \in \mathfrak{m}_R$, so that $x \notin R$. ■

Let K be a field and $A \subset K$ a subring. If a valuation ring R of K contains A we say that R has a centre in A, and the prime ideal $\mathfrak{m}_R \cap A$ of A is called the *centre* of R in A. The set of valuation rings of K having a centre in A is called the *Zariski space* or the *Zariski Riemann surface* of K over A, and written

$\mathrm{Zar}(K, A)$. We will treat $\mathrm{Zar}(K, A)$ as a topological space, introducing a topology as follows.

For $x_1,\ldots,x_n \in K$, set $U(x_1,\ldots,x_n) = \mathrm{Zar}(K, A[x_1,\ldots,x_n])$. Then since

$$U(x_1,\ldots,x_n) \cap U(y_1,\ldots,y_m) = U(x_1,\ldots,x_n,y_1,\ldots,y_m),$$

the collection $\mathscr{F} = \{U(x_1,\ldots,x_n) | n \geqslant 0 \text{ and } x_i \in K\}$ is the basis for the open sets of a topology on $\mathrm{Zar}(K, A)$. That is, we take as open sets the subsets of $\mathrm{Zar}(K, A)$ which can be written as a union of elements of $\mathscr{F}$. As in the case of Spec, this topology is called the Zariski topology.

Theorem 10.5. $\mathrm{Zar}(K, A)$ is quasi-compact.

Proof. It is enough to prove that if $\mathscr{A}$ is a family of closed sets of $\mathrm{Zar}(K, A)$ having the finite intersection property (that is, the intersection of any finite number of elements of $\mathscr{A}$ is non-empty) then the intersection of all the elements of $\mathscr{A}$ is non-empty. By Zorn's lemma there exists a maximal family of closed sets $\mathscr{A}'$ having the finite intersection property and containing $\mathscr{A}$. Since it is then enough to show that the intersection of all the elements of $\mathscr{A}'$ is non-empty, we can take $\mathscr{A} = \mathscr{A}'$. Then it is easy to see that $\mathscr{A}$ has the following properties:

(α) $F_1,\ldots,F_r \in \mathscr{A} \Rightarrow F_1 \cap \cdots \cap F_r \in \mathscr{A}$;

(β) $Z_1,\ldots,Z_n$ are closed sets and $Z_1 \cup \cdots \cup Z_n \in \mathscr{A} \Rightarrow Z_i \in \mathscr{A}$ for some i;

(γ) if a closed set F contains an element of $\mathscr{A}$ then $F \in \mathscr{A}$.

For a subset $F \subset \mathrm{Zar}(K, A)$ we write F^c to denote the complement of F. If $F \in \mathscr{A}$ and $F^c = \bigcup_\lambda U_\lambda$ then $F = \bigcap_\lambda U_\lambda^c$, and moreover if $U(x_1,\ldots,x_n)^c = \bigcup_{i=1}^n U(x_i)^c \in \mathscr{A}$ then by (β) above one of the $U(x_i)^c$ must belong to $\mathscr{A}$. Hence the intersection of all the elements of $\mathscr{A}$ is the same thing as the intersection of the sets of the form $U(x)^c$ belonging to $\mathscr{A}$. Set

$$\Gamma = \{y \in K \,|\, U(y^{-1})^c \in \mathscr{A}.\}.$$

Now since the condition for $R \in \mathrm{Zar}(K, A)$ to belong to $U(y^{-1})^c$ is that $y \in \mathfrak{m}_R$, the intersection of all the elements of $\mathscr{A}$ is equal to

$$\{R \in \mathrm{Zar}(K, A) | \mathfrak{m}_R \supset \Gamma\}.$$

Write I for the ideal of $A[\Gamma]$ generated by Γ. If $1 \notin I$ then by Theorem 2, the above set is non-empty, as required to prove. But if $1 \in I$ then there is a finite subset $\{y_1,\ldots,y_r\} \subset \Gamma$ such that $1 \in \sum y_i A[y_1,\ldots,y_r]$; but then $U(y_1^{-1})^c \cap \cdots \cap U(y_r^{-1})^c = \varnothing$, which contradicts the finite intersection property of $\mathscr{A}$. ∎

When K is an algebraic function field over an algebraically closed field k of characteristic 0 (that is, K is a finitely generated extension of k), Zariski gave a classification of valuation rings of K containing k, and using this and the compactness result above, he succeeded in giving an algebraic proof in characteristic 0 of the resolution of singularities of algebraic varieties of dimension 2 and 3. However, Hironaka's general proof of resolution of singularities in characteristic 0 in all dimensions was obtained by other methods, without the use of valuation theory.

As we saw at the beginning of this section, the ideals of R form a totally ordered set under inclusion. This holds not just for ideals, but for all R-modules contained in K. In particular, if we set

$$G = \{xR \mid x \in K \quad \text{and} \quad x \neq 0\},$$

then G is a totally ordered set under inclusion; we will, however, give G the opposite order to that given by inclusion. That is, we define $\leqslant$ by

$$xR \leqslant yR \Leftrightarrow xR \supset yR.$$

Moreover, G is an Abelian group with product $(xR)\cdot(yR) = xyR$. In general, an Abelian group H written additively, together with a total order relation $\geqslant$ is called an *ordered group* if the axiom

$$x \geqslant y, z \geqslant t \Rightarrow x + z \geqslant y + t$$

holds. This axiom implies

$$(1) \quad x > 0, \quad y \geqslant 0 \Rightarrow x + y > 0, \quad \text{and} \quad (2) \quad x \geqslant y \Rightarrow -y \geqslant -x.$$

We make an ordered set $H \cup \{\infty\}$ by adding to H an element ∞ bigger than all the elements of H, and fix the conventions $\infty + \alpha = \infty$ for $\alpha \in H$ and $\infty + \infty = \infty$. A map $v: K \longrightarrow H \cup \{\infty\}$ from a field K to $H \cup \{\infty\}$ is called an *additive valuation* or just a *valuation* of K if it satisfies the conditions

(1) $v(xy) = v(x) + v(y)$;

(2) $v(x + y) \geqslant \min\{v(x), v(y)\}$;

(3) $v(x) = \infty \Leftrightarrow x = 0$.

If we write K^* for the multiplicative group of K then v defines a homomorphism $K^* \longrightarrow H$; the image is a subgroup of H, called the *value groups* of v. We also set

$$R_v = \{x \in K \mid v(x) \geqslant 0\} \quad \text{and} \quad m_v = \{x \in K \mid v(x) > 0\},$$

obtaining a valuation ring R_v of K with m_v as its maximal ideal, and call R_v the valuation ring of v, and m_v the valuation ideal of v. Conversely, if R is a valuation ring of K, then the group $G = \{xR \mid x \in K^*\}$ described above is an ordered group, and we obtain an additive valuation of K with value group G by defining $v: K \longrightarrow G \cup \{\infty\}$ by $v(0) = \infty$ and $v(x) = xR$ for $x \in K^*$ (there is no real significance in whether or not we rewrite the multiplication in G additively); the valuation ring of v is R. The additive valuation corresponding to a valuation ring R is not quite unique, but if v and v' are two additive valuations of K with value groups H and H' and both having the valuation ring R then there exists an order-isomorphism $\varphi: H \longrightarrow H'$ such that $v' = \varphi v$ (prove this!). Thus we can think of valuation rings and additive valuations as being two aspects of the same thing.

We now give some examples of ordered groups:

(1) the additive group of real numbers $\mathbb{R}$ (this is isomorphic to the multiplicative group of positive reals), or any subgroup of this;

(2) the group $\mathbb{Z}$ of rational integers;

(3) the direct product $\mathbb{Z}^n$ of n copies of $\mathbb{Z}$, with lexicographical order, that is

$$(a_1,\ldots,a_n)<(b_1,\ldots,b_n)\Leftrightarrow\begin{cases}\text{the first non-zero element of}\\ b_1-a_1,\ldots,b_n-a_n \text{ is positive.}\end{cases}$$

An ordered group G is said to be *Archimedean* if it is order-isomorphic to a suitable subgroup of $\mathbb{R}$. The name is explained by the following theorem: the condition in it is known as the Archimedean axiom. (Note that our usage is completely unrelated to that in number theory, where non-Archimedean fields are p-adic fields, as opposed to subfields of $\mathbb{R}$ and $\mathbb{C}$ with the usual 'Archimedean' metrics.)

Theorem 10.6. Let G be an ordered group; then G is Archimedean if and only if the following condition holds:

if a, $b\in G$ with $a>0$, there exists a natural number n such that $na>b$.

Proof. The condition is obviously necessary, and we prove sufficiency. If $G=\{0\}$ then we can certainly embed G in $\mathbb{R}$. Suppose that $G\neq\{0\}$. Fix some $0<x\in G$. For any $y\in G$, there is a well-defined largest integer n such that $nx\leqslant y$ (if $y\geqslant 0$ this is clear by assumption; if $y<0$, let m be the smallest integer such that $-y\leqslant mx$, and set $n=-m$). Let this be n_0. Now set $y_1=y-n_0x$ and let n_1 be the largest integer n such that $nx\leqslant 10\,y_1$; we have $0\leqslant n_1<10$. Set $y_2=10\,y_1-n_1x$ and let n_2 be the largest integer n such that $nx\leqslant 10\,y_2$. Continuing in the same way, we find integers n_0, $n_1, n_2,\ldots$, and set $\varphi(y)=\alpha$, where α is the real number given by the decimal expression $n_0+0.n_1n_2n_3\ldots$. Then it can easily be checked that $\varphi\colon G\longrightarrow\mathbb{R}$ is order-preserving, in the sense that $y<y'$ implies $\varphi(y)\leqslant\varphi(y')$.

We also see that φ is injective. For this, we only need to observe that if $y<y'$ then there exists a natural number r such that $x<10^r(y'-y)$; the details are left to the reader.

Finally we show that φ is a group homomorphism. For $y\in G$, we write $n/10^r$ with $n\in\mathbb{Z}$ to denote the number obtained by taking the first r decimal places of $\varphi(y)$; the numerator n is determined by the property that $nx\leqslant 10^r y<(n+1)x$. For $y'\in G$, if $n'x\leqslant 10^r y'<(n'+1)x$ then we have

$$(n+n')x\leqslant 10^r(y+y')<(n+n'+2)x,$$

and hence

$$\varphi(y+y')-(n+n')\cdot 10^{-r}<2.10^{-r},$$

so that

$$|\varphi(y+y')-\varphi(y)-\varphi(y')|<4.10^{-r},$$

and since r is arbitrary, $\varphi(y+y')=\varphi(y)+\varphi(y')$. ■

A non-zero group G order-isomorphic to a subgroup of $\mathbb{R}$ is said to have *rank 1*. The *rational rank* of an ordered group G of rank 1 is the maximum number of elements of G (viewed as a subgroup of $\mathbb{R}$) which are linearly independent over $\mathbb{Z}$. For example, the additive group $G = \mathbb{Z} + \mathbb{Z}\sqrt{2} \subset \mathbb{R}$ is an ordered group of rank 1 and rational rank 2.

Theorem 10.7. Let R be a valuation ring having value group G. Then G has rank $1 \Leftrightarrow R$ has Krull dimension 1.

Proof. ($\Rightarrow$) Since $G \neq 0$, R is not a field. Suppose that $\mathfrak{p}$ is a prime ideal of R distinct from $\mathfrak{m}_R$. Let $\xi \in \mathfrak{m}_R$ be such that $\xi \notin \mathfrak{p}$, and set $v(\xi) = x$, where v is the additive valuation corresponding to R. Suppose that $0 \neq \eta \in \mathfrak{p}$, and set $y = v(\eta)$; then $y \in G$ and $x > 0$, so that $nx > y$ for some sufficiently large natural number n. This means that $\xi^n/\eta \in R$, so that $\xi^n \in \eta R \subset \mathfrak{p}$; then since $\mathfrak{p}$ is prime we have $\xi \in \mathfrak{p}$, which is a contradiction. Therefore $\mathfrak{p} = (0)$. The only prime ideals of R are $\mathfrak{m}_R$ and (0), which means $\dim R = 1$.

($\Leftarrow$) If $0 \neq \eta \in \mathfrak{m}_R$ then $\mathfrak{m}_R$ is the unique prime ideal containing ηR, and hence $\sqrt{(\eta R)} = \mathfrak{m}_R$. Thus for any $\xi \in \mathfrak{m}_R$ there exists a natural number n such that $\xi^n \in \eta R$. From this one sees easily that G satisfies the Archimedean axiom. ■

Exercises to §10. Prove the following propositions.

10.1. In a valuation ring any finitely generated ideal is principal.

10.2. If R is a valuation ring then an R-module M is flat if and only if it is torsion-free (that is, $a \neq 0$, $x \neq 0 \Rightarrow ax \neq 0$ for $a \in R$, $x \in M$).

10.3. In Theorem 10.4, if A is a local ring then B is the intersection of the valuation rings of K dominating A.

10.4. If R is a valuation ring of Krull dimension $\geqslant 2$ then the formal power series ring $R[\![X]\!]$ is not integrally closed ([B5], Ex. 27, p. 76 and Seidenberg [1]).

10.5. If R is a valuation ring of Krull dimension 1 and K its field of fractions then there do not exist any rings intermediate between R and K. In other words R is maximal among proper subrings of K. Conversely if a ring R, not a field, is a maximal proper subring of a field K then R is a valuation ring of Krull dimension 1.

10.6. If v is an additive valuation of a field K, and if $\alpha, \beta \in K$ are such that $v(\alpha) \neq v(\beta)$ then $v(\alpha + \beta) = \min(v(\alpha), v(\beta))$.

10.7. If v is an additive valuation of a field K and $\alpha_1, \dots, \alpha_n \in K$ are such that $\alpha_1 + \cdots + \alpha_n = 0$ then there exist two indices i, j such that $i \neq j$ and $v(\alpha_i) = v(\alpha_j)$.

10.8. Let $K \subset L$ be algebraic field extension of degree $[L:K] = n$, and let S be a valuation ring of L; set $R = S \cap K$. Write k, k' for the residue fields of S and R, and set $[k:k'] = f$. Now let G be the value group of S, and let G' be the image of K^* under the valuation map $L^* \longrightarrow G$; set $|G:G'| = e$. Then $ef \leqslant n$. (The numbers f and e are called the degree and the ramification index of the valuation ring extension S/R.)

10.9. Let L, K, S and R be as in the previous question, and let $S_1 \neq S$ be a valuation ring of L such that $S_1 \cap K = R$. Then neither of S or S_1 contains the other.

10.10 Let A be an integral domain with field of fractions K, and let H be an ordered group. If a map $v: A \longrightarrow H \cup \{\infty\}$ satisfies conditions (1), (2) and (3) of an additive valuation (on elements of A), then v can be extended uniquely to an additive valuation $K \longrightarrow H \cup \{\infty\}$.

10.11 Let k be a field, X and Y indeterminates, and suppose that α is a positive irrational number. Then the map $v: k[X, Y] \longrightarrow \mathbb{R} \cup \{\infty\}$ defined by taking $\sum c_{n,m} X^n Y^m$ (with $c_{n,m} \in k$) into $v(\sum c_{n,m} X^n Y^m) = \min\{n + m\alpha \mid c_{n,m} \neq 0\}$ determines a valuation of $k(X, Y)$ with value group $\mathbb{Z} + \mathbb{Z}\alpha$.

11 DVRs and Dedekind rings

A valuation ring whose value group is isomorphic to $\mathbb{Z}$ is called a *discrete valuation ring* (DVR). Discrete refers to the fact that the value group is a discrete subgroup of $\mathbb{R}$, and has nothing to do with the $\mathfrak{m}$-adic topology of the local ring being discrete.

Theorem 11.1. Let R be a valuation ring. Then the following conditions are equivalent.

(1) R is a DVR;

(2) R is a principal ideal domain;

(3) R is Noetherian.

Proof. Let K be the field of fractions of R and $\mathfrak{m}$ its maximal ideal.

(1) $\Rightarrow$ (2) Let v_R the additive valuation of R having value group $\mathbb{Z}$; this is called the normalised additive valuation corresponding to R. There exists $t \in \mathfrak{m}$ such that $v_R(t) = 1$. For $0 \neq x \in \mathfrak{m}$, the valuation $v_R(x)$ is a positive integer, say $v_R(x) = n$; then $v_R(x/t^n) = 0$, so that we can write $x = t^n u$ with u a unit of R. In particular $\mathfrak{m} = tR$. Let $I \neq (0)$ be any ideal of R; then $\{v_R(a) \mid 0 \neq a \in I\}$ is a set of non-negative integers, and so has a smallest element, say n. If $n = 0$ then I contains a unit of R, so that $I = R$. If $n > 0$ then there exists an $x \in I$ such that $v_R(x) = n$; then $I = xR = t^n R$. Therefore R is a principal ideal domain, and moreover every non-zero ideal of R is a power of $\mathfrak{m} = tR$.

(2) $\Rightarrow$ (3) is obvious.

(3) $\Rightarrow$ (2) In general, given any two ideals of a valuation ring, one contains the other, so that any finitely generated ideal $a_1 R + \cdots + a_r R$ is equal to one of the $a_i R$, and therefore principal. Hence, if R is Noetherian every ideal of R is principal.

(2)⇒(1) We can write $\mathfrak{m} = xR$ for some x. Now if we set $I = \bigcap_{v=1}^{\infty} x^v R$ then this is also a principal ideal, so that we can write $I = yR$. If we set $y = xz$, then from $y \in x^v R$ we get $z \in x^{v-1} R$, and since this holds for every v, we have $z \in I$, hence we can write $z = yu$. Since $y = xz = xyu$, we have $y(1 - xu) = 0$, but then since $x \in \mathfrak{m}$ we must have $y = 0$, and therefore $I = (0)$. Because of this, for every non-zero element $a \in R$, there is a well-defined integer $v \geqslant 0$ such that $a \in x^v R$ but $a \notin x^{v+1} R$; we then set $v(a) = v$. It is not difficult to see that if $a, b, c, d \in R - \{0\}$ satisfy $a/b = c/d$ then

$$v(a) - v(b) = v(c) - v(d);$$

therefore setting $v(\xi) = v(a) - v(b)$ for $\xi = a/b \in K^*$ gives a map $v: K^* \longrightarrow \mathbb{Z}$ which can easily be seen to be an additive valuation of K whose valuation ring is R. The value group of v is clearly $\mathbb{Z}$, so that R is a DVR. ■

If R is a DVR with maximal ideal $\mathfrak{m}$ then an element $t \in R$ such that $\mathfrak{m} = tR$ is called a *uniformising element of R*.

Remark. A valuation ring S whose maximal ideal $\mathfrak{m}_S$ is principal does not have to be a DVR. To obtain a counter-example, let K be a field, and R a DVR of K; set $k = R/\mathfrak{m}_R$, and suppose that $\mathfrak{R}$ is a DVR of k. Now let S be the composite of R and $\mathfrak{R}$. Let f be a uniformising element of R, and $g \in S$ be any element mapping to a uniformising element $\bar{g}$ of $\mathfrak{R}$. Then $\mathfrak{m}_R = fR \subset \mathfrak{m}_S \subset S \subset R$, and $\mathfrak{m}_S/\mathfrak{m}_R = \bar{g}\mathfrak{R} = \bar{g}(S/\mathfrak{m}_R)$, and so

$$\mathfrak{m}_S = \mathfrak{m}_R + gS.$$

On the other hand $g^{-1} \in R$, so that for any $h \in \mathfrak{m}_R$ we have $h/g \in \mathfrak{m}_R \subset S$, and hence $\mathfrak{m}_R \subset gS$, so that

$$\mathfrak{m}_S = gS.$$

However, $\mathfrak{m}_R = fR$ is not finitely generated as an ideal of S, being generated by $f, fg^{-1}, fg^{-2}, \ldots$. The value group of S is $\mathbb{Z}^2$, with the valuation $v: K^* \longrightarrow \mathbb{Z}^2$ given by

$$v(x) = (n, m), \quad \text{where} \quad n = v_R(x) \quad \text{and} \quad m = v_{\mathfrak{R}}(\varphi(xf^{-n})),$$

where $\varphi: R \longrightarrow R/\mathfrak{m}_R = k$ is the natural homomorphism.

The previous theorem gives a characterisation of DVRs among valuation rings; now we consider characterisations among all rings.

Theorem 11.2. Let R be a ring; then the following conditions are equivalent:

(1) R is a DVR;

(2) R is a local principal ideal domain, and not a field;

(3) R is a Noetherian local ring, $\dim R > 0$ and the maximal ideal $\mathfrak{m}_R$ is principal;

(4) R is a one-dimensional normal Noetherian local ring.

Proof. We saw (1)$\Rightarrow$(2) in the previous theorem; (2)$\Rightarrow$(3) is obvious.

(3)$\Rightarrow$(1) Let xR be the maximal ideal of R. If x were nilpotent then we would have $\dim R = 0$, and hence $x^\nu \neq 0$ for all ν. By the Krull intersection theorem (Theorem 8.10, (i)) we have $\bigcap_{\nu=1}^{\infty} x^\nu R = (0)$, so that for $0 \neq y \in R$ there is a well-determined ν such that $y \in x^\nu R$ and $y \notin x^{\nu+1}R$. If $y = x^\nu u$, then since $u \notin xR$ it must be a unit. Similarly, for $0 \neq z \in R$ we have $z = x^\mu v$, with v a unit. Therefore $yz = x^{\nu+\mu}uv \neq 0$, and so R is an integral domain. Finally, any element t of the fraction field of R can be written $t = x^\nu u$, with u a unit of R and $\nu \in \mathbb{Z}$, and it is easy to see that setting $v(t) = \nu$ defines an additive valuation of the field of fractions of R whose valuation ring is R.

(1)$\Rightarrow$(4) In a DVR the only ideals are (0) and the powers of the maximal ideal, so that the only prime ideals of R are (0) and $\mathfrak{m}_R$, and hence $\dim R = 1$. By the previous theorem R is Noetherian, and it is normal because it is a valuation ring.

(4)$\Rightarrow$(3) By assumption R is an integral domain. Write K for the field of fractions and $\mathfrak{m}$ for the maximal ideal of R. Then $\mathfrak{m} \neq 0$, so that by Theorem 8.10, $\mathfrak{m} \neq \mathfrak{m}^2$; choose some $x \in \mathfrak{m} - \mathfrak{m}^2$. Since $\dim R = 1$ the only prime ideals of R are (0) and $\mathfrak{m}$, so that $\mathfrak{m}$ must be a prime divisor of xR, and there exists $y \in R$ such that $xR{:}y = \mathfrak{m}$. Set $a = yx^{-1}$; then $a \notin R$, but $a\mathfrak{m} \subset R$. Now we set $\mathfrak{m}^{-1} = \{b \in K \mid b\mathfrak{m} \subset R\}$, so that $R \subset \mathfrak{m}^{-1}$, and $R \neq \mathfrak{m}^{-1}$ since $a \in \mathfrak{m}^{-1}$. Consider the ideal $\mathfrak{m}^{-1}\mathfrak{m}$ of R; since $R \subset \mathfrak{m}^{-1}$ we have $\mathfrak{m} \subset \mathfrak{m}^{-1}\mathfrak{m}$. If we had $\mathfrak{m} = \mathfrak{m}^{-1}\mathfrak{m}$ then we would get $a\mathfrak{m} \subset \mathfrak{m}$, and then a would be integral over R by Theorem 2.1, so that $a \in R$, which is a contradiction. Hence we must have $\mathfrak{m}^{-1}\mathfrak{m} = R$. Moreover, $x\mathfrak{m}^{-1} \subset R$ is an ideal, and if $x\mathfrak{m}^{-1} \subset \mathfrak{m}$ then we would have $xR = x\mathfrak{m}^{-1}\mathfrak{m} \subset \mathfrak{m}^2$, contradicting $x \notin \mathfrak{m}^2$. Therefore $x\mathfrak{m}^{-1} = R$, and hence $xR = x\mathfrak{m}^{-1}\mathfrak{m} = \mathfrak{m}$, so that $\mathfrak{m}$ is principal. ■

Quite generally, if R is an integral domain and K its fields of fractions, we say that an R-submodule I of K is a *fractional ideal* of R if $I \neq 0$, and there exists a non-zero element $\alpha \in R$ such that $\alpha I \subset R$ (see Ex. 3.4). As an R-module we have $I \simeq \alpha I$, so that if R is a Noetherian integral domain then any fractional ideal is finitely generated. For I a fractional ideal we set $I^{-1} = \{\alpha \in K \mid \alpha I \subset R\}$; we say that I is *invertible* if $I^{-1}I = R$.

Theorem 11.3. Let R be an integral domain and I a fractional ideal of R. Then the following conditions are equivalent:

(1) I is invertible;

(2) I is a projective R-module;

(3) I is finitely generated, and for every maximal ideal P of R, the fractional ideal $I_P = IR_P$ of R_P is principal.

Proof. (1)$\Rightarrow$(2) If $I^{-1}I = R$ then there exist $a_i \in I$ and $b_i \in I^{-1}$ such that $\sum_1^n a_i b_i = 1$. Then $a_1, \ldots, a_n$ generate I, since for any $x \in I$ we have $\sum (xb_i)a_i =$

x, and $xb_i \in R$. Let $F = Re_1 + \cdots + Re_n$ be the free R-module with basis $e_1, \ldots, e_n$; we define the R-linear map $\varphi: F \longrightarrow I$ by $\varphi(e_i) = a_i$, so that φ is surjective. Then we defined $\psi: I \longrightarrow F$ by writing $\psi_i: I \longrightarrow R$ for the map $\psi_i(x) = b_i x$, and setting $\psi(x) = \sum \psi_i(x) e_i$. We then have $\varphi\psi(x) = x$, so that φ splits, and I is isomorphic to a direct summand of the free module F, and therefore projective.

(2)$\Rightarrow$(1) Every R-linear map from I to R is given by multiplication by some element of K (prove this!). If we let $\varphi: F \longrightarrow I$ be a surjective map from a free module $F = \bigoplus Re_i$, by assumption there exists a splitting $\psi: I \longrightarrow F$ such that $\varphi\psi = 1$. Write $\psi(x) = \sum \lambda_i(x) e_i$ for $x \in I$; then by what we have said, each λ_i determines a $b_i \in K$ such that $\lambda_i(x) = b_i x$, and since for each x that are only finitely many i such that $\lambda_i(x) \neq 0$, we have $b_i = 0$ for all but finitely many i. Letting $b_1, \ldots, b_n$ be the non-zero ones, we have $\sum a_i b_i x = x$ for all $x \in I$, where $a_i = \varphi(e_i)$. Thus $\sum_1^n a_i b_i = 1$. Moreover, since $b_i I = \lambda_i(I) \subset R$ we have $b_i \in I^{-1}$, and therefore $I^{-1} I = R$.

(1)$\Rightarrow$(3) As we have already seen, I is finitely generated. Now if $\sum a_i b_i = 1$ and P is any prime ideal then at least one of $a_i b_i$ must be a unit of R_P, and $I_P = a_i R_P$. Hence I_P is a principal fractional ideal.

(3)$\Rightarrow$(1) If I is finitely generated then $(I^{-1})_P = (I_P)^{-1}$. Indeed, the inclusion $\subset$ holds for any ideal; for $\supset$, if $I = a_1 R + \cdots + a_n R$ and $x \in (I_P)^{-1}$ then $x a_i \in R_P$, so there exist $c_i \in R - P$ such that $x a_i c_i \in R$, so that setting $c = c_1 \ldots c_n$ we have $(cx) a_i \in R$ for all i, which gives $cx \in I^{-1}$ and $x \in (I^{-1})_P$. From the fact that I_P is principal, we get $I_P \cdot (I_P)^{-1} = R_P$. Now if $II^{-1} \neq R$ then we can take a maximal ideal P such that $II^{-1} \subset P$, and then $I_P \cdot (I_P)^{-1} = I_P \cdot (I^{-1})_P \subset PR_P$, which is a contradiction. Thus we must have $II^{-1} = R$. ∎

Theorem 11.4. Let R be a Noetherian integral domain, and P a non-zero prime ideal of R. If P is invertible then $\operatorname{ht} P = 1$ and R_P is a DVR.

Proof. If P is invertible the maximal ideal PR_P of R_P is principal, and condition (3) of Theorem 2 is satisfied; thus R_P is a DVR, and so $\dim R_P = 1$.

Theorem 11.5. Let R be a normal Noetherian domain. Then we have

(i) all the prime divisors of a non-zero principal ideal have height 1;

(ii) $R = \bigcap_{\operatorname{ht} P = 1} R_P$.

Proof. (i) Suppose $0 \neq a \in R$ and that P is one of the prime divisors of aR; then there exists an element $b \in R$ such that $aR{:}b = P$. We set $PR_P = \mathfrak{m}$, and then $aR_P{:}b = \mathfrak{m}$, so that $ba^{-1} \in \mathfrak{m}^{-1}$ and $ba^{-1} \notin R_P$. If $ba^{-1}\mathfrak{m} \subset \mathfrak{m}$ then by the determinant trick ba^{-1} is integral over R_P, which contradicts the fact that R_P is integrally closed. Thus $ba^{-1}\mathfrak{m} = R_P$, so that $\mathfrak{m}^{-1}\mathfrak{m} = R_P$, and then by the previous theorem we get $\operatorname{ht} \mathfrak{m} = \operatorname{ht} P = 1$.

(ii) It is sufficient to prove that for $a, b \in R$ with $a \neq 0$, $b \in aR_P$ for every height 1 prime $P \in \operatorname{Spec} R$ implies $b \in aR$. Let $P_1, \ldots, P_n$ be the prime divisors of aR, and let $aR = \mathfrak{q}_1 \cap \cdots \cap \mathfrak{q}_n$ be a primary decomposition of aR, where $\mathfrak{q}_i$ is a P_i-primary ideal for each i. Then since $\operatorname{ht} P_i = 1$, we have $b \in aR_{P_i} \cap R = \mathfrak{q}_i$ for $i = 1, \ldots, n$, and therefore $b \in \bigcap \mathfrak{q}_i = aR$. ■

Corollary. Let R be a Noetherian domain. The following two conditions are necessary and sufficient for R to be normal:

(*a*) for P a height 1 prime ideal, R_P is a DVR;

(*b*) all the prime divisors of a non-zero principal ideal of R have height 1.

Proof. We have already seen necessity. For sufficiency, note that the proof of (ii) above shows that (b) implies $R = \bigcap_{\operatorname{ht} P = 1} R_P$. Then by (a) each R_P is normal, so that R is normal. ■

Definition. An integral domain for which every non-zero ideal is invertible is called a *Dedekind ring* (sometimes Dedekind domain).

Theorem 11.6. For an integral domain R the following conditions are equivalent:

(1) R is a Dedekind ring;

(2) R is either a field or a one-dimensional Noetherian normal domain;

(3) every non-zero ideal of R can be written as a product of a finite number of prime ideals.

Moreover, the factorisation into primes in (3) is unique.

Proof. (1)⇒(2) Every non-zero ideal is invertible, and therefore finitely generated, so that R is Noetherian. Let P be a non-zero prime ideal of R; then by Theorem 4, the local ring R_P is a DVR and $\operatorname{ht} P = 1$, and therefore either R is a field or $\dim R = 1$. Also, by Theorem 4.7 we know that R is the intersection of the R_P as P runs through all the maximal ideals of R, but since each R_P is a DVR it follows that R is normal.

(2)⇒(1) If R is a field there is no problem. If R is not a field then for every maximal ideal P of R the local ring R_P is a one-dimensional Noetherian local ring and is normal, so that by Theorem 2 it is a principal ideal ring. Thus by Theorem 3, R is a Dedekind ring.

(1)⇒(3) Let I be a non-zero ideal. If $I = R$ then we can view it as the product of zero ideals; if I is itself maximal then it is the product of just one prime ideal. We have already seen that R is Noetherian, so that we can use descending induction on I, that is assume that $I \neq R$ and that every ideal strictly bigger than J is a product of prime ideals. If $I \neq R$ then there is a maximal ideal P containing I, and $I \subset IP^{-1} \subset R$. If $IP^{-1} = I$ then using $P^{-1}P = R$ we would have $I = IP$, and by NAK this would lead to a contradiction. Hence $IP^{-1} \neq I$, so that by induction we can write $IP^{-1} = Q_1 \ldots Q_r$, with $Q_i \in \operatorname{Spec} R$. Multiplying both sides by P gives $I = Q_1 \ldots Q_r P$.

The proof of (3)⇒(1) is a little harder, and we break it up into four steps.

Step 1. In general, any non-zero principal ideal aR of an integral domain R is obviously invertible. Moreover, suppose that I and J are non-zero fractional ideals and $B = IJ$; then obviously I and J invertible implies B invertible, but the converse also holds. To see this, from $I^{-1}J^{-1}B \subset R$ we get $I^{-1}J^{-1} \subset B^{-1}$, and also from $B^{-1}IJ \subset R$ we get $B^{-1}I \subset J^{-1}$ and $B^{-1}J \subset I^{-1}$; now if B is invertible then multiplying the last two inclusions together we get $B^{-1} = B^{-1}B^{-1}IJ \subset I^{-1}J^{-1}$, and hence $B^{-1} = I^{-1}J^{-1}$. Therefore

$$R = BB^{-1} = IJI^{-1}J^{-1} = (II^{-1})(JJ^{-1}),$$

and we must have $II^{-1} = JJ^{-1} = R$.

Step 2. Let P be a non-zero prime ideal. Let us prove that if I is an ideal strictly bigger than P then $IP = P$. For this it is sufficient to show that if $I = P + aR$ with $a \notin P$ then $P \subset IP$. Consider expressions of I^2 and $a^2R + P$ as product of prime ideals, $I^2 = P_1 \dots P_r$ and $a^2R + P = Q_1 \dots Q_s$. Then P_i and Q_j are prime ideals containing I, and so are prime ideals strictly bigger than P. We now set $\bar{R} = R/P$, and write $^-$ to denote the image in $\bar{R}$ of elements or ideals of R. Then we have

$$(*) \quad \bar{P}_1 \dots \bar{P}_r = \bar{a}^2\bar{R} = \bar{Q}_1 \dots \bar{Q}_s,$$

and applying Step 1 to the domain $\bar{R}$ we find that $\bar{P}_i$ and $\bar{Q}_j$ are all invertible, and are prime ideals of $\bar{R}$. We can suppose that $\bar{P}_1$ is a minimal element of the set $\{\bar{P}_1, \dots, \bar{P}_r\}$. Moreover, at least one of $\bar{Q}_1, \dots, \bar{Q}_s$ is contained in $\bar{P}_1$, so that we can assume that $\bar{Q}_1 \subset \bar{P}_1$, and, on the other hand, since $\bar{Q}_1$ is also a prime ideal and $\bar{P}_1 \dots \bar{P}_r \subset \bar{Q}_1$ we must have $\bar{Q}_1 \supset \bar{P}_i$ for some i. Then $\bar{P}_i \subset \bar{Q}_1 \subset \bar{P}_1$, and by the minimality of $\bar{P}_1$ we have $\bar{P}_i = \bar{P}_1 = \bar{Q}_1$. Multiplying through both sides of (*) by $\bar{P}_1^{-1}$ gives

$$\bar{P}_2 \dots \bar{P}_r = \bar{Q}_2 \dots \bar{Q}_s.$$

Proceeding in the same way, we see that $r = s$, and that after reordering the $\bar{Q}_i$ we can assume that $\bar{P}_i = \bar{Q}_i$ for $i = 1, \dots, r$. From this we get $P_i = Q_i$, and $a^2R + P = (P + aR)^2 = P^2 + aP + a^2R$. Thus any element $x \in P$ can be written

$$x = y + az + a^2t \quad \text{with} \quad y \in P^2, \quad z \in P \quad \text{and} \quad t \in R.$$

Since $a \notin P$ we must have $t \in P$, and then as required we have $P \subset P^2 + aP = (P + aR)P$.

Step 3. Let $b \in R$ be a non-zero element; then in the factorisation $bR = P_1 \dots P_r$, every P_i is a maximal ideal of R. Indeed, if I is any ideal strictly greater than P_i then $IP_i = P_i$, and by Step 1 P_i is invertible, so that $I = R$.

Step 4. Let P be a prime ideal of R, and $0 \neq a \in P$. If $aR = P_1 \dots P_r$ with $P_i \in \operatorname{Spec} R$ then P must contain one of the P_i, but from Step 3 we know that P_i is maximal, so that $P = P_i$. We deduce that P is a maximal ideal and is invertible. If every non-zero prime ideal is invertible then any non-zero ideal of R is invertible, since it can be written as a product of primes. This completes the proof of (3)$\Rightarrow$(1).

Finally, if (1), (2) and (3) hold, then as we have seen in Step 2 above, the uniqueness of factorisation into primes is a consequence of the fact that the prime ideals of R are invertible. ∎

Theorem 11.7 (the Krull–Akizuki theorem). Let A be a one-dimensional Noetherian integral domain with field of fractions K, let L be a finite algebraic extension field of K, and B a ring with $A \subset B \subset L$; then B is a Noetherian ring of dimension at most 1, and if J is a non-zero ideal of B then B/J is an A-module of finite length.

Proof. We follow the method of proof of Akizuki [1] in the linear algebra formulation of [B5]. First of all we prove the following lemma.

Lemma. Let A and K be as in the theorem, and let M be a torsion-free A-module (see Ex. 10.2) of rank $r < \infty$. Then for $0 \neq a \in A$ we have

$$l(M/aM) \leqslant r \cdot l(A/aA).$$

Remark. The rank of a module M over an integral domain A is the maximal number of elements of M linearly independent over A; this is equal to the dimension of the K-vector space $M \otimes_A K$.

Proof of the lemma. First we assume that M is finitely generated. Choose elements $\xi_1, \ldots, \xi_r \in M$ linearly independent over A and set $E = \sum A\xi_i$; then for any $\eta \in M$ there exists $t \in A$ with $t \neq 0$ such that $t\eta \in E$. If we set $C = M/E$ then from the assumption on M we see that C is also finitely generated, so that $tC = 0$ for suitable $0 \neq t \in A$. Applying Theorem 6.4 to C, we can find $C = C_0 \supset C_1 \supset \cdots \supset C_m = 0$, such that $C_i/C_{i+1} \simeq A/\mathfrak{p}_i$ with $\mathfrak{p}_i \in \operatorname{Spec} A$. Now $t \in \mathfrak{p}_i$, and since A is one-dimensional each $\mathfrak{p}_i$ is maximal, so that $l(C) = m < \infty$. If $0 \neq a \in A$ then the exact sequence

$$E/a^n E \longrightarrow M/a^n M \longrightarrow C/a^n C \to 0$$

gives

$$(*) \quad l(M/a^n M) \leqslant l(E/a^n E) + l(C) \quad \text{for all} \quad n > 0.$$

Now E and M are both torsion-free A-modules, and one sees easily that $a^i M/a^{i+1} M \simeq M/aM$, and similarly for E. Hence (*) can be written $n \cdot l(M/aM) \leqslant n \cdot l(E/aE) + l(C)$ for all $n > 0$, which gives $l(M/aM) \leqslant l(E/aE)$. Since $E \simeq A^r$ we have $l(E/aE) = r \cdot l(A/aA)$. This completes the proof in the case that M is finitely generated. If M is not finitely generated, take a finitely generated submodule $\bar{N} = A\bar{\omega}_1 + \cdots + A\bar{\omega}_s$ of $\bar{M} = M/aM$. Then choosing an inverse image ω_i in M for each $\bar{\omega}_i$, and setting $M_1 = \sum A\omega_i$, we get

$$l(\textstyle\sum A\bar{\omega}_i) = l(M_1/M_1 \cap aM) \leqslant l(M_1/aM_1) \leqslant r \cdot l(A/aA).$$

The right-hand side is now independent of $\bar{N}$, so that $\bar{M}$ is in fact finitely generated, and $l(\bar{M}) \leqslant r \cdot l(A/aA)$.

We return to the proof of the theorem. We can replace L by the field of fractions of B. Set $[L:K] = r$; then B is a torsion-free A-module of rank r. Hence by the lemma, for any $0 \neq a \in A$ we have $l_A(B/aB) < \infty$. Now if $J \neq 0$ is an ideal of B and $0 \neq b \in J$ then since b is algebraic over A it satisfies a relation

$$a_m b^m + a_{m-1} b^{m-1} + \cdots + a_1 b + a_0 = 0 \quad \text{with} \quad a_i \in A.$$

B is an integral domain, so that we can assume $a_0 \neq 0$. Then $0 \neq a_0 \in J \cap A$ and so

$$l_A(B/J) \leqslant l_A(B/a_0 B) < \infty.$$

Moreover, one sees from $l_B(J/a_0B) \leqslant l_A(J/a_0B) \leqslant l_A(B/a_0B) < \infty$ that J/a_0B is a finite B-module; hence, J itself is a finite B-module, and therefore B is Noetherian. If P is a non-zero prime ideal of B then B/P is an Artinian ring and an integral domain, and therefore a field. Thus P is maximal and $\dim B = 1$.

Corollary. Let A be a one-dimensional Noetherian integral domain, K its field of fractions, and L a finite algebraic extension field of K; write B for the integral closure of A in L. Then B is a Dedekind ring, and for any maximal ideal P of A there are just a finite number of primes of B lying over P.

Proof. By the theorem B is a one-dimensional Noetherian integral domain and is normal by construction; hence it is a Dedekind ring. It is easy to see that if we factorise PB as a product $PB = Q_1^{\alpha_1} \ldots Q_r^{\alpha_r}$ of a finite number of prime ideals, then $Q_1, \ldots Q_r$ are all the prime ideals of B lying over P.

Exercises to §11. Prove the following propositions.

11.1. Let A be a DVR, K its field of fractions, and $\bar{K}$ an algebraic closure of K; then any valuation ring of $\bar{K}$ dominating A is a one-dimensional non-discrete valuation ring.

11.2. Let A be a DVR, K its field of fractions, and L a finite extension field of K; then a valuation ring of L dominating A is a DVR.

11.3. Let A be a DVR and $\mathfrak{m}$ its maximal ideal; then the $\mathfrak{m}$-adic completion $\hat{A}$ of A is again a DVR.

11.4. Let $v: K \longrightarrow \mathbb{R} \cup \{\infty\}$ be an Archimedean additive valuation of a field K, and let c be a real number with $0 < c < 1$. For $\alpha, \beta \in K$, set $d(\alpha, \beta) = c^{v(\alpha - \beta)}$; then d satisfies the axioms for a metric on K (that is $d(\alpha, \beta) \geqslant 0, d(\alpha, \beta) = 0 \Leftrightarrow \alpha = \beta$, $d(\alpha, \beta) = d(\beta, \alpha)$ and $d(\alpha, \gamma) \leqslant d(\alpha, \beta) + d(\beta, \gamma)$), and the topology of K defined by d does not depend on the choice of c. Let R be the valuation ring of v and $\mathfrak{m}$ its valuation ideal; if R is a DVR then the topology determined by d restricts to the $\mathfrak{m}$-adic topology on R.

11.5. Any ideal in a Dedekind ring can be generated by at most two elements.

11.6. Let A be the integral closure of $\mathbb{Z}$ in $\mathbb{Q}(\sqrt{10})$; then A is a Dedekind ring but not a principal ideal ring.

11.7. If a Dedekind ring A is semilocal then it is a principal ideal ring.

11.8. A module over a Dedekind ring is flat if and only if it is torsion-free.

11.9. Let A be an integral domain (not necessarily Noetherian). The following two conditions are equivalent:

(1) A_P is a valuation ring for every maximal ideal P of A;

(2) an A-module is flat if and only if it is torsion-free. (An integral domain satisfying these conditions is called a Prüfer domain.)

11.10. A finite torsion-free module over a Dedekind ring is projective, and is isomorphic to a direct sum of ideals.

12 Krull rings

Let A be an integral domain and K its field of fractions. We write K^* for the multiplicative group of K. We say that A is a *Krull ring* if there is a family $\mathscr{F} = \{R_\lambda\}_{\lambda\in\Lambda}$ of DVRs of K such that the following two conditions hold, where we write v_λ for the normalised additive valuation corresponding to R_λ:

(1) $A = \bigcap_\lambda R_\lambda$;

(2) for every $x \in K^*$ there are at most a finite number of $\lambda\in\Lambda$ such that $v_\lambda(x) \neq 0$.

The family $\mathscr{F}$ of DVRs is said to be a defining family of A. Since DVRs are completely integrally closed (see Ex. 9.5), so are Krull rings. If A is a Krull ring then for any subfield $K' \subset K$ the intersection $A \cap K'$ is again Krull.

Theorem 12.1. If A is a Krull ring and $S \subset A$ a multiplicative set, then A_S is again Krull. If $\mathscr{F} = \{R_\lambda\}_{\lambda\in\Lambda}$ is a defining family of A then the subfamily $\{R_\lambda\}_{\lambda\in\Gamma}$, where $\Gamma = \{\lambda\in\Lambda \mid R_\lambda \supset A_S\}$ is a defining family of A_S.

Proof. Setting $\mathfrak{m}_\lambda$ for the maximal ideal of R_λ we have

$$\lambda\in\Gamma \Leftrightarrow S\cap\mathfrak{m}_\lambda = \varnothing.$$

Let $0 \neq x \in \bigcap_{\lambda\in\Gamma} R_\lambda$; there are at most finitely many $\lambda\in\Lambda$ such that $v_\lambda(x) < 0$; let $\Delta = \{\lambda_1, \ldots, \lambda_n\}$ be the set of these. If $\lambda\in\Delta$ then $\lambda\notin\Gamma$, hence we can find $t_\lambda \in \mathfrak{m}_\lambda \cap S$. Replacing t_λ by a suitable power, we can assume that $v_\lambda(t_\lambda x) \geqslant 0$. We then set $t = \prod_{\lambda\in\Delta} t_\lambda$, so that for every $\lambda\in\Lambda$ we have $v_\lambda(tx) \geqslant 0$, and therefore $tx \in A$; but on the other hand $t \in S$ so that $x \in A_S$ and we have proved that $A_S \supset \bigcap_{\lambda\in\Gamma} R_\lambda$. The opposite inclusion is obvious. The finiteness condition (2) holds for Λ, so also for the subset Γ. ■

Krull rings defined by a finite number of DVRs have a simple structure.

Lemma 1 (Nagata). Let K be a field and $R_1, \ldots, R_n$ valuation rings of K;

set $A = \bigcap R_i$. Then for any given $a \in K$ there exists a natural number $s \geqslant 2$ such that

$$(1 + a + \cdots + a^{s-1})^{-1} \quad \text{and} \quad a \cdot (1 + a + \cdots + a^{s-1})^{-1}$$

both belong to A.

Proof. We consider separately each R_i. Note first that $(1 - a)(1 + a + \cdots + a^{s-1}) = 1 - a^s$. If $a \notin R_i$ then $a^{-1} \in \mathfrak{m}_i$, and any $s \geqslant 2$ will do. If $a \in R_i$ then provided that there does not exist t such that $1 - a^t \in \mathfrak{m}_i$, any $s \geqslant 2$ will do. If $1 - a \in \mathfrak{m}_i$ then any s which is not a multiple of the characteristic of $R_i/\mathfrak{m}_i$ will do. If on the other hand $1 - a \notin \mathfrak{m}_i$ but $1 - a^t \in \mathfrak{m}_i$ for some $t \geqslant 2$, letting t_0 be the smallest value of t for which this happens, we see that $1 - a^s \in \mathfrak{m}_i$ only for s multiples of t_0, so that we only have to avoid these. Thus for each i the bad values of s (if any) are multiples of some number $d_i > 1$, so that choosing s not divisible by any of these d_i we get the result. ∎

Theorem 12.2. Let K be a field and $R_1, \ldots, R_n$ valuation rings of K such that $R_i \not\subset R_j$ for $i \neq j$; set $\mathfrak{m}_i = \mathrm{rad}(R_i)$. Then the intersection $A = \bigcap_{i=1}^n R_i$ is a semilocal ring, having $\mathfrak{p}_i = \mathfrak{m}_i \cap A$ for $i = 1, \ldots, n$ as its only maximal ideals; moreover $A_{\mathfrak{p}_i} = R_i$. If each R_i is a DVR then A is a principal ideal ring.

Proof. The inclusion $A_{\mathfrak{p}_i} \subset R_i$ is obvious. For the opposite inclusion, let $a \in R_i$; choosing $s \geqslant 2$ as in the lemma, and setting $u = (1 + a + \cdots + a^{s-1})^{-1}$ we get $u \in A$ and $au \in A$. Obviously u is a unit of R_i, so that $u \in A - \mathfrak{p}_i$ and $a = (au)/u \in A_{\mathfrak{p}_i}$. This proves that $A_{\mathfrak{p}_i} = R_i$. It follows from this that there are no inclusions among $\mathfrak{p}_1, \ldots, \mathfrak{p}_n$. If I is an ideal of A not contained in any $\mathfrak{p}_i$ then (by Ex. 1.6) there exists $x \in I$ not contained in $\bigcup_{i=1}^n \mathfrak{p}_i$; then x is a unit in each R_i, and hence in A, so that $I = A$. Thus $\mathfrak{p}_1, \ldots, \mathfrak{p}_n$ are all the maximal ideals of A.

If each R_i is a DVR then we have $\mathfrak{m}_i \neq \mathfrak{m}_i^2$, and hence $\mathfrak{p}_i \neq \mathfrak{p}_i^{(2)}$, (where $\mathfrak{p}^{(2)}$ denotes $\mathfrak{p}^2 A_{\mathfrak{p}} \cap A$). Thus there exists $x_i \in \mathfrak{p}_i$ such that $x_i \notin \mathfrak{p}_i^{(2)}$, and $x_i \notin \mathfrak{p}_j$ for $i \neq j$; then $\mathfrak{p}_i = x_i A$. If I is any ideal of A and $IR_i = x_i^{\nu_i} R_i$ for $i = 1, \ldots, n$ then it is easy to see that $I = x_1^{\nu_1} \ldots x_n^{\nu_n} A$. ∎

If a Krull ring A is defined by an infinite number of DVRs then the defining family of DVRs is not necessarily unique, but the following theorem tells us that among them there is a minimal family.

Theorem 12.3. Let A be a Krull ring, K its field of fractions, and $\mathfrak{p}$ a height 1 prime ideal of A; then if $\mathscr{F} = \{R_\lambda\}_{\lambda \in \Lambda}$ is a family of DVRs of K defining A, we must have $A_{\mathfrak{p}} \in \mathscr{F}$. If we set $\mathscr{F}_0 = \{A_{\mathfrak{p}} \mid \mathfrak{p} \in \operatorname{Spec} A \text{ and } \operatorname{ht} \mathfrak{p} = 1\}$ then $\mathscr{F}_0$ is a defining family of A. Thus $\mathscr{F}_0$ is the minimal defining family of DVRs of A.

Proof. By Theorem 1, $A_{\mathfrak{p}}$ is a Krull ring defined by the subfamily $\mathscr{F}_1 = \{R_\lambda \mid A_{\mathfrak{p}} \subset R_\lambda\} \subset \mathscr{F}$; if $A_{\mathfrak{p}} \subset R_\lambda$ then the elements of $A - \mathfrak{p}$ are units of R_λ

so that $\mathfrak{p} \supset \mathfrak{m}_\lambda \cap A$. If $\mathfrak{m}_\lambda \cap A = (0)$ then $R_\lambda \supset K$ which is a contradiction, hence $\mathfrak{m}_\lambda \cap A \neq (0)$; since $\operatorname{ht} \mathfrak{p} = 1$, we must have $\mathfrak{p} = \mathfrak{m}_\lambda \cap A$. Thus if we fix some $0 \neq x \in \mathfrak{p}$, then $v_\lambda(x) > 0$ for all $R_\lambda \in \mathscr{F}_1$, and hence $\mathscr{F}_1$ is a finite set. Now by the previous theorem and Ex. 10.5, the elements of $\mathscr{F}_1$ correspond bijectively with maximal ideals of $A_\mathfrak{p}$, and $\mathscr{F}_1$ has just one element $A_\mathfrak{p}$. Thus $A_\mathfrak{p} \in \mathscr{F}$, in other words $\mathscr{F}_0 \subset \mathscr{F}$.

To prove that $\mathscr{F}_0$ is a defining family of DVRs of A it is enough to show that $A \supset \bigcap_{\operatorname{ht}\mathfrak{p}=1} A_\mathfrak{p}$. That is, it is enough to prove the implication

$$\text{for } a, b \in A \text{ with } a \neq 0,\ b \in aA_\mathfrak{p} \text{ for all } A_\mathfrak{p} \in \mathscr{F}_0 \Rightarrow b \in aA.$$

As one sees easily, this is equivalent to saying that aA can be written as the intersection of height 1 primary ideals. The set of $R \in \mathscr{F}$ such that $aR \neq R$ is finite, so we write $R_1, \ldots, R_t$ for this. If we set

$$aR_i \cap A = \mathfrak{q}_i \quad \text{and} \quad \operatorname{rad}(R_i) \cap A = \mathfrak{p}_i$$

then $\mathfrak{q}_i$ is a primary ideal belonging to $\mathfrak{p}_i$ for each i, and $aA = \mathfrak{q}_1 \ldots \cap \mathfrak{q}_t$. Eliminating redundant terms from this expression, we get an irredundant expression, say $aA = \mathfrak{q}_1 \cap \ldots \cap \mathfrak{q}_r$. It is enough to show that then $\operatorname{ht} \mathfrak{p}_i = 1$ for $1 \leqslant i \leqslant r$. By contradiction, suppose that $\operatorname{ht} \mathfrak{p}_1 > 1$. By Theorem 1, $A_{\mathfrak{p}_1}$ is a Krull ring with defining family $\mathscr{F}' = \{R \in \mathscr{F} \mid A_{\mathfrak{p}_1} \subset R\}$, but is not itself a DVR, and hence by Theorem 2, $\mathscr{F}'$ is infinite. Thus there exists $R' \in \mathscr{F}'$ such that $aR' = R'$; we set $\mathfrak{p}' = \operatorname{rad}(R') \cap A$. We have $a \notin \mathfrak{p}'$, and $A_{\mathfrak{p}_1} \subset R'$ implies that $\mathfrak{p}' \subset \mathfrak{p}_1$. Now by assumption $aA \neq \mathfrak{q}_2 \cap \cdots \cap \mathfrak{q}_r$, and R_1 is a DVR, so that $(\operatorname{rad}(R_1))^\nu \subset aR_1$ for some $\nu > 0$, and hence $\mathfrak{p}_1^\nu \subset \mathfrak{q}_1$. Therefore there exists an $i \geqslant 0$ such that

$$aA \not\supset \mathfrak{p}_1^i \cap \mathfrak{q}_2 \cap \cdots \cap \mathfrak{q}_r \quad \text{and} \quad aA \supset \mathfrak{p}_1^{i+1} \cap \mathfrak{q}_2 \cap \cdots \cap \mathfrak{q}_r.$$

Hence there exist $b \in A$ such that $b \notin aA$ but $b\mathfrak{p}_1 \subset aA$. In particular $b\mathfrak{p}' \subset aA$, but since a is a unit of R' we have

$$(b/a)\mathfrak{p}' \subset A \cap \operatorname{rad}(R') = \mathfrak{p}'.$$

Taking $0 \neq c \in \mathfrak{p}'$ then for every $n > 0$ we have $(b/a)^n c \in \mathfrak{p}' \subset A$, and since A is completely integrally-closed, $b/a \in A$. This is a contradiction, and it proves that $\operatorname{ht} \mathfrak{p}_i = 1$ for $1 \leqslant i \leqslant r$. ∎

Corollary. Let A be a Krull ring and $\mathscr{P}$ the set of height 1 prime ideals of A. For $0 \neq a \in A$ set $v_\mathfrak{p}(a) = n_\mathfrak{p}$; then

$$aA = \bigcap_{\mathfrak{p} \in \mathscr{P}} \mathfrak{p}^{(n_\mathfrak{p})},$$

where $\mathfrak{p}^{(n)}$ denotes the symbolic nth power $\mathfrak{p}^n A_\mathfrak{p} \cap A$.

Proof. According to the theorem we have $aA = \bigcap_{\mathfrak{p} \in \mathscr{P}} (aA_\mathfrak{p} \cap A)$, but $aA_\mathfrak{p} = \mathfrak{p}^{n_\mathfrak{p}} A_\mathfrak{p}$, so that $aA_\mathfrak{p} \cap A = \mathfrak{p}^{(n_\mathfrak{p})}$. ∎

Theorem 12.4. (i) A Noetherian normal domain is a Krull ring.

(ii) Let A be an integral domain, K its field of fractions, and L an extension field of K. If $\{A_i\}_{i \in I}$ is a family of Krull rings contained in L satisfying the two conditions (1) $A=\bigcap A_i$ and (2) given any $0\neq a\in A$ we have $aA_i=A_i$ for all but finitely many i, then A is a Krull ring.

(iii) If A is a Krull ring then so is $A[X]$ and $A[\![X]\!]$.

Proof. (i) This follows from Theorem 11.5 and the fact that for any non-zero $a\in A$ there are only finitely many height 1 prime ideals containing aA (because these are the prime divisors of aA).

(ii) is easy, and we leave it to the reader.

(iii) $K[X]$ is a principal ideal ring and therefore a Krull ring. Moreover, if we let $\mathscr{P}$ be the set of height 1 prime ideals of A then for $\mathfrak{p}\in\mathscr{P}$ the ideal $\mathfrak{p}[X]$ is prime in $A[X]$, and by Theorem 11.2, (3), the local ring $A[X]_{\mathfrak{p}[X]}$ is a DVR of $K(X)$. (If we write v for the additive valuation of K corresponding to the valuation ring $A_\mathfrak{p}$, we can extend v to an additive valuation of $K(X)$ by setting $v(F(X))=\min\{v(a_i)\}$ for a polynomial

$$F(X)=a_0+a_1X+\cdots+a_rX^r \text{ (with } a_i\in K),$$

and $v(F/G)=v(F)-v(G)$ for a rational function $F(X)/G(X)$; then the valuation ring of v in $K(X)$ is $A[X]_{\mathfrak{p}[X]}$.) Now we have $K[X]\cap A[X]_{\mathfrak{p}[X]}=A_\mathfrak{p}[X]$ (prove this!), and so

$$A[X]=K[X]\cap\left(\bigcap_{\mathfrak{p}\in\mathscr{P}}A[X]_{\mathfrak{p}[X]}\right);$$

by (ii) this is a Krull ring.

Now for $A[\![X]\!]$, let $\{R_\lambda\}_{\lambda\in\Lambda}$ be a family of DVRs of K defining A; then inside $K[\![X]\!]$ we have $A[\![X]\!]=\bigcap_\lambda R_\lambda[\![X]\!]$, also by Ex. 9.5, $R_\lambda[\![X]\!]$ is an integrally closed Noetherian ring, and is therefore a Krull ring by (i). However, we cannot use (ii) as it stands, since X is a non-unit of all the rings $R_\lambda[\![X]\!]$, so we set $R_\lambda[\![X]\!][X^{-1}]=B_\lambda$, and note that $A[\![X]\!]=K[\![X]\!]\cap(\bigcap_\lambda B_\lambda)$; now the hypothesis in (ii) is easily verified. Indeed,

$$\varphi(X)=a_rX^r+a_{r+1}X^{r+1}+\cdots\in A[\![X]\!] \quad\text{with}\quad a_r\neq 0$$

is a non-unit of B_λ if and only if a_r is a non-unit of R_λ, and there are only finitely many such λ. Therefore $A[\![X]\!]$ is a Krull ring. ■

Remark 1. Note that the field of fractions of $A[\![X]\!]$ is in general smaller than the field of fractions of $K[\![X]\!]$.

Remark 2. The B_λ occurring above are Euclidean rings ([B7], §1, Ex. 9).

Theorem 12.5. The notions of Dedekind ring and one-dimensional Krull ring coincide.

Proof. A Dedekind ring is a normal Noetherian domain, and therefore a Krull ring. Conversely, if A is a one-dimensional Krull ring, let us prove that A is Noetherian. Let I be a non-zero ideal of A, and let $0\neq a\in I$. If we can prove that A/aA is Noetherian then I/aA is finitely generated, and thus so is

I. By the corollary of Theorem 3 we can write $aA = \mathfrak{q}_1 \cap \cdots \cap \mathfrak{q}_r$, where $\mathfrak{q}_i$ are symbolic powers of prime ideals $\mathfrak{p}_i$ and $\mathfrak{p}_i \neq \mathfrak{p}_j$ if $i \neq j$; now since $\dim A = 1$ each $\mathfrak{p}_i$ is maximal and we have

$$A/aA = A/\mathfrak{q}_1 \times \cdots \times A/\mathfrak{q}_r$$

by Theorem 1.3 and Theorem 1.4. But $A/\mathfrak{q}_i$ is a local ring with maximal ideal $\mathfrak{p}_i/\mathfrak{q}_i$, and hence $A/\mathfrak{q}_i \simeq A_{\mathfrak{p}_i}/\mathfrak{q}_i A_{\mathfrak{p}_i}$; now since each $A_{\mathfrak{p}_i}$ is a DVR, A/aA is Noetherian (in fact even Artinian). Hence A is one-dimensional Noetherian integral domain, and is normal, and is therefore a Dedekind ring. ∎

Theorem 12.6. Let A be a Krull ring, K its field of fractions, and write $\mathscr{P}$ for the set of height 1 prime ideals of A. Suppose given any $\mathfrak{p}_1, \ldots \mathfrak{p}_r \in \mathscr{P}$ and $e_1, \ldots, e_r \in \mathbb{Z}$. Then there exists $x \in K$ satisfying

$$v_i(x) = e_i \quad \text{for} \quad 1 \leqslant i \leqslant r$$

and

$$v_\mathfrak{p}(x) \geqslant 0 \quad \text{for all} \quad \mathfrak{p} \in \mathscr{P} - \{\mathfrak{p}_1, \ldots, \mathfrak{p}_r\}.$$

Here v_i and $v_\mathfrak{p}$ stand for the normalised additive valuations of K corresponding to $\mathfrak{p}_i$ and $\mathfrak{p}$.

Proof. If $y_1 \in A$ is chosen so that $y_1 \in \mathfrak{p}_1$ but $y_1 \notin \mathfrak{p}_1^{(2)} \cup \mathfrak{p}_2 \cup \cdots \cup \mathfrak{p}_r$, then $v_i(y_1) = \delta_{1i}$ for $1 \leqslant i \leqslant r$. Similarly we choose $y_2, \ldots, y_r \in A$ such that $v_i(y_j) = \delta_{ij}$. Then we set

$$y = \prod_{i=1}^{r} y_i^{e_i};$$

let $\mathfrak{p}'_1, \ldots, \mathfrak{p}'_s$ be all the primes $\mathfrak{p} \in \mathscr{P} - \{\mathfrak{p}_1, \ldots, \mathfrak{p}_r\}$ for which $v_\mathfrak{p}(y) < 0$. Then choosing for each $j = 1, \ldots, s$ an element $t_j \in \mathfrak{p}'_j$ not belonging to $\mathfrak{p}_1 \cup \cdots \cup \mathfrak{p}_r$, and taking ν to be sufficiently large, we see that

$$x = y(t_1 \ldots t_s)^\nu$$

satisfies the requirements of the theorem. ∎

Theorem 12.7 (Y. Mori and J. Nishimura). Let A and $\mathscr{P}$ be as in the previous theorem. If $A/\mathfrak{p}$ is Noetherian for every $\mathfrak{p} \in \mathscr{P}$ then A is Noetherian.

Proof (J. Nishimura). As in the proof of Theorem 5, it is enough to show that $A/\mathfrak{p}^{(n)}$ is Noetherian for $\mathfrak{p} \in \mathscr{P}$ and any $n > 0$. If $n = 1$ this holds by hypothesis. For $n > 1$ we proceed as follows. Applying the previous theorem with $r = 1$ and $e = -1$ we can find an element x in the field of fractions of A such that $v_\mathfrak{p}(x) = 1$ and $v_\mathfrak{q}(x) \leqslant 0$ for every other $\mathfrak{q} \in \mathscr{P}$. Set $B = A[x]$. If $y \in \mathfrak{p}$ then $y/x \in A$ so that $y \in xB$, and conversely $B \subset A_\mathfrak{p}$ and $xB \subset \mathfrak{p}A_\mathfrak{p}$, so that $\mathfrak{p} = xB \cap A$. Moreover, $B = A + xB$ and so

$$B/xB \simeq A/\mathfrak{p}.$$

Now $x^iB/x^{i+1}B \simeq B/xB$ for each i, so that by induction on i we see that B/x^iB is a Noetherian B-module for each i, and hence a Noetherian ring. Now we have

$$x^nB \cap A \subset x^nA_\mathfrak{p} \cap A = \mathfrak{p}^{(n)},$$

and B/x^nB is a finite $A/(x^nB \cap A)$-module, being generated by the images of $1, x, \ldots, x^{n-1}$, so that by the Eakin–Nagata theorem (Theorem 3.7), $A/(x^nB \cap A)$ is Noetherian ring; therefore its quotient $A/\mathfrak{p}^{(n)}$ is also Noetherian. ■

Remark. If A is a Noetherian integral domain and K its field of fractions, then the integral closure of A in K is a possibly non-Noetherian Krull ring ([N1], (33.10)). This was proved by Y. Mori (1952) in the local case, and in the general case by M. Nagata (1955). Theorem 12.7 was proved by Mori [1] in 1955 as a theorem on the integral closure of Noetherian rings. His proof was correct (in spite of a number of easily rectifiable inaccuracies), and was an extremely interesting piece of work, but due to its difficulty, and the fact that it appeared in an inaccessible journal, the result was practically forgotten. After Marot [1], (1973) applied it successfully, Mori's work attracted attention once more, and J. Nishimura [1], (1975) reformulated the result as above as a theorem on Krull rings and gave an elegant proof.

More results on Krull rings can be found in [N1], [B7], [F], among others.

Exercises to §12. Prove the following propositions.

12.1. Let $K \subset L$ be a finite extension of fields, and R a valuation ring of K. Then there are a finite number of valuation rings of L dominating R, and if L is a normal extension of K then these are all conjugate to one another under elements of the Galois group $\mathrm{Aut}_K(L)$.

12.2. Let R be a valuation ring of a field K, and let $K \subset L$ be a (possibly infinite) algebraic extension; write $\bar{R}$ for the integral closure of R in L. Then the localisation of $\bar{R}$ at a maximal ideal is a valuation ring dominating R, and conversely every valuation ring of L dominating R is obtained in this way.

12.3. Let A be a Krull ring, K its field of fractions, and $K \subset L$ a finite extension field; if B is the integral closure of A in L, then B is also a Krull ring.

12.4. Let A be an integral domain and K its field of fractions. For I a fractional ideal, write $\tilde{I} = (I^{-1})^{-1}$. If $I = \tilde{I}$ we say that I is divisorial. If A is a Krull ring, then an ideal of A is divisorial if and only if it can be expressed as the intersection of a finite number of height 1 primary ideals.

5

Dimension theory

The dimension theory of Noetherian rings is probably the greatest of Krull's many achievements; with his principal ideal theorem (Theorem 13.5) the theory of Noetherian rings gained in mathematical profundity. Then the theory of multiplicities was first treated rigorously and in considerable generality by Chevalley, and was simplified by Samuel's definition of multiplicity in terms of the Samuel function.

Here we follow the method of EGA, proving Theorem 13.4 via the Samuel function, and deducing the principal ideal theorem as a corollary. The Samuel function is of importance as a measure of singularity in Hironaka's resolution of singularities, but in this book we can only cover its basic properties. In §15 we exploit the notion of systems of parameters to discuss among other things the dimension of the fibres of a ring homomorphism and the dimension formula for finitely generated extension rings.

13 Graded rings, the Hilbert function and the Samuel function

Let G be an Abelian semigroup with identity element 0; (that is, G is a set with an addition law + satisfying associativity $(x+y)+z = x+(y+z)$, commutativity $x+y = y+x$, and such that $0+x = x$). A *graded* (*or* G-*graded*) *ring* is a ring R together with a direct sum decomposition of R as an additive group $R = \bigoplus_{i\in G} R_i$ satisfying $R_iR_j \subset R_{i+j}$. Similarly, a *graded* R-*module* is an R-module M together with a direct sum decomposition $M = \bigoplus_{i\in G} M_i$ satisfying $R_iM_j \subset M_{i+j}$. An element $x\in M$ is *homogeneous* if $x\in M_i$ for some $i\in G$, and i is then called the *degree* of x. A general element $x\in M$ can be written uniquely in the form $x = \sum_{i\in G} x_i$ with $x_i \in M_i$ and only finitely many $x_i \neq 0$; x_i is called the homogeneous term of x of degree i.

A submodule $N \subset M$ is called a *homogeneous submodule* (or graded submodule) if it can be generated by homogeneous elements. This condition is equivalent to either of the following two:

(1) For $x\in M$, if $x\in N$ then each homogeneous term of x is in N;

(2) $N = \sum_{i\in G}(N\cap M_i)$.

For a homogeneous submodule $N \subset M$ we set $N_i = M_i \cap N$; then $M/N = \bigoplus_{i\in G} M_i/N_i$ is again a graded R-module.

One sees from the definition that $R_0 \subset R$ is a subring, and that each graded piece M_i of a graded R-module M is an R_0-module.

The notion of graded ring is most frequently used when G is the semigroup $\{0, 1, 2, \ldots\}$ of non-negative integers, which we denote by $\mathbb{N}$. In this case, we set $R^+ = \sum_{n>0} R_n$; then R^+ is an ideal of R, with $R/R^+ \simeq R_0$.

The polynomial ring $R = R_0[X_1, \ldots, X_n]$ over a ring R_0 is usually made into an $\mathbb{N}$-graded ring by defining the degree of a monomial $X_1^{\alpha_1} \ldots X_n^{\alpha_n}$ as the total degree $\alpha_1 + \cdots + \alpha_n$; however, R has other useful gradings. For example, R has an $\mathbb{N}^n$-grading in which $X_1^{\alpha_1} \cdots X_n^{\alpha_n}$ has degree $(\alpha_1, \ldots, \alpha_n)$; the value of systematically using this grading can be seen in Goto-Watanabe [1]. Alternatively, giving each of the X_i some suitable weight d_i and letting the monomial $X_1^{\alpha_1} \ldots X_n^{\alpha_n}$ have weight $\sum \alpha_i d_i$ defines an $\mathbb{N}$-grading of R. For example, the ring $R_0[X, Y, Z]/(f)$, where $f = a_i X^\alpha + a_2 Y^\beta + a_3 Z^\gamma$ can be graded by giving the images of X, Y, Z the weights $\beta\gamma$, $\alpha\gamma$ and $\alpha\beta$, respectively.

A *filtration* of a ring A is a descending chain $A = J_0 \supset J_1 \supset \cdots$ of ideals such that $J_n J_m \subset J_{n+m}$; the associated graded ring $\mathrm{gr}(A)$ is defined as follows. First of all as a module we set $\mathrm{gr}_n(A) = J_n/J_{n+1}$ for $n \geqslant 0$, and $\mathrm{gr}(A) = \bigoplus_{n\in\mathbb{N}} \mathrm{gr}_n(A)$; then we define the product by

$$(x + J_{n+1})\cdot(y + J_{m+1}) = xy + J_{n+m+1} \quad \text{for} \quad x \in J_n \quad \text{and} \quad y \in J_m.$$

It is easy to see that $\mathrm{gr}(A)$ becomes a graded ring. The filtration $J_1 \supset J_2 \subset \cdots$ defines a linear topology on A (see §8), and the completion $\hat{A}$ of A in this topology has a filtration $J_1^* \supset J_2^* \supset \cdots$ such that $\hat{A}/J_n^* \simeq A/J_n$ for all n, hence $J_n^*/J_{n+1}^* \simeq J_n{}^{n+1}$, and

$$\mathrm{gr}(A) = \mathrm{gr}(\hat{A}).$$

Let A be a ring, I an ideal, and let $B = \bigoplus_{n\geqslant 0} I^n/I^{n+1}$ be the graded ring associated with the filtration $I \supset I^2 \supset \cdots$ of A by powers of I; the various notations $\mathrm{gr}_I(A)$, $\mathrm{gr}^I(A)$ and $G_A(I)$ are used to denote B in the current literature. An element of $B_n = I^n/I^{n+1}$ can be expressed as a linear combination of products of n elements of $B_1 = I/I^2$, so that B is generated over the subring $B_0 = A/I$ by elements of B_1. If $I = Ax_1 + \cdots + Ax_r$ and ξ_i denotes the image of x_i in $B_1 = I/I^2$ then

$$B = \mathrm{gr}_I(A) = (A/I)[\xi_1, \ldots, \xi_r],$$

and B is a quotient of the polynomial ring $(A/I)[X_1, \ldots, X_r]$ as a graded ring.

Theorem 13.1. An $\mathbb{N}$-graded ring $R = \bigoplus_{n\geqslant 0} R_n$ is Noetherian if and only if R_0 is Noetherian and R is finitely generated as a ring over R_0.

Proof. The 'if' is obvious, and we prove the 'only if': suppose that R is Noetherian. Then since $R_0 \simeq R/R^+$, R_0 is Noetherian. R^+ is a homogeneous ideal, and is finitely generated, so that we can suppose that it is generated by homogeneous elements $x_1, \ldots, x_r$. Then it is easy to see that $R = R_0[x_1, \ldots, x_r]$; in fact it is enough to show that $R_n \subset R_0[x_1, \ldots, x_r]$ for every n. Now writing d_i for the degree of x_i we have

$$(*) \quad R_n = x_1 R_{n-d_1} + x_2 R_{n-d_2} + \cdots + x_r R_{n-d_r}.$$

Indeed, for $y \in R_n$, write $y = \sum x_i f_i$ with $f_i \in R$; then setting g_i for the homogeneous term of degree $n - d_i$ of f_i (with $g_i = 0$ if $n - d_i < 0$), we also have $y = \sum x_i g_i$. From (*) it follows by induction that $R_n \subset R_0[x_1, \ldots, x_r]$. ■

Let $R = \bigoplus_{n \geq 0} R_n$ be a Noetherian graded ring; then if $M = \bigoplus_{n \geq 0} M_n$ is a finitely generated graded R-module, each M_n is finitely generated as R_0-module. In fact when $M = R$ this is clear from (*) above. In the general case M can be generated by a finite number of homogeneous elements ω_i: $M = R\omega_1 + \cdots + R\omega_s$. Now letting e_i be the degree of ω_i, we have as above that

$$M_n = R_{n-e_1}\omega_1 + \cdots + R_{n-e_s}\omega_s \quad (\text{where } R_i = 0 \quad \text{for} \quad i < 0),$$

and hence M_n is a finite R_0-module. In particular if R_0 is an Artinian ring, then $l(M_n) < \infty$, where l denotes the length of an R_0-module. In this case we define the *Hilbert series* $P(M, t)$ of M by the formula:

$$P(M, t) = \sum_{n=0}^{\infty} l(M_n) t^n \in \mathbb{Z}[\![t]\!].$$

[In combinatorics it is a standard procedure to associate with a sequence of numbers $a_0, a_1, a_2, \ldots$ the generating function $\sum a_i t^i$.]

Theorem 13.2. Let $R = \bigoplus_{n \geq 0} R_n$ be a Noetherian graded ring with R_0 Artinian, and let M be a finitely generated graded R-module. Suppose that $R = R_0[x_1, \ldots, x_r]$ with x_i of degree d_i, and that $P(M, t)$ is as above. Then $P(M, t)$ is a rational function of t, and can be written

$$P(M, t) = f(t) \Big/ \prod_{i=1}^{r} (1 - t^{d_i}),$$

where $f(t)$ is a polynomial with coefficients in $\mathbb{Z}$.

Proof. By induction on r, the number of generators of R. When $r = 0$, we have $R = R_0$, so that for n sufficiently large, $M_n = 0$, and the power series $P(M, t)$ is a polynomial. When $r > 0$, multiplication by x_r defines an R_0-linear map $M_n \longrightarrow M_{n+d_r}$; writing K_n and L_{n+d_r} for the kernel and cokernel, we get an exact sequence

$$0 \to K_n \longrightarrow M_n \xrightarrow{x_r} M_{n+d_r} \longrightarrow L_{n+d_r} \to 0.$$

Set $K = \oplus K_n$ and $L = \oplus L_n$. Then K is a submodule of M, and $L = M/x_r M$, so that K and L are finite R-modules; moreover $x_r K = x_r L = 0$ so that K

and L can be viewed as R/x_rR-modules, and hence we can apply the induction hypothesis to $P(K,t)$ and $P(L,t)$. Now from the above exact sequence we get

$$l(K_n) - l(M_n) + l(M_{n+d_r}) - l(L_{n+d_r}) = 0.$$

If we multiply this by t^{n+d_r} and sum over n this gives

$$t^{d_r}P(K,t) - t^{d_r}P(M,t) + P(M,t) - P(L,t) = g(t),$$

where $g(t) \in \mathbb{Z}[t]$. The theorem follows at once from this. ∎

A lot of information on the values of $l(M_n)$ can be obtained from the above theorem. Especially simple is the case $d_1 = \cdots = d_r = 1$, so that R is generated over R_0 by elements of degree 1. In this case $P(M,t) = f(t)(1-t)^{-r}$; if $f(t)$ has $(1-t)$ as a factor we can cancel to get P in the form

$$P(M,t) = f(t)(1-t)^{-d} \quad \text{with} \quad f \in \mathbb{Z}[t], \quad d \geqslant 0,$$
$$\text{and if} \quad d > 0 \quad \text{then} \quad f(1) \neq 0.$$

If this holds, we will write $d = d(M)$. Since $(1-t)^{-1} = 1 + t + t^2 + \cdots$, we can repeatedly differentiate both sides to get

$$(1-t)^{-d} = \sum_{n=0}^{\infty} \binom{d+n-1}{d-1} t^n.$$

(This can of course be proved in other ways, for example by induction on d.) If $f(t) = a_0 + a_1 t + \cdots + a_s t^s$ then

$$(*) \qquad l(M_n) = a_0\binom{d+n-1}{d-1} + a_1\binom{d+n-2}{d-1} + \cdots + a_s\binom{d+n-s-1}{d-1};$$

here we set $\binom{m}{d-1} = 0$ for $m < d-1$. The right-hand side of (*) can be formally rearranged as a polynomial in n with rational coefficients, say $\varphi(n)$; then

$$\varphi(X) = \frac{f(1)}{(d-1)!} X^{d-1} + (\text{terms of lower degree}).$$

Since $\binom{m}{d-1}$ coincides with the polynomial $m(m-1)\ldots(m-d+2)/(d-1)!$ for $m \geqslant 0$, this implies the following result.

Corollary. If $d_1 = \cdots = d_r = 1$ in Theorem 2, and $d = d(M)$ is defined as above, then there is a polynomial $\varphi_M(X)$ of degree $d-1$ with rational coefficients such that for $n \geqslant s+1-d$ we have $l(M_n) = \varphi_M(n)$. Here s is the degree of the polynomial $(1-t)^d P(M,t)$.

The polynomial φ_M appearing here is called the *Hilbert polynomial* of the graded module M. The numerical function $l(M_n)$ itself is called the *Hilbert function* of M; by the degree of a Hilbert function we mean the degree of the corresponding Hilbert polynomial.

Remark. For general $d_1, \ldots, d_r$, it is no longer necessarily the case that $l(M_n)$ can be represented by one polynomial.

Example 1. When $R = R_0[X_0, X_1, \ldots, X_r]$, the number of monomials of degree n is $\binom{n+r}{r}$, so that

$$l(R_n) = l(R_0) \cdot \binom{n+r}{r}$$

holds for every $n \geqslant 0$, and the right-hand side is $\varphi_R(n)$. Thus $\varphi_R(X) = (l(R_0)/r!)(X+r)(X+r-1)\cdots(X+1)$.

Example 2. Let k be a field, and $F(X_0, \ldots, X_r)$ a homogeneous polynomial of degree s; set $R = k[X_0, \ldots, X_r]/(F(X))$. Then for $n \geqslant s$,

$$l(R_n) = \binom{n+r}{r} - \binom{n-s+r}{r},$$

and hence, setting $\binom{n+r}{r} = (1/r!)n^r + a_1 n^{r-1} + \cdots$, we have

$$\varphi_R(X) = \frac{1}{r!}[X^r - (X-s)^r] + a_1[X^{r-1} - (X-s)^{r-1}] + \cdots$$

$$= \frac{s}{(r-1)!} X^{r-1} + \text{(terms of lower degree)}.$$

Example 3. Let k be a field, and $R = k[X_1, \ldots, X_r]/P = k[\xi_1, \ldots, \xi_r]$, where P is a homogeneous prime ideal. Let t be the transcendence degree of R over k, and suppose that $\xi_1, \ldots, \xi_t$ are algebraically independent over k; then there are $\binom{n+t-1}{t-1}$ monomials of degree n in the $\xi_1, \ldots, \xi_t$, and these are linearly independent over k, so that $l(R_n) \geqslant \binom{n+t-1}{t-1}$, from which it follows that $d \geqslant t$. In fact we will prove later (Theorem 8) that $d = t$.

A homogeneous ideal of the polynomial ring $k[X_0, \ldots, X_r]$ over a field k defines an algebraic variety in r-dimensional projective space $\mathbb{P}^r$, and the Hilbert polynomial plays an important role in algebraic geometry. For example, note that the numerator of the leading term of φ_R in Example 2 is equal to the degree of F. This holds in more generality, but we must leave details of this to textbooks on algebraic geometry [Ha].

The idea of using the construction of $\mathrm{gr}_\mathfrak{m}(A)$ to relate the study of a general Noetherian local ring $(A, \mathfrak{m})$ to the theory of ideals in a polynomial ring over a field was one of the crucial ideas introduced by Krull in his article 'Dimension theory of local rings' [6], a work of monumental significance for the theory of Noetherian rings. If $\mathfrak{m}$ is generated by r elements then $\mathrm{gr}_{\mathfrak{m}}(A)$ is of the form $k[X_1, \ldots, X_r]/I$, where $k = A/\mathfrak{m}$ and I is a homogeneous ideal. However, the Hilbert function of this graded ring was first used in the study of the multiplicity of A by P. Samuel (1951).

Samuel functions

In a little more generality, let A be a Noetherian semilocal ring, and $\mathfrak{m}$ the Jacobson radical of A. If I is an ideal of A such that for some $\nu > 0$ we have $\mathfrak{m}^\nu \subset I \subset \mathfrak{m}$, we call I an *ideal of definition*; the I-adic and $\mathfrak{m}$-adic topologies then coincide, so that 'ideal of definition' means 'ideal defining the $\mathfrak{m}$-adic topology'. Let M be a finite A-module. If we set

$$\mathrm{gr}_I(M) = \bigoplus_{n \geqslant 0} I^n M / I^{n+1} M$$

then $\mathrm{gr}_I(M)$ is in a natural way a graded module over $\mathrm{gr}_I(A) = \bigoplus I^n/I^{n+1}$. For brevity write $\mathrm{gr}_I(A) = A'$ and $\mathrm{gr}_I(M) = M'$. Then the ring $A'_0 = A/I$ is Artinian, and if $I = \sum_1^r x_i A$, and ξ_i is the image of x_i in I/I^2, then $A' = A'_0[\xi_1, \ldots, \xi_r]$. If also $M = \sum_1^s A\omega_i$ then $M' = \sum A'\bar{\omega}_i$ (where $\bar{\omega}_i$ is the image of ω_i in $M'_0 = M/IM$), so that we can apply Theorem 2 and its corollary to M'. Noting that $l(M'_n) = l(I^n M/I^{n+1} M)$ (where on the left-hand side l is the length as an A'_0-module, on the right-hand side as an A-module), we have

$$\sum_{i=0}^{n} l(M'_i) = l(M/I^{n+1}M).$$

We now set $\chi_M^I(n) = l(M/I^{n+1}M)$. In particular we abbreviate $\chi_M^{\mathfrak{m}}(n)$ to $\chi_M(n)$, and call it the Samuel function of the A-module M.

Repeatedly using the well-known formula $\binom{m}{n} = \binom{m-1}{n-1} + \binom{m-1}{n}$ we get

$$\sum_{\nu=0}^{n} \binom{d+\nu-1}{d-1} = \binom{d+n}{d},$$

so that from formula (*) on p. 95 we get

$$\chi_M^I(n) = a_0\binom{d+n}{d} + a_1\binom{d+n-1}{d} + \cdots + a_s\binom{d+n-s}{d},$$

with $a_i \in \mathbb{Z}$. When $n \geqslant s$ this is a polynomial in n of degree d. This degree d is determined by M, and does not depend on I; to see this, if I and J are both ideals of definition of A then there exist natural numbers a and b

such that $I^a \subset J$, $J^b \subset I$, so that

$$\chi_M^I(an+a-1) \geqslant \chi_M^J(n) \quad \text{and} \quad \chi_M^J(bn+b-1) \geqslant \chi_M^I(n).$$

We thus write $d = d(M)$. It is natural to think of $d(M)$ as a measure of the size of M.

Theorem 13.3. Let A be a semilocal Noetherian ring, and $0 \to M' \longrightarrow M \longrightarrow M'' \to 0$ an exact sequence of finite A-modules; then

$$d(M) = \max(d(M'),\ d(M''))$$

If I is any ideal of definition of A, then $\chi_M^I - \chi_{M''}^I$ and $\chi_{M'}^I$ have the same leading coefficient.

Proof. We can assume $M'' = M/M'$. Then since $M''/I^nM'' = M/(M' + I^nM)$ we have

$$l(M/I^nM) = l(M/M' + I^nM) + l(M' + I^nM/I^nM)$$
$$= l(M''/I^nM'') + l(M'/M' \cap I^nM).$$

Thus setting $\varphi(n) = l(M'/M' \cap I^{n+1}M)$, we have $\chi_M^I = \chi_{M''}^I + \varphi$. Since moreover both $\chi_{M''}^I$ and φ take on only positive values, $d(M)$ coincides with whichever is the greater of $d(M'')$ and $\deg \varphi$. However, by the Artin–Rees lemma, there is a $c > 0$ such that

$$n > c \Rightarrow I^{n+1}M' \subset M' \cap I^{n+1}M \subset I^{n-c+1}M',$$

and hence

$$\chi_{M'}^I(n) \geqslant \varphi(n) \geqslant \chi_{M'}^I(n-c);$$

therefore φ and $\chi_{M'}^I$ have the same leading coefficient. ■

We now define a further measure $\delta(M)$ of the size of M: let $\delta(M)$ be the smallest value of n such that there exist $x_1, \ldots, x_n \in \mathfrak{m}$ for which $l(M/x_1M + \cdots + x_nM) < \infty$. When $l(M) < \infty$ we interpret this as $\delta(M) = 0$. If I is any ideal of definition of A then $l(M/IM) < \infty$, so that $\delta(M) \leqslant$ number of generators of I. Conversely, in the case that A is a local ring and $M = A$, then $l(A/I) < \infty$ implies that I is an $\mathfrak{m}$-primary ideal. Therefore in this case $\delta(A)$ is the minimum of the number of generators of $\mathfrak{m}$-primary ideals.

We have now arrived at the fundamental theorem of dimension theory.

Theorem 13.4. Let A be a semilocal Noetherian ring and M a finite A-module; then we have

$$\dim M = d(M) = \delta(M).$$

Proof.

Step 1. Each of $d(M)$ and $\delta(M)$ are finite, but the finiteness of $\dim M$ has not yet been established. First of all, let us prove that $d(A) \geqslant \dim A$ for the case $M = A$, by induction on $d(A)$. Set $\mathfrak{m} = \operatorname{rad}(A)$. If $d(A) = 0$ then $l(A/\mathfrak{m}^n)$ is constant for $n \gg 0$, so that for some n we have $\mathfrak{m}^n = \mathfrak{m}^{n+1}$ and by NAK, $\mathfrak{m}^n = 0$. Hence any prime ideal of A is maximal and $\dim A = 0$.

Next suppose that $d(A)>0$; if $\dim A=0$ then we're done. If $\dim A>0$, consider a strictly increasing sequence $\mathfrak{p}_0\subset\mathfrak{p}_1\subset\cdots\subset\mathfrak{p}_e$ of prime ideals of A, choose some element $x\in\mathfrak{p}_1-\mathfrak{p}_0$ and set $B=A/(\mathfrak{p}_0+xA)$; then by the previous theorem applied to the exact sequence

$$0\to A/\mathfrak{p}_0\xrightarrow{x}A/\mathfrak{p}_0\longrightarrow B\to 0,$$

we have $d(B)<d(A)$, and so by induction

$$\dim B\leqslant d(B)\leqslant d(A)-1.$$

(The values of $d(B)$ and $\dim B$ are independent of whether we consider B as an A-module or as a B-module, as is clear from the definitions.) In B, the image of $\mathfrak{p}_1\subset\cdots\subset\mathfrak{p}_e$ provides a chain of prime ideals of length $e-1$, so that

$$e-1\leqslant\dim B\leqslant d(A)-1;$$

hence $e\leqslant d(A)$. Since this holds for any chain of prime ideals of A, this proves $\dim A\leqslant d(A)$. For general M, by Theorem 6.4, there are submodules M_i such that $0=M_0\subset M_1\subset\cdots\subset M_q=M$ with

$$M_i/M_{i-1}\simeq A/\mathfrak{p}_i\quad\text{and}\quad\mathfrak{p}_i\in\operatorname{Spec}A.$$

Since for an exact sequence $0\to M'\longrightarrow M\longrightarrow M''\to 0$ of finite A-modules we have

$$\operatorname{Supp}(M)=\operatorname{Supp}(M')\cup\operatorname{Supp}(M'')$$

and

$$\dim M=\max(\dim M',\dim M''),$$

it is easy to see that

$$d(M)=\max\{d(A/\mathfrak{p}_i)\}\geqslant\max\{\dim(A/\mathfrak{p}_i)\}=\dim M.$$

Step 2. We show that $\delta(M)\geqslant d(M)$. If $\delta(M)=0$ then $l(M)<\infty$ so that $\chi_M(n)$ is bounded, hence $d(M)=0$. Next suppose that $\delta(M)=s>0$, choose $x_1,\dots,x_s\in\mathfrak{m}$ such that $l(M/x_1M+\cdots+x_sM)<\infty$, and set $M_i=M/x_1M+\cdots+x_iM$; then clearly $\delta(M_i)=\delta(M)-i$. On the other hand,

$$\begin{aligned}l(M_1/\mathfrak{m}^nM_1)&=l(M/x_1M+\mathfrak{m}^nM)\\&=l(M/\mathfrak{m}^nM)-l(x_1M/x_1M\cap\mathfrak{m}^nM)\\&=l(M/\mathfrak{m}^nM)-l(M/(\mathfrak{m}^nM:x_1))\\&\geqslant l(M/\mathfrak{m}^nM)-l(M/\mathfrak{m}^{n-1}M),\end{aligned}$$

so that $d(M_1)\geqslant d(M)-1$. Repeating this, we get $d(M_s)\geqslant d(M)-s$, but since $\delta(M_s)=0$ we have $d(M_s)=0$, so that $s\geqslant d(M)$.

Step 3. We show that $\dim M\geqslant\delta(M)$, by induction on $\dim M$. If $\dim M=0$ then $\operatorname{Supp}(M)\subset\text{m-Spec}\,A=V(\mathfrak{m})$ so that for large enough n we have $\mathfrak{m}^n\subset\operatorname{ann}M$, and $l(M)<\infty$, therefore $\delta(M)=0$. Next suppose that $\dim M>0$, and let $\mathfrak{p}_i$ for $1\leqslant i\leqslant t$ be the minimal prime divisors of $\operatorname{ann}(M)$

with coht $\mathfrak{p}_i = \dim M$; then the $\mathfrak{p}_i$ are not maximal ideals, so do not contain $\mathfrak{m}$. Hence we can choose $x_1 \in \mathfrak{m}$ not contained in any $\mathfrak{p}_i$. Setting $M_1 = M/x_1M$ we get $\dim M_1 < \dim M$. Therefore by the inductive hypothesis $\delta(M_1) \leqslant \dim M_1$; but obviously $\delta(M) \leqslant \delta(M_1) + 1$, so that $\delta(M) \leqslant \dim M_1 + 1 \leqslant \dim M$. ■

Theorem 13.5. Let A be a Noetherian ring, and $I = (a_1, \ldots, a_r)$ an ideal generated by r elements; then if $\mathfrak{p}$ is a minimal prime divisor of I we have $\operatorname{ht} \mathfrak{p} \leqslant r$. Hence the height of a proper ideal of A is always finite.
Proof. The ideal $IA_{\mathfrak{p}} \subset A_{\mathfrak{p}}$ is a primary ideal belonging to the maximal ideal, so that $\operatorname{ht} \mathfrak{p} = \dim A_{\mathfrak{p}} = \delta(A_{\mathfrak{p}}) \leqslant r$. ■

Remark. Krull proved this theorem by induction on r; the case $r = 1$ is then the hardest part of the proof. Krull called the $r = 1$ case the *principal ideal theorem* (Hauptidealsatz), and the whole of Theorem 5 is sometimes known by this name. Here Theorem 5 is merely a corollary of Theorem 4, but one can also deduce the statement $\dim M = \delta(M)$ of Theorem 4 from it. As far as proving Theorem 5 is concerned, Krull's proof, which does not use the Samuel function, is easier. For this proof, see [N1] or [K]. More elementary proofs of the principal ideal theorem can be found in Rees [3] and Caruth [1].

The definition of height is abstract, and even when one can find a lower bound, one cannot expect an upper bound just from the definition, so that this theorem is extremely important. The principal ideal theorem corresponds to the familiar and obvious-looking proposition of geometrical and physical intuition (which is strictly speaking not always true) that 'adding one equation can decrease the dimension of the space of solutions by at most one'.

Theorem 13.6. Let P be a prime ideal of height r in a Noetherian ring A. Then

(i) P is a minimal prime divisor of some ideal $(a_1, \ldots, a_r)$ generated by r elements;

(ii) if $b_1, \ldots, b_s \in P$ we have $\operatorname{ht} P/(b_1, \ldots, b_s) \geqslant r - s$;

(iii) if $a_1, \ldots, a_r$ are as in (i) we have

$$\operatorname{ht} P/(a_1, \ldots, a_i) = r - i \quad \text{for} \quad 1 \leqslant i \leqslant r.$$

Proof. (i) A_P is an r-dimensional local ring, so that by Theorem 4 we can choose r elements $a_1, \ldots, a_r \in PA_P$ such that $(a_1, \ldots, a_r)A_P$ is PA_P-primary. Each a_i is of the form an element of P times a unit of A_P, so that without loss of generality we can assume that $a_i \in P$. Then P is a minimal prime divisor of $(a_1, \ldots, a_r)A$.

(ii) Set $\bar{A} = A/(b_1, \ldots, b_s)$, $\bar{P} = P/(b_1, \ldots, b_s)$ and $\operatorname{ht} \bar{P} = t$. Then by (i), there exist $c_1, \ldots, c_t \in P$ such that $\bar{P}$ is a minimal prime divisor of $(\bar{c}_1, \ldots, \bar{c}_t)\bar{A}$. Then P is a minimal prime divisor of $(b_1, \ldots, b_s, c_1, \ldots, c_t)$, and hence $r \leqslant s + t$ by Theorem 5.

(iii) The ideal $P/(a_1, \ldots, a_i)$ is a minimal prime divisor of $(\bar{a}_{i+1}, \ldots, \bar{a}_r)$ in $A/(a_1, \ldots, a_i)$, hence $\operatorname{ht} P/(a_1, \ldots, a_i) \leqslant r - i$. The opposite inequality was proved in (ii). ■

Theorem 13.7. Let $A = \bigoplus_{n \geqslant 0} A_n$ be a Noetherian graded ring.

(i) If I is a homogeneous ideal and P is a prime divisor of I then P is also homogeneous.

(ii) If P is a homogeneous prime ideal of height r then there exists a sequence $P = P_0 \supset P_1 \supset \cdots \supset P_r$ of length r consisting of homogeneous prime ideals.

Proof. (i) P can be expressed in the form $P = \operatorname{ann}(x)$ for a suitable element x of the graded A-module A/I. Let $a \in P$, and let $x = x_0 + x_1 + \cdots + x_r$ and $a = a_p + a_{p+1} + \cdots + a_q$ be decompositions into homogeneous terms. Then since $ax = 0$,

$$a_p x_0 = 0, \quad a_p x_1 + a_{p+1} x_0 = 0, \quad a_p x_2 + a_{p+1} x_1 + a_{p+2} x_0 = 0, \ldots,$$

from which we get $a_p^2 x_1 = 0$, $a_p^3 x_2 = 0, \ldots$, and finally $a_p^{r+1} x = 0$. It follows that $a_p^{r+1} \in P$, but since P is prime, $a_p \in P$. Thus $a_{p+1} + \cdots + a_q \in P$, so that in turn $a_{p+1} \in P$. Proceeding in the same way, we see that all the homogeneous terms of a are in P, so that P is a homogeneous ideal.

(ii) First of all note that we can assume that A is an integral domain. To see this, if we take a chain $P = \mathfrak{p}_0 \supset \cdots \supset \mathfrak{p}_r$ of prime ideals of length r then $\mathfrak{p}_r$ is a minimal prime divisor of (0), and so by (i) is a homogeneous ideal; so we can replace A by $A/\mathfrak{p}_r$. Now choose a homogeneous element $0 \neq b_1 \in P$; then by Theorem 6, $\operatorname{ht}(P/b_1 A) = r - 1$, and so there is a minimal prime divisor Q of $b_1 A$ such that $\operatorname{ht}(P/Q) = r - 1$; since $Q \neq (0)$, it is a height 1 homogeneous prime ideal. By the inductive hypothesis on r applied to P/Q there exists a chain $P = P_0 \supset P_1 \supset \cdots \supset P_{r-1} = Q$ of homogeneous prime ideals of length $r - 1$, and adding on (0) we get a chain of length r. ■

Let us investigate more closely the relation between local rings and graded rings.

Theorem 13.8. Let k be a field, and $R = k[\xi_1, \ldots, \xi_r]$ a graded ring generated by elements $\xi_1, \ldots, \xi_r$ of degree 1; set $M = \sum \xi_i R$, $A = R_M$ and $\mathfrak{m} = MA$.

(i) Let χ be the Samuel function of the local ring A, and φ the Hilbert function of the graded ring R; then $\varphi(n) = \chi(n) - \chi(n-1)$;

(ii) $\dim R = ht\ M = \dim A = \deg \varphi + 1$;

(iii) $\operatorname{gr}_{\mathfrak{m}}(A) \simeq R$ as graded rings.

Proof. M is a maximal ideal of R so that

$$\mathfrak{m}^n/\mathfrak{m}^{n+1} \simeq M^n/M^{n+1} \simeq R_n;$$

hence $\chi(n) - \chi(n-1) = l(\mathfrak{m}^n/\mathfrak{m}^{n+1}) = l(R_n) = \varphi(n)$, and so $\dim A = \deg \chi = 1 + \deg \varphi$. Then since $A = R_M$, we have $\dim A = \operatorname{ht} M$. After this it is enough to prove that $\dim R = \operatorname{ht} M$. First of all, assume that R is an integral domain, so that by Example 3 in the section on Hilbert functions and by Theorem 5.6, we have

$$1 + \deg \varphi \geqslant \operatorname{tr.deg}_k R = \dim R \geqslant \operatorname{ht} M;$$

putting this together with $\operatorname{ht} M = \dim A = 1 + \deg \varphi$, we get $\dim R = \operatorname{ht} M$. Next for general R, let $P_1, \ldots, P_t$ be the minimal prime ideals of R; then by Theorem 7, these are all homogeneous ideals, and each R/P_i is a graded ring. Choosing P_1 such that $\dim R = \dim R/P_1$ and using the above result, we get

$$\dim R = \dim R/P_1 = \operatorname{ht} M/P_1 \leqslant \operatorname{ht} M \leqslant \dim R,$$

so that $\dim R = \operatorname{ht} M$ as required. We have $R_n \subset M^n \subset \mathfrak{m}^n$ with $\mathfrak{m}^n/\mathfrak{m}^{n+1} \simeq R_n$, and so taking an element x of R_n into its image in $\mathfrak{m}^n/\mathfrak{m}^{n+1}$ we obtain a canonical one-to-one map $R \xrightarrow{\sim} \operatorname{gr}_{\mathfrak{m}} A$, and it is clear from the definition that this is a ring isomorphism. ■

Theorem 13.9. Let $(A, \mathfrak{m}, k)$ be a Noetherian local ring, and set $G = \operatorname{gr}_{\mathfrak{m}} A$; then $\dim A = \dim G$.

Proof. Letting φ be the Hilbert polynomial of G, we have $\dim A = 1 + \deg \varphi$ (by Theorem 4), and by the previous theorem this is equal to $\dim G$. ■

In fact the following more general theorem holds: for I a proper ideal in a Noetherian local ring A, set $G = \operatorname{gr}_I(A)$; then $\dim A = \dim G$. This will be proved a little later (Theorem 15.7).

Exercises to §13. Prove the following propositions.

13.1. Let $R = R_0 + R_1 + \cdots$ be a graded ring, and u a unit of R_0. Then the map T_u defined by $T_u(x_0 + x_1 + \cdots + x_n) = x_0 + x_1 u + \cdots + x_n u^n$ (where $x_i \in R_i$) is an automorphism of R. If R_0 contains an infinite field k, then an ideal I of R is homogeneous if and only if $T_\alpha(I) = I$ for every $\alpha \in k$.

13.2. Let $R = R_0 + R_1 + \cdots$ be a graded ring, I an ideal of R and t an indeterminate over R. Set $R' = R[t, t^{-1}]$ and consider R' as a graded ring where t has degree 0 (that is, $R'_n = R_n[t, t^{-1}]$). Then an ideal I of R is homogeneous if and only if $T_t(IR') = IR'$.

13.3. Let A be a Noetherian ring having an embedded associated prime. If $a \in A$ is a non-zero divisor satisfying $\bigcap_{n=1}^{\infty} a^n A = (0)$, then $A/(a)$ also has an embedded associated prime.

13.4. Let $R = \bigoplus_{n \in \mathbb{Z}} R_n$ be a $\mathbb{Z}$-graded ring. For an ideal I of R, let I^* denote the greatest homogeneous ideal of R contained in I, that is the ideal of R generated by all the homogeneous elements of I.

(i) If P is prime so is P^*.

(ii) If P is a homogeneous prime ideal and Q is a P-primary ideal then Q^* is again P-primary.

13.5. Let R be a $\mathbb{Z}$-graded integral domain; write S for the multiplicative set consisting of all non-zero homogeneous elements of R. Then R_S is a graded ring, and its component of degree 0 is a field $(R_S)_0 = K$; if $R \neq R_0$ then $R_S \simeq K[X, X^{-1}]$, where the degree of X is the greatest common divisor of the degrees of elements of S.

13.6. Let R be a $\mathbb{Z}$-graded ring and P an inhomogeneous prime ideal of R; then there are no prime ideals contained between P^* and P. If $\operatorname{ht} P < \infty$ then $\operatorname{ht} P = \operatorname{ht} P^* + 1$ (Matijevic–Roberts [1]).

Appendix to §13. Determinantal ideals (after Eagon–Northcott [1])

Let $M = (a_{ij})$ be an $r \times s$ matrix ($r \leqslant s$) with elements a_{ij} in a Noetherian ring A, and let I_t be the ideal of A generated by the $t \times t$ minors (that is subdeterminants) of M. When $t = r$ and A is a polynomial ring $k[X_1, \ldots, X_n]$ over a field k, Macaulay proved that all the prime divisors of I_t have height $\leqslant s - r + 1$ ([Mac], p. 54). In his Ph.D. thesis, Eagon generalised this result as follows: for an arbitrary Noetherian ring A, every minimal prime divisor of I_t has height $\leqslant (r - t + 1)(s - t + 1)$. The following ingeneous proof is taken from Eagon–Northcott [1]. We begin with some preliminary observations.

The following operations on a matrix M with elements in a ring A are called elementary row operations: (1) permutation of the rows; (2) replacing C_i by $uC_i + vC_j$, where C_i and $C_j (i \neq j)$ are two distinct rows of M, u is a unit of A and v is an element of A; elementary column operations are defined similarly. The ideal I_t does not change under these operations. Now, if an element of M is a unit in A, we can transform M by a finite number of elementary row and column operations to the following form:

$$\begin{pmatrix} 1 & 0 & \cdots & 0 \\ 0 & & & \\ \vdots & & N & \\ 0 & & & \end{pmatrix}$$

and I_t is equal to the ideal of A generated by the $(t-1) \times (t-1)$ minors of N.

Lemma. Let (A, P) be a Noetherian local ring and set $B = A[X]$. Let J be a P-primary ideal of A and J' an ideal of B such that $J' \subset PB$ and $J' + XB = JB + XB$. Then PB is a minimal prime divisor of J'.

Proof. $PB+XB$ is a maximal ideal of B, and is the radical of $JB+XB=J'+XB$. Thus, in the ring B/J' we have that $(PB+XB)/J'$ is a minimal prime divisor of the principal ideal $(J'+XB)/J'$. Hence $\mathrm{ht}((PB+XB)/J')=1$. Since PB/J' is a prime ideal in B/J', we have $\mathrm{ht}(PB/J')=0$. ∎

Theorem 13.10 (Eagon). Let A be a Noetherian ring and M be an $r\times s$ matrix ($r\leqslant s$) of elements of A. Let I_t be the ideal of A generated by the $t\times t$ minors of M. If P is a minimal prime divisor of I_t then we have

$$\mathrm{ht}\,P\leqslant(r-t+1)(s-t+1).$$

Proof. Induction on r. When $r=1$ we have $t=1$, and so $(r-t+1)(s-t+1)=s$. The ideal I_1 is generated by s elements, so that the assertion is just the principal ideal theorem (Theorem 5) in this case. Next assume that $r>1$. Localising at P we may assume that A is a local ring with maximal ideal P, and that I_t is P-primary.

If $t=1$, then I_t is generated by rs elements and $(r-t+1)(s-t+1)=rs$, so our assertion holds also for this case. Therefore we assume $t>1$. If at least one of the elements of M is a unit of A, then by what we said above, I_t is generated by $(t-1)\times(t-1)$ minors of a $(r-1)\times(s-1)$ matrix, and again we are done. Therefore we assume that all the elements of M are in P. Now comes the brilliant idea. Let M' be the matrix with elements in $B=A[X]$ obtained from M by replacing a_{11} by $a_{11}+X$, and let I' be the ideal of B generated by the $t\times t$ minors of M'. Since $t>1$ and $a_{ij}\in P$ for all i and j we have $I'\subset PB$. We also have $I'+XB=I_tB+XB$ since both sides have the same image in $B/XB=A$. Therefore PB is a minimal prime divisor of I' by the lemma. Since the element $a_{11}+X$ of M' is not in PB, we have $\mathrm{ht}\,PB\leqslant(r-t+1)(s-t+1)$ by our previous argument. Since $\mathrm{ht}\,PB=\mathrm{ht}\,P$, as we can see by Theorems 4 and 5, we are done. ∎

14 Systems of parameters and multiplicity

Let $(A,\mathfrak{m})$ be an r-dimensional Noetherian local ring; by Theorem 13.4, there exists an $\mathfrak{m}$-primary ideal generated by r elements, but none generated by fewer. If $a_1,\dots,a_r\in\mathfrak{m}$ generate an $\mathfrak{m}$-primary ideal, $\{a_1,\dots,a_r\}$ is said to be a *system of parameters* of A (sometimes abbreviated to s.o.p.). If M is a finite A-module with $\dim M=s$, there exist $y_1,\dots,y_s\in\mathfrak{m}$ such that $l(M/(y_1,\dots,y_s)M)<\infty$, and then $\{y_1,\dots,y_s\}$ is said to be a *system of parameters of M*.

If we set $A/\mathfrak{m}=k$, the smallest number of elements needed to generate $\mathfrak{m}$ itself is equal to $\mathrm{rank}_k\,\mathfrak{m}/\mathfrak{m}^2$; (here rank_k is the rank of a free module over k, that is the dimension of $\mathfrak{m}/\mathfrak{m}^2$ as k-vector space). This number is called the *embedding dimension* of A, and is written emb dim A. In general

$$\dim A\leqslant\mathrm{emb\,dim}\,A,$$

and equality holds when $\mathfrak{m}$ can be generated by r elements; in this case A is said to be a *regular local ring*, and a system of parameters generating $\mathfrak{m}$ is called a *regular system of parameters.*

Theorem 14.1. Let $(A, \mathfrak{m})$ be a Noetherian local ring, and $x_1, \ldots, x_r$ a system of parameters. Then

(i) $\dim A/(x_1, \ldots, x_i) = r - i$ for $1 \leqslant i \leqslant r$.

(ii) although it is not true that $\operatorname{ht}(x_1, \ldots, x_i) = i$ for all i for an arbitrary system of parameters, there exists a choice of $x_1, \ldots, x_r$ such that every subset $F \subset \{x_1, \ldots, x_r\}$ generates an ideal of A of height equal to the number of elements of F.

Proof. (i) is contained in Theorem 13.6. We now prove the second half of (ii). If $r \leqslant 1$ the assertion is obvious; suppose that $r > 1$. Let $\mathfrak{p}_{0j}$ (for $1 \leqslant j \leqslant e_0$) be the prime ideals of A of height 0. Choosing $x_1 \in \mathfrak{m}$ not contained in any $\mathfrak{p}_{0j}$, we have $\operatorname{ht}(x_1) = 1$. Next letting $\mathfrak{p}_{1j}$ (for $1 \leqslant j \leqslant e_1$) be the minimal prime divisors of (x_1), so that $\operatorname{ht} \mathfrak{p}_{1j} = 1$, and choosing $x_2 \in \mathfrak{m}$ not contained in any $\mathfrak{p}_{0j}$ or any $\mathfrak{p}_{1j}$, we have $\operatorname{ht}(x_2) = 1$, $\operatorname{ht}(x_1, x_2) = 2$; if $r = 2$ we're done. If $r > 2$ we choose $x_3 \in \mathfrak{m}$ not contained in any minimal prime divisor of (0), (x_1), (x_2), (x_1, x_2), and proceed in the same way to obtain the result.

We now give an example where $\operatorname{ht}(x_1, \ldots, x_i) < i$. Let k be a field and set $R = k[\![X, Y, Z]\!]$; let $I = (X) \cap (Y, Z)$, and write $A = R/I$, and x, y, z for the images in A of X, Y, Z. The minimal prime ideals of A are (x) and (y, z); now $A/(x) \simeq R/(X) \simeq k[\![Y, Z]\!]$ is two-dimensional and $A/(y, z) \simeq R/(Y, Z) \simeq k[\![X]\!]$ is one-dimensional, so that $\dim A = 2$. $\{y, x + z\}$ is a system of parameters of A; in fact $xy = xz = 0$ so that $x^2 = x(x + z) \in (y, x + z)$ and $z^2 = z(x + z) \in (y, x + z)$. However, y is contained in the minimal prime ideal (y, z) of A, and hence $\operatorname{ht}(y) = 0$. ■

Theorem 14.2. Let $(R, \mathfrak{m})$ be an n-dimensional regular local ring, and $x_1, \ldots, x_i$ elements of $\mathfrak{m}$. Then the following conditions are equivalent:

(1) $x_1, \ldots, x_i$ is a subset of a regular system of parameters of R;

(2) the images in $\mathfrak{m}/\mathfrak{m}^2$ of $x_1, \ldots, x_i$ are linearly independent over $R/\mathfrak{m}$;

(3) $R/(x_1, \ldots, x_i)$ is an $(n - i)$-dimensional regular local ring.

Proof. (1)$\Rightarrow$(2) If $x_1, \ldots, x_i, x_{i+1}, \ldots, x_n$ is a regular system of parameters then their images generate $\mathfrak{m}/\mathfrak{m}^2$ over $k = R/\mathfrak{m}$, and since $\operatorname{rank}_k \mathfrak{m}/\mathfrak{m}^2 = n$ they must be linearly independent over k.

(1)$\Rightarrow$(3) We know that $\dim R/(x_1, \ldots, x_i) = n - i$, and the images of $x_{i+1}, \ldots, x_n$ generate the maximal ideal of $R/(x_1, \ldots, x_i)$.

(3)$\Rightarrow$(1) If the maximal ideal $\mathfrak{m}/(x_1, \ldots, x_i)$ of $R/(x_1, \ldots, x_i)$ is generated by the images of $y_1, \ldots, y_{n-i} \in \mathfrak{m}$ then $\mathfrak{m}$ is generated by $x_1, \ldots, x_i$, $y_1, \ldots, y_{n-i}$.

Remark. The hypothesis that R is regular is not needed for (3)$\Rightarrow$(1).

(2)$\Rightarrow$(1) Using $\operatorname{rank}_k \mathfrak{m}/\mathfrak{m}^2 = n$, if we choose $x_{i+1},\ldots,x_n \in \mathfrak{m}$ such that the images of $x_1,\ldots,x_n$ in $\mathfrak{m}/\mathfrak{m}^2$ form a basis then $x_1,\ldots,x_n$ generate $\mathfrak{m}$ and so forms a regular system of parameters. ■

Theorem 14.3. A regular local ring is an integral domain.

Proof. Let $(R,\mathfrak{m})$ be an n-dimensional regular local ring; we proceed by induction on n. If $n=0$ then $\mathfrak{m}$ is an ideal generated by 0 elements, so that $\mathfrak{m}=(0)$. This in turn means that R is a field. Thus a zero-dimensional regular local ring is just a field by another name.

When $n=1$, the maximal ideal $\mathfrak{m}=xR$ is principal and $\operatorname{ht}\mathfrak{m}=1$, so that there exists a prime ideal $\mathfrak{p}\neq\mathfrak{m}$ with $\mathfrak{m}\supset\mathfrak{p}$. If $y\in\mathfrak{p}$ we can write $y=xa$ with $a\in R$, and since $x\notin\mathfrak{p}$ we have $a\in\mathfrak{p}$; hence $\mathfrak{p}=x\mathfrak{p}$, and by NAK, $\mathfrak{p}=(0)$. This proves that R is an integral domain. (There is a slightly different proof in the course of the proof of Theorem 11.2; as proved there, a one-dimensional regular local ring is just a DVR by another name.)

When $n>1$, let $\mathfrak{p}_1,\ldots,\mathfrak{p}_r$ be the minimal prime ideals of R; then since $\mathfrak{m}\not\subset\mathfrak{m}^2$ and $\mathfrak{m}\not\subset\mathfrak{p}_i$ for all i, there exists an element $x\in\mathfrak{m}$ not contained in any of $\mathfrak{m}^2,\mathfrak{p}_1,\ldots,\mathfrak{p}_r$ (see Ex. 1.6). Then the image of x in $\mathfrak{m}/\mathfrak{m}^2$ is non-zero, so that by the previous theorem R/xR is an $(n-1)$-dimensional regular local ring. By the induction hypothesis, R/xR is an integral domain, in other words, xR is a prime ideal of R. If $\mathfrak{p}_1$ is one of the minimal prime ideals contained in xR then since $x\notin\mathfrak{p}_1$, the same argument as in the $n=1$ case shows that $\mathfrak{p}_1=x\mathfrak{p}_1$, and hence $\mathfrak{p}_1=(0)$. ■

Theorem 14.4. Let $(A,\mathfrak{m},k)$ be a d-dimensional regular local ring; then

$$\operatorname{gr}_{\mathfrak{m}}(A)\simeq k[X_1,\ldots,X_d],$$

and if $\chi(n)$ is the Samuel function of A then

$$\chi(n)=\binom{n+d}{d}\quad\text{for all } n\geqslant 0.$$

Proof. Since $\mathfrak{m}$ is generated by d elements, $\operatorname{gr}_{\mathfrak{m}}(A)$ is of the form $k[X_1,\ldots,X_d]/I$, where I is a homogeneous ideal. Now if $I\neq(0)$, let $f\in I$ be a non-zero homogeneous element of degree r; then for $n>r$ the homogeneous piece of $k[X]/I$ of degree n has length at most $\binom{n+d-1}{d-1}-\binom{n-r+d-1}{d-1}$, which is a polynomial of degree $d-2$ in n. This implies that the Samuel function of A is of degree at most $d-1$, and contradicts $\dim A=d$. Hence $I=(0)$; the second assertion follows from the first. ■

Let $(A,\mathfrak{m})$ be a Noetherian local ring. Elements $y_1,\ldots,y_r\in\mathfrak{m}$ are said to be *analytically independent* if they have the following property; for every homogeneous form $F(Y_1,\ldots,Y_r)$ with coefficients in A,

$$F(y_1,\ldots,y_r)=0 \Rightarrow \text{the coefficients of } F \text{ are in } \mathfrak{m}.$$

If $y_1,\ldots,y_r$ are analytically independent and A contains a field k, then $F(y)\neq 0$ for any non-zero homogeneous form $F(Y)\in k[Y_1,\ldots,Y_r]$.

Theorem 14.5. Let $(A,\mathfrak{m})$ be a d-dimensional Noetherian local ring and $x_1,\ldots,x_d$ a system of parameters of A; then $x_1,\ldots,x_d$ are analytically independent.

Proof. Set $\mathfrak{q}=\sum x_iA$. Since $\mathfrak{q}$ is an ideal of definition of A, by Theorem 13.4, $\chi_A^{\mathfrak{q}}(n)=l(A/\mathfrak{q}^n)$ is a polynomial of degree d in n for $n\gg 0$. Set $A/\mathfrak{m}=k$; we say that a homogeneous form $f(X)\in k[X_1,\ldots,X_d]$ of degree n is a *null-form* of $\mathfrak{q}$ if $F(x_1,\ldots,x_d)\in\mathfrak{q}^n\mathfrak{m}$ for any homogeneous form $F(X)\in A[X_1,\ldots,X_d]$ which reduces to $f(X)$ modulo $\mathfrak{m}$. Write $\mathfrak{n}$ for the ideal of $k[X_1,\ldots,X_d]$ generated by the null-forms of $\mathfrak{q}$. Then

$$k[X]/\mathfrak{n}\simeq\bigoplus\mathfrak{q}^n/\mathfrak{q}^n\mathfrak{m},$$

and writing φ for the Hilbert polynomial of $k[X]/\mathfrak{n}$, we have $\varphi(n)=l(\mathfrak{q}^n/\mathfrak{q}^n\mathfrak{m})$ for $n\gg 0$. The right-hand side is just the number of elements in a minimal basis of $\mathfrak{q}^n$, so that $\varphi(n)\cdot l(A/\mathfrak{q})\geqslant l(\mathfrak{q}^n/\mathfrak{q}^{n+1})$. Now

$$l(\mathfrak{q}^n/\mathfrak{q}^{n+1})=\chi_A^{\mathfrak{q}}(n)-\chi_A^{\mathfrak{q}}(n-1)$$

is a polynomial in n of degree $d-1$, so that $\deg\varphi\geqslant d-1$, but if $\mathfrak{n}\neq(0)$ this is impossible. Thus $\mathfrak{n}=(0)$, and the statement in the theorem follows at once. ■

Multiplicity

Let $(A,\mathfrak{m})$ be a d-dimensional Noetherian local ring, M a finite A-module, and $\mathfrak{q}$ an ideal of definition of A (that is, an $\mathfrak{m}$-primary ideal). As we saw in §13, the Samuel function $l(M/\mathfrak{q}^{n+1}M)=\chi_M^{\mathfrak{q}}(n)$ can be expressed for $n\gg 0$ as a polynomial in n with rational coefficients, and degree equal to $\dim M$, and therefore at most d. In addition, this polynomial can only take integer values for $n\gg 0$, so it is easy to see by induction on d (using the fact that $\chi(n+1)-\chi(n)$ has the same property) that

$$\chi_M^{\mathfrak{q}}(n)=\frac{e}{d!}n^d+(\text{terms of lower order}),$$

with $e\in\mathbb{Z}$. This integer will be written $e(\mathfrak{q},M)$. By definition we have the following property.

Formula 14.1. $e(\mathfrak{q},M)=\lim_{n\to\infty}\frac{d!}{n^d}l(M/\mathfrak{q}^nM)$, and in particular, if $d=0$ then $e(\mathfrak{q},M)=l(M)$.

From this we see easily the following:

Formula 14.2. $e(\mathfrak{q},M)>0$ if $\dim M=d$, and $e(\mathfrak{q},M)=0$ if $\dim M<d$;

Formula 14.3. $e(\mathfrak{q}^r,M)=e(\mathfrak{q},M)r^d$;

Formula 14.4. If $\mathfrak{q}$ and $\mathfrak{q}'$ are both $\mathfrak{m}$-primary ideals and $\mathfrak{q} \supset \mathfrak{q}'$ then $e(\mathfrak{q}, M) \leqslant e(\mathfrak{q}', M)$.

We set $e(\mathfrak{q}, A) = e(\mathfrak{q})$, and define this to be the *multiplicity* of $\mathfrak{q}$. In addition, we will refer to the multiplicity $e(\mathfrak{m})$ of the maximal ideal as the multiplicity of the local ring A, and sometimes write $e(A)$ for it. For example, if A is a regular local ring then by Theorem 4, we can see that $e(A) = 1$.

Theorem 14.6. Let $0 \to M' \longrightarrow M \longrightarrow M'' \to 0$ be an exact sequence of finite A-modules. Then

$$e(\mathfrak{q}, M) = e(\mathfrak{q}, M') + e(\mathfrak{q}, M'').$$

Proof. We view M' as a submodule of M. Then

$$l(M/\mathfrak{q}^n M) = l(M''/\mathfrak{q}^n M'') + l(M'/M' \cap \mathfrak{q}^n M),$$

and obviously $\mathfrak{q}^n M' \subset M' \cap \mathfrak{q}^n M$. On the other hand by Artin–Rees, there exists $c > 0$ such that

$$M' \cap \mathfrak{q}^n M \subset \mathfrak{q}^{n-c} M' \quad \text{for all} \quad n > c.$$

Hence

$$l(M'/\mathfrak{q}^{n-c} M') \leqslant l(M'/M' \cap \mathfrak{q}^n M) \leqslant l(M'/\mathfrak{q}^n M').$$

From this and Formula 14.1 it follows easily that

$$e(\mathfrak{q}, M) - e(\mathfrak{q}, M'') = \lim_{n \to \infty} \frac{d!}{n^d} l(M'/M' \cap \mathfrak{q}^n M) = e(\mathfrak{q}, M').$$

Theorem 14.7. Let $\{\mathfrak{p}_1, \ldots, \mathfrak{p}_t\}$ be all the minimal prime ideals of A such that $\dim A/\mathfrak{p} = d$; then

$$e(\mathfrak{q}, M) = \sum_{i=1}^{t} e(\bar{\mathfrak{q}}_i, A/\mathfrak{p}_i) l(M_{\mathfrak{p}_i}),$$

where $\bar{\mathfrak{q}}_i$ denotes the image of $\mathfrak{q}$ in $A/\mathfrak{p}_i$ and $l(M_\mathfrak{p})$ stands for the length of $M_\mathfrak{p}$ as $A_\mathfrak{p}$-module.

Proof (taken from Nagata [N1]). We write $\sigma = \sum_i l(M_{\mathfrak{p}_i})$ and proceed by induction on σ. If $\sigma = 0$ then $\dim M < d$, so that the left-hand side is 0, and the right-hand side is obviously 0; now suppose $\sigma > 0$. Now there is some $\mathfrak{p} \in \{\mathfrak{p}_1, \ldots, \mathfrak{p}_t\}$ for which $M_\mathfrak{p} \neq 0$; then $\mathfrak{p}$ is a minimal element of $\operatorname{Supp}(M)$. Hence $\mathfrak{p} \in \operatorname{Ass}(M)$, that is M contains a submodule N isomorphic to $A/\mathfrak{p}$. Then

$$e(\mathfrak{q}, M) = e(\mathfrak{q}, N) + e(\mathfrak{q}, M/N).$$

On the other hand, $N_\mathfrak{p} \simeq A_\mathfrak{p}/\mathfrak{p}A_\mathfrak{p}$ and $N_{\mathfrak{p}_i} = 0$ for $\mathfrak{p}_i \neq \mathfrak{p}$, so that $l(N_\mathfrak{p}) = 1$, and the value of σ for M/N has decreased by one, so that the theorem holds for M/N. However, from the definition

$$e(\mathfrak{q}, N) = e(\mathfrak{q}, A/\mathfrak{p}) = e(\bar{\mathfrak{q}}, A/\mathfrak{p}), \quad \text{where} \quad \bar{\mathfrak{q}} = (\mathfrak{q} + \mathfrak{p})/\mathfrak{p}.$$

Putting this together, we see that the theorem also holds for M. ∎

Theorem 7 allows us to reduce the study of $e(\mathfrak{q}, M)$ to the case that A is an integral domain and $M = A$. In particular, if A is an integral domain then $l(M_{(0)})$ is just the rank of M, so that we obtain the following theorem.

Theorem 14.8. Let A be a Noetherian local integral domain, $\mathfrak{q}$ an ideal of definition of A and M a finite A-module; then

$$e(\mathfrak{q}, M) = e(\mathfrak{q})\cdot s, \quad \text{where} \quad s = \operatorname{rank} M.$$

Theorem 14.9. Let $(A, \mathfrak{m})$ be a Noetherian local ring, $\mathfrak{q}$ an ideal of definition of A, and $x_1, \ldots, x_d$ a system of parameters of A contained in $\mathfrak{q}$. Suppose that $x_i \in \mathfrak{q}^{v_i}$ for $1 \leqslant i \leqslant d$. Then for a finite A-module M and $s = 1, \ldots, d$ we have

$$e(\mathfrak{q}/(x_1, \ldots, x_s), M/(x_1, \ldots, x_s)M) \geqslant v_1 v_2 \ldots v_s e(\mathfrak{q}, M).$$

In particular if $s = d$, we have

$$l(M/(x_1, \ldots, x_d)M) \geqslant v_1 v_2 \ldots v_d e(\mathfrak{q}, M).$$

Proof. It is enough to prove the case $s = 1$. We set $A' = A/x_1 A, \mathfrak{q}' = \mathfrak{q}/x_1 A$, $M' = M/x_1 M$ and $v = v_1$. By Theorem 1, we have $\dim A' = d - 1$. On the other hand,

$$\begin{aligned} l(M'/\mathfrak{q}'^n M') &= l(M/x_1 M + \mathfrak{q}^n M) \\ &= l(M/\mathfrak{q}^n M) - l(x_1 M + \mathfrak{q}^n M/\mathfrak{q}^n M). \end{aligned}$$

In addition, in view of $(x_1 M + \mathfrak{q}^n M)/\mathfrak{q}^n M \simeq x_1 M/x_1 M \cap \mathfrak{q}^n M \simeq M/(\mathfrak{q}^n M : x_1)$ and $\mathfrak{q}^{n-v} M \subset \mathfrak{q}^n M : x_1$, we have

$$-l(x_1 M + \mathfrak{q}^n M/\mathfrak{q}^n M) \geqslant -l(M/\mathfrak{q}^{n-v} M),$$

and therefore

$$l(M'/\mathfrak{q}'^n M') \geqslant l(M/\mathfrak{q}^n M) - l(M/\mathfrak{q}^{n-v} M).$$

When $n \gg 0$ the right-hand side is of the form

$$\begin{aligned} &\frac{e(\mathfrak{q}, M)}{d!}[n^d - (n - v)^d] + (\text{polynomial of degree } d-2 \text{ in } n) \\ &= \frac{e(\mathfrak{q}, M)}{(d-1)!} v \cdot n^{d-1} + (\text{polynomial of degree } d-2 \text{ in } n), \end{aligned}$$

so that the assertion is clear. ∎

A case of the above theorem which is particularly simple, but important, is the following.

Theorem 14.10. Let $(A, \mathfrak{m})$ be a d-dimensional Noetherian local ring, let $x_1, \ldots, x_d$ be a system of parameters of A, and set $\mathfrak{q} = (x_1, \ldots, x_d)$; then

$$l(A/\mathfrak{q}) \geqslant e(\mathfrak{q}),$$

and if in addition $x_i \in \mathfrak{m}^v$ for all i then $l(A/\mathfrak{q}) \geqslant v^d e(\mathfrak{m})$.

Theorem 14.11. Let A, $\mathfrak{m}$, x_i and $\mathfrak{q}$ be as above. Let M be a finite A-module, and set $A' = A/x_1 A$, $M' = M/x_1 M$ and $\mathfrak{q}' = \mathfrak{q}/x_1 A = \sum_2^d x_i A'$. Then if x_1 is a non-zero-divisor of M, we have the following equality

$$e(\mathfrak{q}, M) = e(\mathfrak{q}', M').$$

Proof. Since $l(M'/\mathfrak{q}'^{n+1}M') = l(M/x_1M + \mathfrak{q}^{n+1}M)$ we have

$$\begin{aligned} l(M/\mathfrak{q}^{n+1}M) - l(M'/\mathfrak{q}'^{n+1}M') &= l(x_1M + \mathfrak{q}^{n+1}M/\mathfrak{q}^{n+1}M) \\ &= l(x_1M/x_1M \cap \mathfrak{q}^{n+1}M) = l(M/(\mathfrak{q}^{n+1}M : x_1)) \\ &= l(M/\mathfrak{q}^nM) - l((\mathfrak{q}^{n+1}M : x_1)/\mathfrak{q}^nM). \end{aligned}$$

On the other hand, setting $\mathfrak{a} = \sum_2^d x_iA$ we have $\mathfrak{q} = x_1A + \mathfrak{a}$ and $\mathfrak{q}^{n+1} = x_1\mathfrak{q}^n + \mathfrak{a}^{n+1}$, and therefore

$$\mathfrak{q}^{n+1}M : x_1 = \mathfrak{q}^nM + (\mathfrak{a}^{n+1}M : x_1).$$

Moreover, by Artin–Rees, there is a $c > 0$ such that for $n > c$ we have $\mathfrak{a}^{n+1}M \cap x_1M = \mathfrak{a}^{n-c}(\mathfrak{a}^{c+1}M \cap x_1M)$, and therefore $\mathfrak{a}^{n+1}M : x_1 \subset \mathfrak{a}^{n-c}M$. Thus

$$\begin{aligned} (\mathfrak{q}^{n+1}M : x_1)/\mathfrak{q}^nM &= (\mathfrak{q}^nM + (\mathfrak{a}^{n+1}M : x_1))/\mathfrak{q}^nM \\ &\subset (\mathfrak{q}^nM + \mathfrak{a}^{n-c}M)/\mathfrak{q}^nM \\ &\simeq \mathfrak{a}^{n-c}M/\mathfrak{a}^{n-c}M \cap \mathfrak{q}^nM. \end{aligned}$$

Now $\mathfrak{a}^{n-c}M/\mathfrak{a}^{n-c}M \cap \mathfrak{q}^nM$ is a module over $A/\mathfrak{q}^c$, and since $\mathfrak{a}$ is generated by $d-1$ elements, $\mathfrak{a}^{n-c}$ is generated by $\binom{n-c+d-2}{d-2}$ elements. Thus for $n > c$ we have

$$l(\mathfrak{a}^{n-c}M/\mathfrak{a}^{n-c}M \cap \mathfrak{q}^nM) \leqslant \binom{n-c+d-2}{d-2} \cdot l(A/\mathfrak{q}^c)m,$$

where m is the number of generators of M. The right-hand side is a polynomial of degree $d-2$ in n, so that

$$\begin{aligned} e(\mathfrak{q}', M') &= (d-1)! \lim_{n\to\infty} l(M'/\mathfrak{q}'^{n+1}M')/n^{d-1} \\ &= (d-1)! \lim_{n\to\infty} [l(M/\mathfrak{q}^{n+1}M) - l(M/\mathfrak{q}^nM)]/n^{d-1} \\ &= e(\mathfrak{q}, M). \quad \blacksquare \end{aligned}$$

Theorem 14.12 (Lech's lemma). Let A be a d-dimensional Noetherian local ring, and $x_1, \ldots, x_d$ a system of parameters of A; set $\mathfrak{q} = (x_1, \ldots, x_d)$, and suppose that M is a finite A-module. Then

$$e(\mathfrak{q}, M) = \lim_{\min(\nu_i)\to\infty} \frac{l(M/(x_1^{\nu_1}, \ldots, x_d^{\nu_d})M)}{\nu_1 \ldots \nu_d}.$$

Proof. If $d = 0$ then both sides are equal to $l(M)$. If $d = 1$ then the right-hand side is exactly Formula 14.1 which defines $e(\mathfrak{q}, M)$. For $d > 1$ we use induction on d.

Setting $N_j = \{m \in M \mid x_1^j m = 0\}$ we have $N_1 \subset N_2 \subset \cdots$ so that there is a $c > 0$ such that $N_c = N_{c+1} = \cdots$. If we set $M' = x_1^cM$ then x_1 is a non-zero-divisor for M', and there is an exact sequence $0 \to N_c \longrightarrow M \longrightarrow M' \to 0$. Since N_c is a module over A/x_1^cA we have $\dim N_c < d$, and therefore

$e(\mathfrak{q}, M) = e(\mathfrak{q}, M')$. On the other hand,

$$\begin{aligned} &l(M/(x_1^{\nu_1},\dots,x_d^{\nu_d})M) - l(M'/(x_1^{\nu_1},\dots,x_d^{\nu_d})M') \\ &\quad = l(N_c + (x_1^{\nu_1},\dots,x_d^{\nu_d})M/(x_1^{\nu_1},\dots,x_d^{\nu_d})M) \\ &\quad = l(N_c/N_c \cap (x_1^{\nu_1},\dots,x_d^{\nu_d})M) \\ &\quad \leqslant l(N_c/(x_1^{\nu_1},\dots,x_d^{\nu_d})N_c). \end{aligned}$$

If $\nu_1 > c$ then $x_1^{\nu_1}N_c = 0$, and N_c is a module over the $(d-1)$-dimensional local ring A/x_1^cA, so that by induction there is a constant C such that as $\min(\nu_i) \to \infty$ we have

$$l(N_c/(x_1^{\nu_1},\dots,x_d^{\nu_d})N_c) = l(N_c/(x_2^{\nu_2},\dots,x_d^{\nu_d})N_c) < C\cdot\nu_2\dots\nu_d.$$

Therefore,

$$\lim\,[l(M/(x_1^{\nu_1},\dots,x_d^{\nu_d})M) - l(M'/(x_1^{\nu_1},\dots,x_d^{\nu_d})M')]/\nu_1\dots\nu_d = 0.$$

This means that we can replace M by M' in the theorem, and so we can assume that x_1 is a non-zero-divisor in M. Then by the previous theorem we have $e(\mathfrak{q}, M) = e(\bar{\mathfrak{q}}, \bar{M})$, with $\bar{\mathfrak{q}} = \mathfrak{q}/x_1A$ and $\bar{M} = M/x_1M$. If we furthermore set

$$E = (x_2^{\nu_2},\dots,x_d^{\nu_d})M \quad \text{and} \quad F = M/E$$

then by Theorem 9, we have

$$\begin{aligned} e(\mathfrak{q}, M)\cdot\nu_1\dots\nu_d &\leqslant l(M/(x_1^{\nu_1},\dots,x_d^{\nu_d})M) = l(F/x_1^{\nu_1}F) \\ &= \sum_{i=1}^{\nu_1} l(x_1^{i-1}F/x_1^iF) \leqslant \nu_1 l(F/x_1F) = \nu_1 l(M/x_1M + E) \\ &= \nu_1 l(\bar{M}/(x_2^{\nu_2},\dots,x_d^{\nu_d})\bar{M}). \end{aligned}$$

Then by induction on d we have

$$\begin{aligned} \lim l(M/(x_1^{\nu_1},\dots,x_d^{\nu_d})M)/\nu_1\dots\nu_d &= \lim l(\bar{M}/(x_2^{\nu_2},\dots,x_d^{\nu_d})\bar{M})/\nu_2\dots\nu_d \\ &= e(\mathfrak{q}, M). \quad \blacksquare \end{aligned}$$

Although we will not use it in this book, we state here without proof a remarkable result of Serre which shows that multiplicity can be expressed as the Euler characteristic of the homology groups of the Koszul complex (discussed in §16).

Theorem. Let A be a d-dimensional Noetherian local ring, and $x_1,\dots,x_d$ a system of parameters of A; set $\mathfrak{q} = (x_1,\dots,x_d)$ and let M be a finite A-module. Then

$$e(\mathfrak{q}, M) = \sum(-1)^i l(H_i(x, M)).$$

For a proof, see for example Auslander and Buchsbaum [2].

As we have seen in several of the above theorems, the multiplicity of ideals generated by systems of parameters enjoy various nice properties. We are now going to see that in a certain sense the general case can be reduced to this one. We follow the method of Northcott and Rees [1].

Quite generally, let A be a ring and $\mathfrak{a}$ an ideal. We say that an ideal $\mathfrak{b}$ is a *reduction* of $\mathfrak{a}$ if it satisfies the following condition:

$$\mathfrak{b} \subset \mathfrak{a}, \quad \text{and for some} \quad r > 0 \quad \text{we have} \quad \mathfrak{a}^{r+1} = \mathfrak{b}\mathfrak{a}^r.$$

If $\mathfrak{b}$ is a reduction of $\mathfrak{a}$ and $\mathfrak{a}^{r+1} = \mathfrak{b}\mathfrak{a}^r$ then for any $n > 0$ we have $\mathfrak{a}^{r+n} = \mathfrak{b}^n\mathfrak{a}^r$.

Theorem 14.13. Let $(A, \mathfrak{m})$ be a Noetherian local ring, $\mathfrak{q}$ an $\mathfrak{m}$-primary ideal and $\mathfrak{b}$ a reduction of $\mathfrak{q}$; then $\mathfrak{b}$ is also $\mathfrak{m}$-primary, and for any finite A-module M we have

$$e(\mathfrak{q}, M) = e(\mathfrak{b}, M).$$

Proof. If $\mathfrak{q}^{r+1} = \mathfrak{b}\mathfrak{q}^r$ then $\mathfrak{q}^{r+1} \subset \mathfrak{b} \subset \mathfrak{q}$, hence $\mathfrak{b}$ is also $\mathfrak{m}$-primary. Moreover,

$$l(M/\mathfrak{b}^{n+r}M) \geqslant l(M/\mathfrak{q}^{n+r}M) = l(M/\mathfrak{b}^n\mathfrak{q}^r) \geqslant l(M/\mathfrak{b}^n M),$$

so that $e(\mathfrak{q}, M) = e(\mathfrak{b}, M)$ follows easily.

Theorem 14.14. Let $(A, \mathfrak{m})$ be a d-dimensional Noetherian local ring, and suppose that $A/\mathfrak{m}$ is an infinite field; let $\mathfrak{q} = (u_1, \ldots, u_s)$ be an $\mathfrak{m}$-primary ideal. Then if $y_i = \sum a_{ij}u_j$ for $1 \leqslant i \leqslant d$ are d 'sufficiently general' linear combinations of $u_1, \ldots, u_s$, the ideal $\mathfrak{b} = (y_1, \ldots, y_d)$ is a reduction of $\mathfrak{q}$ and $\{y_1, \ldots, y_d\}$ is a system of parameters of A.

Proof. If $d = 0$ then $\mathfrak{q}^r = (0)$ for some $r > 0$, hence (0) is a reduction of $\mathfrak{q}$ so that the result holds. We suppose below that $d > 0$.

Step 1. Set $A/\mathfrak{m} = k$ and consider the polynomial ring $k[X_1, \ldots, X_s]$ (or $k[X]$ for short). For a homogeneous form $\varphi(X) = \varphi(X_1, \ldots, X_s) \in A[X]$ of degree n, we write $\bar{\varphi}(X) \in k[X]$ for the polynomial obtained by reducing the coefficients of φ modulo $\mathfrak{m}$. As in the proof of Theorem 5 we say that $\bar{\varphi}(X) \in k[X]$ is a null-form of $\mathfrak{q}$ if $\varphi(u_1, \ldots, u_s) \in \mathfrak{q}^n\mathfrak{m}$; this notion depends not just on $\mathfrak{q}$, but also on $u_1, \ldots, u_s$. However, for fixed $\bar{\varphi}$ it does not depend on the choice of φ. We write Q for the ideal of $k[X]$ generated by all the null-forms of $\mathfrak{q}$, and call Q the *ideal of null-forms* of $\mathfrak{q}$. One sees easily that all the homogeneous elements of Q are null-forms of $\mathfrak{q}$, and that the graded ring $k[X]/Q$ has graded component of degree n isomorphic to $\mathfrak{q}^n/\mathfrak{q}^n\mathfrak{m}$, so that we have

$$k[X]/Q \simeq \bigoplus_{n \geqslant 0} \mathfrak{q}^n/\mathfrak{q}^n\mathfrak{m} = \mathrm{gr}_{\mathfrak{q}}(A) \otimes_{A/\mathfrak{q}} k.$$

Write $\varphi(n)$ for the Hilbert function of $k[X]/Q$; then

$$\varphi(n) = l(\mathfrak{q}^n/\mathfrak{q}^n\mathfrak{m}) \leqslant l(\mathfrak{q}^n/\mathfrak{q}^{n+1}) \leqslant \varphi(n) \cdot l(A/\mathfrak{q})$$

(see the proof of Theorem 5). We know that for $n \gg 0$, the function $l(\mathfrak{q}^n/\mathfrak{q}^{n+1})$ is a polynomial in n of degree $d - 1$ (where $d = \dim A$). Thus from the above inequality, φ is also a polynomial of degree $d - 1$, so that by Theorem 13.8, (ii), we have $\dim k[X]/Q = d$.

Now set $V = \sum_1^s kX_i$, and let $P_1, \ldots, P_t$ be the minimal prime divisors of

Q. By the assumption that $d>0$, we have $P_i \not\supset V$, so that $P_i \cap V$ is a proper vector subspace of V. Since k is an infinite field,

$$V \neq \bigcup_{i=1}^{t} (V \cap P_i).$$

Hence we can take a linear form $l_1(X) \in V$ not belonging to any P_i. If $d>1$ then similarly we can take $l_2(X) \in V$ such that $l_2(X)$ is not contained in any minimal prime divisor of $(Q, l_1(X))$, and, proceeding in the same way, we get $l_1(X), \ldots, l_d(X) \in V$ such that $(Q, l_1, \ldots, l_d)$ is a primary ideal belonging to $(X_1, \ldots, X_s)$.

Step 2. We let $\mathfrak{b}$ be the ideal of A generated by t linear combinations $L_i(u) = \sum a_{ij} u_j$ (for $1 \leqslant i \leqslant t$) of $u_1, \ldots, u_s$ with coefficients in A. Then if we set $l_i(X) = \bar{L}_i(X) = \sum \bar{a}_{ij} X_j$, a necessary and sufficient condition for $\mathfrak{b}$ to be a reduction of $\mathfrak{q}$ is that the ideal $(Q, l_1, \ldots, l_t)$ of $k[X]$ is $(X_1, \ldots, X_s)$-primary.

Proof of necessity. Suppose that $\mathfrak{b}\mathfrak{q}^r = \mathfrak{q}^{r+1}$. Then if $M = M(X)$ is a monomial of degree $r+1$ in $X_1, \ldots, X_s$, we can write

$$M(u) = \sum_1^t L_i(u) F_i(u),$$

where the $F_i(X)$ are homogeneous forms of degrees r with coefficients in A. Thus

$$\bar{M}(X) - \sum l_i(X) \bar{F}_i(X) \in Q.$$

Hence

$$(X_1, \ldots, X_s)^{r+1} \subset (Q, l_1, \ldots, l_t).$$

Proof of sufficiency. We go through the same argument in reverse: if $\bar{M} - \sum l_i \bar{F}_i \in Q$ then

$$M(u) - \sum L_i(u) F_i(u) \in \mathfrak{q}^{r+1}\mathfrak{m},$$

so that $\mathfrak{q}^{r+1} \subset \mathfrak{b}\mathfrak{q}^r + \mathfrak{q}^{r+1}\mathfrak{m}$; thus by NAK, $\mathfrak{q}^{r+1} = \mathfrak{b}\mathfrak{q}^r$.

Step 3. Putting together Steps 1 and 2 we see that $\mathfrak{q}$ has a reduction $\mathfrak{b} = (y_1, \ldots, y_d)$ generated by d elements. Both $\mathfrak{q}$ and its reduction $\mathfrak{b}$ are $\mathfrak{m}$-primary ideals, so that $y_1, \ldots, y_d$ is a system of parameters of A. We are going to prove that there exists a finite number of polynomials $D_\alpha(Z_{ij})$ for $1 \leqslant \alpha \leqslant v$ in sd indeterminates Z_{ij} (for $1 \leqslant i \leqslant d$ and $1 \leqslant j \leqslant s$) such that d linear combinations $y_i = \sum a_{ij} u_j$ (for $1 \leqslant i \leqslant d$) generate a reduction ideal of $\mathfrak{q}$ if and only if at least one of $D_\alpha(\bar{a}_{ij}) \neq 0$. (The expression '$d$ sufficiently general linear combinations' in the statement of the theorem is quite vague, but in the present case it has a precise interpretation as above.)

Let $G_1(X), \ldots, G_m(X)$ be a set of generators of Q, with G_j homogeneous of degree e_j. For any sd elements α_{ij} of k (for $1 \leqslant i \leqslant d$ and $1 \leqslant j \leqslant s$), set $l_i(X) = \sum \alpha_{ij} X_j$. We write I_n for the homogeneous component of degree n

of a homogeneous ideal $I \subset k[X_1,\ldots,X_s]$, and in particular we write $(X_1,\ldots,X_s)_n = V_n$, so that

$$(X_1,\ldots,X_s)^n \subset (Q,l_1,\ldots,l_d) \Leftrightarrow V_n = (Q,l_1,\ldots,l_d)_n.$$

Set $c_n = \dim_k V_n$. We have

$$(Q,l_1,\ldots,l_d)_n = \{\sum l_i F_i + \sum G_j H_j | F_i \in V_{n-1} \text{ and } H_j \in V_{n-e_j}\}.$$

Let $K_1,\ldots,K_w$ be the elements obtained as $\sum l_i F_i + \sum G_j H_j$ as the F_i run through a basis of V_{n-1} and the H_j run independently through a basis of V_{n-e_j}; it is clear that they span $(Q,l_1,\ldots,l_d)_n$. Each of $K_1,\ldots,K_w$ is a linear combination of the c_n monomials of degree n in the X_i, with linear functions in the α_{ij} as coefficients; we write out these coefficients in a $c_n \times w$ matrix. If $\varphi_{n\nu}(\alpha_{ij})$ for $1 \leqslant \nu \leqslant p_n$ are the $c_n \times c_n$ minors of this matrix then the necessary and sufficient condition for $(X_1,\ldots,X_s)^n \subset (Q,l_1,\ldots,l_d)$ to hold is that at least one of the $\varphi_n(\alpha_{ij})$ is non-zero. Therefore the ideal $(Q,l_1,\ldots,l_d)$ will fail to be $(X_1,\ldots,X_s)$-primary if and only if the quantities α_{ij} satisfy $\varphi_{n\nu}(\alpha_{ij}) = 0$ for all n and all ν. However, the ring $k[Z_{ij}]$ is Noetherian, so that the ideal of $k[Z_{ij}]$ generated by all of the $\varphi_{n\nu}(Z_{ij})$ is generated by finitely many elements $D_\alpha(Z_{ij})$ for $1 \leqslant \alpha \leqslant v$. These D_α clearly meet our requirements. ∎

Remark. The polynomials $D_\alpha(Z_{ij})$ obtained above are in fact the necessary and sufficient conditions on the coefficients α_{ij} for the system of homogeneous equations $l_1(X) = \cdots = l_d(X) = G_1(X) = \cdots = G_m(X) = 0$ to have a non-trivial solution, and as such they are known as a system of resultants. Here we have avoided appealing to the classical theory of resultants by following a method given in Shafarevich [Sh].

If $k = A/\mathfrak{m}$ is a finite field then Theorem 14 cannot to be used as it stands, but we can use the following trick. Let x be an indeterminate over A, and set $S = A[x] - \mathfrak{m}[x]$; then S consists of polynomials having a unit of A among their coefficients, and so the composite of the canonical maps $A \longrightarrow A[x] \longrightarrow A[x]_S$ is injective. (In fact S does not contain any zero-divisors of $A[x]$, so that $A \subset A[x] \subset A[x]_S$; for this see [AM], Chap. 1, Ex. 2.) Following Nagata [N1] we write $A(x)$ for $A[x]_S$. This is a Noetherian local ring containing A, with maximal ideal $\mathfrak{m}A(x)$, and the residue class field $A(x)/\mathfrak{m}A(x)$ is the field of fractions of $A[x]/\mathfrak{m}[x] = k[x]$, that is, the field $k(x)$ of rational functions over k; this is an infinite field. If $\mathfrak{q}$ is an $\mathfrak{m}$-primary ideal of A then $\mathfrak{q}A(x)$ is a primary ideal belonging to $\mathfrak{m}A(x)$. Moreover, since $A(x)$ is flat over A, we see that quite generally if $I \supset I'$ are ideals of A such that $I/I' \simeq k$, then

$$IA(x)/I'A(x) \simeq (I/I') \otimes_A A(x) \simeq k \otimes A(x) = A(x)/\mathfrak{m}A(x).$$

This gives $l_A(A/\mathfrak{q}^n) = l_{A(x)}(A(x)/\mathfrak{q}^n A(x))$, so that

$$\dim A = \dim A(x) \quad \text{and} \quad e(\mathfrak{q}) = e(\mathfrak{q}A(x)).$$

Thus there are many instances when we can discuss properties of $e(\mathfrak{q})$ in terms of $A(x)$, to which Theorem 14 applies.

Exercises to §14. Prove the following propositions.

14.1. Let $(A, \mathfrak{m})$ be a Noetherian local ring and set $G = \mathrm{gr}_{\mathfrak{m}}(A)$.
(i) If G is an integral domain then so is A (hence Theorem 3 also follows from Theorem 4).
(ii) Let k be a field, and $A = k[\![X, Y]\!]/(Y^2 - X^3)$; then A is an integral domain, but G has nilpotents.

14.2. Let $(A, \mathfrak{m})$ and G be as above. For $a \in A$, suppose that $a \in \mathfrak{m}^i$ but $a \notin \mathfrak{m}^{i+1}$, and write a^* for the image of a in $\mathfrak{m}^i/\mathfrak{m}^{i+1}$, viewed as an element of G; define a^* to be the *leading term* of a. Set $0^* = 0$. Then
(i) if $a^*b^* \neq 0$ then $(ab)^* = a^*b^*$;
(ii) if a^* and b^* have the same degree and $a^* + b^* \neq 0$ then $(a+b)^* = a^* + b^*$;
(iii) let $I \subset \mathfrak{m}$ be an ideal of A. Write I^* for the ideal of G generated by all the leading terms of elements of I; then setting $B = A/I$ and $\mathfrak{n} = \mathfrak{m}/I$, we have $\mathrm{gr}_{\mathfrak{n}}(B) = G/I^*$.

14.3. In the above notation, if G is an integral domain and $I = aA$ then $I^* = a^*G$. If $I = (a_1, \ldots, a_r)$ with $r > 1$ then it can happen that $I^* \neq (a_1^*, \ldots, a_r^*)$. Construct an example.

14.4. Let $(A, \mathfrak{m})$ be a regular local ring, and K its field of fractions.
(i) For $0 \neq a \in A$, set $v(a) = i$ if $a \in \mathfrak{m}^i$ but $a \notin \mathfrak{m}^{i+1}$; then v extends to an additive valuation of K.
(ii) Let R be the valuation ring of v; then R is a DVR of K dominating A. Let $x_1, \ldots, x_d$ be a regular system of parameters of A, and set $B = A[x_2/x_1, \ldots, x_d/x_1]$ and $P = x_1B$; then P is a prime ideal of B and $R = B_P$.

14.5. In the above notation, if $0 \neq f \in \mathfrak{m}$ then $v(f)$ is equal to the multiplicity of $A/(f)$.

14.6. (Associativity formula for multiplicities.) Let A be a d-dimensional Noetherian local ring, $x_1, \ldots, x_d$ a system of parameters of A, $\mathfrak{q} = (x_1, \ldots, x_d)$, and for $s \leqslant d$ let $\mathfrak{a} = (x_1, \ldots, x_s)$. Write Γ for the set of all prime divisors of $\mathfrak{a}$ satisfying $\mathrm{ht}\,\mathfrak{p} = s$, $\mathrm{coht}\,\mathfrak{p} = d - s$. Let M be a finite A-module. Use Lech's lemma to prove the following formula:

$$e(\mathfrak{q}, M) = \sum_{\mathfrak{p} \in \Gamma} e(\mathfrak{q} + \mathfrak{p}/\mathfrak{p}) \cdot e(\mathfrak{a}A_{\mathfrak{p}}, M_{\mathfrak{p}});$$

(in particular, it follows that $\Gamma \neq \varnothing$).

Remark. The name of the formula comes from its connection with the associativity of intersection product in algebraic geometry. For details, see [S3], pp. 84–5.

14.7. Let $(A, \mathfrak{m})$ be an n-dimensional Noetherian local integral domain, with $n > 1$. If $0 \neq f \in \mathfrak{m}$ then A_f is a Jacobson ring (see p. 34).

15 The dimension of extension rings

1. Fibres

Let $\varphi: A \longrightarrow B$ be a ring homomorphism, and for $\mathfrak{p} \in \operatorname{Spec} A$, write $\kappa(\mathfrak{p}) = A_\mathfrak{p}/\mathfrak{p}A_\mathfrak{p}$; then $\operatorname{Spec}(B \otimes \kappa(\mathfrak{p}))$ is called the *fibre* of φ *over* $\mathfrak{p}$. As we saw in §7, it can be identified with the inverse image in $\operatorname{Spec} A$ of $\mathfrak{p}$ under the map ${}^a\varphi: \operatorname{Spec} B \longrightarrow \operatorname{Spec} A$ induces by φ. The ring $B \otimes \kappa(\mathfrak{p})$ will be called the fibre ring over $\mathfrak{p}$. When $(A, \mathfrak{m})$ is a local ring, $\mathfrak{m}$ is the unique closed point of $\operatorname{Spec} A$, and so the spectrum of $B \otimes \kappa(\mathfrak{m}) = B/\mathfrak{m}B$ is called the *closed fibre* of φ. If A is an integral domain and K its field of fractions then the spectrum of $B \otimes_A K = B \otimes_A \kappa(0)$ is called the *generic fibre* of φ.

Theorem 15.1. Let $\varphi: A \longrightarrow B$ be a homomorphism of Noetherian rings, and P a prime ideal of B; then setting $\mathfrak{p} = P \cap A$, we have

(i) $\operatorname{ht} P \leqslant \operatorname{ht} \mathfrak{p} + \dim B_P/\mathfrak{p}B_P$;

(ii) if φ is flat, or more generally if the going-down theorem holds between A and B, then equality holds in (i).

Proof. We can replace A and B by $A_\mathfrak{p}$ and B_P, and assume that $(A, \mathfrak{m})$ and $(B, \mathfrak{n})$ are local rings, with $\mathfrak{m}B \subset \mathfrak{n}$. Rewriting (i) in the form

$$\dim B \leqslant \dim A + \dim B/\mathfrak{m}B$$

makes clear the geometrical content. To prove this, take a system of parameters $x_1, \dots, x_r$ of A, and choose $y_1, \dots, y_s \in B$ such that their images in $B/\mathfrak{m}B$ form a system of parameters of $B/\mathfrak{m}B$. Then for ν, μ large enough we have $\mathfrak{n}^\nu \subset \mathfrak{m}B + \sum y_i B$ and $\mathfrak{m}^\mu \subset \sum x_j A$, giving $\mathfrak{n}^{\nu\mu} \subset \sum y_i B + \sum x_j B$. Hence $\dim B \leqslant r + s$.

(ii) Let $\dim B/\mathfrak{m}B = s$, and let $\mathfrak{n} = P_0 \supset P_1 \supset \cdots \supset P_s$ be a strictly decreasing chain of prime ideals of B between $\mathfrak{n}$ and $\mathfrak{m}B$. Obviously we have $P_i \cap A = \mathfrak{m}$ for $0 \leqslant i \leqslant s$. Now set $\dim A = r$ and let $\mathfrak{m} = \mathfrak{p}_0 \supset \mathfrak{p}_1 \supset \cdots \supset \mathfrak{p}_r$ be a strictly decreasing chain of prime ideals of A; by the going-down theorem, we can construct a strictly decreasing chain of prime ideals of B

$$P_s \supset P_{s+1} \supset \cdots \supset P_{s+r} \quad \text{such that} \quad P_{s+i} \cap A = \mathfrak{p}_i.$$

Thus $\dim B \geqslant r + s$, and putting this together with (i) gives equality. ■

Theorem 15.2. Let $\varphi: A \longrightarrow B$ be a homomorphism of Noetherian rings, and suppose that the going-up theorem holds between A and B. Then if $\mathfrak{p}$ and $\mathfrak{q}$ are prime ideals of A such that $\mathfrak{p} \supset \mathfrak{q}$, we have

$$\dim B \otimes \kappa(\mathfrak{p}) \geqslant \dim B \otimes \kappa(\mathfrak{q}).$$

Proof. Set $r = \dim B \otimes \kappa(\mathfrak{q})$ and $s = \operatorname{ht}(\mathfrak{p}/\mathfrak{q})$. We choose a strictly increasing chain $Q_0 \subset Q_1 \subset \cdots \subset Q_r$ of prime ideals of B lying over $\mathfrak{q}$ and a strictly increasing chain $\mathfrak{q} = \mathfrak{p}_0 \subset \mathfrak{p}_1 \subset \cdots \subset \mathfrak{p}_s = \mathfrak{p}$ of prime ideals of A. By the going-up theorem there exists a chain $Q_r \subset Q_{r+1} \subset \cdots \subset Q_{r+s}$ of prime ideals of B such that $Q_{r+i} \cap A = \mathfrak{p}_i$. We set $P = Q_{r+s}$; then

$$\mathrm{ht}(P/\mathfrak{q}B) \geqslant r+s \quad \text{and} \quad P \cap A = \mathfrak{p}.$$

Thus applying the previous theorem to the homomorphism $A/\mathfrak{q} \longrightarrow B/\mathfrak{q}B$ induced by φ we get $r+s \leqslant \mathrm{ht}(P/\mathfrak{q}B) \leqslant s + \dim B_P/\mathfrak{p}B_P$, and therefore

$$r \leqslant \dim B_P/\mathfrak{p}B_P \leqslant \dim B \otimes \kappa(\mathfrak{p}). \quad \blacksquare$$

Theorem 15.3. Let $\varphi: A \to B$ be a homomorphism of Noetherian rings, and suppose that the going-down theorem holds between A and B. If $\mathfrak{p}$ and $\mathfrak{q}$ are prime ideals of A with $\mathfrak{p} \supset \mathfrak{q}$ then

$$\dim B \otimes \kappa(\mathfrak{p}) \leqslant \dim B \otimes \kappa(\mathfrak{q}).$$

Proof. We may assume that $\mathrm{ht}(\mathfrak{p}/\mathfrak{q}) = 1$, and it is enough to prove that, given a chain $P_0 \subset P_1 \subset \ldots \subset P_r$ of prime ideals of B lying over $\mathfrak{p}$ such that $\mathrm{ht}(P_i/P_{i-1}) = 1$ we can construct a chain of prime ideals $Q_0 \subset Q_1 \subset \ldots \subset Q_r$ of B lying over q such that

$$Q_i \subset P_i \ (0 \leqslant i \leqslant r) \text{ and } \mathrm{ht}(Q_i/Q_{i-1}) = 1 \ (0 < i \leqslant r).$$

We can find Q_0 by going down. If $r \geqslant 1$ then take $x \in \mathfrak{p} - \mathfrak{q}$ and let $T_1, \ldots, T_s$ be the minimal prime divisors of $Q_0 + xB$. Then $\mathrm{ht}(T_i/Q_0) = 1$, while $\mathrm{ht}(P_1/Q_0) \geqslant 2$, hence we can choose

$$y \in P_1 - (\textstyle\bigcup_i T_i).$$

Let Q_1 be a minimal prime divisor of $Q_0 + yB$ contained in P_1. Then $\mathrm{ht}(Q_1/Q_0) = 1$, and $Q_1 \neq T_i$ for all i, hence $\varphi(x) \notin Q_1$.

Therefore $Q_1 \cap A \neq \mathfrak{p}$, and since $\mathrm{ht}(\mathfrak{p}/\mathfrak{q}) = 1$ we must have $Q_1 \cap A = \mathfrak{q}$. By the same method we can successively construct $Q_1, Q_2, \ldots, Q_r$. $\blacksquare$

2. *Polynomial and formal power series rings*

Theorem 15.4. Let A be a Noetherian ring, and $X_1, \ldots, X_n$ indeterminates over A. Then

$$\dim A[X_1, \ldots, X_n] = \dim A[\![X_1, \ldots, X_n]\!] = \dim A + n.$$

Proof. It is enough to consider the case $n = 1$. For any $\mathfrak{p} \in \operatorname{Spec} A$, the ring $A[X] \otimes_A \kappa(\mathfrak{p}) = \kappa(\mathfrak{p})[X]$ is a principal ideal ring, and therefore one-dimensional; also $A[X]$ is a free A-module, hence faithfully flat, so by Theorem 1, (ii), $\dim A[X] = \dim A + 1$.

For $A[\![X]\!]$ it is not true in general that $A[\![X]\!] \otimes_A \kappa(\mathfrak{p})$ and $\kappa(\mathfrak{p})[\![X]\!]$ coincide; however, if $\mathfrak{m}$ is a maximal ideal of A we have

$$A[\![X]\!] \otimes \kappa(\mathfrak{m}) = A[\![X]\!] \otimes (A/\mathfrak{m}) = (A/\mathfrak{m})[\![X]\!],$$

and this fibre ring is one-dimensional. Also, as we saw on p. 4, every maximal ideal $\mathfrak{M}$ of $A[\![X]\!]$ is of the form $\mathfrak{M} = (\mathfrak{m}, X)$, where $\mathfrak{m} = \mathfrak{M} \cap A$ is a maximal ideal of A. Thus for a maximal ideal $\mathfrak{M}$ of $A[\![X]\!]$ we have

$$\operatorname{ht} \mathfrak{M} = \operatorname{ht}(\mathfrak{M} \cap A) + 1;$$

conversely, if $\mathfrak{m}$ is a maximal ideal of A then $\operatorname{ht}(\mathfrak{m}, X) = \operatorname{ht} \mathfrak{m} + 1$, and putting these together gives $\dim A[\![X]\!] = \dim A + 1$. ∎

Remark 1. It is not necessarily true that a maximal ideal of $A[X]$ lies over a maximal ideal of A. For example, if A is a DVR and t a uniformising element then $A[t^{-1}] = K$ is the field of fractions of A, so that $A[X]/(tX - 1) \simeq K$, and $(tX - 1)$ is a maximal ideal of $A[X]$; however, $(tX - 1) \cap A = (0)$.

Remark 2. It is quite common for fibre rings of $A \longrightarrow A[\![X_1, \ldots, X_n]\!]$ to have dimension strictly greater than n. For example, let k be a field and set $A = k[Y, Z]$. It is well-known that the field of fractions of $k[\![X]\!]$ has infinite transcendence degree over $k(X)$ (see [ZS], vol. II, p. 220). Let $u(X)$, $v(X) \in k[\![X]\!]$ be two elements algebraically independent over $k(X)$, and define a k-homomorphism (continuous for the X-adic topology)

$$\varphi: A[\![X]\!] \longrightarrow k[\![X]\!]$$

by $\varphi(X) = X, \varphi(Y) = u(X), \varphi(Z) = v(X)$. If we set $\operatorname{Ker} \varphi = P$ then $P \cap A = (0)$, and $A[\![X]\!]/P \simeq k[\![X]\!]$ is one-dimensional. Now every maximal ideal of $A[\![X]\!]$ has height 3, and, as we will see later, $A[\![X]\!]$ is catenary, so that $\operatorname{ht} P = 2$. Thus we see that the generic fibre of $A \longrightarrow A[\![X]\!]$ is two-dimensional.

3. The dimension inequality

We say that a ring A is *universally catenary* if A is Noetherian and every finitely generated A-algebra is catenary. Since any A-algebra generated by n elements is a quotient of $A[X_1, \ldots, X_n]$, and since a quotient of a catenary ring is again catenary, a necessary and sufficient condition for a Noetherian ring A to be universally catenary is that $A[X_1, \ldots, X_n]$ is catenary for every $n \geqslant 0$. (In fact it is known that it is sufficient for $A[X_1]$ to be catenary, compare Theorem 31.7.)

Theorem 15.5 (I. S. Cohen [3]). Let A be a Noetherian integral domain, and

B an extension ring of A which is an integral domain. Let $P \in \operatorname{Spec} B$ and $p = P \cap A$; then we have

(*) $\operatorname{ht} P + \operatorname{tr.deg}_{\kappa(p)} \kappa(P) \leqslant \operatorname{ht} p + \operatorname{tr.deg}_A B$,

where $\operatorname{tr.deg}_A B$ is the transcendence degree of the field of quotients of B over that of A.

Proof. We may assume that B is finitely generated over A. For if the right hand side is finite and m and t are non-negative integers such that $m \leqslant \operatorname{ht} P$ and $t \leqslant \operatorname{tr.deg}_{\kappa(p)} \kappa(P)$, then there is a prime ideal chain $P = P_0 \supset P_1 \supset \cdots \supset P_m$ in B. Take $a_i \in P_i - P_{i+1}$, $0 \leqslant i < m$, and let $c_1, \ldots, c_t \in B$ be such that their images modulo P are algebraically independent over A/p. Set $C = A[\{a_i\}, \{c_j\}]$. If the theorem holds for C, then we have $m + t \leqslant \operatorname{ht} p + \operatorname{tr.deg}_A C \leqslant \operatorname{ht} p + \operatorname{tr.deg}_A B$. Letting m and t vary we see the validity of (*).

We may furthermore assume, by induction, that B is generated over A by a single element: $B = A[x]$. We can replace A by A_p and B by $B_p = A_p[x]$, and hence assume that A is local and p its maximal ideal. Set $k = A/p$ and write $B = A[X]/Q$. If $Q = (0)$ then $B = A[X]$ and by Theorem 1 we have $\operatorname{ht} P = \operatorname{ht} p + \operatorname{ht}(P/pB)$, and since $B/pB = k[X]$ we have either $P = pB$ or $\operatorname{ht}(P/pB) = 1$. In both cases the equality holds in (*).

If $Q \neq (0)$ then $\operatorname{tr.deg}_A B = 0$. Since A is a subring of B we have $Q \cap A = (0)$, so that writing K for the field of fractions of A we have $\operatorname{ht} Q = \operatorname{ht} QK[X] = 1$. Let P^* be the inverse image of P in $A[X]$. Then $P = P^*/Q, \kappa(P) = \kappa(P^*)$, and $\operatorname{ht} P \leqslant \operatorname{ht} P^* - \operatorname{ht} Q = \operatorname{ht} P^* - 1 = \operatorname{ht} p + 1 - \operatorname{tr.deg}_{\kappa(p)} \kappa(P^*) - 1 = \operatorname{ht} p - \operatorname{tr.deg}_{\kappa(p)} \kappa(P)$. ∎

Definition. Suppose that A and B satisfy the conditions of the previous theorem. We refer to (*) as the *dimension inequality*, and if the equality in (*) holds for every $P \in \operatorname{Spec} B$, we say that the *dimension formula* holds between A and B. The above proof shows that dimension formula holds between A and $A[X_1, \ldots, X_n]$.

Theorem 15.6 (Ratliff). A Noetherian ring A is universally catenary if and only if the dimension formula holds between $A/\mathfrak{p}$ and B for every prime ideal $\mathfrak{p}$ of A and every finitely generated extension ring B of $A/\mathfrak{p}$ which is an integral domain.

Proof of 'only if'. If A is universally catenary then so is $A/\mathfrak{p}$, so that we need only consider the case that A is an integral domain, and B is a finitely generated extension ring which is an integral domain. If $B = A[X_1, \ldots, X_n]/Q$ and $P = P^*/Q$, then since $A[X_1, \ldots, X_n]$ is catenary we have $\operatorname{ht} P = \operatorname{ht} P^* - \operatorname{ht} Q$, and an easy calculation proves our assertion.

Proof of 'if'. We suppose that A is not universally catenary, so that there exists a finitely generated A-algebra B which is not catenary; without loss of generality we can assume that B is an integral domain. Write $\mathfrak{p}$ for the kernel of the homomorphism $A \longrightarrow B$. There exist prime ideals P and Q of B such that

$$P \subset Q, \quad \mathrm{ht}(Q/P) = d \quad \text{but} \quad \mathrm{ht}\, Q > \mathrm{ht}\, P + d.$$

We write $h = \mathrm{ht}\, P$, choose $a_1, \ldots, a_h \in P$ such that $\mathrm{ht}(a_1, \ldots, a_h) = h$, and set $I = (a_1, \ldots, a_h)$, so that P is a minimal prime divisor of I. Let

$$I = \mathfrak{q}_1 \cap \cdots \cap \mathfrak{q}_r$$

be a shortest primary decomposition of I, with P the minimal prime divisor of $\mathfrak{q}_1$. Then for $b \in Q\mathfrak{q}_2 \ldots \mathfrak{q}_r - P$ we have

$$I : b^\nu B = \mathfrak{q}_1 \quad \text{for} \quad \nu = 1, 2, \ldots.$$

We set $y_i = a_i/b$ for $1 \leqslant i \leqslant h$,

$$C = B[y_1, \ldots, y_h], \quad J = (y_1, \ldots, y_h)C \text{ and } M = J + QC = J + Q.$$

Every element of C can be written in the form u/b^k for suitable k, with $u \in (I + bB)^k$, so that if $z \in J \cap B$ then $zb^\nu \in I$ holds for sufficiently large ν. Hence $z \in I : b^\nu = \mathfrak{q}_1$. The converse inclusion $\mathfrak{q}_1 \subset J \cap B$ is obvious, hence $J \cap B = \mathfrak{q}_1$. Thus

$$M \cap B = (J + Q) \cap B = (J \cap B) + Q = Q,$$

$$C/J \simeq B/\mathfrak{q}_1 \quad \text{and} \quad C/M \simeq B/Q.$$

Therefore, $C_M/JC_M = B_Q/\mathfrak{q}_1 B_Q$ is a d-dimensional local ring, and J is generated by h elements, so that

$$\mathrm{ht}\, M = \dim C_M \leqslant h + d < \mathrm{ht}\, Q.$$

Now C and B have the same field of fractions, and $\kappa(M) = \kappa(Q)$, so that this inequality implies that the dimension formula does not hold between B and C. This is a contradiction, since we are assuming that the dimension formula holds between $A/\mathfrak{p}$ and B and between $A/\mathfrak{p}$ and C, and one sees easily that it must then hold between B and C. ■

4. The Rees ring and $\mathrm{gr}_I(A)$

Let A be a ring, I an ideal of A and t an indeterminate over A. Consider $A[t]$ as a graded ring in the usual way. We obtain a graded ring $R_+ \subset A[t]$ by setting

$$R_+ = R_+(A, I) = \{\textstyle\sum c_n t^n \mid c_n \in I^n\} = \bigoplus_n I^n t^n \subset A[t].$$

If $I = (a_1, \ldots, a_r)$ then R_+ can be written $R_+ = A[a_1 t, \ldots, a_r t]$, so that R_+ is Noetherian if A is.

R_+ is related to the graded ring $\mathrm{gr}_I(A)$ associated with A and I by the fact that

$$\mathrm{gr}_I(A) = \bigoplus_n I^n/I^{n+1} \simeq R_+/IR_+.$$

Now let $u=t^{-1}$, and consider $A[t,u]=A[t,t^{-1}]$ as a $\mathbb{Z}$-graded ring in the obvious way. The Rees ring $R(A,I)$ is the graded subring

$$R=R(A,I)=R_+[u]=\left\{\sum c_nt^n \middle| \begin{array}{l} c_n\in I^n \text{ for } n\geqslant 0 \\ c_n\in A \text{ for } n\leqslant 0 \end{array}\right\}\subset A[t,t^{-1}].$$

Since

$$uR=\left\{\sum c_nt^n \middle| \begin{array}{l} c_n\in I^{n+1} \text{ for } n\geqslant 0 \\ c_n\in A \text{ for } n\leqslant -1 \end{array}\right\}$$

we have $\mathrm{gr}_I(A)\simeq R/uR$.

Set $S=\{1,u,u^2,\ldots\}$. Then $R_S=R[u^{-1}]=R[t]=A[t^{-1},t]$, and $R_S/(1-u)R_S=A[t^{-1},t]/(1-t)=A$. But $R_S/(1-u)R_S=(R/(1-u)R)_{\bar{S}}$, where $\bar{S}$ is the image of S in $R/(I-u)R$, and since $\bar{S}=1$, we see that $R_S/(1-u)R_s=R/(1-u)R$. Thus we have

$$R/(1-u)R=A \quad\text{and}\quad R/uR=\mathrm{gr}_I(A),$$

so that the graded ring $\mathrm{gr}_I(A)$ is a 'deformation' of the original ring A, with R as 'total space of the deformation', in the sense that R contains a parameter u such that the values $u=1$ and 0 correspond to A and $\mathrm{gr}_I(A)$, respectively.

We also have

$$u^nR\cap A=I^n \quad\text{for all}\quad n\geqslant 0$$

and this property is often used to reduce problems about powers of I to the corresponding problems for powers of the principal ideal uR.

We conclude this section by applying the dimension inequality to the study of the dimension of the Rees ring and $\mathrm{gr}_I(A)$.

Let A be a Noetherian ring, $I=\sum_1^r a_iA$ a proper ideal of A, and t an indeterminate over A. We set

$$u=t^{-1},\quad R=R(A,I)=A[u,a_1t,\ldots,a_rt] \quad\text{and}\quad G=\mathrm{gr}_I(A).$$

We have $R\subset A[t,u]$ and $R/uR\simeq G$. For any ideal $\mathfrak{a}$ of A, set

$$\mathfrak{a}'=\mathfrak{a}A[t,u]\cap R.$$

That $\mathfrak{a}'\cap A=\mathfrak{a}A[t,u]\cap A=\mathfrak{a}$, so that for $\mathfrak{a}_1\neq\mathfrak{a}_2$ we have $\mathfrak{a}_1'\neq\mathfrak{a}_2'$. Moreover, if $\mathfrak{p}$ is a prime ideal of A then $\mathfrak{p}'$ is prime in R, and the same thing goes for primary ideals. If $(0)=\mathfrak{q}_1\cap\cdots\cap\mathfrak{q}_n$ is a primary decomposition of (0) in A then $(0)=\mathfrak{q}_1'\cap\cdots\cap\mathfrak{q}_n'$ is a primary decomposition of (0) in R. Hence if $\mathfrak{p}_{0i}$ (for $1\leqslant i\leqslant m$) are all the minimal prime ideals of A then $\{\mathfrak{p}_{0i}'\}_{1\leqslant i\leqslant m}$ is the set of all minimal prime ideals of R. Let $\mathfrak{p}$ be a prime ideal of A with $\mathrm{ht}\,\mathfrak{p}=h$, and let $\mathfrak{p}=\mathfrak{p}_0\supset\mathfrak{p}_1\supset\cdots\supset\mathfrak{p}_h$ be a strictly decreasing chain of prime ideals of A; then $\mathfrak{p}'\supset\mathfrak{p}_1'\supset\cdots\supset\mathfrak{p}_h'$ is a strictly descending chain of prime ideals of R, so that

$$\mathrm{ht}\,\mathfrak{p}\leqslant\mathrm{ht}\,\mathfrak{p}'.$$

Conversely, suppose that $P\in\mathrm{Spec}\,R$ and $P\cap A=\mathfrak{p}$. Let $\mathfrak{p}_{0i}'$ be a minimal prime of R contained in P and such that $\mathrm{ht}\,P=\mathrm{ht}\,(P/\mathfrak{p}_{0i}')$; then $R/\mathfrak{p}_{0i}'\supset$

$A/\mathfrak{p}_{0i}$, so that by the dimension inequality

$$\begin{aligned}\operatorname{ht} P = \operatorname{ht}(P/\mathfrak{p}'_{0i}) &\leqslant \operatorname{ht}(\mathfrak{p}/\mathfrak{p}_{0i}) + 1 - \operatorname{tr.deg}_{\kappa(\mathfrak{p})}\kappa(P)\\ &\leqslant \operatorname{ht}\mathfrak{p} + 1.\end{aligned}$$

Hence $\dim R \leqslant \dim A + 1$. On the other hand $A[u,t] = R[u^{-1}]$ is a localisation of R so that $\dim R \geqslant \dim A[u,t] = \dim A + 1$, so that finally

$$\dim R = \dim A + 1.$$

Moreover, for any $\mathfrak{p} \in \operatorname{Spec} A$ we set $\alpha_i = a_i \bmod \mathfrak{p}$, so that $R/\mathfrak{p}' = (A/\mathfrak{p})[u, \alpha_1 t, \ldots, \alpha_r t]$, and hence $\operatorname{tr.deg}_{\kappa(\mathfrak{p})}\kappa(\mathfrak{p}') = 1$; carrying out the above calculation using the dimension inequality with $\mathfrak{p}'$ in place of P we get $\operatorname{ht}\mathfrak{p}' \leqslant \operatorname{ht}\mathfrak{p}$, and so

$$\operatorname{ht}\mathfrak{p} = \operatorname{ht}\mathfrak{p}'.$$

We now choose a maximal ideal $\mathfrak{m}$ of A containing I; then since $R/\mathfrak{m}' = (A/\mathfrak{m})[u]$ we see that $\mathfrak{M} = (\mathfrak{m}', u)$ is a maximal ideal of R and $\mathfrak{M} \neq \mathfrak{m}'$, so that $\operatorname{ht}\mathfrak{M} > \operatorname{ht}\mathfrak{m}'$. However, by the dimension inequality, we have $\operatorname{ht}\mathfrak{M} \leqslant \operatorname{ht}\mathfrak{m} + 1 = \operatorname{ht}\mathfrak{m}' + 1$. Thus

$$\operatorname{ht}\mathfrak{M} = \operatorname{ht}\mathfrak{m}' + 1 = \operatorname{ht}\mathfrak{m} + 1.$$

The element u is a non-zero-divisor of R so that considering a system of parameters gives $\operatorname{ht}(\mathfrak{M}/uR) = \operatorname{ht}\mathfrak{M} - 1 = \operatorname{ht}\mathfrak{m}$. Thus providing that there exists a maximal ideal such that $\operatorname{ht}\mathfrak{m} = \dim A$ containing I, (in particular if A is local), then we have

$$\dim G = \dim(R/uR) = \dim A.$$

We summarise the above in the following theorem.

Theorem 15.7. Let A be a Noetherian ring and I a proper ideal; then setting $R = R(A, I)$ and $G = \operatorname{gr}_I(A)$ we have

$$\dim R = \dim A + 1, \quad \dim G \leqslant \dim A.$$

If in addition A is local, then

$$\dim G = \dim A.$$

Exercises to §15. Let k be a field.

15.1. Let $A = k[X, Y] \subset B = k[X, Y, X/Y]$, and $P = (Y, X/Y)B, \mathfrak{p} = (X, Y)A$; then check that $P \cap A = \mathfrak{p}$, $\operatorname{ht} P = \operatorname{ht}\mathfrak{p} = 2$, and $\dim B_P/\mathfrak{p}B_P = 1$, and hence that

$$\operatorname{ht} P < \operatorname{ht}\mathfrak{p} + \dim B_P/\mathfrak{p}B_P.$$

Show also by a concrete example that the going-down theorem does not hold between A and B.

15.2. Does the going-up theorem hold between A and B, where $A = k[X] \subset B = k[X, Y]$?

15.3. In Theorem 15.7, construct an example where $\dim G < \dim A$.

6

Regular sequences

In the 1950s homological algebra was introduced into commutative ring theory, opening up new avenues of study. In this chapter we run through some fundamental topics in this direction.

In §16 we define regular sequences, depth and the Koszul complex. The notion of depth is not very geometric, and rather hard to grasp, but is an extremely important invariant. It can be treated either in terms of Ext's, or by means of the Koszul complex, and we give both versions. We discuss the relation between regular and quasi-regular sequences in a transparent treatment due to Rees. §17 contains the definition and principal properties of Cohen–Macaulay (CM) rings. The theorem that quotients of CM rings are always catenary is of great significance in dimension theory. In §18 we treat a distinguished subclass of CM rings having even nicer properties, the Gorenstein rings. In the famous paper of H. Bass [1], Gorenstein rings are discussed using Matlis' theory of injective modules. But here we give an elementary treatment of Gorenstein rings following Greco before going through Matlis' theory.

16 Regular sequences and the Koszul complex

Let A be a ring and M an A-module. An element $a \in A$ is said to be *M-regular* if $ax \neq 0$ for all $0 \neq x \in M$. A sequence $a_1, \ldots, a_n$ of elements of A is an *M-sequence* (or an M-regular sequence) if the following two conditions hold:

(1) a_1 is M-regular, a_2 is (M/a_1M)-regular,..., a_n is $(M/\sum_1^{n-1} a_iM)$-regular;

(2) $M/\sum_1^n a_iM \neq 0$.

Note that, after permutation, the elements of an M-sequence may no longer form an M-sequence.

Theorem 16.1. If $a_1, \ldots, a_n$ is an M-sequence then so is $a_1^{\nu_1}, \ldots, a_n^{\nu_n}$ for any positive integers $\nu_1, \ldots, \nu_n$.

Proof. It is sufficient to prove that if $a_1, \ldots, a_n$ is an M-sequence then so is $a_1^{\nu}, a_2, \ldots, a_n$. Indeed, assuming this, we have in turn that $a_1^{\nu_1}, a_2, \ldots, a_n$

is an M-sequence, then setting $M_1 = M/a_1^{\nu_1}M$ that $a_2, a_3, \ldots, a_n$ and hence also $a_2^{\nu_2}, a_3, \ldots, a_n$ is an M_1-sequence, and so on. Also, the second condition $M \neq \sum_1^n a_i^{\nu_i}M$ is obvious.

Let us now prove by induction on n that if $b_1, \ldots, b_n$ is an M-sequence, and if $b_1\xi_1 + \cdots + b_n\xi_n = 0$ with $\xi_i \in M$ then $\xi_i \in b_1M + \cdots + b_nM$ for all i. First of all from the condition that b_n is not a zero-divisor modulo $b_1, \ldots, b_{n-1}$ we can write

$$\xi_n = \sum_1^{n-1} b_i\eta_i, \quad \text{with} \quad \eta_i \in M.$$

Therefore $\sum_1^{n-1} b_i(\xi_i + b_n\eta_i) = 0$, so that by induction we have

$$\xi_i + b_n\eta_i \in b_1M + \cdots + b_{n-1}M \quad \text{for} \quad 1 \leqslant i \leqslant n-1,$$

giving $\xi_i \in b_1M + \cdots + b_nM$ for $1 \leqslant i \leqslant n-1$. The condition for ξ_n is already known

Now assuming $\nu > 1$ we prove by induction on ν that $a_1^\nu, a_2, \ldots, a_n$ is an M-sequence. Since a_1 is M-regular, so is a_1^ν. For $i > 1$, suppose that for some $\omega \in M$ we have

$$a_i\omega = a_1^\nu\xi_1 + a_2\xi_2 + \cdots + a_{i-1}\xi_{i-1} \quad \text{with} \quad \xi_j \in M.$$

Then since $a_1^{\nu-1}, a_2, \ldots, a_i$ is an M-sequence, we can write

$$\omega = a_1^{\nu-1}\eta_1 + \cdots + a_{i-1}\eta_{i-1} \quad \text{with} \quad \eta_j \in M.$$

Hence we get

$$0 = a_1^{\nu-1}(a_1\xi_1 - a_i\eta_1) + a_2(\xi_2 - a_i\eta_2) + \cdots + a_{i-1}(\xi_{i-1} - a_i\eta_{i-1}).$$

The above assertion gives $a_1\xi_1 - a_i\eta_1 \in a_1^{\nu-1}M + a_2M + \cdots + a_{i-1}M$, and hence $a_i\eta_1 \in a_1M + a_2M + \cdots + a_{i-1}M$. Therefore $\eta_1 \in a_1M + \cdots + a_{i-1}M$, and so as required we have $\omega \in a_1^\nu M + a_2M + \cdots + a_{i-1}M$. ■

Let A be a ring, $X_1, \ldots, X_n$ indeterminates over A, and M an A-module. We can view elements of $M \otimes_A A[X_1, \ldots, X_n]$ as polynomials in the X_i with coefficients in M,

$$F(X) = F(X_1, \ldots, X_n) = \sum \xi_{(\alpha)} X_1^{\alpha_1} \ldots X_n^{\alpha_n}, \quad \text{with} \quad \xi_{(\alpha)} \in M.$$

For this reason we write $M[X_1, \ldots, X_n]$ for $M \otimes_A A[X_1, \ldots, X_n]$; we can consider this either as an A-module or as an $A[X_1, \ldots, X_n]$-module. For $a_1, \ldots, a_n \in A$ and $F \in M[X_1, \ldots, X_n]$, we can substitute the a_i for X_i to get $F(a_1, \ldots, a_n) \in M$.

Definition. Let $a_1, \ldots, a_n \in A$, set $I = \sum_1^n a_iA$, and let M be an A-module with $IM \neq M$. We say that $a_1, \ldots, a_n$ is an *M-quasi-regular sequence* if the following condition holds for each ν:

(*) $F(X_1, \ldots, X_n) \in M[X_1, \ldots, X_n]$ is homogeneous of degree ν and $F(a) \in I^{\nu+1}M$ implies that all the coefficients of F are in IM.

This notion is obviously independent of the order of $a_1, \ldots, a_n$.

In the above definition it would not make any difference if we replaced the condition that $F(a)\in I^{\nu+1}M$ by the condition $F(a)=0$. Indeed, if F is homogeneous of degree ν and $F(a)\in I^{\nu+1}M$ then there exist a homogeneous element $G(X)\in M[X_1,\ldots,X_n]$ of degree $\nu+1$ such that $F(a)=G(a)$. Then write $G(X)=\sum_1^n X_iG_i(X)$ with each G_i homogeneous of degree ν, and set $F^*(X)=F(X)-\sum a_iG_i(X)$, so that F^* is homogeneous of degree ν and $F^*(a)=0$. Moreover, if F^* has coefficients in IM then so does F.

We can define a map $\varphi:(M/IM)[X_1,\ldots,X_n]\longrightarrow \mathrm{gr}_I M=\bigoplus_{\nu\geq 0} I^\nu M/I^{\nu+1}M$ as follows: taking a homogeneous element $F(X)\in M[X]$ of degree ν into the class of $F(a)$ in $I^\nu M/I^{\nu+1}M$ provides a homomorphism (of additive groups) from $M[X]$ into $\mathrm{gr}_I M$ which preserves degrees. Since $IM[X]$ is in the kernel, this induces a homomorphism

$$\varphi:M[X]/IM[X]=(M/IM)[X]\longrightarrow \mathrm{gr}_I M,$$

which is obviously surjective. Then $a_1,\ldots,a_n$ is a quasi-regular sequence precisely when φ is injective, and hence an isomorphism.

Theorem 16.2. Let A be a ring, M an A-module, and $a_1,\ldots,a_n\in A$; set $I=(a_1,\ldots,a_n)A$. Then we have the following:

(i) if $a_1,\ldots,a_n$ is an M-sequence then it is M-quasi-regular;

(ii) if $a_1,\ldots,a_n$ is an M-quasi-regular sequence, and if $x\in A$ satisfies $IM:x=IM$ then $I^\nu M:x=I^\nu M$ for any $\nu>0$.

Proof (taken from Rees [5]). First of all we prove (ii) by induction on ν. The case $\nu=1$ is just the assumption; suppose that $\nu>1$. For $\xi\in M$, if $x\xi\in I^\nu M$ then also $x\xi\in I^{\nu-1}M$, so that by the inductive hypothesis $\xi\in I^{\nu-1}M$, and hence we can write $\xi=F(a)$ with $F=F(X)\in M[X_1,\ldots,X_n]$ homogeneous of degree $\nu-1$. Now $x\xi=xF(a)\in I^\nu M$, so that by definition of quasi-regular sequence each coefficient of $xF(X)$ belongs to IM. Using $IM:x=IM$ once more we find that the coefficients of $F(X)$ also belong to IM, and therefore $\xi=F(a)\in I^\nu M$.

Now we prove (i) by induction on n. The case $n=1$ can easily be checked. Suppose that $n>1$, and that the statement holds up to $n-1$, so that in particular $a_1,\ldots,a_{n-1}$ is M-quasi-regular. Now let $F(X)\in M[X_1,\ldots,X_n]$ be homogeneous of degree ν, such that $F(a)=0$. We prove by induction on ν that the coefficients of F belong to IM. We separate out $F(X)$ into terms containing X_n and not containing X_n, writing

$$F(X)=G(X_1,\ldots,X_{n-1})+X_nH(X_1,\ldots,X_n).$$

Here G is homogeneous of degree ν and H of degree $\nu-1$. Then, as we proved in (ii),

$$H(a)\in(a_1,\ldots,a_{n-1})^\nu M:a_n=(a_1,\ldots,a_{n-1})^\nu M\subset I^\nu M,$$

and hence, by induction on ν, the coefficients of $H(X)$ belong to IM. Moreover, by the above formula there is a homogeneous polynomial $h(X_1,\ldots,X_{n-1})$ of degree ν with coefficients in M such that $H(a) = h(a_1,\ldots,a_{n-1})$, and so setting

$$G(X_1,\ldots,X_{n-1}) + a_n h(X_1,\ldots,X_{n-1}) = g(X),$$

since $a_1,\ldots,a_{n-1}$ is M-quasi-regular, we get that the coefficients of g belong to $(a_1,\ldots,a_{n-1})M$; therefore the coefficients of G belong to $(a_1,\ldots,a_n)M$. ■

This theorem holds for any A and M, but as we will see in the next theorem, under some conditions we can say that conversely, quasi-regular implies regular. Then the notions of regular and quasi-regular sequences for M coincide, and so reordering an M-sequence gives again an M-sequence.

Theorem 16.3. Let A be a Noetherian ring, $M \neq 0$ an A-module, and $a_1,\ldots,a_n \in A$; set $I = (a_1,\ldots,a_n)A$. Under the condition

(*) each of $M, M/a_1M,\ldots,M/(a_1,\ldots,a_{n-1})M$ is I-adically separated, if $a_1,\ldots,a_n$ is M-quasi-regular it is an M-sequence.

Remark. The hypothesis (*) holds in either of the following cases:

(α) M is finite and $I \subset \operatorname{rad}(A)$;

(β) A is an $\mathbb{N}$-graded ring, M an $\mathbb{N}$-graded module, and each a_i is homogeneous of positive degree.

However, for a non-Noetherian ring A there are examples where the theorem fails (Dieudonné [1]) even if A is local, $M = A$ and $I \subset \operatorname{rad}(A)$.

Proof. We prove first that a_1 is M-regular. If $\xi \in M$ with $a_1\xi = 0$ then by hypothesis $\xi \in IM$. Then setting $\xi = \sum a_i\eta_i$ we get $0 = \sum a_1a_i\eta_i$, so that $\eta_i \in IM$. Proceeding in the same way we get $\xi \in \bigcap I^\nu M = (0)$.

Now set $M_1 = M/a_1M$; if we prove that $a_2,\ldots,a_n$ is an M_1-quasi-regular sequence then the theorem follows by induction on n. (If M is I-adically separated and $M \neq 0$ then $M \neq IM$.) So let $f(X_2,\ldots,X_n)$ be a homogeneous polynomial of degree ν with coefficients in M_1 such that $f(a_2,\ldots,a_n) = 0$. If $F(X_2,\ldots,X_n)$ is a homogeneous polynomial of degree ν with coefficients in M which reduces to f modulo a_1M, then $F(a_2,\ldots,a_n) \in a_1M$. Set $F(a_2,\ldots,a_n) = a_1\omega$; suppose that $\omega \in I^iM$, so that we can write $\omega = G_i(a)$ with $G_i(X) \in M[X_1,\ldots,X_n]$ homogeneous of degree i. Then

$$F(a_2,\ldots,a_n) = a_1G_i(a_1,\ldots,a_n),$$

and if $i < \nu - 1$ it follows that the coefficients of G_i belong to IM, so that $\omega \in I^{i+1}M$; repeating this argument we see that $\omega \in I^{\nu-1}M$. Setting $i = \nu - 1$ in the above formula, then since X_1 does not appear in F, we can apply the definition of quasi-regular sequence to $F(X_2,\ldots,X_n) -$

$X_1G_{\nu-1}(X_1,\ldots,X_n)$ to deduce that the coefficients of F belong to IM. Hence, the coefficients of f belong to IM_1. ■

Corollary. Let A be a Noetherian ring, M and A-module and $a_1,\ldots,a_n$ an M-sequence. If conditions (α) or (β) of the above remark hold then any permutation of $a_1,\ldots,a_n$ is again an M-sequence.

Here is an example where a permutation of an M-sequence fails to be an M-sequence: let k be a field, $A=k[X, Y, Z]$ and set $a_1=X(Y-1)$, $a_2=Y$, $a_3=Z(Y-1)$. Then $(a_1, a_2, a_3)A=(X, Y, Z)A\neq A$, and a_1, a_2, a_3 is an A-sequence, whereas a_1, a_3, a_2 is not.

The Koszul complex

Given a ring A and $x_1,\ldots,x_n\in A$, we define a complex $K_.$ as follows: set $K_0=A$, and $K_p=0$ if p is not in the range $0\leqslant p\leqslant n$. For $1\leqslant p\leqslant n$, let $K_p=\bigoplus Ae_{i_1\ldots i_p}$ be the free A-module of rank $\binom{n}{p}$ with basis $\{e_{i_1\ldots i_p}|1\leqslant i_1<\cdots<i_p\leqslant n\}$. The differential d: $K_p\longrightarrow K_{p-1}$ is defined by setting

$$\mathrm{d}(e_{i_1\ldots i_p})=\sum_{r=1}^{p}(-1)^{r-1}x_{i_r}e_{i_1\ldots\hat{i}_r\ldots i_p};$$

(for $p=1$, set $\mathrm{d}(e_i)=x_i$). One checks easily that $\mathrm{dd}=0$. This complex is called the Koszul complex, and written $K_.(x_1,\ldots,x_n)$ (alternatively, $K_.(\underline{x})$ or $K_{.x,1\ldots n}$). For an A-module M we set $K_.(\underline{x}, M)=K_.(\underline{x})\otimes_A M$. Moreover, for a complex $C_.$ of A-modules we set $C_.(\underline{x})=C_.\otimes K_.(\underline{x})$. In particular, for $n=1$ the complex $K_.(x)$ is just

$$\cdots\to 0\to 0\to A\xrightarrow{x}A\to 0,$$

and it is easy to check that $K_.(x_1,\ldots,x_n)=K_.(x_1)\otimes\cdots\otimes K_.(x_n)$. Since the tensor product of complexes satisfies $L_.\otimes M_.\simeq M_.\otimes L_.$, the Koszul complex is invariant (up to isomorphism) under permutation of $x_1,\ldots,x_n$. The Koszul complex $K_.(\underline{x}, M)$ has homology groups $H_p(K_.(\underline{x}, M))$, which we abbreviate to $H_p(\underline{x}, M)$. Quite generally we have

$$H_0(\underline{x}, M)\simeq M/\underline{x}M,$$

where $\underline{x}M$ stands for $\sum x_iM$, and

$$H_n(\underline{x}, M)\simeq\{\xi\in M|x_1\xi=\cdots=x_n\xi=0\}.$$

Theorem 16.4. Let $C_.$ be a complex of A-modules and $x\in A$. Then we obtain an exact sequence of complexes

$$0\to C_.\longrightarrow C_.(x)\longrightarrow C'_.\to 0,$$

where $C'_.$ is the complex obtained by shifting the degrees in $C_.$ up by 1 (that is $C'_{p+1}=C_p$ and the differential of $C'_.$ is that of $C_.$). The homology long exact sequence obtained from this is

$$\cdots \longrightarrow H_p(C.) \longrightarrow H_p(C.(x)) \longrightarrow H_{p-1}(C.) \xrightarrow{(-1)^{p-1}x} H_{p-1}(C.) \longrightarrow \cdots;$$

we have $x \cdot H_p(C.(x)) = 0$ for all p.

Proof. From the fact that $K_1(x) = Ae_1$ and $K_0(x) = A$ and the definition of tensor product of complexes, we can identify $C_p(x)$ with $C_p \oplus C_{p-1}$, and for $\xi \in C_p$, $\eta \in C_{p-1}$ we have

$$d(\xi, \eta) = (d\xi + (-1)^{p-1}x\eta, d\eta).$$

The first assertion is clear from this. Moreover, $H_p(C'.) = H_{p-1}(C.)$ is also clear, and if $\eta \in C'_p = C_{p-1}$ satisfies $d\eta = 0$ then in $C.(x)$ we have $d(0, \eta) = ((-1)^{p-1}x\eta, 0)$, so that the long exact sequence has the form indicated in the theorem. Finally, if $d(\xi, \eta) = 0$ then $d\eta = 0$ and $d\xi = (-1)^p x\eta$, so that $x \cdot (\xi, \eta) = d(0, (-1)^p \xi) \in dC_{p+1}(x)$, and therefore $x \cdot H_p(C.(x)) = 0$. ∎

Applying this theorem to $K.(\underline{x}, M)$ and using the commutativity of tensor product of complexes, we see that the ideal $(\underline{x}) = (x_1, \ldots, x_n)$ generated by $\underline{x}$ annihilates the homology groups $H_p(\underline{x}, M)$:

$$(\underline{x}) \cdot H_p(\underline{x}, M) = 0 \quad \text{for all} \quad p.$$

Theorem 16.5.

(i) Let A be a ring, M an A-module, and $x_1, \ldots, x_n$ an M-sequence; then

$$H_p(\underline{x}, M) = 0 \quad \text{for} \quad p > 0 \quad \text{and} \quad H_0(\underline{x}, M) = M/\underline{x}M.$$

(ii) Suppose that one of the following two conditions (α) or (β) holds:

(α) $(A, \mathfrak{m})$ is a local ring, $x_1, \ldots, x_n \in \mathfrak{m}$ and M is a finite A-module;

(β) A is an $\mathbb{N}$-graded ring, M is an $\mathbb{N}$-graded A-module, and $x_1, \ldots, x_n$ are homogeneous elements of degree > 0.

Then the converse of (i) holds in the following strong form: if $H_1(\underline{x}, M) = 0$ and $M \neq 0$ then $x_1, \ldots, x_n$ is an M-sequence.

Proof. We use induction on n.

(i) When $n = 1$ we have $H_1(x, M) = \{\xi \in M \mid x\xi = 0\} = 0$, so that the assertion holds. When $n > 1$, for $p > 1$ the previous theorem provides an exact sequence

$$0 = H_p(x_1, \ldots, x_{n-1}, M) \longrightarrow H_p(x_1, \ldots, x_n, M) \longrightarrow H_{p-1}(x_1, \ldots, x_{n-1}, M) = 0.$$

so that $H_p(x_1, \ldots, x_n, M) = 0$. For $p = 1$, setting $M_i = M/(x_1, \ldots, x_i)M$ we have an exact sequence

$$0 \to H_1(\underline{x}, M) \longrightarrow H_0(x_1, \ldots, x_{n-1}, M) = M_{n-1} \xrightarrow{\pm x_n} M_{n-1} \to \cdots,$$

and since x_n is M_{n-1}-regular we have $H_1(\underline{x}, M) = 0$.

(ii) Assuming either (α) or (β), $M \neq 0$ implies that $M_i \neq 0$ for $1 \leqslant i \leqslant n$. By hypothesis and by the previous theorem,

$$H_1(x_1,\ldots,x_{n-1},M) \xrightarrow{\pm x_n} H_1(x_1,\ldots,x_{n-1},M) \longrightarrow H_1(\underline{x},M)=0;$$

but quite generally $H_p(\underline{x},M)$ is a finite A-module in case (α), or a $\mathbb{N}$-graded A-module in case (β), so that by NAK, $H_1(x_1,\ldots,x_{n-1},M)=0$. Thus by induction $x_1,\ldots,x_{n-1}$ is an M-sequence. Now by the same exact sequence as in the case $p=1$ of (i), we see that x_n is M_{n-1}-regular, and therefore $x_1,\ldots,x_n$ is an M-sequence. ■

Let A be a ring, M an A-module and I an ideal of A. If $a_1,\ldots,a_r$ are elements of I, we say that they form a maximal M-sequence in I if $a_1,\ldots,a_r$ is an M-sequence, and $a_1,\ldots,a_r$, b is not an M-sequence for any $b \in I$. If $a_1,\ldots,a_r$ is an M-sequence then $a_1M,(a_1,a_2)M,\ldots,(a_1,\ldots,a_r)M$ is strictly increasing, so that the chain of ideals $(a_1) \subset (a_1,a_2) \subset \ldots$ is also strictly increasing. If A is Noetherian this cannot continue indefinitely, so that any M-sequence can be extended until we arrive at a maximal M-sequence.

Remark. In Theorems 6–8 below, the hypothesis that M is a finite A-module can be weakened to the statement that M is a finite B-module for a homomorphism $A \longrightarrow B$ of Noetherian rings, as one sees on inspecting the proof. The reason for this is that, if we set $\mathrm{Ass}_B(M)=\{P_1,\ldots,P_r\}$ and $P_i \cap A = \mathfrak{p}_i$, then any ideal of A consisting entirely of zero-divisors of M is contained in $\bigcup \mathfrak{p}_i$, and therefore contained in one of the $\mathfrak{p}_i$. Note that according to [M], (9.A), we have $\mathrm{Ass}_A(M)=\{\mathfrak{p}_1,\ldots,\mathfrak{p}_r\}$.

Theorem 16.6. Let A be a Noetherian ring, M a finite A-module and I an ideal of A; suppose that $IM \neq M$. For a given integer $n>0$ the following conditions are equivalent;

(1) $\mathrm{Ext}^i_A(N,M)=0$ for all $i<n$ and for any finite A-module N with $\mathrm{Supp}(N) \subset V(I)$;

(2) $\mathrm{Ext}^i_A(A/I,M)=0$ for all $i<n$;

(2′) $\mathrm{Ext}^i_A(N,M)=0$ for all $i<n$ and for some finite A-module N with $\mathrm{Supp}(N)=V(I)$;

(3) there exists an M-sequence of length n contained in I.

Proof. (1)⇒(2)⇒(2′) is obvious. For (2′)⇒(3), if I consists only of zero-divisors of M then there exists an associated prime P of M containing I (this is where we need the finiteness of M). Hence there is an injective map $A/P \longrightarrow M$. Localising at P, we see that $\mathrm{Hom}_{A_P}(k,M_P) \neq 0$, where $k=(A/P)_P=A_P/PA_P$. Now $P \in V(I)=\mathrm{Supp}(N)$, so that $N_P \neq 0$, and hence by NAK, $N_P/PN_P = N \otimes_A k \neq 0$. Thus $N \otimes k$ is a non-zero vector space over k, and $\operatorname{Hom}_k(N\otimes k, k)\neq 0$. Putting together what we have said, we can follow the composite $N_P \longrightarrow N\otimes k \longrightarrow k \longrightarrow M_P$ to show that $\operatorname{Hom}_{A_P}(N_P, M_P)\neq 0$. The left-hand side is equal to $(\operatorname{Hom}_A(N, M))_P$, so that $\operatorname{Hom}_A(N, M)\neq 0$. But this contradicts (2′). Hence I contains an M-regular element f. By assumption, $M/IM\neq 0$, and if $n=1$ then we are done. If $n>1$ we set $M_1=M/fM$; then from the exact sequence

$$0\rightarrow M \xrightarrow{f} M \longrightarrow M_1 \rightarrow 0$$

we get $\operatorname{Ext}^i_A(N, M)=0$ for $i<n-1$, so that by induction I contains an M_1-sequence $f_2,\ldots,f_n$.

For the proof of (3)⇒(1) we do not need to assume that A is Noetherian or M finite. Let $f_1,\ldots,f_n\in I$ be an M-sequence; we have the exact sequence

$$0\rightarrow M \xrightarrow{f_1} M \longrightarrow M_1 \rightarrow 0,$$

and if $n>1$ the inductive hypothesis $\operatorname{Ext}^i_A(N, M_1)=0$ for $i<n-1$, so that

$$0\rightarrow \operatorname{Ext}^i_A(N, M) \xrightarrow{f_1} \operatorname{Ext}^i_A(N, M)$$

is exact for $i<n$. But $\operatorname{Ext}^i_A(N, M)$ is annihilated by elements of $\operatorname{ann}(N)$. Since $\operatorname{Supp}(N)=V(\operatorname{ann}(N))\subset V(I)$, we have $I\subset\sqrt{(\operatorname{ann}(N))}$, and a sufficiently large power of f_1 annihilates $\operatorname{Ext}^i_A(N, M)$. Therefore, $\operatorname{Ext}^i_A(N, M)=0$ for $i<n$. ∎

Let M and I be as in the above theorem, and $a_1,\ldots,a_n$ an M-sequence in I. For $1\leqslant i\leqslant n$, set $M_i=M/(a_1,\ldots,a_i)M$; then it is easy to see that $\operatorname{Hom}_A(A/I, M_n)\cong \operatorname{Ext}^1_A(A/I, M_{n-1})\cong\cdots\cong \operatorname{Ext}^n_A(A/I, M)$. Therefore, if $\operatorname{Ext}^n_A(A/I, M)=0$ we can find another element $a_{n+1}\in I$ such that $a_1,\ldots,a_{n+1}$ is an M-sequence. Hence if $a_1,\ldots,a_n$ is a maximal M-sequence in I we must have $\operatorname{Ext}^n_A(A/I, M)\neq 0$. We thus obtain the following theorem.

Theorem 16.7. Let A be a Noetherian ring, I an ideal of A and M a finite A-module such that $M\neq IM$; then the length of a maximal M-sequence in I is a well-determined integer n, and n is determined by

$$\operatorname{Ext}^i_A(A/I, M)=0 \quad \text{for} \quad i<n \quad \text{and} \quad \operatorname{Ext}^n_A(A/I, M)\neq 0.$$

We write $n=\operatorname{depth}(I, M)$, and call n the *I-depth* of M. (If $M=IM$, the I-depth is by convention ∞.) Theorem 7 takes the form

$$\operatorname{depth}(I, M)=\inf\{i\,|\,\operatorname{Ext}^i_A(A/I, M)\neq 0\}.$$

In particular for a Noetherian local ring $(A, \mathfrak{m}, k)$, we call $\operatorname{depth}(\mathfrak{m}, M)$ simply the depth of M, and write $\operatorname{depth} M$ or $\operatorname{depth}_A M$:

$$\operatorname{depth} M=\inf\{i\,|\,\operatorname{Ext}^i_A(k, M)\neq 0\}.$$

From Theorem 6 we see that if $V(I)=V(I')$ then $\operatorname{depth}(I, M)=\operatorname{depth}(I', M)$; this also follows easily from Theorem 1.

If $\operatorname{ann}(M)=\mathfrak{a}$ and we set $A/\mathfrak{a}=\bar{A}$ then M is also an $\bar{A}$-module. Writing

$\bar{a}$ or $\bar{I}$ for the image of an element a or an ideal I of A under the natural homomorphism $A \longrightarrow \bar{A}$ we clearly have that $a_1, \ldots, a_r$ is an M-sequence if and only if $\bar{a}_1, \ldots, \bar{a}_r$ is. Thus $\operatorname{depth}(I, M) = \operatorname{depth}(\bar{I}, M)$, and if we set $I + \mathfrak{a} = J$, then since $\bar{I} = \bar{J}$ we also have $\operatorname{depth}(I, M) = \operatorname{depth}(J, M)$.

We can also prove that the length of a maximal M-sequence is well-determined by means of the Koszul complex.

Theorem 16.8. Let A be a Noetherian ring, $I = (y_1, \ldots, y_n)$ an ideal of A, and M a finite A-module such that $M \neq IM$. If we set

$$q = \sup\{i \mid H_i(\underline{y}, M) \neq 0\},$$

then any maximal M-sequence in I has length $n - q$.

Proof. Let $x_1, \ldots, x_s$ be a maximal M-sequence in I; we argue by induction on s. If $s = 0$ then every element of I is a zero-divisor of M, so that there exists $P \in \operatorname{Ass}(M)$ containing I. By definition of Ass, there exists $0 \neq \xi \in M$ such that $P = \operatorname{ann}(\xi)$, and hence $I\xi = 0$. Thus $\xi \in H_n(\underline{y}, M)$ so that $q = n$, and the assertion holds in this case.

If $s > 0$ we set $M_1 = M/x_1 M$; then from the exact sequence

$$0 \rightarrow M \xrightarrow{x_1} M \longrightarrow M_1 \rightarrow 0$$

and from the fact that $IH_i(\underline{y}, M) = 0$ (by Theorem 4), it follows that

$$0 \rightarrow H_i(\underline{y}, M) \longrightarrow H_i(\underline{y}, M_1) \longrightarrow H_{i-1}(\underline{y}, M) \rightarrow 0$$

is exact for every i. Thus $H_{q+1}(\underline{y}, M_1) \neq 0$ and $H_i(\underline{y}, M_1) = 0$ for $i > q + 1$; but $x_2, \ldots, x_s$ is a maximal M_1-sequence in I, so that by induction we have $q + 1 = n - (s - 1)$, and therefore $q = n - s$. ■

In other words, $\operatorname{depth}(I, M)$ is the number of successive zero terms from the left in the sequence

$$H_n(\underline{y}, M), H_{n-1}(\underline{y}, M), \ldots, H_0(\underline{y}, M) = M/IM \neq 0.$$

This fact is sometimes referred to as the 'depth sensitivity' of the Koszul complex.

Corollary. In the situation of the theorem, $y_1, \ldots, y_n$ is an M-sequence if and only if $\operatorname{depth}(I, M) = n$.

Proof. $\operatorname{depth}(I, M) = n \Leftrightarrow H_i(\underline{y}, M) = 0$ for all $i > 0 \Leftrightarrow \underline{y}$ is an M-sequence.

Grade

A little before Auslander and Buchsbaum [2], Rees [5] introduced and developed the theory of another notion related to regular sequences, that of grade. Let A be a Noetherian ring and $M \neq 0$ a finite A-module. Then Rees made the definition

$$\operatorname{grade} M = \inf\{i \mid \operatorname{Ext}^i_A(M, A) \neq 0\}.$$

For a proper ideal J of A we also call grade (A/J) the grade of the ideal J, and write grade J. If we set $\mathfrak{a} = \operatorname{ann}(M)$ then since $\operatorname{Supp}(M) = V(\mathfrak{a})$, Theorem 6 gives $\operatorname{grade} M = \operatorname{depth}(\mathfrak{a}, A)$. Moreover, if $g = \operatorname{grade} M$ then $\operatorname{Ext}^g_A(M, A) \neq 0$, so that

$$\operatorname{grade} M \leqslant \operatorname{proj\,dim} M.$$

If I is an ideal then $\operatorname{grade} I (= \operatorname{grade}(A/I)) = \operatorname{depth}(I, A)$ is the length of a maximal A-sequence in I, but in general if $a_1, \ldots, a_r$ is an A-sequence then one sees easily from Theorem 13.5 that $\operatorname{ht}(a_1, \ldots, a_r) = r$. Thus if $a_1, \ldots, a_r$ is a maximal A-sequence in I, we have $r = \operatorname{ht}(a_1, \ldots, a_r) \leqslant \operatorname{ht} I$. Hence for an ideal I we have $\operatorname{grade} I \leqslant \operatorname{ht} I$.

Theorem 16.9. Let A be a Noetherian ring, and M, N finite A-modules; suppose $M \neq 0$, $\operatorname{grade} M = k$ and $\operatorname{proj\,dim} N = l < k$. Then

$$\operatorname{Ext}^i_A(M, N) = 0 \quad \text{for} \quad i < k - l.$$

Proof. We use induction on l. If $l = 0$ then N is a direct summand of some free module A^n, so that we need only say what happens for $N = A$, but then the assertion is just the definition of grade. If $l > 0$ we choose an exact sequence

$$0 \to N_1 \longrightarrow L_0 \longrightarrow N \to 0$$

with L_0 a finite free module; then $\operatorname{proj\,dim} N_1 = l - 1$, so that by induction

$$\operatorname{Ext}^i_A(M, L_0) = 0 \quad \text{for} \quad i < k \quad \text{and}$$
$$\operatorname{Ext}^{i+1}_A(M, N_1) = 0 \quad \text{for} \quad i < k - l;$$

the assertion follows from this. ■

Exercises to §16. Prove the following propositions.

16.1. Let $(A, \mathfrak{m})$ be a Noetherian local ring, $M \neq 0$ a finite A-module, and $a_1, \ldots, a_r \in \mathfrak{m}$ an M-sequence. Set $M' = M/(a_1, \ldots, a_r)M$. Then $\dim M' = \dim M - r$.

16.2. Let A be a Noetherian ring, $\mathfrak{a}$ and $\mathfrak{b}$ ideals of A; then if grade $\mathfrak{a}$ $> \operatorname{projdim} A/\mathfrak{b}$ we have $\mathfrak{b}:\mathfrak{a} = \mathfrak{b}$.

16.3. Let A be a Noetherian ring. A proper ideal I of A is called a *perfect ideal* if $\operatorname{grade} I = \operatorname{proj\,dim} A/I$. If I is a perfect ideal of grade k then all the prime divisors of I have grade k.

Remark. Quite generally, we have $\operatorname{grade} I (= \operatorname{grade}(A/I)) \leqslant \operatorname{proj\,dim} A/I$. If A is a regular local ring and $P \in \operatorname{Spec} A$ then as we will see in Theorems 19.1 and 19.2, P is perfect $\Leftrightarrow A/P$ is Cohen–Macaulay.

16.4. Let $f: A \longrightarrow B$ be a flat ring homomorphism, M an A-module, and $a_1, \ldots, a_r \in A$ an M-sequence; if $(M/(a_1, \ldots, a_r)M) \otimes B \neq 0$ then $f(a_1), \ldots, f(a_r)$ is an $M \otimes B$-sequence.

16.5. Let A be a Noetherian local ring, M a finite A-module, and P a prime ideal of A; show that $\operatorname{depth}(P, M) \leqslant \operatorname{depth}_{A_P} M_P$, and construct an example where the inequality is strict.

16.6. Let A be a ring and $a_1, \ldots, a_n \in A$ an A-quasi-regular sequence. If A contains a field k then $a_1, \ldots, a_n$ are algebraically independent over k.

16.7. Let $(A, \mathfrak{m})$ and $(B, \mathfrak{n})$ be Noetherian local rings, and suppose that $A \subset B$, $\mathfrak{n} \cap A = \mathfrak{m}$ and that $\mathfrak{m}B$ is an $\mathfrak{n}$-primary ideal. Then for a finite B-module M we have

$$\operatorname{depth}_B M = \operatorname{depth}_A M.$$

16.8. Let A be a ring, $P_1, \ldots, P_r$ prime ideals, I an ideal, and x an element of A. If $xA + I \not\subset P_1 \cup \cdots \cup P_r$ then there is a $y \in I$ such that $x + y \notin P_1 \cup \cdots \cup P_r$ (E. Davis).

16.9. Use the previous question to show the following: let A be a Noetherian ring, and suppose that $I \neq A$ is an ideal generated by n elements; then $\operatorname{grade} I \leqslant n$, and if $\operatorname{grade} I = n$ then I can be generated by an A-sequence ([K], Th. 125).

16.10. Let A be a Noetherian ring, and suppose that P is a height $r > 0$ prime ideal generated by r elements $a_1, \ldots, a_r$.

(i) Suppose either that A is local, or that A is $\mathbb{N}$-graded and the a_i are homogeneous of positive degree. Then A is an integral domain, and for $1 \leqslant i \leqslant r$ the ideal $(a_1, \ldots, a_i)$ is prime; hence $a_1, \ldots, a_r$ is an A-sequence.

(ii) In general $a_1, \ldots, a_r$ does not have to be an A-sequence, but P can in any case be generated by an A-sequence (E. Davis).

17 Cohen–Macaulay rings

Theorem 17.1 (Ischebeck). Let $(A, \mathfrak{m})$ be a Noetherian local ring, M and N non-zero finite A-modules, and suppose that $\operatorname{depth} M = k$, $\dim N = r$. Then

$$\operatorname{Ext}^i_A(N, M) = 0 \quad \text{for} \quad i < k - r.$$

Proof. By induction on r; if $r = 0$ then $\operatorname{Supp}(N) = \{\mathfrak{m}\}$ and the assertion holds by Theorem 16.6. Suppose $r > 0$. By Theorem 6.4, there exists a chain

$$N = N_0 \supset N_1 \supset \cdots \supset N_n = (0) \quad \text{with} \quad N_j/N_{j+1} \simeq A/P_j$$

of submodules N_j, where $P_j \in \operatorname{Spec} A$. It is easy to see that if $\operatorname{Ext}^i_A(N_j/N_{j+1}, M) = 0$ for each j then $\operatorname{Ext}^i_A(N, M) = 0$, and since $\dim N_j/N_{j+1} \leqslant \dim N = r$ it is enough to prove that $\operatorname{Ext}^i_A(N, M) = 0$ for $i < k - r$ in the case $N = A/P$ with $P \in \operatorname{Spec} A$ and $\dim N = r$. Since $r > 0$ we can take an element $x \in \mathfrak{m} - P$ and get the exact sequence

$$0 \to N \xrightarrow{x} N \longrightarrow N' \to 0,$$

where $N' = A/(P, x)$; then $\dim N' < r$ so that by induction we have $\operatorname{Ext}^i_A(N', M) = 0$ for $i < k - r + 1$. Thus for $i < k - r$ we have an exact sequence

$$0 \to \mathrm{Ext}^i_A(N,M) \xrightarrow{x} \mathrm{Ext}^i_A(N,M) \longrightarrow \mathrm{Ext}^{i+1}_A(N',M)=0.$$

We have $x \in \mathfrak{m}$ so that by NAK, $\mathrm{Ext}^i_A(N,M)=0$. ∎

Theorem 17.2. Let A be a Noetherian local ring, M a finite A-module, and assume that $P \in \mathrm{Ass}(M)$; then $\dim(A/P) \geqslant \operatorname{depth} M$.

Proof. If $P \in \mathrm{Ass}(M)$ then $\mathrm{Hom}_A(A/P, M) \neq 0$, so that by the previous theorem we cannot have $\dim A/P < \operatorname{depth} M$. ∎

Definition. Let $(A, \mathfrak{m}, k)$ be a Noetherian local ring, and M a finite A-module. We say that M is a *Cohen–Macaulay module* (abbreviated to *CM module*) if $M \neq 0$ and $\operatorname{depth} M = \dim M$, or if $M=0$. If A itself is a CM module we say that A is a *CM ring* or a *Macaulay ring.*

Theorem 17.3. Let A be a Noetherian local ring and M a finite A-module.

(i) If M is a CM module then for any $P \in \mathrm{Ass}(M)$ we have $\dim(A/P) = \dim M = \operatorname{depth} M$. Hence M has no embedded associated primes.

(ii) If $a_1, \ldots, a_r \in \mathfrak{m}$ is an M-sequence and we set $M' = M/(a_1, \ldots, a_r)$ then

$$M \text{ is a CM module} \Leftrightarrow M' \text{ is a CM module}$$

(iii) If M is a CM module then M_P is a CM module over A_P for every $P \in \operatorname{Spec} A$, and if $M_P \neq 0$ then

$$\operatorname{depth}(P, M) = \operatorname{depth}_{A_P} M_P.$$

Proof. (i) Quite generally, we have

$$\begin{aligned}\dim M &= \sup\{\dim A/P \mid P \in \mathrm{Ass}\, M\}\\ &\geqslant \inf\{\dim A/P \mid P \in \mathrm{Ass}\, M\} \geqslant \operatorname{depth} M,\end{aligned}$$

so that this is clear.

(ii) By definition $\operatorname{depth} M' = \operatorname{depth} M - r$, and by Ex. 16.1, $\dim M' = \dim M - r$, so that this is clear.

(iii) It is enough to consider the case $M_P \neq 0$, when $P \supset \operatorname{ann}(M)$. Then quite generally we have $\dim M_P \geqslant \operatorname{depth} M_P \geqslant \operatorname{depth}(P, M)$, so that we need only show that

$$\dim M_P = \operatorname{depth}(P, M).$$

We prove this by induction on $\operatorname{depth}(P, M)$. If $\operatorname{depth}(P, M) = 0$ then P is contained in an associated prime of M, but in view of $P \supset \operatorname{ann}(M)$ and the fact that by (i) all the associated primes of M are minimal, it follows that P is itself an associated prime of M; therefore $\dim M_P = 0$. If $\operatorname{depth}(P, M) > 0$ then we can take an M-regular element $a \in P$, and set $M' = M/aM$. Then

$$\operatorname{depth}(P, M') = \operatorname{depth}(P, M) - 1,$$

and M' is a CM module with $M'_P \neq 0$, so that by induction $\dim M'_P = \operatorname{depth}(P, M')$. However, a is M_P-regular as an element of A_P, and $M'_P = M_P/aM_P$, so that using Ex. 16.1 once more, we have $\dim M'_P = \dim M_P - 1$. Putting these together gives $\operatorname{depth}(P, M) = \dim M_P$. ∎

Theorem 17.4. Let $(A, \mathfrak{m})$ be a CM local ring.

(i) For a proper ideal I of A we have

$$\operatorname{ht} I = \operatorname{depth}(I, A) = \operatorname{grade} I, \quad \text{and} \quad \operatorname{ht} I + \dim A/I = \dim A.$$

(ii) A is catenary.

(iii) For any sequence $a_1, \ldots, a_r \in \mathfrak{m}$ the following four conditions are equivalent:

(1) $a_1, \ldots, a_r$ is an A-sequence;

(2) $\operatorname{ht}(a_1, \ldots, a_i) = i$ for $1 \leqslant i \leqslant r$;

(3) $\operatorname{ht}(a_1, \ldots, a_r) = r$;

(4) $a_1, \ldots, a_r$ is part of a system of parameters of A.

Proof. (iii) The implication (1)⇒(2) follows from Theorem 13.5, together with the fact that from the definition of A-sequence we have $0 < \operatorname{ht}(a_1) < \operatorname{ht}(a_1, a_2) < \cdots$.

(2)⇒(3) is trivial.

(3)⇒(4) If $\dim A = r$ this is obvious; if $\dim A > r$ then $\mathfrak{m}$ is not a minimal prime divisor of $(a_1, \ldots, a_r)$, so that we can choose $a_{r+1} \in \mathfrak{m}$ not contained in any minimal prime divisor of $(a_1, \ldots, a_r)$, and then $\operatorname{ht}(a_1, \ldots, a_{r+1}) = r + 1$. Proceeding in the same way we arrive at a system of parameters of A. (Up to now we have not used the CM assumption.)

(4)⇒(1) It is enough to show that any system of parameters $x_1, \ldots, x_n$ (with $n = \dim A$) is an A-sequence. If $P \in \operatorname{Ass}(A)$ then by Theorem 3, (i), $\dim A/P = n$, so that $x_1 \notin P$. Thus x_1 is A-regular. Therefore if we set $A' = A/x_1A$ we have by the previous theorem that A' is an $(n-1)$-dimensional CM ring, and the images of $x_2, \ldots, x_n$ form a system of parameters of A'. Thus by induction on n we see that $x_1, \ldots, x_n$ is an A-sequence.

(i) If $\operatorname{ht} I = r$ then we can take $a_1, \ldots, a_r \in I$ such that $\operatorname{ht}(a_1, \ldots, a_i) = i$ for $1 \leqslant i \leqslant r$. Thus by (iii), $a_1, \ldots, a_r$ is an A-sequence. Thus $r \leqslant \operatorname{grade} I$. Conversely if $b_1, \ldots, b_s \in I$ is an A-sequence then $\operatorname{ht}(b_1, \ldots, b_s) = s \leqslant \operatorname{ht} I$, and hence $r \geqslant \operatorname{grade} I$, so that equality must hold. For the second equality, letting S be the set of minimal prime divisors of I, we have

$$\operatorname{ht} I = \inf\{\operatorname{ht} P \mid P \in S\}$$
$$\text{and } \dim(A/I) = \sup\{\dim A/P \mid P \in S\},$$

and so it is enough to show that $\operatorname{ht} P = \dim A - \dim A/P$ for every $P \in S$. Set $\operatorname{ht} P = \dim A_P = r$ and $\dim A = n$. By Theorem 3, (iii), A_P is a CM ring and $r = \operatorname{depth}(P, A)$. Now if we take an A-sequence $a_1, \ldots, a_r \in P$ then by Theorem 3, (ii), $A/(a_1,\dots,a_r)$ is an $(n-r)$-dimensional CM ring, and from the fact that $\operatorname{ht}(a_1,\dots,a_r)=r=\operatorname{ht}P$ we see that P is a minimal prime divisor of $(a_1,\dots,a_r)$; thus by Theorem 3, (i), $\dim A/P=\dim A/(a_1,\dots,a_r)=n-r$.

(ii) Let $P\supset Q$ be prime ideals of A. Then since A_P is a CM ring, (i) above gives $\dim A_P=\operatorname{ht}QA_P+\dim A_P/QA_P$; in other words $\operatorname{ht}P-\operatorname{ht}Q=\operatorname{ht}(P/Q)$. ∎

If one system of parameters of a Noetherian local ring A is an A-sequence then depth $A=\dim A$, so that A is a CM ring, and therefore, by the above theorem, every system of parameters of A is an A-sequence.

Theorem 17.5. Let A be a Noetherian local ring and $\hat{A}$ its completion; then

(i) depth $A=$ depth $\hat{A}$;

(ii) A is CM $\Leftrightarrow \hat{A}$ is CM.

Proof. (i) This comes for example from the fact that $\operatorname{Ext}^i_A(A/\mathfrak{m},A)\otimes\hat{A}=\operatorname{Ext}^i_{\hat{A}}(\hat{A}/\mathfrak{m}\hat{A},\hat{A})$ for all i. (ii) follows from (i) and the fact that $\dim A=\dim\hat{A}$.

Definition. A proper ideal I in a Noetherian ring A is said to be *unmixed* if the heights of its prime divisors are all equal. We say that the *unmixedness theorem* holds for A if for every $r\geqslant 0$, every height r ideal I of A generated by r elements is unmixed. This includes as the case $r=0$ the statement that (0) is unmixed. By Theorem 13.5, if I is an ideal satisfying the hypotheses of this proposition, then all the minimal prime divisors of I have height r, so that to say that I is unmixed is to say that I does not have embedded prime divisors.

A Noetherian ring A is said to be a CM ring if $A_\mathfrak{m}$ is a CM local ring for every maximal ideal $\mathfrak{m}$ of A. By Theorem 3, (iii), a localisation $S^{-1}A$ of a CM ring A is again CM.

Theorem 17.6. A necessary and sufficient condition for a Noetherian ring A to be a CM ring is that the unmixedness theorem holds for A.

Proof. First suppose that A is a CM ring and that $I=(a_1,\dots,a_r)$ is an ideal of A with $\operatorname{ht}A=r$. We assume that P is an embedded prime divisor of I and derive a contradiction. Localising at P we can assume that A is a CM local ring; then by Theorem 4, (iii), $a_1,\dots,a_r$ is an A-sequence, and hence A/I is also a CM local ring. But then I does not have embedded prime divisors, and this is a contradiction. Next we suppose that the unmixedness theorem holds for A. If $P\in\operatorname{Spec}A$ with $\operatorname{ht}P=r$ then we can choose $a_1,\dots,a_r\in P$ such that

$$\operatorname{ht}(a_1,\dots,a_i)=i\quad\text{for}\quad 1\leqslant i\leqslant r.$$

Then by the unmixedness theorem, all the prime divisors of $(a_1,\ldots,a_i)$ have height i, and therefore do not contain a_{i+1}. Hence a_{i+1} is an $A/(a_1,\ldots,a_i)$-regular element; in other words, $a_1,\ldots,a_r$ is an A-sequence. Therefore $\operatorname{depth} A_P = r = \dim A_P$, so that A_P is a CM local ring; P was any element of $\operatorname{Spec} A$, so that A is a CM ring. ■

The unmixedness theorem for polynomial rings over a field was a brilliant early result of Macaulay in 1916; for regular local rings, the unmixedness theorem was proved by I. S. Cohen [1] in 1946. This explains the term Cohen–Macaulay. Having come this far, we are now in a position to give easy proofs of these two theorems.

Theorem 17.7. If A is a CM ring then so is $A[X_1,\ldots,X_n]$.
Proof. We need only consider the case $n=1$. Set $B=A[X]$ and let P be a maximal ideal of B. Set $P\cap A=\mathfrak{m}$; then B_P is also a localisation of $A_{\mathfrak{m}}[X]$, so that replacing A by $A_{\mathfrak{m}}$ we have a local CM ring A with maximal ideal $\mathfrak{m}$, and we need only prove that B_P is CM. Setting $A/\mathfrak{m}=k$ we get

$$B/\mathfrak{m}B = k[X],$$

so that $P/\mathfrak{m}B$ is a principal ideal of $k[X]$ generated by an irreducible monic polynomial $\varphi(X)$. If we let $f(X)\in A[X]$ be a monic polynomial of $A[X]$ which reduces to $\varphi(X)$ modulo $\mathfrak{m}B$ then $P=(\mathfrak{m},f)$. We choose a system of parameters $a_1,\ldots,a_n$ for A, so that $a_1,\ldots,a_n$, f is a system of parameters of B_P. Since B is flat over A the A-sequence $a_1,\ldots,a_n$ is also a B-sequence. We set $A/(a_1,\ldots,a_n)=A'$; then the image of f in $A'[X]$ is a monic polynomial, and therefore $A'[X]$-regular, so that $a_1,\ldots,a_n$, f is a B-sequence, and

$$\operatorname{depth} B_P \geqslant \operatorname{depth}(P,B) \geqslant n+1 = \dim B_P.$$

Therefore B_P is a CM ring. ■

Remark. If A is a CM local ring, then a similar (if anything, rather easier) method can be used to prove that $A[\![X]\!]$ is also CM. The statement also holds for a non-local CM ring, but the proof is a little more complicated, and we leave it to §23.

Theorem 17.8 A regular local ring is a CM ring.
Proof. Let $(A,\mathfrak{m})$ be an n-dimensional regular local ring, and $x_1,\ldots,x_n$ a regular system of parameters. By Theorems 14.2 and 14.3, (x_1), $(x_1,x_2),\ldots,$ $(x_1,\ldots,x_n)$ is a strictly increasing chain of prime ideals; therefore $x_1,\ldots,x_n$ is an A-sequence. ■

Theorem 17.9. Any quotient of a CM ring is universally catenary.
Proof. Clear from Theorems 7 and 4. ■

Theorem 17.10 A necessary and sufficient condition for a Noetherian local ring $(A, \mathfrak{m}, k)$ to be a regular ring is that $\mathrm{gr}_{\mathfrak{m}}(A)$ is isomorphic as a graded k-algebra to a polynomial ring over k.

Proof. If A is regular, let $x_1, \ldots, x_r$ be a regular system of parameters, that is a minimal basis of $\mathfrak{m}$; then $x_1,\ldots,x_r$ is an A-sequence, so that by Theorem 16.2 (see also Theorem 14.4 for another proof) $\mathrm{gr}_{\mathfrak{m}}(A) \simeq k[X_1,\ldots, X_r]$. Conversely, if $\mathrm{gr}_{\mathfrak{m}}(A) \simeq k[X_1,\ldots,X_r]$, then comparing the homogeneous components of degree 1, we see that $\mathfrak{m}/\mathfrak{m}^2 \simeq kX_1 + \cdots + kX_r$. On the other hand, the homogeneous component of degree n of $k[X_1,\ldots,X_r]$ is a vector space over k of dimension $\binom{n+r-1}{r-1}$, so that the Samuel function is

$$\chi_A(n) = l(A/\mathfrak{m}^{n+1}) = \sum_{i=0}^{n} \binom{i+r-1}{r-1} = \binom{n+r}{r},$$

and $\dim A = r$. Therefore A is regular. ■

We can also characterise CM local rings in terms of properties of multiplicities. Let A be a Noetherian local ring. An ideal of A is said to be a *parameter ideal* if it can be generated by a system of parameters. By Theorem 14.10, if $\mathfrak{q}$ is a parameter ideal then $l(A/\mathfrak{q}) \geqslant e(\mathfrak{q})$. As we are about to see, equality here is characteristic of CM rings.

Theorem 17.11. The following three conditions on a Noetherian local ring $(A, \mathfrak{m})$ are equivalent:

(1) A is a CM ring;

(2) $l(A/\mathfrak{q}) = e(\mathfrak{q})$ for any parameter ideal $\mathfrak{q}$ of A;

(3) $l(A/\mathfrak{q}) = e(\mathfrak{q})$ for some parameter ideal $\mathfrak{q}$ of A.

Proof. (1)⇒(2). If $x_1,\ldots,x_d$ is a system of parameters of A and $\mathfrak{q} = (x_1,\ldots,x_d)$ then by Theorem 16.2, $\mathrm{gr}_{\mathfrak{q}}(A) \simeq (A/\mathfrak{q})[X_1,\ldots,X_d]$, so that as in the proof of the previous theorem, $\chi_A^{\mathfrak{q}}(n) = l(A/\mathfrak{q})\cdot\binom{n+d}{d}$ so that $e(\mathfrak{q}) = l(A/\mathfrak{q})$.

(2)⇒(3) is obvious.

(3)⇒(1) Suppose that $\mathfrak{q} = (x_1,\ldots,x_d)$ is a parameter ideal satisfying $e(\mathfrak{q}) = l(A/\mathfrak{q})$. We set $B = (A/\mathfrak{q})[X_1,\ldots,X_d]$; then there is a homogeneous ideal $\mathfrak{b}$ of B such that $\mathrm{gr}_{\mathfrak{q}}(A) \simeq B/\mathfrak{b}$. We write $\varphi_B(n)$ and $\varphi_{\mathfrak{b}}(n)$ for the Hilbert polynomials of B and $\mathfrak{b}$ (see §13); then

$$\varphi_B(n) = l(A/\mathfrak{q})\binom{n+d-1}{d-1},$$

and for $n \gg 0$ we have $l(\mathfrak{q}^n/\mathfrak{q}^{n+1}) = \varphi_B(n) - \varphi_{\mathfrak{b}}(n)$. The left-hand side is a polynomial in n of degree $d-1$, and the coefficient of n^{d-1} is $e(\mathfrak{q})/(d-1)!$. By hypothesis $e(\mathfrak{q}) = l(A/\mathfrak{q})$, so that $\varphi_{\mathfrak{b}}(n)$ must be a polynomial in n of degree at most $d-2$. However, if $\mathfrak{b} \neq (0)$ then we can take a non-zero homogeneous element $f(X) \in \mathfrak{b}$. If $\mathfrak{m}^r \subset \mathfrak{q}$ and we set $\mathfrak{m}/\mathfrak{q} = \bar{\mathfrak{m}}$ then in B we have $\bar{\mathfrak{m}}^r = (0)$, and therefore replacing f by the product of f with a suitable element of $\bar{\mathfrak{m}}$, we can assume that $f \neq 0$ but $\bar{\mathfrak{m}} f = 0$. Then

$$\mathfrak{b} \supset fB \simeq (A/\mathfrak{m})[X_1, \ldots, X_d],$$

and therefore if $\deg f = p$ then $\varphi_{\mathfrak{b}}(n) \geqslant \binom{n-p+d-1}{d-1}$, the length of the homogeneous component of degree $n-p$ in $(A/\mathfrak{m})[X_1, \ldots, X_d]$. This contradicts $\deg \varphi_{\mathfrak{b}} < d-1$. Hence $\mathfrak{b} = (0)$, and

$$\mathrm{gr}_{\mathfrak{q}}(A) \simeq B = (A/\mathfrak{q})[X_1, \ldots, X_d],$$

so that by Theorem 16.3, $\{x_1, \ldots, x_d\}$ is an A-sequence. Therefore A is a CM ring. ■

Exercises to §17. Prove the following propositions.

17.1. (*a*) A zero-dimensional Noetherian ring is a CM ring.
(*b*) A one-dimensional ring is CM provided that it is reduced (= no nilpotent elements); also, construct an example of a one-dimensional ring which is not CM.

17.2. Let k be a field, x, y indeterminates over k, and set $A = k[x^3, x^2y, xy^2, y^3] \subset k[x, y]$ and $P = (x^3, x^2y, xy^2, y^3)A$. Is $R = A_P$ a CM ring? How about $k[x^4, x^3y, xy^3, y^4]$?

17.3. A two-dimensional normal ring is CM.

17.4. Let A be a CM ring, $a_1, \ldots, a_n$ an A-sequence, and set $J = (a_1, \ldots, a_n)$. Then for every integer ν the ring A/J^ν is CM, and therefore J^ν is unmixed.

17.5. Let A be a Noetherian local ring and $P \in \operatorname{Spec} A$. Then
(i) $\operatorname{depth} A \leqslant \operatorname{depth}(P, A) + \dim A/P$;
(ii) call $\dim A - \operatorname{depth} A$ the *codepth* of A. Then $\operatorname{codepth} A \geqslant \operatorname{codepth} A_P$.

17.6. Let A be a Noetherian ring, $P \in \operatorname{Spec} A$ and set $G = \mathrm{gr}_P(A)$. If G is an integral domain then $P^n = P^{(n)}$ for all $n > 0$. (This observation is due to Robbiano. One sees from it that if P is a prime ideal generated by an A-sequence then $P^n = P^{(n)}$.)

18 Gorenstein rings

Lemma 1. Let A be a ring, M an A-module, and $n \geqslant 0$ a given integer. Then

$$\operatorname{inj\,dim} M \leqslant n \Leftrightarrow \operatorname{Ext}_A^{n+1}(A/I, M) = 0 \text{ for all ideals } I.$$

If A is Noetherian, then we can replace 'for all ideals' by 'for all prime ideals' in the right-hand condition.

Proof. ($\Rightarrow$) This is clear on calculating the Ext by an injective resolution of M.

($\Leftarrow$) If $n=0$ then from the exact sequence $0\to I\longrightarrow A\longrightarrow A/I\to 0$ and from the fact that $\operatorname{Ext}^1_A(A/I,M)=0$ we get that $\operatorname{Hom}(A,M)\longrightarrow \operatorname{Hom}(I,M)\to 0$ is exact. Since this holds for every I, Theorem B3 of Appendix B implies that M is injective. Suppose then that $n>0$.

There exists an exact sequence

$$0\to M\longrightarrow Q^0\longrightarrow Q^1\longrightarrow\cdots\longrightarrow Q^{n-1}\longrightarrow C\to 0,$$

with each Q^i injective. (We can obtain this by taking an injective resolution of M up to Q^{n-1} and setting C for the cokernel of $Q^{n-2}\longrightarrow Q^{n-1}$.) One sees easily that $\operatorname{Ext}^{n+1}_A(A/I,M)\simeq\operatorname{Ext}^1_A(A/I,C)$, so that by the argument used in the $n=0$ case, C is injective, and so $\operatorname{inj\,dim} M\leqslant n$.

If A is Noetherian then by Theorem 6.4, any finite A-module N has a chain $N=N_0\supset N_1\supset\cdots\supset N_{r+1}=0$ of submodules such that $N_j/N_{j+1}\simeq A/P_j$ with $P_j\in\operatorname{Spec}A$. Using this, if $\operatorname{Ext}^i_A(A/P,M)=0$ for all prime ideals P then we also have $\operatorname{Ext}^i_A(N,M)=0$ for all finite A-modules N. Now we just have to apply this with $i=n+1$ and $N=A/I$. ∎

Lemma 2. Let A be a ring, M and N two A-modules, and $x\in A$; suppose that x is both A-regular and M-regular, and that $xN=0$. Set $B=A/xA$ and $\bar M=M/xM$. Then

(i) $\operatorname{Hom}_A(N,M)=0$, and $\operatorname{Ext}^{n+1}_A(N,M)\simeq\operatorname{Ext}^n_B(N,\bar M)$ for all $n\geqslant 0$;

(ii) $\operatorname{Ext}^n_A(M,N)\simeq\operatorname{Ext}^n_B(\bar M,N)$ for all $n\geqslant 0$;

(iii) $\operatorname{Tor}^A_n(M,N)\simeq\operatorname{Tor}^B_n(\bar M,N)$ for all $n\geqslant 0$.

Proof. (i) The first formula is obvious. For the second, set $T^n(N)=\operatorname{Ext}^{n+1}_A(N,M)$, and view T^n as a contravariant functor from the category of B-modules to that of Abelian groups. Then first of all, the exact sequence

$$0\to M\xrightarrow{\;x\;}M\longrightarrow\bar M\to 0$$

gives $T^0(N)=\operatorname{Hom}_A(N,\bar M)=\operatorname{Hom}_B(N,\bar M)$. Moreover, since x is A-regular we have $\operatorname{proj\,dim}_A B=1$, and therefore $T^n(B)=0$ for $n>0$, so that $T^n(L)=0$ for $n>0$ and every projective B-module L. Finally, for any short exact sequence $0\to N'\longrightarrow N\longrightarrow N''\to 0$ of B-modules, there is a long exact sequence

$$\begin{aligned}0\to T^0(N'')&\longrightarrow T^0(N)\longrightarrow T^0(N')\\ &\longrightarrow T^1(N'')\longrightarrow T^1(N)\longrightarrow T^1(N')\to\cdots.\end{aligned}$$

This proves that T^i is the derived functor of $\operatorname{Hom}_B(-,\bar M)$, and therefore coincides with $\operatorname{Ext}^i_B(-,M)$.

(ii) We first prove $\operatorname{Tor}^A_n(M,B)=0$ for $n>0$. For $n>1$ this follows from $\operatorname{proj\,dim}_A B = 1$. For $n = 1$, consider the long exact sequence $0 \to \operatorname{Tor}_1^A(M, B) \longrightarrow M \xrightarrow{x} M \longrightarrow \bar{M} \to 0$ associated with the short exact sequence $0 \to A \xrightarrow{x} A \longrightarrow B \to 0$. Since x is M-regular we have $\operatorname{Tor}_1^A(M, B) = 0$.

Now let $L_{.} \longrightarrow M \to 0$ be a free resolution of the A-module M. Then $L_{.} \otimes_A B \longrightarrow M \otimes_A B \to 0$ is exact by what we have just proved, so that $L_{.} \otimes B$ is a free resolution of the B-module $M \otimes B = \bar{M}$. Then $\operatorname{Ext}_A^n(M, N) = H^n(\operatorname{Hom}_A(L_{.}, N)) = H^n(\operatorname{Hom}_B(L_{.} \otimes_A B, N)) = \operatorname{Ext}_B^n(\bar{M}, N)$ by Formula 9 of Appendix A.

(iii) Using the same notation as above, we have $\operatorname{Tor}_n^A(M, N) = H_n(L_{.} \otimes_A N) = H_n((L_{.} \otimes_A B) \otimes_B N) = \operatorname{Tor}_n^B(\bar{M}, N)$. ■

Lemma 3. Let $(A, \mathfrak{m}, k)$ be a Noetherian local ring, M a finite A-module, and $P \in \operatorname{Spec} A$; suppose that $\operatorname{ht}(\mathfrak{m}/P) = 1$. Then

$$\operatorname{Ext}_A^{i+1}(k, M) = 0 \Rightarrow \operatorname{Ext}_A^i(A/P, M) = 0.$$

Proof. Choose $x \in \mathfrak{m} - P$; then $0 \to A/P \xrightarrow{x} A/P \longrightarrow A/(P + Ax) \to 0$ is an exact sequence, and $P + Ax$ is an $\mathfrak{m}$-primary ideal, so that if we let $N = A/(P + Ax)$, there exists a chain of submodules of N

$$N = N_0 \supset N_1 \supset \cdots \supset N_r = 0 \quad \text{such that} \quad N_i/N_{i+1} \simeq k.$$

Hence from $\operatorname{Ext}_A^{i+1}(k, M) = 0$ we get $\operatorname{Ext}_A^{i+1}(A/(P + Ax), M) = 0$, and

$$\operatorname{Ext}_A^i(A/P, M) \xrightarrow{x} \operatorname{Ext}_A^i(A/P, M) \to 0.$$

is exact, so that by NAK $\operatorname{Ext}_A^i(A/P, M) = 0$. ■

Lemma 4. Let $(A, \mathfrak{m}, k)$ be a Noetherian local ring, M a finite A-module, and $P \in \operatorname{Spec} A$; suppose that $\operatorname{ht}(\mathfrak{m}/P) = d$. Then

$$\operatorname{Ext}_A^{i+d}(k, M) = 0 \Rightarrow \operatorname{Ext}_{A_P}^i(\kappa(P), M_P) = 0.$$

Proof. Let $\mathfrak{m} = P_0 \supset P_1 \supset \cdots \supset P_d = P$, with $P_i \in \operatorname{Spec} A$ and $\operatorname{ht}(P_i/P_{i+1}) = 1$. Then by Lemma 3,

$$\operatorname{Ext}_A^{i+d-1}(A/P_1, M) = 0,$$

and localising at P_1 we get

$$\operatorname{Ext}_{A_{P_1}}^{i+d-1}(\kappa(P_1), M_{P_1}) = 0.$$

Proceeding in the same way gives the result. ■

Theorem 18.1. Let $(A, \mathfrak{m}, k)$ be an n-dimensional Noetherian local ring. Then the following conditions are equivalent:

(1) $\operatorname{inj\,dim} A < \infty$;

(1′) $\operatorname{inj\,dim} A = n$;

(2) $\operatorname{Ext}_A^i(k, A) = 0$ for $i \neq n$ and $\simeq k$ for $i = n$;

(3) $\operatorname{Ext}^i_A(k, A) = 0$ for some $i > n$;
(4) $\operatorname{Ext}^i_A(k, A) = 0$ for $i < n$ and $\simeq k$ for $i = n$;
(4′) A is a CM ring and $\operatorname{Ext}^n_A(k, A) \simeq k$;
(5) A is a CM ring, and every parameter ideal of A is irreducible;
(5′) A is a CM ring and there exists an irreducible parameter ideal.

Recall that an ideal I is irreducible if $I = J \cap J'$ implies either $I = J$ or $I = J'$ (see §6).

Definition. A Noetherian local ring for which the above equivalent conditions hold is said to be *Gorenstein*.

Proof of (1)⇒(1′). Set $\operatorname{inj\,dim} A = r$. If P is a minimal prime ideal of A such that $\operatorname{ht}(\mathfrak{m}/P) = \dim A = n$ then $PA_P \in \operatorname{Ass}(A_P)$, so that $\operatorname{Hom}(\kappa(P), A_P) \neq 0$; hence, by Lemma 4, $\operatorname{Ext}^n_A(k, A) \neq 0$, therefore $r \geqslant n$. If $r = 0$ this means that $n = 0$, and we are done. If $r > 0$, set $\operatorname{Ext}^r_A(-, A) = T$; then this is a right-exact contravariant functor, and by Lemma 1, there is a prime ideal P such that $T(A/P) \neq 0$. Now if $P \neq \mathfrak{m}$ and we take $x \in \mathfrak{m} - P$, the exact sequence

$$0 \to A/P \xrightarrow{x} A/P$$

leads to an exact sequence

$$T(A/P) \xrightarrow{x} T(A/P) \to 0;$$

but then by NAK, $T(A/P) = 0$, which is a contradiction. Thus $P = \mathfrak{m}$, and so $T(k) \neq 0$. We have $\mathfrak{m} \neq \operatorname{Ass}(A)$, since otherwise there would exist an exact sequence $0 \to k \longrightarrow A$, and hence an exact sequence

$$T(A) = \operatorname{Ext}^r_A(A, A) = 0 \longrightarrow T(k) \to 0,$$

which is a contradiction. Hence $\mathfrak{m}$ contains an A-regular element x. If we set $B = A/xA$ then by Lemma 2, $\operatorname{Ext}^i_B(N, B) = \operatorname{Ext}^{i+1}_A(N, A)$ for every B-module N, so that $\operatorname{inj\,dim} B = r - 1$. By induction on r we have $r - 1 = \dim B = n - 1$, and hence $r = n$.

Proof of (1′)⇒(2). When $n = 0$ we have $\mathfrak{m} \in \operatorname{Ass}(A)$, so there exists an exact sequence $0 \to k \longrightarrow A$, and since $\operatorname{inj\,dim} A = 0$,

$$A = \operatorname{Hom}(A, A) \longrightarrow \operatorname{Hom}(k, A) \to 0$$

is exact. Therefore $\operatorname{Hom}(k, A)$ is generated by one element. But $\operatorname{Hom}(k, A) \neq 0$, so that we must have $\operatorname{Hom}(k, A) \simeq k$. By assumption, A is an injective module, so that $\operatorname{Ext}^i_A(k, A) = 0$ for $i > 0$; thus we are done in the case $n = 0$. If $n > 0$ then, as we have seen above, $\mathfrak{m}$ contains an A-regular element x, and if we set $B = A/xA$ then $\dim B = \operatorname{inj\,dim} B = n - 1$, so that by Lemma 2 and induction on n we have

$$\operatorname{Ext}^i_A(k, A) = \operatorname{Ext}^{i-1}_B(k, B) = \begin{cases} 0 & \text{if } 0 < i \neq n \\ k & \text{if } i = n, \end{cases}$$

and

$$\operatorname{Hom}_A(k, A) = 0.$$

(2)⇒(3) is trivial.

Proof of (3)⇒(1). We use induction on n. Assume that for some $i > n$ we have $\mathrm{Ext}^i_A(k, A) = 0$. If $n = 0$ then $\mathfrak{m}$ is the unique prime ideal of A, so that by Lemma 1, $\mathrm{inj\,dim}\, A < i < \infty$. If $n > 0$ let P be a prime ideal distinct from $\mathfrak{m}$ and set $d = \mathrm{ht}(\mathfrak{m}/P)$ and $B = A_P$; then by Lemma 4 we have $\mathrm{Ext}^{i-d}_B(\kappa(P), B) = 0$. Moreover, $\dim B \leqslant n - d < i - d$, so that by induction $\mathrm{inj\,dim}\, B < \infty$. Thus for any finite A-module M we have

$$(\mathrm{Ext}^i_A(M, A))_P = \mathrm{Ext}^i_B(M_P, B) = 0$$

(since $i > n > \dim B = \mathrm{inj\,dim}\, B$). Therefore, setting $T(M) = \mathrm{Ext}^i_A(M, A)$ we get $\mathrm{Supp}(T(M)) \subset \{\mathfrak{m}\}$, and since $T(M)$ is a finite A-module, $l(T(M)) < \infty$. Using this, we now prove that $T(A/P) = 0$ for every prime ideal P. If $T(A/P) \neq 0$ for some P, choose a maximal P with this property. By assumption $T(k) = 0$, so that $P \neq \mathfrak{m}$, so that we can take $x \in \mathfrak{m} - P$ and form the exact sequence

$$0 \to A/P \xrightarrow{x} A/P \longrightarrow A/(P + Ax) \to 0.$$

Then write $A/(P + Ax) = M_0 \supset M_1 \supset \cdots \supset M_s = 0$ with $M_i/M_{i+1} \simeq A/P_i$; each P_i is strictly bigger than P, so that $T(A/(P + Ax)) = 0$. Therefore

$$0 \to T(A/P) \xrightarrow{x} T(A/P)$$

is exact, so that multiplication by x in $T(A/P)$ is injective; but since $l(T(A/P)) < \infty$, injective implies surjective. Hence by NAK, $T(A/P) = 0$, which is contradiction. Therefore $T(A/P) = 0$ for every $P \in \mathrm{Spec}\, A$, so that by Lemma 1, $\mathrm{inj\,dim}\, A < i$.

So far we have proved that (1), (1′), (2) and (3) are equivalent. Now we prove the equivalence of (2), (4), (4′), (5) and (5′).

(2)⇒(4) is obvious. (4)⇔(4′) comes at once from the fact that A is CM if and only if $\mathrm{Ext}^i_A(k, A) = 0$ for all $i < n$ (the implication (2)⇔(3) of Theorem 16.6).

Proof of (4′)⇒(5). A system of parameters $x_1, \ldots, x_n$ in a CM ring A is an A-sequence, so that setting $B = A/\sum_1^n x_i A$, we have

$$\mathrm{Hom}_B(k, B) \simeq \mathrm{Ext}^n_A(k, A) \simeq k.$$

Now B is an Artinian ring, and any minimal non-zero ideal of B is isomorphic to k, so that the above formula says that B has just one such minimal ideal, say I_0. If I_1 and I_2 are any non-zero ideals of B then both of them must contain I_0, so that $I_1 \cap I_2 \neq (0)$. Lifting this up to A, this means that $(x_1, \ldots, x_n)$ is an irreducible ideal.

(5)⇒(5′) is obvious.

Proof of (5′)⇒(2). If A is CM we already have $\mathrm{Ext}^i_A(k, A) = 0$ for $i < n$. If $\mathfrak{q}$ is an irreducible parameter ideal and we set $B = A/\mathfrak{q}$ then, in the same way as above,

$$\mathrm{Ext}^{n+i}_A(k, A) \simeq \mathrm{Ext}^i_B(k, B),$$

so that it is enough to prove that in an Artinian ring B, (0) is irreducible implies that

$$\operatorname{Hom}_B(k, B) \simeq k \quad \text{and} \quad \operatorname{Ext}^i_B(k, B) = 0 \quad \text{for} \quad i > 0.$$

The statement for Hom is easy: first of all, B is Artinian, so that $\operatorname{Hom}_B(k, B) \neq 0$; for non-zero $f, g \in \operatorname{Hom}_B(k, B)$ we must have $f(k) = g(k)$, since otherwise $f(k) \cap g(k) = (0)$, which contradicts the irreducibility of (0). Hence $f(1) = g(\alpha)$ for some $\alpha \in k$, and $f = \alpha g$, so that $\operatorname{Hom}_B(k, B) \simeq k$.

Now consider the $\operatorname{Ext}^i_B(k, B)$. Choose a chain of ideals $(0) = N_0 \subset N_1 \subset \cdots \subset N_r = B$ such that $N_i/N_{i-1} \simeq k$, and consider the exact sequences

$$0 \to N_1 \longrightarrow N_2 \longrightarrow k \to 0$$
$$0 \to N_2 \longrightarrow N_3 \longrightarrow k \to 0$$
$$\vdots$$
$$0 \to N_{r-1} \longrightarrow B \longrightarrow k \to 0.$$

From the long exact sequence

$$0 \to \operatorname{Hom}_B(k, B) \longrightarrow \operatorname{Hom}_B(N_{i+1}, B) \longrightarrow \operatorname{Hom}_B(N_i, B) \xrightarrow{\delta_i} \operatorname{Ext}^1_B(k, B) \longrightarrow \cdots$$

and an easy induction (using $N_1 \simeq k$ and $\operatorname{Hom}_B(k, B) \simeq k$), we get that $l(\operatorname{Hom}_B(N_i, B)) \leqslant i$, with equality holding if and only if $\delta_1, \ldots, \delta_{i-1}$ are all zero. However,

$$l(\operatorname{Hom}_B(N_r, B)) = l(\operatorname{Hom}_B(B, B)) = l(B) = r,$$

so that we must have $\delta_1 = \cdots = \delta_{r-1} = 0$. Then from

$$0 \to N_{r-1} \longrightarrow B \longrightarrow k \to 0$$

we get the exact sequence

$$0 \to \operatorname{Ext}^1_B(k, B) \longrightarrow \operatorname{Ext}^1_B(B, B) = 0,$$

and therefore $\operatorname{Ext}^1_B(k, B) = 0$. Now from Lemma 1, B is an injective B-module, so that $\operatorname{Ext}^i_B(k, B) = 0$ for all $i > 0$. ∎

Lemma 5. Let A be a Noetherian ring, $S \subset A$ a multiplicative set, and I an injective A-module; then I_S is an injective A_S-module.

Proof. Every ideal of A_S is the localisation $\mathfrak{a}_S$ of an ideal $\mathfrak{a}$ of A. From $0 \to \mathfrak{a} \longrightarrow A$ we get the exact sequence $\operatorname{Hom}_A(A, I) \longrightarrow \operatorname{Hom}_A(\mathfrak{a}, I) \to 0$, and, since $\mathfrak{a}$ is finitely generated

$$\operatorname{Hom}_{A_S}(A_S, I_S) \longrightarrow \operatorname{Hom}_{A_S}(\mathfrak{a}_S, I_S) \to 0$$

is exact. This proves that I_S is an injective A_S-module.

Theorem 18.2. If A is a Gorenstein local ring and $P \in \operatorname{Spec} A$ then A_P is also Gorenstein.

Proof. If $0 \to A \longrightarrow I^0 \longrightarrow I^1 \longrightarrow \cdots \longrightarrow I^n \to 0$ is an injective resolution of A then

$$0 \to A_P \longrightarrow (I^0)_P \longrightarrow \cdots \longrightarrow (I^n)_P \to 0$$

is an injective resolution of A_P, so that inj dim $A_P < \infty$. ∎

Definition. A Noetherian ring A is *Gorenstein* if its localisation at every maximal ideal is a Gorenstein local ring. (By the previous theorem, it then follows that A_P is Gorenstein for every $P \in \operatorname{Spec} A$.)

Theorem 18.3. Let A be a Noetherian local ring and $\hat{A}$ its completion. Then A is Gorenstein $\Leftrightarrow \hat{A}$ is Gorenstein.

Proof. We have $\dim A = \dim \hat{A}$, and since $\hat{A}$ is faithfully flat over A, $\operatorname{Ext}^i_A(k, A) \otimes_A \hat{A} = \operatorname{Ext}^i_{\hat{A}}(k, \hat{A})$, so that we only need to use condition (3) of Theorem 1. ∎

Closely related to the theory of Gorenstein rings is Matlis' theory of injective modules over Noetherian rings. We now discuss the main results of Matlis [1].

Let A be a Noetherian ring, and E an injective A-module. If E is a submodule of an A-module M then since we can extend the identity map $E \longrightarrow E$ to a linear map $f: M \longrightarrow E$, we have $M = E \oplus F$ (with $F = \operatorname{Ker} f$). Say that an A-module N is *indecomposable* if N cannot be written as a direct sum of two submodules. We write $E_A(N)$ or $E(N)$ for the injective hull of an A-module E (see Appendix B).

Theorem 18.4. Let A be a Noetherian ring and $P, Q \in \operatorname{Spec} A$.

(i) $E(A/P)$ is indecomposable.

(ii) Any indecomposable injective A-module is of the form $E(A/P)$ for some $P \in \operatorname{Spec} A$.

(iii) If $x \in A - P$, multiplication by x induces an automorphism of $E(A/P)$.

(iv) $P \neq Q \Rightarrow E(A/P) \not\cong E(A/Q)$.

(v) For any $\xi \in E(A/P)$ there exists a positive integer ν (depending on ξ) such that $P^\nu \xi = 0$.

(vi) If $Q \subset P$ then $E(A/Q)$ is an A_P-module, and is an injective hull of $(A/Q)_P = A_P/QA_P$, that is

$$E_A(A/Q) = E_{A_P}(A_P/QA_P).$$

Proof. (i) If I_1 and I_2 are non-zero ideals of A/P then $0 \neq I_1 I_2 \subset I_1 \cap I_2$. Now $E(A/P)$ is an essential extension of A/P (see Appendix B), so that for any two non-zero submodules N_1, N_2 of $E(A/P)$ we have $N_i \cap (A/P) \neq 0$, so that

$$N_1 \cap N_2 \supset (N_1 \cap A/P) \cap (N_2 \cap A/P) \neq 0.$$

(ii) Let $N \neq 0$ be an indecomposable injective A-module and choose $P \in \operatorname{Ass}(N)$. Then A/P can be embedded into N, and so $E(A/P)$ can also; but an injective submodule is always a direct summand, and since N is indecomposable, $N = E(A/P)$.

(iii) Write φ for multiplication by x in $E(A/P)$; then $\mathrm{Ker}(\varphi)\cap(A/P)=0$, so that $\mathrm{Ker}(\varphi)=0$, and $\mathrm{Im}\,\varphi$ is isomorphic to $E(A/P)$. Hence $\mathrm{Im}\,\varphi$ is injective, and is therefore a direct summand of $E(A/P)$, so that by (i), $\mathrm{Im}\,\varphi = E(A/P)$.

(iv) If $P \not\subset Q$ and $x \in P - Q$ then multiplication by x is injective in $E(A/Q)$ but not in $E(A/P)$.

(v) By the proof of (ii) together with (iv), $\mathrm{Ass}\,(E(A/P)) = \{P\}$, so that the submodule $A\xi \simeq A/\mathrm{ann}(\xi)$ also has $\mathrm{Ass}\,(A\xi) = \{P\}$. Hence $\mathrm{ann}\,(\xi)$ is a P-primary ideal.

(vi) By (iii), we can view $E(A/Q)$ as an A_P-module; hence it contains $(A/Q)_P$. Since $E(A/Q)$ is an essential extension of A/Q and $A/Q \subset (A/Q)_P \subset E(A/Q)$, it is also an essential extension of $(A/Q)_P$. For A_P-modules M and N, any A-linear map $M \longrightarrow N$ is also A_P-linear, and of course conversely, so that for an A_P-module, being injective as an A_P-module is the same as being injective as an A-module. Thus $E(A/Q)$ is an injective hull of the A_P-module $(A/Q)_P$. ■

Example 1. If A is an integral domain and K its field of fractions, $K = E(A)$ (prove this !).

Example 2. If A is a DVR with uniformising element x and field of fractions K, and $k = A/xA$, then $E(k) = K/A$. Indeed, if I is a non-zero ideal of A we can write $I = x^r A$, and if $f: I \longrightarrow K/A$ is a given map, let $f(x^r) = \alpha \bmod A$ for some $\alpha \in K$; then f can be extended to a map $f: A \longrightarrow K/A$ by setting $f(1) = (\alpha/x^r) \bmod A$. Therefore K/A is injective. We have $(x^{-1}A)/A \simeq A/xA = k$, and it is easy to see that K/A is an essential extension of $x^{-1}A/A$. Thus K/A can be thought of as $E(k)$.

Theorem 18.5. We consider modules over a Noetherian ring A.

(i) A direct sum of any number of injective modules is injective.

(ii) Every injective module is a direct sum of indecomposable injective modules.

(iii) The direct sum decomposition in (ii) is unique, in the sense that if

$$M = \bigoplus M_i \text{ (with indecomposable } M_i\text{)}$$

then for any $P \in \mathrm{Spec}\,A$, the sum $M(P)$ of all the M_i isomorphic to $E(A/P)$ depends only on M and P, and not on the decomposition $M = \bigoplus M_i$. Moreover, the number of M_i isomorphic to $E(A/P)$ is equal to

$$\dim_{\kappa(P)} \mathrm{Hom}_{A_P}(\kappa(P), M_P), \text{ (where } \kappa(P) = A_P/PA_P\text{)},$$

so that this also is independent of the decomposition.

Proof. (i) Let M_λ for $\lambda \in \Lambda$ be injective modules. It is enough to prove that for an ideal I of A, any linear map $\varphi: I \longrightarrow \bigoplus M_\lambda$ can be extended to a linear map from the whole of A. Since I is finitely generated, $\varphi(I)$ is contained in a direct sum of a finite number of the M_λ. If $\varphi(I)\subset M_1\oplus\cdots\oplus M_n$ and we write $\varphi_i(a)$ for the component of $\varphi(a)$ in M_i, then $\varphi_i: I\longrightarrow M_i$ extends to $\psi_i: A\longrightarrow M_i$. Defining $\psi: A\longrightarrow\bigoplus_\lambda M_\lambda$ by $\psi(1)=\psi_1(1)+\cdots+\psi_n(1)$ extends φ to A.

(ii) Say that a family $\mathscr{F}=\{E_\lambda\}$ of indecomposable injective submodules of M is *free* if the sum in M of the E_λ is direct, that is if, for any finite number $E_{\lambda_1},\dots,E_{\lambda_n}$ of them,

$$E_{\lambda_1}\cap(E_{\lambda_2}+\cdots+E_{\lambda_n})=0.$$

Let $\mathfrak{M}$ be the set of all free families $\mathscr{F}$, ordered by inclusion. Then by Zorn's lemma $\mathfrak{M}$ has a maximal element, say $\mathscr{F}_0$. Write $N=\sum_{E\in\mathscr{F}_0}E$; then by (i), N is injective, hence a direct summand of M, and $M=N\oplus N'$. If $N'\neq 0$ then since it is a direct summand of M it must be injective, and for $P\in\operatorname{Ass}(N')$, the proof of Theorem 4, (ii), shows that N' contains a direct summand E' isomorphic to $E(A/P)$. Thus $\mathscr{F}_0\cup\{E'\}$ is a free family, contradicting the maximality of $\mathscr{F}_0$. Hence $N'=0$ and $M=N$.

(iii) If we can show that $M(P)$ has the property that every submodule E of M isomorphic to $E(A/P)$ is contained in $M(P)$, then $M(P)$ is the submodule of M generated by all such E, and therefore is determined by M and P only. To prove this, take any $\xi\in E$; we can write $\xi=\xi_1+\cdots+\xi_r$ with $\xi_i\in M(P_i)$, where $P_1,\dots,P_r$ are distinct prime ideals and $P=P_1$. Setting $\xi_1-\xi=\eta_1$ and $\xi_i=\eta_i$ for $2\leqslant i\leqslant r$ we have $\eta_1+\cdots+\eta_r=0$, with $\eta_1\in M(P_1)+E$ and $\eta_i\in M(P_i)$ for $i\geqslant 2$. We need only prove that in this case each $\eta_i=0$. Suppose that P_r is minimal among $P_1,\dots,P_r$; then for any m we have $(P_1\dots P_{r-1})^m\not\subset P_r$, so that taking $a\in(P_1\dots P_{r-1})^m-P_r$ and m large enough, we get $a\eta_1=\cdots=a\eta_{r-1}=0$. Then also $a\eta_r=0$, but multiplication by a is an automorphism of $M(P_r)$, so that $\eta_r=0$. By induction on r we get $\eta_i=0$ for all i.

We now prove that if $M(P)=M_1\oplus\cdots\oplus M_s$ with $M_i\simeq E(A/P)$ then

$$s=\dim_{\kappa(P)}\operatorname{Hom}_{A_P}(\kappa(P),M_P).$$

(We are writing this as if s were finite, but, as one can see from the proof below, the same works for any cardinal number.) By Theorem 4, (vi), both sides of $M(P)=M_1\oplus\cdots\oplus M_s$ are A_P-modules, and $M_i\simeq E(\kappa(P))$. Moreover, by Theorem 4, (v), $E(A/Q)_P=0$ if $Q\not\subset P$, so that

$$M_P=\bigoplus_{Q\subset P}M(Q)_P=\bigoplus_{Q\subset P}M(Q).$$

Hence we can replace A by A_P, and assume that A is a local ring with P its maximal ideal; set $k=\kappa(P)$. If $Q\neq P$ then any $x\in P-Q$ gives an automorphism of $M(Q)$, but $x\cdot k=0$, so that $\operatorname{Hom}_A(k,M(Q))=0$. Hence $\operatorname{Hom}_A(k,M)=\operatorname{Hom}_A(k,M(P))$, so that there is no loss of generality in assuming that $M = M(P)$. For any A-module N, we can identify $\mathrm{Hom}_A(k, N)$ with the submodule $\{\xi \in N \mid P\xi = 0\}$, but since $E(k)$ is an essential extension of k we must have $\dim_k \mathrm{Hom}_A(k, E(k)) = 1$, so that if $M = M_1 \oplus \cdots \oplus M_s$ with $M_i \simeq E(k)$ then $s = \dim_k \mathrm{Hom}_A(k, M)$. ∎

Theorem 18.6. Let $(A, \mathfrak{m}, k)$ be a Noetherian local ring, and $E = E_A(k)$ the injective hull of k. For each A-module M set $M' = \mathrm{Hom}_A(M, E)$.

(i) If M is an A-module and $0 \neq x \in M$, then there exists $\varphi \in M'$ such that $\varphi(x) \neq 0$. In other words the canonical map $\theta: M \longrightarrow M''$ defined by $\theta(x)(\varphi) = \varphi(x)$ for $x \in M$ and $\varphi \in M'$ is injective.

(ii) If M is an A-module of finite length, then $l(M) = l(M')$ and the canonical map $M \longrightarrow M''$ is an isomorphism.

(iii) Let $\hat{A}$ be the completion of A; then E is also an $\hat{A}$-module, and is an injective hull of k as $\hat{A}$-module.

(iv) $\mathrm{Hom}_A(E, E) = \mathrm{Hom}_{\hat{A}}(E, E) = \hat{A}$. In other words, each endomorphism of the A-module E is multiplication by a unique element of $\hat{A}$.

(v) E is Artinian as an A-module and also as an $\hat{A}$-module. Assume now that A is complete, and write $\mathscr{N}$ (resp. $\mathscr{A}$) for the category of Noetherian (respectively Artinian) A-modules. Then if $M \in \mathscr{N}$ we have $M' \in \mathscr{A}$ and $M \simeq M''$; if $M \in \mathscr{A}$ we have $M' \in \mathscr{N}$ and $M \simeq M''$.

Proof. (i) Let $f: Ax \longrightarrow E$ be the composite of the canonical maps $Ax \simeq A/\mathrm{ann}(x)$, $A/\mathrm{ann}(x) \longrightarrow A/\mathfrak{m} = k$ and $k \longrightarrow E$. Then $f(x) \neq 0$. Since E is injective we can extend f to $\varphi: M \longrightarrow E$.

(ii) If $l(M) = n < \infty$ then M has a submodule M_1 of length $n - 1$, and $0 \to M_1 \longrightarrow M \longrightarrow k \to 0$ is exact, so that $0 \to k' \longrightarrow M' \longrightarrow M_1' \to 0$ is exact. However

$$k' = \mathrm{Hom}(k, E) = \mathrm{Hom}(k, k) \simeq k,$$

so that by induction on n we get $l(M) = n = l(M')$. The canonical map $M \longrightarrow M''$ is injective by (*i*), and $l(M) = l(M') = l(M'')$, hence it must be an isomorphism.

(iii) Each element of E is annihilated by some power of $\mathfrak{m}$, so that the canonical map $E \longrightarrow E \otimes_A \hat{A}$ is surjective. However, since $\hat{A}$ is faithfully flat over A it is also injective, so that $E \simeq E \otimes_A \hat{A}$, and we can view E as an $\hat{A}$-module. Let F be the injective hull of E as an $\hat{A}$-module. Then F is also the $\hat{A}$-injective hull of k, so that every element of F is annihilated by some power of $\mathfrak{m}\hat{A}$. As an A-module F splits into a direct sum of E and an A-module C. If $x \in C$, and if $\mathfrak{m}^r Ax = 0$, then for each $a^* \in \hat{A}$ we can find $a \in A$ such that $a^* \equiv a \mod \mathfrak{m}^r \hat{A}$ and hence $a^* x = ax \in C$. Therefore C is an $\hat{A}$-module. But F is indecomposable as an $\hat{A}$-module. Hence $C = 0$ and $E = F$.

(iv) For $\nu > 0$ set $E_\nu = \{x \in E \mid \mathfrak{m}^\nu x = 0\}$. Then we have $(A/\mathfrak{m}^\nu)' = \mathrm{Hom}_A(A/\mathfrak{m}^\nu, E) \simeq E_\nu$, and $\mathrm{Hom}_A(E_\nu, E_\nu) = \mathrm{Hom}_A(E_\nu, E) = E_\nu' = (A/\mathfrak{m}^\nu)'' \simeq$

$A/\mathfrak{m}$. Now $E_1 \subset E_2 \subset \cdots$ and $E = \bigcup_\nu E_\nu$ by Theorem 4, (v), hence $E = \varinjlim E_\nu$. Therefore $\mathrm{Hom}_A(E, E) = \mathrm{Hom}_A(\varinjlim E_\nu, E) = \varprojlim \mathrm{Hom}_A(E_\nu, E) = \varprojlim A/\mathfrak{m}^\nu = \hat{A}$.

(v) If an A-module M is Artinian and $x \in M$, then $Ax \simeq A/\mathrm{ann}(x)$ is also Artinian and consequently $\mathfrak{m}^\nu \subset \mathrm{ann}(x)$ for some ν. Therefore M can be viewed as an $\hat{A}$-module, and its $\hat{A}$-submodules are precisely its A-submodules. It is also clear that if an $\hat{A}$-module M is Artinian then we have the same conclusion. Therefore to prove (v) we may assume that A is complete.

If M is a submodule of E set $M^\perp = \{a \in A \mid aM = 0\}$. If I is an ideal of A set $I^\perp = \{x \in E \mid Ix = 0\}$. Then clearly $M^{\perp\perp} \supset M$. If $x \in E - M$ there exist $\varphi \in (E/M')$ satisfying $\varphi(x \bmod M) \neq 0$ by (i), and if we identify $E' = \mathrm{Hom}_A(E, E)$ with A then $(E/M)'$ is identified with $M^\perp$. Thus $\varphi(x \bmod M) = ax$ for some $a \in M^\perp$, and $x \notin M^{\perp\perp}$. Therefore $M^{\perp\perp} = M$. Similarly, if $a \in A - I$ then there exists $\varphi \in (A/I)'$ such that $\varphi(a \bmod I) \neq 0$, and $(A/I)'$ is identified with the submodule $I^\perp$ of $E = A'$. Thus, setting $x = \varphi(1 \bmod I)$ we have $x \in I^\perp$ and $ax = \varphi(a \bmod I) \neq 0$. This proves $a \notin I^{\perp\perp}$, so that $I = I^{\perp\perp}$. Thus $M \longmapsto M^\perp$ is an order-reversing bijection from the set of submodules of E onto the set of ideals of A. Since A is Noetherian, it follows that E is Artinian. By Theorem 3.1 finite direct sums E^n of E are also Artinian for all $n > 0$.

If $M \in \mathcal{N}$ then there is a surjection $A^n \longrightarrow M$ for some n, and so there is an injection $M' \longrightarrow (A^n)' = E^n$. Hence M' is Artinian. On the other hand, if $M \in \mathcal{A}$ there is an injection $M \longrightarrow E^n$ for some n. This can be seen as follows: consider all linear maps $M \longrightarrow E^n$, where n is not fixed, and take one $\varphi: M \longrightarrow E^n$ whose kernel is minimal among the kernels of those maps. Then, using (i) we can easily see that $\mathrm{Ker}(\varphi) = 0$. Now, from $0 \to M \longrightarrow E^n$ we have $(E^n)' = A^n \longrightarrow M' \to 0$ exact, hence $M' \in \mathcal{N}$. Now the assertion $M \simeq M''$ for $M \in \mathcal{N}$ or $\mathcal{A}$ can easily be checked using (iv) if $M = A$ or E, and the general case follows from this and from (i). ∎

Lemma 6. Let A be a Noetherian ring, $S \subset A$ a multiplicative set, M an A-module and $N \subset M$ a submodule. Assume that M is an essential extension of N; then M_S is an essential extension of N_S.

Proof. For $\xi \in M$ we write $\xi_S = \xi/1 \in M_S$; then any element of M_S can be written $u \cdot \xi_S$ (with u a unit of A_S and $\xi \in M$), so that it is enough to show that for any non-zero ξ_S we have $N_S \cap A_S \cdot \xi_S \neq 0$. Suppose that $\mathrm{ann}(t_0\xi)$ is a maximal element of the set of ideals $\{\mathrm{ann}(t\xi) \mid t \in S\}$; then if we set $\eta = t_0\xi$, we have $\xi_S = t_0^{-1}\eta_S$, and hence $\eta \neq 0$. Now let $\mathfrak{b} = \{a \in A \mid a\eta \in N\}$; by assumption,

$$\mathfrak{b}\eta = A\eta \cap N \neq 0.$$

Suppose that $\mathfrak{b} = (b_1, \ldots, b_r)$; if $b_1\eta_S = \cdots = b_r\eta_S = 0$ then there is a $t \in S$

such that $tb_i\eta = 0$ for all i. Then $tb\eta = 0$, but by choice of η we have $\mathrm{ann}(\eta) = \mathrm{ann}(t\eta)$, so that $b\eta = 0$, which is a contradiction. Thus $b_i\eta_S \neq 0$ for some i, and

$$b_i\eta_S \in A_S \cdot \eta_S \cap N_S = A_S\xi_S \cap N_S,$$

as required. ■

By Lemmas 5 and 6, if M is an injective hull of N then the A_S-module M_S is an injective hull of N_S. Hence if $0 \to M \longrightarrow I^0 \longrightarrow I^1 \longrightarrow \cdots$ is a minimal injective resolution of an A-module M, then $0 \to M_S \longrightarrow I^0_S \longrightarrow I^1_S \longrightarrow \cdots$ is a minimal injective resolution of the A_S-module M_S. The I^i are determined uniquely up to isomorphism by M. We can therefore define $\mu_i(P, M)$ to be the number of summands isomorphic to $E(A/P)$ appearing in a decomposition of I^i as a direct sum of indecomposable modules. We can write symbolically

$$I^i = \bigoplus_{P \in \mathrm{Spec}\, A} \mu_i(P, M) E(A/P).$$

From what we have just proved, for a multiplicative set $S \subset A$,

$$\mu_i(P, M) = \mu_i(PA_S, M_S) \quad \text{if } P \cap S = \emptyset.$$

Theorem 18.7. Let A be a Noetherian ring, M an A-module, and $P \in \mathrm{Spec}\, A$. Then

$$\mu_i(P, M) = \dim_{\kappa(P)} \mathrm{Ext}^i_{A_P}(\kappa(P), M_P) = \dim_{\kappa(P)} (\mathrm{Ext}^i_A(A/P, M))_P.$$

In particular, if M is a finite A-module then $\mu_i(P, M) < \infty$.

Proof. Replacing A and M by A_P and M_P we can assume that (A, P, k) is a local ring. Let $0 \to M \longrightarrow I^0 \xrightarrow{\mathrm{d}} I^1 \xrightarrow{\mathrm{d}} \cdots$ be a minimal injective resolution of M, so that $\mathrm{Ext}^i_A(k, M)$ is obtained as the homology of the complex

$$\cdots \longrightarrow \mathrm{Hom}_A(k, I^{i-1}) \longrightarrow \mathrm{Hom}_A(k, I^i) \longrightarrow \mathrm{Hom}_A(k, I^{i+1}) \longrightarrow \cdots$$

We can identify $\mathrm{Hom}_A(k, I^i)$ with the submodule $T^i = \{x \in I^i \mid Px = 0\} \subset I^i$. By construction of the minimal injective resolution, I^i is an essential extension of $\mathrm{d}(I^{i-1})$, so that for $x \in T^i$ the submodule $Ax \simeq k$ intersects $\mathrm{d}(I^{i-1})$, and $x \in \mathrm{d}(I^{i-1})$. Therefore, $T^i \subset \mathrm{d}(I^{i-1})$, and $\mathrm{d}\, T^{i-1} = \mathrm{d}\, T^i = 0$, so that $\mathrm{Ext}^i_A(k, A) = T^i$. Also,

$$\dim_k T^i = \dim_k \mathrm{Hom}_A(k, I^i),$$

and by Theorem 4, (iii), this is equal to $\mu_i(P, M)$. ■

Theorem 18.8. A necessary and sufficient condition for a ring A to be Gorenstein is that a minimal injective resolution $0 \to A \longrightarrow I^0 \longrightarrow I^1 \longrightarrow \cdots$ of A satisfies

$$I^i = \bigoplus_{\mathrm{ht}\, P = i} E(A/P),$$

or, in other words, $\mu_i(P, A) = \delta_{i,\mathrm{ht}P}$ (the Kronecker δ) for every $P \in \operatorname{Spec} A$.
Proof. By Theorem 7 and condition (2) of Theorem 1 we have

$$A_P \text{ is Gorenstein} \Leftrightarrow \mu_i(P, A) = \delta_{i,\mathrm{ht}P}.$$

Theorem 18.9. Let $(A, \mathfrak{m})$ be a Noetherian local ring, and M a finite A-module. Then

$$\operatorname{inj\,dim} M < \infty \Rightarrow \operatorname{inj\,dim} M = \operatorname{depth} A.$$

Proof. Suppose that $\operatorname{inj\,dim} M = r < \infty$. If P is a prime ideal distinct from $\mathfrak{m}$, choose $x \in \mathfrak{m} - P$. Then

$$0 \to A/P \xrightarrow{x} A/P,$$

together with the right-exactness of $\operatorname{Ext}^r_A(-, M)$ gives an exact sequence

$$\operatorname{Ext}^r_A(A/P, M) \xrightarrow{x} \operatorname{Ext}^r_A(A/P, M) \to 0,$$

so that by NAK $\operatorname{Ext}^r_A(A/P, M) = 0$. Putting this together with Lemma 1, we get $\operatorname{Ext}^r_A(k, M) \neq 0$. Set $t = \operatorname{depth} A$, and let $x_1, \ldots, x_t \in \mathfrak{m}$ be a maximal A-sequence; then setting $A/(x_1, \ldots, x_t) = N$ we have $\mathfrak{m} \in \operatorname{Ass}(N)$. Hence there exists an exact sequence $0 \to k \longrightarrow N$, and we must have $\operatorname{Ext}^r_A(N, M) \neq 0$. The Koszul complex $K(x_1, \ldots, x_t)$ is a projective resolution of $N = A/(x_1, \ldots, x_t)$, so computing Ext by means of it we see that

$$\operatorname{Ext}^t_A(N, M) \simeq M/(x_1, \ldots, x_t)M,$$

and by NAK this is non-zero. Thus $\operatorname{proj\,dim} N = t$, and from $\operatorname{Ext}^t_A(N, M) \neq 0$ we get $t \leqslant r$, whereas from $\operatorname{Ext}^r_A(N, M) \neq 0$ we get $t \geqslant r$. Hence $t = r$. ∎

Remark (the Bass conjecture and the intersection theorem). Let (A, m, k) be a Noetherian local ring of dimension d. H. Bass [1] conjectured the following:

(B) if there exists a finite A-module M $(\neq 0)$ of finite injective dimension, then A is a CM ring.

According to Theorem 9 this is equivalent to asking that $\operatorname{inj\,dim} M = d$. The converse of the Bass conjecture is true. Indeed, if A is CM, taking a maximal A-sequence $x_1, \ldots, x_d$ and setting $B = A/(x_1, \ldots, x_d)$ and $E = E(k)$ we have $l_A(B) < \infty$. By Theorem 6, $M = \operatorname{Hom}_A(B, E)$ is also of finite length, hence is finitely generated. We prove $\operatorname{inj\,dim}_A M \leqslant d$; the Koszul complex $0 \to A \to A^d \to \cdots \to A^d \to A \to B \to 0$ with respect to $x_1, \ldots, x_d$ provides an A-free resolution of B. Now applying the exact functor $\operatorname{Hom}_A(-, E)$ to this gives the exact sequence $0 \to M \to E \to E^d \to \cdots \to E^d \to E \to 0$. This proves $\operatorname{inj\,dim}_A M \leqslant d$.

(B) is a special case of the following theorem.

(C) If $I^\bullet\colon 0 \to I^0 \to \cdots \to I^d \to 0$ is a complex of injective modules such that $H^i(I^\bullet)$ is finitely generated for all i and $I^\bullet$ is not exact, then $I^d \neq 0$.

Using the theory of dualizing complexes (see [Rob]) one can prove that (C) is equivalent to the following

Intersection Theorem. If $F.: 0 \to F_d \to \cdots \to F_0 \to 0$ is a complex of finitely generated free modules such that $H_i(F)$ has finite length for all i and $F.$ is not exact, then $F_d \neq 0$.

(B) was proved by Peskine and Szpiro [1] in some important cases, and by Hochster [H] in the equal characteristic case (i.e. when A contains a field) as a corollary of his existence theorem for the 'big CM module', i.e. a (not necessarily finite) A-module with depth = dim A, see [H] p.10 and p.70. The intersection theorem was conjectured by Peskine–Szpiro [3] and by P. Roberts independently. They pointed out that it was also a consequence of Hochster's theorem. Finally, P. Roberts [3] settled the remaining unequal characteristic case of the intersection theorem by using the advanced technique of algebraic geometry developed by W. Fulton ([Ful]). Therefore (B), which was known as Bass's conjecture for 24 years, is now a theorem. Some other conjectures listed in [H] are still open.

Exercises to § 18. Prove the following propositions.

18.1. Let $(A, \mathfrak{m})$ be a Noetherian local ring, $x_1, \ldots, x_r$ an A-sequence, and set $B = A/(x_1, \ldots, x_r)$; then A is Gorenstein $\Leftrightarrow B$ is Gorenstein.

18.2. Use the result of Ex. 18.1 to give another proof of Theorem 3.

18.3. If A is Gorenstein then so is the polynomial ring $A[X]$.

18.4. Is the ring R of Ex. 17.2 Gorenstein?

18.5. Let $(A, \mathfrak{m}, k)$ be a local ring; then $E = E_A(k)$ is a faithful A-module (that is $0 \neq a \in A \Rightarrow aE \neq 0$).

18.6. Let $(A, \mathfrak{m}, k)$ be a complete Noetherian local ring and M an A-module. If M is a faithful A-module and is an essential extension of k then $M \cong E_A(k)$.

18.7. Let k be a field, $S = k[X_1, \ldots, X_n]$ and $P = (X_1, \ldots, X_n)$; set $A = S_P$, $\hat{A} = k[\![X_1, \ldots, X_n]\!]$ and $E = k[X_1^{-1}, \ldots, X_n^{-1}]$. We make E into an A-module by the following multiplication: if $X^\alpha = X_1^{\alpha_1} \ldots X_n^{\alpha_n}$ and $X^{-\beta} = X_1^{-\beta_1} \ldots X_n^{-\beta_n}$, the product $X^\alpha X^{-\beta}$ is defined to be $X^{\alpha-\beta}$ if $\alpha_i \leqslant \beta_i$ for all i and 0 otherwise. Then $E = E_S(S/P) = E_A(k)$. (Use the preceding question; see also Northcott [8] for further results. The elements of this A-module E are called *inverse polynomials*; they were defined and used by Macaulay [Mac] as early as 1916.)

18.8. Let k be a field and t an indeterminate. Consider the subring $A = k[\![t^3, t^5, t^7]\!]$ of $k[\![t]\!]$ and show that A is a one-dimensional CM ring which is not Gorenstein. How about $k[\![t^3, t^4, t^5]\!]$ and $k[\![t^4, t^5, t^6]\!]$?

7

Regular rings

Regular local rings have already been mentioned several times, and in this chapter we are going to carry out a study of them using homological algebra. Serre's Theorem 19.2, characterising regular local rings as Noetherian local rings of finite global dimension, is the really essential result, and from this one can deduce at once, for example, that a localisation of a regular local ring is again regular (Theorem 19.3); this is a result which ideal theory on its own was only able to prove with difficulty in special cases. §20 on UFDs is centred around the theorem that a regular local ring is a UFD, another important achievement of homological methods; we only cover the basic topics. This section was written referring to the early parts of Professor M. Narita's lectures at Tokyo Metropolitan University. In §21 we give a simple discussion of the most elementary results on complete intersection rings. This is an area where the homology theory of M. André plays an essential role, but we are only able to mention this in passing.

19 Regular rings

Minimal free resolutions. Let $(A, \mathfrak{m}, k)$ be a local ring, M and N finite A-modules. An A-linear map $\varphi: M \longrightarrow N$ induces a k-linear map $M \otimes k \longrightarrow N \otimes k$, which we denote $\bar{\varphi}$; then one sees easily that

$$\bar{\varphi} \text{ is an isomorphism} \Leftrightarrow \varphi \text{ is surjective and } \operatorname{Ker} \varphi \subset \mathfrak{m}M.$$

In particular for free modules M and N, if $\bar{\varphi}$ is an isomorphism then rank $M =$ rank N, and writing φ as a matrix we have $\det \varphi \notin \mathfrak{m}$, so that

$$\bar{\varphi} \text{ is an isomorphism} \Leftrightarrow \varphi \text{ is an isomorphism.}$$

Let M be a finite A-module. An exact sequence

$$(*) \quad \cdots \longrightarrow L_i \xrightarrow{d_i} L_{i-1} \xrightarrow{d_{i-1}} \cdots \longrightarrow L_1 \xrightarrow{d_1} L_0 \xrightarrow{\varepsilon} M \to 0,$$

(or the complex $L_\cdot$) is called *a minimal (free) resolution* of M if it satisfies the three conditions (1) each L_i is a finite free A-modules, (2) $\bar{d}_i = 0$, or in other words $d_i L_i \subset \mathfrak{m} L_{i-1}$ for all i, and (3) $\bar{\varepsilon}: L_0 \otimes k \longrightarrow M \otimes k$ is an isomorphism. Breaking up (*) into short exact sequences $0 \to K_1 \longrightarrow L_0 \longrightarrow M \to 0$, $0 \to K_2 \longrightarrow L_1 \longrightarrow K_1 \to 0, \ldots,$ we have $L_0 \otimes k \xrightarrow{\sim} M \otimes k$, $L_1 \otimes k \xrightarrow{\sim} K_1 \otimes k, \ldots$. Any two minimal resolutions of M are isomorphic as complexes (prove this!).

Example. Let $x_1, \ldots, x_n \in \mathfrak{m}$ be an A-sequence, and let $K_\bullet = K_\bullet(x_1, \ldots, x_n)$ be the Koszul complex

$$0 \to K_n \longrightarrow K_{n-1} \longrightarrow \cdots \longrightarrow K_0 \longrightarrow A/(x_1, \ldots, x_n) \to 0;$$

then $K_\bullet$ is a minimal resolution of $A/(x_1, \ldots, x_n)$ over A.

Let $(A, \mathfrak{m}, k)$ be a Noetherian local ring; then a finite A-module M always has a minimal resolution. Construction: let $\{\omega_1, \ldots, \omega_p\}$ be a minimal basis of M, let $L_0 = Ae_1 + \cdots + Ae_p$ be a free module, and define $\varepsilon: L_0 \longrightarrow M$ by $\varepsilon(e_i) = \omega_i$; taking K_1 to be the kernel of ε we get $0 \to K_1 \longrightarrow L_0 \longrightarrow M \to 0$ with $L_0 \otimes k \simeq M \otimes k$. Now K_1 is again a finite A-module, so that we need only proceed as before.

Lemma 1. Let $(A, \mathfrak{m}, k)$ be a local ring, and M a finite A-module. Suppose that $L_\bullet$ is a minimal resolution of M; then

(i) $\dim_k \operatorname{Tor}_i^A(M, k) = \operatorname{rank} L_i$ for all i,

(ii) $\operatorname{proj\,dim} M = \sup\{i \mid \operatorname{Tor}_i^A(M, k) \neq 0\} \leqslant \operatorname{proj\,dim}_A k$,

(iii) if $M \neq 0$ and $\operatorname{proj\,dim} M = r < \infty$ then for any finite A-module $N \neq 0$ we have $\operatorname{Ext}_A^r(M, N) \neq 0$.

Proof. (i) We have $\operatorname{Tor}_i^A(M, k) = H_i(L_\bullet \otimes k)$, but from the definition of minimal resolution, $\bar{d}_i = 0$, and hence $H_i(L_\bullet \otimes k) = L_i \otimes k$, and the dimension of this as a k-vector space is equal to $\operatorname{rank}_A L_i$.

(ii) follows from (i).

(iii) Since $L_{r+1} = 0$ and $L_r \neq 0$, $\operatorname{Ext}_A^r(M, N)$ is the cokernel of d_r^*: $\operatorname{Hom}(L_r, N) \longleftarrow \operatorname{Hom}(L_{r-1}, N)$, but since L_i is free, $\operatorname{Hom}(L_i, N)$ is just a direct sum of a number of copies of N; we can write $d_r: L_r \longrightarrow L_{r-1}$ as a matrix with entries in $\mathfrak{m}$, and then d_r^* is given by the same matrix, so that $\operatorname{Im}(d_r^*) \subset \mathfrak{m} \operatorname{Hom}(L_r, N)$, and by NAK $\operatorname{Ext}_A^r(M, N) \neq 0$. ∎

Remark. One sees from the above lemma that $\operatorname{Tor}_i(M, k) = 0$ implies that $L_i = 0$, and therefore $\operatorname{proj\,dim} M < i$, so that $\operatorname{Tor}_j(M, k) = 0$ for $j > i$. It is conjectured that this holds in more generality, or more precisely:

Rigidity conjecture. Let R be a Noetherian ring, M and N finite R-modules; suppose that $\operatorname{proj\,dim} M < \infty$. Then $\operatorname{Tor}_i^R(M, N) = 0$ implies that $\operatorname{Tor}_j^R(M, N) = 0$ for all $j > i$.

This has been proved by Lichtenbaum [1] if R is a regular ring, but is unsolved in general.

The following theorem is not an application of Lemma 1, but is proved by a similar technique.

Theorem 19.1 (Auslander and Buchsbaum). Let A be a Noetherian local ring and $M \neq 0$ a finite A-module. Suppose that $\operatorname{proj\,dim} M < \infty$; then

$$\operatorname{proj\,dim} M + \operatorname{depth} M = \operatorname{depth} A.$$

Proof. Set $\operatorname{proj\,dim} M = h$; we work by induction on h. If $h = 0$ then M is a free A-module, so that the assertion is trivial. If $h = 1$, let

$$(\dagger) \quad 0 \to A^m \xrightarrow{\varphi} A^n \xrightarrow{\varepsilon} M \to 0$$

be a minimal resolution of M. We can write φ as an $m \times n$ matrix with entries in $\mathfrak{m}$. From (†) we obtain the long exact sequence

$$\cdots \longrightarrow \operatorname{Ext}^i_A(k, A^m) \xrightarrow{\varphi_*} \operatorname{Ext}^i_A(k, A^n) \xrightarrow{\varepsilon_*} \operatorname{Ext}^i_A(k, M) \longrightarrow \cdots,$$

and writing out $\operatorname{Ext}^i_A(k, A^m) = \operatorname{Ext}^i_A(k, A)^m$ and $\operatorname{Ext}^i_A(k, A^n) = \operatorname{Ext}^i_A(k, A)^n$, we can express φ_* by the same matrix as φ. However, the entries of φ are elements of $\mathfrak{m}$, and therefore annihilate $\operatorname{Ext}^i_A(k, A)$, so that $\varphi_* = 0$, and we have an exact sequence

$$0 \to \operatorname{Ext}^i_A(k, A)^n \longrightarrow \operatorname{Ext}^i_A(k, M) \longrightarrow \operatorname{Ext}^{i+1}_A(k, A)^m \to 0$$

for every i. Since $\operatorname{depth} M = \inf\{i \mid \operatorname{Ext}^i_A(k, M) \neq 0\}$ we have $\operatorname{depth} M = \operatorname{depth} A - 1$ and the theorem holds if $h = 1$. If $h > 1$ then taking any exact sequence

$$0 \to M' \longrightarrow A^n \longrightarrow M \to 0,$$

we have $\operatorname{proj\,dim} M' = h - 1$, so that an easy induction completes the proof. ■

Lemma 2. Let A be a ring and $n \geqslant 0$ a given integer. Then the following conditions are equivalent.

(1) $\operatorname{proj\,dim} M \leqslant n$ for every A-module M;

(2) $\operatorname{proj\,dim} M \leqslant n$ for every finite A-module M;

(3) $\operatorname{inj\,dim} N \leqslant n$ for every A-module N;

(4) $\operatorname{Ext}^{n+1}_A(M, N) = 0$ for all A-modules M and N.

Proof. (1)⇒(2) is trivial.

(2)⇒(3) For any ideal I, the A-module A/I is finite, so that $\operatorname{Ext}^{n+1}_A(A/I, N) = 0$, so that by §18, Lemma 1, $\operatorname{inj\,dim} N \leqslant n$.

(3)⇒(4) is trivial, and (4)⇒(1) is well-known (see p. 280). ■

We define the *global dimension* of a ring A by

$$\operatorname{gl\,dim} A = \sup\{\operatorname{proj\,dim} M \mid M \text{ is an } A\text{-module}\}.$$

According to Lemma 2 above, this is also equal to the maximum projective dimension of all finite A-modules. If $(A, \mathfrak{m}, k)$ is a Noetherian local ring then by Lemma 1, $\operatorname{gl\,dim} A = \operatorname{proj\,dim}_A k$.

We have defined regular local rings (see §14) as Noetherian local rings for which $\dim A = \operatorname{emb}\dim A$, and we have seen that they are integral domains (Theorem 14.3) and CM rings (Theorem 17.8). A regular local ring is Gorenstein (Theorem 18.1, (5′)). A necessary and sufficient condition for a Noetherian local ring $(A, \mathfrak{m}, k)$ to be regular is that $\operatorname{gr}_{\mathfrak{m}}(A)$ is a polynomial ring over k (Theorem 17.10). The following theorem gives another important necessary and sufficient condition.

Theorem 19.2 (Serre). Let A be a Noetherian local ring. Then

$$A \text{ is regular} \Leftrightarrow \operatorname{gl}\dim A = \dim A \Leftrightarrow \operatorname{gl}\dim A < \infty.$$

Proof. (I) Suppose that $(A, \mathfrak{m}, k)$ is an n-dimensional regular local ring. Let $x_1, \ldots, x_n$ be a regular system of parameters; then since this is an A-sequence, the Koszul complex $K_{.}(x_1, \ldots, x_n)$ is a minimal free resolution of $A/(x_1, \ldots, x_n) = k$, and $K_n \neq 0$, $K_{n+1} = 0$, so that as we have already seen, $\operatorname{gl}\dim A = \operatorname{proj}\dim k = n$.

(II) Conversely, suppose that $\operatorname{gl}\dim A = r < \infty$ and $\operatorname{emb}\dim A = s$. We prove that A is regular by induction on s; we can assume that $s > 0$, that is $\mathfrak{m} \neq 0$. Then $\mathfrak{m} \notin \operatorname{Ass}(A)$: for if $0 \neq a \in A$ is such that $\mathfrak{m}a = 0$, consider a minimal resolution

$$0 \to L_r \longrightarrow L_{r-1} \longrightarrow \cdots \longrightarrow L_0 \longrightarrow k \to 0$$

of k (with $r \geqslant 0$); then $L_r \subset \mathfrak{m}L_{r-1}$, but then $aL_r = 0$, which contradicts the assumption that L_r is a free module. Thus we can choose $x \in \mathfrak{m}$ not contained in $\mathfrak{m}^2$ or in any associated prime of A. Then x is A-regular, hence also $\mathfrak{m}$-regular, so that if we set $B = A/xA$ then according to Lemma 2 of §18, $\operatorname{Ext}^i_A(\mathfrak{m}, N) = \operatorname{Ext}^i_B(\mathfrak{m}/x\mathfrak{m}, N)$ for all B-modules N, and hence we obtain $\operatorname{proj}\dim_B \mathfrak{m}/x\mathfrak{m} \leqslant r$.

Now we prove that the natural map $\mathfrak{m}/x\mathfrak{m} \longrightarrow \mathfrak{m}/xA$ splits, so that $\mathfrak{m}/xA$ is isomorphic to a direct summand of $\mathfrak{m}/x\mathfrak{m}$. Since $x \notin \mathfrak{m}^2$, we can take a minimal basis $x_1 = x, x_2, \ldots, x_s$ of $\mathfrak{m}$ starting with x (here $s = \operatorname{emb}\dim A$). We set $\mathfrak{b} = (x_2, \ldots, x_s)$, so that by the minimal basis condition, $\mathfrak{b} \cap xA \subset x\mathfrak{m}$, and therefore there exists a chain

$$\mathfrak{m}/xA = (\mathfrak{b} + xA)/xA \simeq \mathfrak{b}/(\mathfrak{b} \cap xA) \longrightarrow \mathfrak{m}/x\mathfrak{m} \longrightarrow \mathfrak{m}/xA$$

of natural maps, whose composite is the identity. This proves the above claim. Now clearly,

$$\operatorname{proj}\dim_B \mathfrak{m}/xA \leqslant \operatorname{proj}\dim_B \mathfrak{m}/x\mathfrak{m} \leqslant r.$$

Taking a minimal B-projective resolution of $\mathfrak{m}/xA$ and patching it together with the exact sequence $0 \to \mathfrak{m}/xA \longrightarrow B \longrightarrow k \to 0$ gives a projective resolution of k of length $\leqslant r + 1$, and hence $\operatorname{gl}\dim B = \operatorname{proj}\dim_B k \leqslant r + 1$, so that by induction, B is a regular local ring. Since x is not contained in any associated prime of A we have $\dim B = \dim A - 1$, and therefore A is regular. ■

Theorem 19.3 (Serre). Let A be a regular local ring and P a prime ideal; then A_P is again regular.
Proof. Since $\operatorname{proj\,dim}_A A/P \leqslant \operatorname{gl\,dim} A < \infty$, as an A-module A/P has a projective resolution $L_{\bullet}$ of finite length. Then $L_{\bullet} \otimes_A A_P$ is a projective resolution of $(A/P) \otimes_A A_P = A_P/PA_A = \kappa(P)$ as an A_P-module, so that $\kappa(P)$ has a projective resolution of finite length as an A_P-module, which means that A_P has finite global dimension; thus by the previous theorem, A_P is regular. ■

Definition. A *regular ring* is a Noetherian ring such that the localisation at every prime is a regular local ring. By the previous theorem, it is sufficient for the localisation at every maximal ideal to be regular.

Theorem 19.4. A regular ring is normal.
Proof. The definition of normal is local, so that it is enough to show that a regular local ring is normal. We show that the conditions of the corollary of Theorem 11.5 are satisfied. (*a*) The localisation at a height 1 prime ideal is a DVR by the previous theorem and Theorem 11.2. (*b*) All the prime divisors of a non-zero principal ideal have height 1 by Theorem 17.8 (the implication regular $\Rightarrow$ CM). ■

Theorem 19.5. If A is regular then so are $A[X]$ and $A[\![X]\!]$.
Proof. For $A[X]$, let P be a maximal ideal of $A[X]$ and set $P \cap A = \mathfrak{m}$. $A[X]_P$ is a localisation of $A_{\mathfrak{m}}[X]$, so that replacing A by $A_{\mathfrak{m}}$ we can assume that A is a regular local ring. Then setting $A/\mathfrak{m} = k$ we have $A[X]/\mathfrak{m}[X] = k[X]$, so that there is a monic polynomial $f(X)$ with coefficients in A such that $P = (\mathfrak{m}, f(X))$, and such that f reduces to an irreducible polynomial $\bar{f} \in k[X]$ modulo $\mathfrak{m}$. Then by Theorem 15.1, we clearly have

$$\dim A[X]_P = \operatorname{ht} P = 1 + \operatorname{ht} \mathfrak{m} = 1 + \dim A;$$

on the other hand $\mathfrak{m}$ is generated by $\dim A$ elements, so that $P = (\mathfrak{m}, f)$ is generated by $\dim A + 1$ elements, and therefore $A[X]_P$ is regular.

For $A[\![X]\!]$, set $B = A[\![X]\!]$ and let M be a maximal ideal of B; then $X \in M$ by Theorem 8.2, (i). Therefore $M \cap A = \mathfrak{m}$ is a maximal ideal of A. Now although we cannot say that B_M contains $A_{\mathfrak{m}}[\![X]\!]$, the two have the same completion, $(B_M)\hat{} = (A_{\mathfrak{m}})\hat{}\,[\![X]\!]$. A Noetherian local ring is regular if and only if its completion is regular (since both the dimension and embedding dimension remain the same on taking the completion). Thus if we replace A by $(A_{\mathfrak{m}})\hat{}$, the maximal ideal of $B = A[\![X]\!]$ is $M = (\mathfrak{m}, X)$, and $\operatorname{ht} M = \operatorname{ht} \mathfrak{m} + 1$, so that B is also regular. ■

Next we discuss the properties of modules which have finite free resolutions; (the definition is given below).

Lemma 3 (Schanuel). Let A be a ring and M an A-module. Suppose that

$$0 \to K \longrightarrow P \longrightarrow M \to 0 \quad \text{and} \quad 0 \to K' \longrightarrow P' \longrightarrow M \to 0$$

are exact sequences with P and P' projective. Then $K \oplus P' \simeq K' \oplus P$.

Proof. From the fact that P and P' are projective, there exist $\lambda: P \longrightarrow P'$ and $\lambda': P' \longrightarrow P$, giving the diagram:

$$\begin{array}{ccccccc} 0 \to K \longrightarrow & P & \xrightarrow{\alpha} & M & \to 0 & & \\ & {\scriptstyle\lambda'}\uparrow\downarrow{\scriptstyle\lambda} & & \| & & & \text{with } \alpha'\lambda = \alpha \quad \text{and} \quad \alpha\lambda' = \alpha'. \\ 0 \to K' \longrightarrow & P' & \xrightarrow{\alpha'} & M & \to 0 & & \end{array}$$

We add in harmless summands P' and P to the two exact rows, and line up the middle terms:

$$\begin{array}{ccccc} 0 \to K \oplus P' \longrightarrow & P \oplus P' & \xrightarrow{(\alpha, 0)} & M & \to 0 \\ & {\scriptstyle\psi}\uparrow\downarrow{\scriptstyle\varphi} & & \| & \\ 0 \to P \oplus K' \longrightarrow & P \oplus P' & \xrightarrow{(0, \alpha')} & M & \to 0. \end{array}$$

Here $\varphi: P \oplus P' \longrightarrow P \oplus P'$ is defined by

$$\varphi\begin{pmatrix} x \\ x' \end{pmatrix} = \begin{pmatrix} 1 & -\lambda' \\ \lambda & 1 - \lambda\lambda' \end{pmatrix}\begin{pmatrix} x \\ x' \end{pmatrix} \quad \text{for} \quad x \in P, \quad x' \in P',$$

and satisfies

$$(0, \alpha')\begin{pmatrix} 1 & -\lambda' \\ \lambda & 1 - \lambda\lambda' \end{pmatrix} = (\alpha, 0),$$

and similarly ψ is defined by $\begin{pmatrix} 1 - \lambda'\lambda & \lambda' \\ -\lambda & 1 \end{pmatrix}$ and satisfies $(\alpha, 0)\psi = (0, \alpha')$. Moreover, by matrix computation we see that $\varphi\psi = 1$ and $\psi\varphi = 1$, so that φ is an isomorphism and $\psi = \varphi^{-1}$. Therefore φ induces an isomorphism $K \oplus P' \xrightarrow{\sim} P \oplus K'$. ■

Lemma 4 (generalised Schanuel lemma). Let A, M be as above, and suppose that $0 \to P_n \longrightarrow \cdots \longrightarrow P_1 \longrightarrow P_0 \longrightarrow M \to 0$ and $0 \to Q_n \longrightarrow \cdots \longrightarrow Q_1 \longrightarrow Q_0 \longrightarrow M \to 0$ are exact sequences with P_i and Q_i projective for $0 \leqslant i \leqslant n-1$. Then

$$P_0 \oplus Q_1 \oplus P_2 \oplus \cdots \simeq Q_0 \oplus P_1 \oplus Q_2 \oplus \cdots.$$

Proof. Write K for the kernel of $P_0 \longrightarrow M$ and K' for the kernel of $Q_0 \longrightarrow M$; then, by the previous lemma, $K \oplus Q_0 \simeq P_0 \oplus K'$. Now add in harmless summands Q_0 and P_0 to $0 \to P_n \longrightarrow \cdots \longrightarrow P_1 \longrightarrow K \to 0$ and $Q \to Q_n \longrightarrow \cdots \longrightarrow Q_1 \longrightarrow K' \to 0$ respectively, to obtain

$$0 \to P_n \longrightarrow \cdots \longrightarrow P_2 \longrightarrow P_1 \oplus Q_0 \longrightarrow K \oplus Q_0 \to 0$$
$$\downarrow \wr$$
$$0 \to Q_n \longrightarrow \cdots \longrightarrow Q_2 \longrightarrow P_0 \oplus Q_1 \longrightarrow P_0 \oplus K' \to 0.$$

Induction on n now gives

$$(P_1 \oplus Q_0) \oplus Q_2 \oplus P_3 \oplus \cdots \simeq (P_0 + Q_1) \oplus P_2 \oplus Q_3 \oplus \cdots. \quad \blacksquare$$

Definition. A *finite free resolution* (or FFR for short) of an A-module M is an exact sequence $0 \to F_n \longrightarrow \cdots \longrightarrow F_1 \longrightarrow F_0 \longrightarrow M \to 0$ (of finite length n) such that each F_i is a finite free module. If M has an FFR we set $\chi(M) = \sum(-1)^i \operatorname{rank} F_i$, and call $\chi(M)$ the *Euler number* of M. By Lemma 4, this is independent of the choice of FFR. Moreover, since for any prime ideal P of A

$$0 \to (F_n)_P \longrightarrow \cdots \longrightarrow (F_1)_P \longrightarrow (F_0)_P \longrightarrow M_P \to 0$$

is an FFR of the A_P-module M_P we have $\chi(M) = \chi(M_P)$. If M is itself free then one sees easily from Lemma 4 that $\chi(M) = \operatorname{rank} M$.

Theorem 19.6. Let $(A, \mathfrak{m})$ be a local ring, and suppose that for any finite subset $E \subset \mathfrak{m}$ there exists $0 \neq y \in A$ such that $yE = 0$; then the only A-modules having an FFR are the free modules.

Remark. If A is Noetherian then the assumption on $\mathfrak{m}$ is equivalent to $\mathfrak{m} \in \operatorname{Ass}(A)$, or depth $A = 0$. In this case the theorem is a special case of Theorem 19.1.

Proof. Suppose that $0 \to F_n \longrightarrow F_{n-1} \longrightarrow \cdots \longrightarrow F_0 \longrightarrow M \to 0$ is an FFR of M, and set $N = \operatorname{coker}(F_n \longrightarrow F_{n-1})$; if we prove that N is free then we can decrease n by 1, so that we only need consider the case $0 \to F_1 \longrightarrow F_0 \longrightarrow M \to 0$. Now let $0 \to L_1 \longrightarrow L_0 \longrightarrow M \to 0$ be a minimal free resolution of M; then since L_0 and F_1 are finitely generated, by Schanuel's lemma (or by Theorem 2.6), L_1 is also finite. Considering bases of L_0 and L_1, we can write down a set of generators of L_1 as a submodule of $\mathfrak{m}L_0$ using only a finite number of elements of $\mathfrak{m}$. Then by assumption, there exists $0 \neq y \in A$ such that $yL_1 = 0$. Since L_1 is a free module, we must have $L_1 = 0$, so that $M \simeq L_0$, and is free. $\blacksquare$

Theorem 19.7. Let A be any ring; if M is an A-module having an FFR then $\chi(M) \geqslant 0$.

Proof. Choose a minimal prime ideal P of A; since $\chi(M) = \chi(M_P)$, we can replace A by A_P, and then A is a local ring with maximal ideal $\mathfrak{m}$ equal to $\operatorname{nil}(A)$. Then the assumption of the previous theorem is satisfied: for given $x_1, \ldots, x_r \in \mathfrak{m}$, we can assume by induction on r that there is a $z \neq 0$ such that $zx_1 = \cdots = zx_{r-1}$; but x_r is nilpotent, so that there is an $i \geqslant 0$

such that $zx_r^i \neq 0$ but $zx_r^{i+1} = 0$, and we can take $y = zx_r^i$. Therefore by the previous theorem M is a free module, and $\chi(M) = \operatorname{rank} M \geqslant 0$. ∎

Theorem 19.8 (Auslander and Buchsbaum [2]). Let A be a Noetherian ring and M an A-module, and suppose that M has an FFR. Then the following three conditions are equivalent:

(1) $\operatorname{ann}(M) \neq 0$;

(2) $\chi(M) = 0$,

(3) $\operatorname{ann}(M)$ contains an A-regular element.

Proof. (1)$\Rightarrow$(2) Suppose that $\chi(M) > 0$; then for any $P \in \operatorname{Ass}(A)$ we have $\chi(M_P) > 0$, and hence $M_P \neq 0$. By Theorem 6, M_P is a free A_P-module, so that setting $I = \operatorname{ann}(M)$ we have $I_P = \operatorname{ann}(M_P) = 0$. If we set $J = \operatorname{ann}(I)$ then this is equivalent to $J \not\subset P$. Since this holds for every $P \in \operatorname{Ass}(A)$ we see that J contains an A-regular element, but then $J \cdot I = 0$ implies that $I = 0$.

(2)$\Rightarrow$(3) If $\chi(M) = 0$ then by Theorem 6, $M_P = 0$ for every $P \in \operatorname{Ass}(A)$. This means that $\operatorname{ann}(M) \not\subset P$, so that $\operatorname{ann}(M)$ contains an A-regular element.

(3)$\Rightarrow$(1) is obvious. ∎

Theorem 19.9 (Vasconcelos [1]). Let A be a Noetherian local ring, and I a proper ideal of A; assume that $\operatorname{proj\,dim} I < \infty$. Then

I is generated by an A-sequence $\Leftrightarrow$ I/I^2 is a free module over A/I.

Proof. ($\Rightarrow$) is already known (Theorem 16.2). In fact, $I^\nu/I^{\nu+1}$ is a free A/I-module for $\nu = 1, 2, \ldots$.

($\Leftarrow$) We can assume that $I \neq 0$. Since I has finite projective dimension over A so has A/I, and since A is local, A/I has an FFR. Now $\operatorname{ann}(A/I) = I$, so that by the previous theorem I is not contained in any associated prime of A, and therefore we can choose an element $x \in I$ such that x is not contained in $\mathfrak{m}I$ or in any associated prime of A. Then x is A-regular, and $\bar{x} = x \bmod I^2$ is a member of a basis of I/I^2 over A/I; let x, $y_2, \ldots, y_n \in I$ be such that their images form a basis of I/I^2. Then if we set $B = A/xA$, we see by the same argument as in (II) of the proof of Theorem 2 that $\operatorname{proj\,dim}_B I/xI < \infty$, and that I/xA is isomorphic to a direct summand of I/xI. We now set $I^* = I/xA$, so that $\operatorname{proj\,dim}_B I^* < \infty$. But on the other hand on sees easily that I^*/I^{*2} is a free module over B/I^*, and an induction on the number of generators of I completes the proof.

Remark. In Lech [1], a set $x_1, \ldots, x_n$ of elements of A is defined to be *independent* if

$$\sum a_i x_i = 0 \quad \text{for} \quad a_i \in A \Rightarrow a_i \in (x_1, \ldots, x_n) \quad \text{for all } i.$$

If we set $I = (x_1, \ldots, x_n)$ then this condition is equivalent to saying that the images of $x_1,\ldots,x_n$ in I/I^2 form a basis of I/I^2 over A/I. Then if A and I satisfy the hypotheses of the previous theorem, the theorem tells us that $I=(y_1,\ldots,y_n)$ with $y_1,\ldots,y_n$ an A-sequence. Setting $x_i=\sum a_{ij}y_j$ we see that the matrix (a_{ij}) is invertible when considered in A/I; this means that the determinant of (a_{ij}) is not in the maximal ideal of A, and so (a_{ij}) itself is invertible. Thus $x_1,\ldots,x_n$ is an A-quasi-regular sequence, hence an A-sequence. In particular, we get the following corollary.

Corollary. Let $(A,\mathfrak{m})$ be a regular local ring. Then if $x_1,\ldots,x_n\in\mathfrak{m}$ are independent in the sense of Lech, they form an A-sequence.

However, if we try to prove this corollary as it stands, the induction does not go through. The key to success with Vasconcelos' theorem is to strengthen the statement so that induction can be used effectively. Now as Kaplansky has also pointed out, the main part of Theorem 2 (the implication $\operatorname{gl\,dim} A<\infty\Rightarrow$ regular) follows at once from Theorem 9, because if $\mathfrak{m}$ is generated by an A-sequence then $\operatorname{emb\,dim} A\leqslant \operatorname{depth} A\leqslant \dim A$.

Exercises to §19.

19.1. Let k be a field and $R=R_0+R_1+R_2+\cdots$ a Noetherian graded ring with $R_0=k$; set $\mathfrak{m}=R_1+R_2+\cdots$. Show that if $R_\mathfrak{m}$ is an n-dimensional regular local ring then R is a polynomial ring $R=k[y_1,\ldots,y_n]$ with y_i homogeneous of positive degree.

19.2. Let A be a ring and M an A-module. Say that M is *stably free* if there exist finite free modules F and F' such that $M\oplus F\simeq F'$. Obviously a stably free A-module M is a finite projective A-module, and has an FFR $0\to F\longrightarrow F'\longrightarrow M\to 0$. Prove that, conversely, a finite projective module having an FFR is stably free.

19.3. Prove that if every finite projective module over a Noetherian ring A is stably free then every finite A-module of finite projective dimension has an FFR.

19.4. Prove that if every finite module over a Noetherian ring A has an FFR then A is regular.

20 UFDs

This section treats UFDs, which we have already touched on in §1; note that the Bourbaki terminology for UFD is 'factorial ring'. First of all, we have the following criterion for Noetherian rings.

Theorem 20.1. A Noetherian integral domain A is a UFD if and only if every height 1 prime ideal is principal.

Proof of 'only if'. Suppose that A is a UFD and that P is a height 1 prime ideal. Take any non-zero $a \in P$, and express a as a product of prime elements, $a = \prod \pi_i$. Then at least one of the π_i belongs to P; if $\pi_i \in P$ then $(\pi_i) \subset P$, but (π_i) is a non-zero prime ideal and $\operatorname{ht} P = 1$, hence $P = (\pi_i)$.
Proof of 'if'. Since A is Noetherian, every element $a \in A$ which is neither 0 nor a unit can be written as a product of finitely many irreducibles. Hence it will be enough to prove that an irreducible element a is a prime element. Let P be a minimal prime divisor of (a); then by the principal ideal theorem (Theorem 13.5), $\operatorname{ht} P = 1$, so that by assumption we can write $P = (b)$. Thus $a = bc$, and since a is irreducible, c is a unit, so that $(a) = (b) = P$, and a is a prime element. ∎

Theorem 20.2. Let A be a Noetherian integral domain, Γ a set of prime elements of A, and let S be the multiplicative set generated by Γ. If A_S is a UFD then so is A.
Proof. Let P be a height 1 prime ideal of A. If $P \cap S \neq \emptyset$ then P contains an element $\pi \in \Gamma$, and since πA is a non-zero prime ideal we have $P = \pi A$. If $P \cap S = \emptyset$ then PA_S is a height 1 prime ideal of A_S, so that $PA_S = aA_S$ for some $a \in P$. Among all such a choose one such that aA is maximal; then a is not divisible by any $\pi \in \Gamma$. Now if $x \in P$ we have $sx = ay$ for some $s \in S$ and $y \in A$. Let $s = \pi_1 \dots \pi_r$ with $\pi_i \in \Gamma$; then $a \notin \pi_i A$, so that $y \in \pi_i A$, and an induction on r shows that $y \in sA$, so that $x \in aA$. Hence $P = aA$. ∎

Lemma 1. Let A be an integral domain, and $\mathfrak{a}$ an ideal of A such that $\mathfrak{a} \oplus A^n \simeq A^{n+1}$; then $\mathfrak{a}$ is principal.
Proof. Fix the basis $e_0, \dots, e_n$ of A^{n+1}, and viewing $\mathfrak{a} \oplus A^n \subset A \oplus A^n$, fix $f_0, \dots, f_n$ such that f_0 is a basis of A and $f_1, \dots, f_n$ a basis of A^n. Then the isomorphism $\varphi: A^{n+1} \longrightarrow \mathfrak{a} \oplus A^n$ can be given in the form $\varphi(e_i) = \sum_{j=0}^n a_{ij} f_j$. Write d_i for the $(i, 0)$th cofactor of the matrix (a_{ij}), and d for the determinant, so that, since φ is injective, $d \neq 0$, and $\sum a_{i0} d_i = d$, $\sum a_{ij} d_i = 0$ if $j \neq 0$. Hence if we set $e'_0 = \sum_0^n d_i e_i$ we have $\varphi(e'_0) = d f_0$. Moreover, since the image of φ includes $f_1, \dots, f_n$, there exist $e'_1, \dots, e'_n \in A^{n+1}$ such that $\varphi(e'_j) = f_j$. Now define a matrix (c_{jk}) by $e'_j = \sum_{k=0}^n c_{jk} e_k$ for $j = 0, \dots, n$ (so $c_{0k} = d_k$). Then we have

$$(c_{jk})(a_{ij}) = \begin{pmatrix} d & 0 & \cdots & 0 \\ 0 & 1 & & 0 \\ \vdots & & \ddots & \\ 0 & 0 & & 1 \end{pmatrix},$$

so that by comparing the determinants of both sides we get $\det(c_{jk}) = 1$.

Therefore $e'_0, \ldots, e'_n$ is another basis of A^{n+1}, and $\mathfrak{a}f_0 = \varphi(Ae'_0) = dA\bar{f}_0$, so that $\mathfrak{a} = dA$. ∎

Let K be the field of fractions of the integral domain A; for a finite A-module M, the dimension of $M \otimes_A K$ as a vector space over K is called the *rank* of M. A torsion-free finite A-module of rank 1 is isomorphic to an ideal of A. Lemma 1 can be formulated as saying that for an integral domain A, a stably free rank 1 module is free (see Ex. 19.2). The elementary proof given above is taken from a lecture by M. Narita in 1971.

Theorem 20.3 (Auslander and Buchsbaum [3]). A regular local ring is a UFD.

Proof. Let $(A, \mathfrak{m})$ be a regular local ring; the proof works by induction on $\dim A$. If $\dim A = 0$ then A is a field and therefore (trivially) a UFD. If $\dim A = 1$ then A is a DVR, and therefore a UFD. We suppose that $\dim A > 1$ and choose $x \in \mathfrak{m} - \mathfrak{m}^2$; then since xA is a prime ideal, applying Theorem 2 to $\Gamma = \{x\}$, we need only show that A_x is a UFD (where $A_x = A[x^{-1}]$ is as on p. 22). Let P be a height 1 prime ideal of A_x and set $\mathfrak{p} = P \cap A$; we have $P = \mathfrak{p}A_x$. Since A is a regular local ring, the A-module $\mathfrak{p}$ has an FFR, so that the A_x-module P has an FFR. For any prime ideal Q of A_x, the ring $(A_x)_Q = A_{Q \cap A}$ is a regular local ring of dimension less than that of A, so by induction is a UFD. Thus P_Q is free as an $(A_x)_Q$-module, so that by Theorem 7.12, the A_x-module P is projective; hence by Ex. 19.2, P is stably free, and therefore by the previous lemma, P is a principal ideal of A_x. ∎

The above proof is due to Kaplansky. Instead of our Lemma 1, he used the following more general proposition, which he had previously proved: if A is an integral domain, and I_i, J_i are ideals of A for $1 \leqslant i \leqslant r$ such that $\bigoplus_{i=1}^r I_i \simeq \bigoplus_{i=1}^r J_i$, then $I_1 \ldots I_r \simeq J_1 \ldots J_r$. This is an interesting property of ideals, and we have given a proof in Appendix C.

Theorem 20.4. Let A be a Noetherian integral domain. Then if any finite A-module has an FFR, A is a UFD.

Proof. By Ex. 19.4, A is a regular ring. Let P be a height 1 prime ideal of A. Then $A_\mathfrak{m}$ is a regular local ring for any $\mathfrak{m} \in \operatorname{Spec} A$, so by the previous theorem, the ideal $P_\mathfrak{m}$ is principal, and is therefore a free $A_\mathfrak{m}$-module. Hence by Theorem 7.12, P is projective. Therefore by Ex. 19.2, P is stably free, and so by Lemma 1 is principal. ∎

Let A be an integral domain; for any two non-zero elements $a, b \in A$, the notion of *greatest common divisor* (g.c.d.) and *least common multiple* (l.c.m.) are defined as in the ring of integers. That is, d is a g.c.d. of a and b if d divides both a and b, and any element x dividing both a and b

divides d; and e is an l.c.m. of a and b if e is divisible by both a and b, and any y divisible by a and b is divisible by e; this condition is equivalent to $(e) = (a) \cap (b)$.

Lemma 2. If an l.c.m. of a and b exists then so does a g.c.d.
Proof. If $(a) \cap (b) = (e)$ then there exists d such that $ab = ed$. From $e \in (a)$ we get $b \in (d)$ and similarly $a \in (d)$, so that $(a, b) \subset (d)$. Now if x is a common divisor of a and b then $a = xt$ and $b = xs$, so that xst is a common multiple of a, b, and is hence divisible by e. Then from $ed = ab = x \cdot xst$ we get that d is divisible by x. Therefore, d is a g.c.d. of a and b. ■

Remark 1. If A is a Noetherian integral domain which is not a UFD then A has an irreducible element a which is not prime. If $xy \in (a)$ but $x \notin (a)$, $y \notin (a)$ then the only common divisors of a and x are units, so that 1 is a g.c.d. of a and x. However, $xy \in (a) \cap (x)$, but $xy \notin (ax)$, so that $(a) \cap (x) \neq (ax)$, and there does not exist any l.c.m. of a and x. Thus the converse of Lemma 2 does not hold in general.
Remark 2. If A is a UFD then an intersection of an arbitrary collection of principal ideals is again principal (we include (0)). Indeed, if $\bigcap_{i \in I} a_i A \neq 0$, then factorise each a_i as a product of primes:

$$a_i = u_i \cdot \prod_\alpha p_\alpha^{r(i,\alpha)},$$

with u_i units, and p_α prime elements such that $p_\alpha A \neq p_\beta A$ for $\alpha \neq \beta$. Then $\bigcap a_i A = dA$, where $d = \prod p_\alpha^{\max\{r(i,\alpha) | i \in I\}}$. (We could even allow the a_i to be elements of the field of fractions of A.)

Theorem 20.5. An integral domain A is a UFD if and only if the ascending chain condition holds for principal ideals, and any two elements of A have an l.c.m.
Proof. The 'only if' is already known, and we prove the 'if'. From the first condition it follows that every element which is neither 0 nor a unit can be written as a product of a finite number of irreducible elements, so that we need only prove that an irreducible element is prime. Let a be an irreducible element, and let $xy \in (a)$ and $x \notin (a)$. By assumption we can write $(a) \cap (x) = (z)$; now 1 is a g.c.d. of a and x, so that one sees from the proof of Lemma 2 that $(z) = (ax)$, and then $xy \in (a) \cap (x) = (ax)$ implies that $y \in (a)$. Therefore (a) is prime. ■

Theorem 20.6. Let A be a regular ring and $u, v \in A$. Then $uA \cap vA$ is a projective ideal.
Proof. $A_\mathfrak{m}$ is a UFD for every maximal ideal $\mathfrak{m}$, so that $(uA \cap vA)A_\mathfrak{m} = uA_\mathfrak{m} \cap vA_\mathfrak{m}$ is a principal ideal, and hence a free module. ■

Theorem 20.7. If A is a UFD then a projective ideal is principal.

Proof. By Theorem 11.3, it is equivalent to say that a non-zero ideal $\mathfrak{a}$ is projective or invertible. Hence if we set K for the field of fractions of A, then there exist $u_i \in K$ such that $u_i\mathfrak{a} \subset A$ and $a_i \in \mathfrak{a}$ such that $\sum u_i a_i = 1$. We have $\mathfrak{a} \subset \bigcap u_i^{-1}A$, and conversely if $x \in \bigcap u_i^{-1}A$ then $x = \sum (xu_i)a_i \in \mathfrak{a}$, and hence $\mathfrak{a} = \bigcap u_i^{-1}A$; now since A is a UFD, the intersection of principal fractional ideals is again principal. ∎

Theorem 20.8. If A is a regular UFD then so is $A[\![X]\!]$.

Proof. Set $B = A[\![X]\!]$. By Theorem 5, it is enough to prove that $uB \cap vB$ is principal for $u, v \in B$; set $\mathfrak{a} = uB \cap vB$. Then by Theorem 6 and Theorem 19.5, $\mathfrak{a}$ is projective, so that

$$\mathfrak{a} \otimes_B A = \mathfrak{a} \otimes_B (B/XB) = \mathfrak{a}/X\mathfrak{a}$$

is projective as an A-module. Suppose that $\mathfrak{a} = X^r\mathfrak{b}$ with $\mathfrak{b} \not\subset XB$; then $\mathfrak{a}/X\mathfrak{a} \simeq \mathfrak{b}/X\mathfrak{b}$, so that $\mathfrak{b}$ is isomorphic to $\mathfrak{a}$, hence projective, and therefore locally principal. B is a regular ring, so that the prime divisors of $\mathfrak{b}$ all have height 1. Since XB is also a height 1 prime ideal and $\mathfrak{b} \not\subset XB$ we have $\mathfrak{b}:XB = \mathfrak{b}$, hence $\mathfrak{b} \cap XB = X\mathfrak{b}$. Therefore since we can view $\mathfrak{b}/X\mathfrak{b}$ as $\mathfrak{b}/X\mathfrak{b} = \mathfrak{b}/\mathfrak{b} \cap XB \subset B/XB = A$, by Theorem 7 it is principal, hence $\mathfrak{b} = yB + X\mathfrak{b}$ for some $y \in \mathfrak{b}$; then by NAK, $\mathfrak{b} = yB$, so that $\mathfrak{a} = X^r yB$. ∎

Remark. There are examples where A is a UFD but $A[\![X]\!]$ is not.

It is easy to see that a UFD is a Krull ring. For any Krull ring A we can define the divisor class group of A, which should be thought of as a measure of the extent to which A fails to be a UFD. We can give the definition in simple terms as follows: let $\mathscr{P}$ be the set of height 1 prime ideals of the Krull ring A, and $D(A)$ the free Abelian group on $\mathscr{P}$. That is, $D(A)$ consists of formal sums $\sum_{\mathfrak{p}\in\mathscr{P}} n_\mathfrak{p}\cdot\mathfrak{p}$ (with $n_\mathfrak{p} \in \mathbb{Z}$ and all but finitely many $n_\mathfrak{p} = 0$), with addition defined by

$$\left(\sum n_\mathfrak{p}\cdot\mathfrak{p}\right) + \left(\sum n'_\mathfrak{p}\cdot\mathfrak{p}\right) = \sum (n_\mathfrak{p} + n'_\mathfrak{p})\mathfrak{p}.$$

Let K be the field of fractions of A, and K^* the multiplicative group of non-zero elements of K, and for $a \in K^*$ set $\operatorname{div}(a) = \sum_{\mathfrak{p}\in\mathscr{P}} v_\mathfrak{p}(a)\cdot\mathfrak{p}$, where $v_\mathfrak{p}$ is the normalised additive valuation of K corresponding to $\mathfrak{p}$. Then $\operatorname{div}(ab) = \operatorname{div}(a) + \operatorname{div}(b)$, so that div is a homomorphism from K^* to $D(A)$. We write $F(A)$ for the image of K^*; this is a subgroup of $D(A)$, so that we can define $C(A) = D(A)/F(A)$ to be the *divisor class group* of A. Obviously, if A is a UFD then each $\mathfrak{p} \in \mathscr{P}$ is principal, and if $\mathfrak{p} = aA$ then as an element of $D(A)$ we have $\mathfrak{p} = \operatorname{div}(a)$, so that $C(A) = 0$. Conversely, if $C(A) = 0$ then each $\mathfrak{p} \in \mathscr{P}$ is a principal ideal, and putting this together with the corollary of Theorem 12.3, one sees easily that A is a UFD. Hence

$$A \text{ is a UFD} \Leftrightarrow C(A) = 0.$$

Now let A be any ring, and M a finite projective A-module. For each $P \in \operatorname{Spec} A$, the localisation M_P is a free module over A_P, and we write $n(P)$ for its rank. Then n is a function on $\operatorname{Spec} A$, and is constant on every connected component (since $n(P) = n(Q)$ if $P \supset Q$). This function n is called the *rank* of M. If the rank is a constant r over the whole of $\operatorname{Spec} A$ then we say that M is a projective module of rank r. We write $\operatorname{Pic}(A)$ for the set of isomorphism classes of finite projective A-modules of rank 1; $\operatorname{cl}(M)$ denotes the isomorphism class of M. If M and N are finite projective rank 1 module then so is $M \otimes_A N$; this is clear on taking localisations. Thus we can define a sum in $\operatorname{Pic}(A)$ by setting

$$\operatorname{cl}(M) + \operatorname{cl}(N) = \operatorname{cl}(M \otimes N).$$

We set $M^* = \operatorname{Hom}_A(M, A)$, and define $\varphi: M \otimes M^* \longrightarrow A$ by

$$\varphi\left(\sum m_i \otimes f_i\right) = \sum f_i(m_i);$$

then φ is an isomorphism (taking localisations and using the corollary to Theorem 7.11 reduces to the case $M = A$, which is clear). Hence $\operatorname{cl}(M^*) = -\operatorname{cl}(M)$, and $\operatorname{Pic}(A)$ becomes an Abelian group, called the Picard group of A. If A is local then $\operatorname{Pic}(A) = 0$.

If A is an integral domain with field of fractions K, then $M_{(0)} = M \otimes K$, so that the rank we have just defined coincides with the earlier definition (after Lemma 1). If M is a finite projective rank 1 module, then since M is torsion-free we have $M \subset M_{(0)} \simeq K$, so that M is isomorphic as an A-module to a fractional ideal; for fractional ideals, by Theorem 11.3, projective and invertible are equivalent conditions, so that for an integral domain A, we can consider $\operatorname{Pic}(A)$ as a quotient of the group of invertible fractional ideals under multiplication. A fractional ideal I is isomorphic to A as an A-module precisely when I is principal, so that

$$\operatorname{Pic}(A) = \left\{ \begin{matrix} \text{invertible frac-} \\ \text{tional ideals} \end{matrix} \right\} \Big/ \left\{ \begin{matrix} \text{principal} \\ \text{ideals} \end{matrix} \right\}.$$

Suppose in addition that A is a Krull ring. Then we can view $\operatorname{Pic}(A)$ as a subgroup of $C(A)$. To prove this, for $\mathfrak{p} \in \mathscr{P}$ and I a fractional ideal, set

$$v_{\mathfrak{p}}(I) = \min \{ v_{\mathfrak{p}}(x) \mid x \in I \};$$

this is zero for all but finitely many $\mathfrak{p} \in \mathscr{P}$ (check this!), so that we can set

$$\operatorname{div}(I) = \sum_{\mathfrak{p} \in \mathscr{P}} v_{\mathfrak{p}}(I) \cdot \mathfrak{p} \in D(A).$$

For a principal ideal $I = \alpha A$ we have $\operatorname{div}(I) = \operatorname{div}(\alpha)$. One sees easily that $\operatorname{div}(II') = \operatorname{div}(I) + \operatorname{div}(I')$, and that $\operatorname{div}(A) = 0$, so that if I is invertible, $\operatorname{div}(I) = -\operatorname{div}(I^{-1})$.

For invertible I we have $(I^{-1})^{-1} = I$: indeed, $I \subset (I^{-1})^{-1}$ from the definition, and $I = I \cdot A \supset I(I^{-1}(I^{-1})^{-1}) \supset (I^{-1})^{-1}$. If I is invertible and $\operatorname{div}(I) = 0$ then $\operatorname{div}(I^{-1}) = 0$, so that $I \subset A$, $I^{-1} \subset A$; hence $A \subset (I^{-1})^{-1} = I$, and $I = A$. It follows that if I, I' are invertible, $\operatorname{div}(I) = \operatorname{div}(I')$ implies $I = I'$. Thus we can view the group of invertible fractional ideals as a subgroup of $D(A)$, and $\operatorname{Pic}(A)$ as a subgroup of $C(A)$.

If A is a regular ring then as we have seen, $\mathfrak{p} \in \mathscr{P}$ is a locally free module, and so is invertible. Clearly from the definition, $\operatorname{div}(\mathfrak{p}) = \mathfrak{p}$. Hence, in the case of a regular ring, $D(A)$ is identified with the group of invertible fractional ideals, and $C(A)$ coincides with $\operatorname{Pic}(A)$.

The notions of $D(A)$ and $\operatorname{Pic}(A)$ originally arise in algebraic geometry. Let V be an algebraic variety, supposed to be irreducible and normal. We write $\mathscr{P}$ for the set of irreducible codimension 1 subvarieties of V, and define the group of divisors $D(V)$ of V to be the free Abelian group on $\mathscr{P}$; a divisor (or Weil divisor) is an element of $D(V)$. Corresponding to a rational function f on V and an element $W \in \mathscr{P}$, let $v_W(f)$ denote the order of zero of f along W, or minus the order of the pole if f has a pole along W. Write $\operatorname{div}(f) = \sum_{W \in \mathscr{P}} v_W(f) \cdot W$ for the divisor of f on V (or just (f)). For $W \in \mathscr{P}$, the local ring $\mathcal{O}_W$ of W on V is a DVR of the function field of V, and v_W is the corresponding valuation. We say that two divisors $M, N \in D(V)$ are linearly equivalent if their difference $M - N$ is the divisor of a function, and write $M \sim N$. The quotient group of $D(V)$ by $\sim$, that is the quotient by the subgroup of divisors of functions, is the divisor class group of V (up to linear equivalence), and we write $C(V)$ for this. (In addition to linear equivalence one also considers other equivalence relations with certain geometric significance (algebraic equivalence, numerical equivalence, ...), and divisor class groups, quotients of $D(V)$ by the corresponding subgroups.)

A divisor M on V is said to be a Cartier divisor if it is the divisor of a function in a neighbourhood of every point of V. From a Cartier divisor one constructs a line bundle over V, and two Cartier divisors give rise to isomorphic line bundles if and only if they are linearly equivalent. Cartier divisors form a subgroup of $D(V)$, and their class group up to linear equivalence is written $\operatorname{Pic}(V)$; this can also be considered as the group of isomorphism classes of line bundles over V (with group law defined by tensor product). If V is smooth then (by Theorem 3) there is no distinction between Cartier and Weil divisors, and $C(V) = \operatorname{Pic}(V)$.

The reader familiar with algebraic geometry will know that the divisor class group and Picard group of a Krull ring are an exact translation of the corresponding notions in algebraic geometry. If V is an affine variety, with coordinate ring $k[V] = A$ then $C(V) = C(A)$ and $\operatorname{Pic}(V) = \operatorname{Pic}(A)$. In this case, to say that A is a UFD expresses the fact that every codimension 1 subvariety of V can be defined as the intersection of V with a hypersurface. If $V \subset \mathbb{P}^n$ is a projective algebraic variety, defined by a prime ideal $I \subset k[X_0, \ldots, X_n]$, and we set $A = k[X]/I = k[\xi_0, \ldots, \xi_n]$ (with ξ_i the class of X_i) then A is the so-called homogeneous coordinate ring of V. If A is integrally closed we say that V is projectively normal (also arithmetically normal). This condition is stronger than saying that V is normal (the local ring of any point of V is normal). If A is a UFD then every codimension 1 subvariety of V can be given as the intersection in $\mathbb{P}^n$ of V with a hypersurface. Let $\mathfrak{m} = (\xi_0, \ldots, \xi_n)$ be the homogeneous maximal ideal of A, and write $R = A_\mathfrak{m}$ for the localisation. The above statement holds if we just assume that R is a UFD; see Ex. 20.6. All the information about V is contained in the local ring R.

Thus $C(A)$, $\mathrm{Pic}(A)$ and the UFD condition are notions with important geometrical meaning, and methods of algebraic geometry can also be used in their study. For example, in this way Grothendieck [G5] was able to prove the following theorem conjectured by Samuel: let R be a regular local ring, P a prime ideal generated by an R-sequence, and set $A = R/P$; if $A_\mathfrak{p}$ is a UFD for every $\mathfrak{p} \in \mathrm{Spec}\, A$ with $\mathrm{ht}\, \mathfrak{p} \leqslant 3$ then A is a UFD.

We do not have the space to discuss $C(A)$ and $\mathrm{Pic}(A)$ in detail, and we just mention the following two theorems as examples:

(1) If A is a Krull ring then $C(A) \simeq C(A[X])$.

This generalises the well-known theorem (see Ex. 20.2) that if A is a UFD then so is $A[X]$.

(2) If A is a regular ring then $C(A) \simeq C(A[\![X]\!])$.

This generalises Theorem 8.

Finally we give an example. Let k be a field of characteristic 0, and set $A = k[X, Y, Z]/(Z^n - XY) = k[x, y, z]$ for some $n > 1$. Then $A/(z, x) \simeq k[X, Y, Z]/(X, Z) \simeq k[Y]$, so that $\mathfrak{p} = (x, z)$ is a height 1 prime ideal of A. In $D(A)$ we have $n\mathfrak{p} = \mathrm{div}(x)$, and it can be proved that $C(A) \simeq \mathbb{Z}/n\mathbb{Z}$ (see [S2], p. 58). The relation $xy = z^n$ shows that A is not a UFD.

For those wishing to know more about UFDs, consult [K], [S2] and [F].

Exercises to §20. Prove the following propositions.

20.1. (Gauss' lemma) Let A be a UFD, and $f(X) = a_0 + a_1 X + \cdots + a_n X^n \in A[X]$; say that f is ***primitive*** if the g.c.d. of the coefficients $a_0, \ldots, a_n$ is 1. Then if $f(X)$ and $g(X)$ are primitive, so is $f(X)g(X)$.

20.2. If A is a UFD so is $A[X]$ (use the previous question).

20.3. If A is a UFD and $\mathfrak{q}_1, \ldots, \mathfrak{q}_r$ are height 1 primary ideals then $\mathfrak{q}_1 \cap \cdots \cap \mathfrak{q}_r$ is a principal ideal.

20.4. Let A be a Zariski ring (see §8) and $\hat{A}$ the completion of A. Then if $\hat{A}$ is a UFD so is A (there are counter-examples to the converse).

20.5. Let A be an integral domain. We say that A is *locally UFD* if $A_\mathfrak{m}$ is a UFD for every maximal ideal $\mathfrak{m}$. If A is a semilocal integral domain and A is locally UFD, then A is a UFD.

20.6. Let $R = \bigoplus_{n\geq 0} R_n$ be a graded ring, and suppose that R_0 is a field. Set $\mathfrak{m} = \bigoplus_{n>0} R_n$. If I is a homogeneous ideal of R such that $IR_\mathfrak{m}$ is principal then there is a homogeneous element $f \in I$ such that $I = fR$.

21 Complete intersection rings

Let $(A, \mathfrak{m}, k)$ be a Noetherian local ring; we choose a minimal basis $x_1, \ldots, x_n$ of $\mathfrak{m}$, where $n = \operatorname{emb}\dim A$ is the embedding dimension of A (see §14). Set $E_. = K_{x,1\ldots n}$ for the Koszul complex. The complex $E_.$ is determined by A up to isomorphism. Indeed, if $x'_1, \ldots, x'_n$ is another minimal basis of $\mathfrak{m}$ then by Theorem 2.3, there is an invertible $n \times n$ matrix (a_{ij}) over A such that $x'_i = \sum a_{ij} x_j$. It is proved in Appendix C that $K_{x,1\ldots n}$ can be thought of as the exterior algebra $\bigwedge (Ae_1 + \cdots + Ae_n)$ with differential defined by $\mathrm{d}(e_i) = x_i$. Similarly,

$$K_{x',1\ldots n} = \bigwedge (Ae'_1 + \cdots + Ae'_n) \quad \text{with} \quad \mathrm{d}(e'_i) = x'_i.$$

Now $f(e'_i) = \sum a_{ij} e_j$ defines an isomorphism from the free A-module $Ae'_1 + \cdots + Ae'_n$ to $Ae_1 + \cdots + Ae_n$, which extends to an isomorphism f of the exterior algebra; f commutes with the differential d, since for a generator e'_i of $\bigwedge (Ae'_1 + \cdots + Ae'_n)$ we have $\mathrm{d}f(e'_i) = \sum a_{ij} x_j = x'_i = f\mathrm{d}(e'_i)$. Therefore $f: K_{x',1\ldots n} \xrightarrow{\sim} K_{x,1\ldots n}$ is an isomorphism of complexes.

Since $\mathfrak{m} H_p(E) = 0$ by Theorem 16.4, $H_p(E_.)$ is a vector space over $k = A/\mathfrak{m}$. Set

$$\varepsilon_p = \dim_k H_p(E_.) \quad \text{for} \quad p = 0, 1, 2, \ldots;$$

then these are invariants of the local ring A. In view of $H_0(E_.) = A/(\underline{x}) = A/\mathfrak{m} = k$, we have $\varepsilon_0 = 1$. In this section we are concerned with ε_1. If A is regular then $x_1, \ldots, x_n$ is an A-sequence, so that $\varepsilon_1 = \cdots = \varepsilon_n = 0$, and conversely by Theorem 16.5, $\varepsilon_1 = 0$ implies that A is regular.

Let us consider the case when A can be expressed as a quotient of a regular local ring R; let $A = R/\mathfrak{a}$, and write $\mathfrak{n}$ for the maximal ideal of R. If $\mathfrak{a} \not\subset \mathfrak{n}^2$, we can take $x \in \mathfrak{a} - \mathfrak{n}^2$; then $R' = R/xR$ is again a regular local ring, and $A = R'/\mathfrak{a}'$, so that we can write A as a quotient of a ring R' of dimension smaller than R. In this way we see that there exist an expression $A = R/\mathfrak{a}$ of A as a quotient of a regular local ring $(R, \mathfrak{n})$ with $\mathfrak{a} \subset \mathfrak{n}^2$. Then we have $\mathfrak{m} = \mathfrak{n}/\mathfrak{a}$ and $\mathfrak{m}/\mathfrak{m}^2 = \mathfrak{n}/(\mathfrak{a} + \mathfrak{n}^2) = \mathfrak{n}/\mathfrak{n}^2$, so that $\dim R = n = \operatorname{emb}\dim A$. Conversely, equality here implies that $\mathfrak{a} \subset \mathfrak{n}^2$.

Let $(R, \mathfrak{n})$ be a regular local ring and $A = R/\mathfrak{a}$ with $\mathfrak{a} \subset \mathfrak{n}^2$; choose a regular system of parameters (that is a minimal basis of $\mathfrak{n}$) $\xi_1,\dots,\xi_n$. Then the images x_i of ξ_i in A form a minimal basis $x_1,\dots,x_n$ of $\mathfrak{m}$. Let

$$K_{\xi,1\dots n}: 0 \to L_n \longrightarrow L_{n-1} \longrightarrow \cdots \longrightarrow L_1 \longrightarrow L_0 \to 0$$

be the Koszul complex of R and $\underline{\xi}$. By Theorem 16.5, we know that this becomes exact on adding $\cdots \to L_0 \longrightarrow k \to 0$ to the right-hand end, so that $K_{\xi,1\dots n}$ is a projective resolution of k as an R-module. Taking the tensor product with $A = R/\mathfrak{a}$, we get the complex $E_. = K_{x,1\dots n}$ of A-modules. Thus we have

$$H_p(E_.) = H_p(K_{\xi,1\dots n} \otimes_R A) = \operatorname{Tor}_p^R(k, A) \quad \text{for all } p \geqslant 0.$$

However, from the exact sequence of R-modules $0 \to \mathfrak{a} \longrightarrow R \longrightarrow A \to 0$ we get the long exact sequence

$$0 = \operatorname{Tor}_1^R(k, R) \longrightarrow \operatorname{Tor}_1^R(kl, A) \xrightarrow{\sim} k \otimes_R \mathfrak{a} \longrightarrow k \otimes_R R$$
$$\longrightarrow k \otimes_R A \to 0;$$

at the right-hand end we have $k \otimes R \xrightarrow{\sim} k \otimes A = k$, so that

$$\operatorname{Tor}_1^R(k, A) \simeq k \otimes_R \mathfrak{a} = \mathfrak{a}/\mathfrak{n}\mathfrak{a}.$$

Quite generally, we write $\mu(M)$ for the minimum number of generators of an R-module M. Then we see that

$$\mu(\mathfrak{a}) = \dim_k H_1(E_.) = \varepsilon_1(A).$$

Theorem 21.1. Let $(A, \mathfrak{m}, k)$ be a Noetherian local ring, and $\hat{A}$ its completion.

(i) $\varepsilon_p(A) = \varepsilon_p(\hat{A})$ for all $p \geqslant 0$.

(ii) $\varepsilon_1(A) \geqslant \operatorname{emb\,dim} A - \dim A$.

(iii) If R is a regular local ring, $\mathfrak{a}$ an ideal of R and $A \simeq R/\mathfrak{a}$, then

$$\mu(\mathfrak{a}) = \dim R - \operatorname{emb\,dim} A + \varepsilon_1(A).$$

Proof. (i) is clear from the fact that a minimal basis of $\mathfrak{m}$ is a minimal basis of $\mathfrak{m}\hat{A}$, so that applying $\otimes_A \hat{A}$ to the complex $E_.$ made from A gives that made from $\hat{A}$. Then since $\hat{A}$ is A-flat, $H_p(E_.) \otimes \hat{A} = H_p(E_. \otimes \hat{A})$, and $\mathfrak{m}H_p(E_.) = 0$ gives $H_p(E_.) \otimes \hat{A} = H_p(E_.)$.

(ii) If A is a quotient of a regular local ring, then as we have seen above, there exists a regular ring $(R, \mathfrak{n})$ such that $A = R/\mathfrak{a}$ with $\mathfrak{a} \subset \mathfrak{n}^2$, so that $\varepsilon_1(A) = \mu(\mathfrak{a}) \geqslant \operatorname{ht} \mathfrak{a} = \dim R - \dim A = \operatorname{emb\,dim} A - \dim A$, where the equality for $\operatorname{ht} \mathfrak{a}$ comes from Theorem 17.4, (i). Now A itself is not necessarily a quotient of a regular local ring, but we will prove later (see §29) that $\hat{A}$ always is, and we admit this in the section. Having said this, the two sides of (ii) are unaltered on replacing A by $\hat{A}$, and the inequality holds for $\hat{A}$.

(iii) Set $\mathfrak{n} = \operatorname{rad}(R)$. If $\mathfrak{a} \subset \mathfrak{n}^2$ then, as we have seen above, $\mu(\mathfrak{a}) = \varepsilon_1(A)$, and $\dim R = \operatorname{emb\,dim} A$, so that we are done. If $\mathfrak{a} \not\subset \mathfrak{n}^2$, take $x \in \mathfrak{a} - \mathfrak{n}^2$;

when we pass to R/xR and $\mathfrak{a}/xR$, each of $\dim R$ and $\mu(\mathfrak{a})$ decreases by 1, so that induction completes the proof. ■

Definition. A Noetherian local ring A is a *complete intersection ring* (abbreviated to c.i. ring) if $\varepsilon_1(A) = \operatorname{emb dim} A - \dim A$.

Theorem 21.2. Let A be a Noetherian local ring.

(i) A is c.i. $\Leftrightarrow \hat{A}$ is c.i.

(ii) Let A be a c.i. ring and R a regular local ring such that $A = R/\mathfrak{a}$; then $\mathfrak{a}$ is generated by an R-sequence. Conversely, if $\mathfrak{a}$ is an ideal generated by an R-sequence then $R/\mathfrak{a}$ is a c.i. ring.

(iii) A necessary and sufficient condition for A to be a c.i. ring is that the completion $\hat{A}$ should be a quotient of a complete regular local ring R by an ideal generated by an R-sequence.

Proof. (i) is obvious.

(ii) By Theorem 1, (iii), $\mu(\mathfrak{a}) = \dim R - \operatorname{emb dim} A + \varepsilon_1(A)$, and by Theorem 17.4, (i), $\operatorname{ht} \mathfrak{a} = \dim R - \dim A$, so that A is a c.i. ring is equivalent to $\operatorname{ht} \mathfrak{a} = \mu(\mathfrak{a})$. But by Theorem 17.4, (iii), this is equivalent to $\mathfrak{a}$ being generated by an R-sequence.

(iii) The sufficiency is clear from (i) and (ii). Necessity follows from the fact that $\hat{A}$ is a quotient of a complete regular local ring (see §29), together with (i) and (ii). ■

Theorem 21.3. A c.i. ring is Gorenstein.

Proof. If A is c.i. then so is $\hat{A}$, and if $\hat{A}$ is Gorenstein then so is A, so that we can assume that A is complete. Then we can write $A = R/\mathfrak{a}$, where R is a regular local ring and $\mathfrak{a}$ is an ideal generated by a regular sequence. Since R is Gorenstein, A is also by Ex. 18.1. ■

Thus we have the following chain of implications for Noetherian local rings:

$$\text{regular} \Rightarrow \text{c.i.} \Rightarrow \text{Gorenstein} \Rightarrow \text{CM}.$$

Let A be a c.i. ring, and $\mathfrak{p}$ a prime ideal of A. If A is of the form $A = R/(x_1, \ldots, x_r)$, where R is regular and $x_1, \ldots, x_r$ is an R-sequence, then since $A_\mathfrak{p}$ can be written $A_\mathfrak{p} = R_P/(x_1, \ldots, x_r)$, where R_P is regular and $x_1, \ldots, x_r$ is an R_P-sequence, it follows that $A_\mathfrak{p}$ is again a c.i. ring. The question of deciding whether $A_\mathfrak{p}$ is still a c.i. ring even if A is not a quotient of a regular local ring remained unsolved for some time, but was answered affirmatively by Avramov [1], making use of André's homology theory [An 1, 2]. This theory defines homology and cohomology groups $H_n(A, B, M)$ and $H^n(A, B, M)$ for $n \geqslant 0$ associated with a ring A, an A-algebra B and a B-module M. The definition is complicated, but in any case these are B-modules having various nice functorial properties. If A is a

Noetherian local ring with residue field k then

$$A \text{ is regular} \Leftrightarrow H_2(A,k,k)=0,$$

and

$$A \text{ is c.i.} \Leftrightarrow H_3(A,k,k)=0;$$

for $n \geqslant 3$ the statements $H_3(A,k,k)=0$ and $H_n(A,k,k)=0$ are equivalent. Thus André homology is particularly relevant to the study of regular and c.i. rings.

Exercises to § 21. Prove the following propositions.

21.1. Let R be a regular ring, I an ideal of R, and let $A = R/I$; then the subset $\{\mathfrak{p} \in \operatorname{Spec} A \mid A_\mathfrak{p} \text{ is c.i.}\}$ is open is $\operatorname{Spec} A$ (use Theorem 19.9).

21.2. Let A be a Noetherian local ring with $\operatorname{emb\,dim} A = \dim A + 1$; if A is CM then it is c.i.

21.3. Let k be a field, and set $A = k[\![X, Y, Z]\!]/(X^2 - Y^2, Y^2 - Z^2, XY, YZ, ZX)$; then A is Gorenstein but not c.i.

8

Flatness revisited

The main theme of this chapter is flatness over Noetherian rings. In §22 we prove a number of theorems known as the 'local flatness criterion' (the main result is Theorem 22.3). Together with Theorem 23.1 in the following section, this is extremely useful in applications.

In §23 we consider a flat morphism $A \longrightarrow B$ of Noetherian local rings, and investigate the remarkable relationships holding between A, B and the fibre ring $F = B/\mathfrak{m}_A B$. Roughly speaking, good properties of B are usually inherited by A, and sometimes by F. Conversely, in order for B to inherit good properties of A one also requires F to be good.

In §24 we discuss the so-called generic freeness theorem in the improved form due to Hochster and Roberts (Theorem 24.1), and investigate, following the ideas of Nagata, the openness of loci of points at which various properties hold, arising out of Theorem 24.3, which states that the set of points of flatness is open.

22 The local flatness criterion

Theorem 22.1. Let A be a ring, B a Noetherian A-algebra, M a finite B-module, and J an ideal of B contained in rad (B); set $M_n = M/J^{n+1}M$ for $n \geqslant 0$. If M_n is flat over A for every $n \geqslant 0$, then M is also flat over A.

Proof. According to Theorem 7.7, we need only show that for a finitely generated ideal I of A, the standard map $u: I \otimes_A M \longrightarrow M$ is injective. Set $I \otimes M = M'$; then M' is also a finite B-module, and hence is separated for the J-adic topology. Let $x \in \operatorname{Ker}(u)$; we prove that $x \in \bigcap J^n M' = 0$. For any $n \geqslant 0$, $M'_n = M'/J^{n+1}M' = (I \otimes_A M) \otimes_B B/J^{n+1} = I \otimes_A M_n$, and the induced map $M'_n \longrightarrow M_n$ is injective, by the assumption that M_n is flat. Then we deduce that $x \in J^{n+1}M'$ from the commutative diagram

$$\begin{array}{ccc} M' & \xrightarrow{u} & M \\ \downarrow & & \downarrow \\ M'_n & \longrightarrow & M_n. \end{array} \quad \blacksquare$$

Theorem 22.2. Let A be a ring, B a Noetherian A-algebra, and M

a finite B-module; suppose that b is an M-regular element of $\operatorname{rad}(B)$. Then if M/bM is flat over A, so is M.

Proof. For each $i>0$ the sequence $0 \to M/b^iM \xrightarrow{b} M/b^{i+1}M \longrightarrow M/bM \to 0$ is exact, so that by Theorem 7.9 and an induction on i, every M/b^iM is flat over A. Thus we can just apply the previous theorem. ∎

Definition. Let A be a ring and I an ideal of A; an A-module M is said to be *I-adically ideal-separated* if $\mathfrak{a} \otimes M$ is separated for the I-adic topology for every finitely generated ideal $\mathfrak{a}$ of A.

For example, if B is a Noetherian A-algebra and $IB \subset \operatorname{rad}(B)$ then a finite B-module M is I-adically ideal-separated as an A-module.

Let A be a ring, I an ideal of A and M an A-module. Set $A_n = A/I^{n+1}$, $M_n = M/I^{n+1}M$ for $n \geqslant 0$ and $\operatorname{gr}(A) = \bigoplus_{n\geqslant 0} I^n/I^{n+1}$, $\operatorname{gr}(M) = \bigoplus_{n\geqslant 0} I^nM/I^{n+1}M$. There exist standard maps

$$\gamma_n : (I^n/I^{n+1}) \otimes_{A_0} M_0 \longrightarrow I^nM/I^{n+1}M \quad \text{for } n \geqslant 0,$$

and we can put together the γ_n into a morphism of $\operatorname{gr}(A)$-modules

$$\gamma : \operatorname{gr}(A) \otimes_{A_0} M_0 \longrightarrow \operatorname{gr}(M).$$

Theorem 22.3. In the above notation, suppose that one of the following two conditions is satisfied:

(α) I is a nilpotent ideal;

or (β) A is a Noetherian ring and M is I-adically ideal-separated. Then the following conditions are equivalent.

(1) M is flat over A;

(2) $\operatorname{Tor}_1^A(N, M) = 0$ for every A_0-module N;

(3) M_0 is flat over A_0 and $I \otimes_A M = IM$;

(3′) M_0 is flat over A_0 and $\operatorname{Tor}_1^A(A_0, M) = 0$;

(4) M_0 is flat over A_0 and γ_n is an isomorphism for every $n \geqslant 0$;

(4′) M_0 is flat over A_0 and γ is an isomorphism;

(5) M_n is flat over A_n for every $n \geqslant 0$.

In fact, the implications (1)⇒(2)⇔(3)⇔(3′)⇒(4)⇒(5) hold without any assumption on M.

Proof. First of all, let M be arbitrary.

(1)⇒(2) is trivial.

(2)⇒(3) If N is an A_0-module then we have

$$N \otimes_A M = (N \otimes_{A_0} A_0) \otimes_A M = N \otimes_{A_0} M_0,$$

and hence for an exact sequence $0 \to N_1 \longrightarrow N_2 \longrightarrow N_3 \to 0$ of A_0-modules we get an exact sequence

$$0 = \operatorname{Tor}_1^A(N_3, M) \longrightarrow N_1 \otimes_{A_0} M_0 \longrightarrow N_2 \otimes_{A_0} M_0 \longrightarrow N_3 \otimes_{A_0} M_0 \to 0;$$

therefore M_0 is flat over A_0. Also, from the exact sequence $0 \to I \longrightarrow A \longrightarrow A_0 \to 0$ we get an exact sequence

$$0 = \mathrm{Tor}_1^A(A_0, M) \longrightarrow I \otimes M \longrightarrow M \longrightarrow M_0 \to 0,$$

so that $I \otimes M = IM$.

(3)⇔(3′) is easy.

(3′)⇒(2) If N is an A_0-module, we can choose an exact sequence of A_0-modules $0 \to R \longrightarrow F_0 \longrightarrow N \to 0$ with F_0 a free A_0-module. From this we get the exact sequence

$$\mathrm{Tor}_1^A(F_0, M) = 0 \longrightarrow \mathrm{Tor}_1^A(N, M) \longrightarrow R \otimes_{A_0} M_0 \longrightarrow F_0 \otimes_{A_0} M_0,$$

and since M_0 is flat over A_0 the final arrow is injective, so that $\mathrm{Tor}_1^A(N, M) = 0$.

(3)⇒(4) By (2) we have $\mathrm{Tor}_1^A(I/I^2, M) = 0$, so that from $0 \to I^2 \longrightarrow I \longrightarrow I/I^2 \to 0$, the sequence $0 \to I^2 \otimes M \longrightarrow I \otimes M \longrightarrow (I/I^2) \otimes M \to 0$ is exact. From $I \otimes M = IM$ we get $I^2 \otimes M = I^2 M$ and $(I/I^2) \otimes M \simeq IM/I^2M$. Proceeding similarly, from $0 \to I^{n+1} \longrightarrow I^n \longrightarrow I^n/I^{n+1} \to 0$ we get by induction $I^{n+1} \otimes M = I^{n+1}M$ and $(I^n/I^{n+1}) \otimes M \simeq I^nM/I^{n+1}M$. (4′) is just a restatement of (4).

(4)⇒(5) We fix an $n > 0$ and prove that M_n is flat over A_n. For $i \leqslant n$ we have a commutative diagram

$$\begin{array}{ccccc} (I^{i+1}/I^{n+1}) \otimes M & \longrightarrow & (I^{i+1}/I^n) \otimes M \longrightarrow (I^i/I^{i+1}) \otimes M \to 0 \\ \alpha_{i+1}\downarrow & & \alpha_i\downarrow \qquad\qquad \gamma_i\downarrow \\ 0 \to I^{i+1}/M_n = I^{i+1}M/I^{n+1}M & \longrightarrow & I^i/M_n = I^i/I^{n+1}M \longrightarrow I^iM/I^{i+1}M \to 0 \end{array}$$

with exact rows. By assumption γ_i is an isomorphism, and since α_{n+1} is an isomorphism (from 0 to 0), by downwards induction on i we see that α_n, $\alpha_{n-1}, \ldots, \alpha_1$ are isomorphisms. In particular,

$$\alpha_1 : (I/I^{n+1}) \otimes_A M = IA_n \otimes_{A_n} M_n \xrightarrow{\simeq} IM_n,$$

so that the conditions in (3) are satisfied by A_n, M_n and I/I^{n+1}. Therefore by (2)⇔(3), we have $\mathrm{Tor}_1^{A_n}(N, M_n) = 0$ for every A_0-module N. Now if N is an A_i-module then IN and N/IN are both A_{i-1}-modules, and $0 \to IN \longrightarrow N \longrightarrow N/IN \to 0$ is exact, so that by induction on i we get finally that $\mathrm{Tor}_1^{A_n}(N, M_n) = 0$ for all A_n-modules N. Therefore M_n is a flat A_n-module.

Next, assuming either (α) or (β) we prove (5)⇒(1). In case (α) we have $A = A_n$ and $M = M_n$ for large enough n, so that this is clear. In case (β), by Theorem 7.7, it is enough to prove that the standard map $J : \mathfrak{a} \otimes M \longrightarrow M$ is injective for any ideal $\mathfrak{a}$ of A. By hypothesis we have $\bigcap_n I^n(\mathfrak{a} \otimes M) = 0$, so that we need only prove that $\mathrm{Ker}(j) \subset I^n(\mathfrak{a} \otimes M)$ for all $n > 0$. For a fixed n, by the Artin–Rees lemma, $I^k \cap \mathfrak{a} \subset I^n\mathfrak{a}$ for sufficiently large $k > n$. We now consider the natural map

$$\mathfrak{a}\otimes M \xrightarrow{f} (\mathfrak{a}/I^k\cap\mathfrak{a})\otimes M \xrightarrow{g} (\mathfrak{a}/I^n\mathfrak{a})\otimes M = (\mathfrak{a}\otimes M)/I^n(\mathfrak{a}\otimes M).$$

Since M_{k-1} is flat over $A_{k-1} = A/I^k$, the map

$$(\mathfrak{a}/I^k\cap\mathfrak{a})\otimes_A M = (\mathfrak{a}/I^k\cap\mathfrak{a})\otimes_{A_{k-1}} M_{k-1} \longrightarrow M_{k-1}$$

is injective, so that from the commutative diagram

$$\begin{array}{ccc} \mathfrak{a}\otimes M & \xrightarrow{f} & (\mathfrak{a}/I^k\cap\mathfrak{a})\otimes M \\ {\scriptstyle j}\downarrow & & \downarrow \\ M & \longrightarrow & M_{k-1} \end{array}$$

we get $\mathrm{Ker}\,(j)\subset \mathrm{Ker}(f)\subset \mathrm{Ker}(gf) = I^n(\mathfrak{a}\otimes M)$. This is what we needed to prove. ■

This theorem is particularly effective when A is a Noetherian local ring and I is the maximal ideal, since if A_0 is a field, M_0 is automatically flat over A_0 in (3)–(4'). Also, in this case, requiring M_n to be flat over A_n in (5) is the same as requiring it to be a free A_n-module, by Theorem 7.10.

We now discuss some applications of the above theorem.

Theorem 22.4. Let $(A,\mathfrak{m})$ and $(B,\mathfrak{n})$ be Noetherian local rings, $\hat{A}$ and $\hat{B}$ their respective completions, and $A\longrightarrow B$ a local homomorphism.

(i) For M a finite B-module, set $\hat{M} = M\otimes_B \hat{B}$; then

$$M \text{ is flat over } A \Leftrightarrow \hat{M} \text{ is flat over } A \Leftrightarrow \hat{M} \text{ is a flat over } \hat{A}.$$

(ii) Writing M^* for the $(\mathfrak{m}B)$-adic completion of M we have

$$M \text{ is flat over } A \Leftrightarrow M^* \text{ is flat over } A \Leftrightarrow M^* \text{ is flat over } \hat{A}.$$

Proof. (i) The first equivalence comes from the transitivity law for flatness, together with the fact that $\hat{B}$ is faithfully flat over B; the second, from the fact that both sides are equivalent to $\hat{M}/\mathfrak{m}^n\hat{M}$ being flat over $A/\mathfrak{m}^n$ for all $n>0$.

(ii) All three conditions are equivalent to $M/\mathfrak{m}^nM$ being flat over $A/\mathfrak{m}^n$ for all n.

Theorem 22.5. Let $(A,\mathfrak{m},k)$ and $(B,\mathfrak{n},k')$ be Noetherian local rings, $A\longrightarrow B$ a local homomorphism, and $u: M\longrightarrow N$ a morphism of finite B-modules. Then if N is flat over A, the following two conditions are equivalent:

(1) u is injective and $N/u(M)$ is flat over A;

(2) $\bar{u}: M\otimes_A k\longrightarrow N\otimes_A k$ is injective.

Proof. (1)⇒(2) is easy, so we only give the proof of (2)⇒(1). Suppose that $x\in M$ is such that $u(x)=0$; then $\bar{u}(\bar{x})=0$, so that $\bar{x}=0$, in other words, $x\in\mathfrak{m}M$. Now assuming $x\in\mathfrak{m}^nM$, we will deduce $x\in\mathfrak{m}^{n+1}M$. Let $\{a_1,\ldots,a_r\}$ be a minimal basis of the A-module $\mathfrak{m}^n$, and write $x=\sum a_iy_i$ with $y_i\in M$; then $0=\sum a_iu(y_i)$. Since N is flat over A, by Theorem 7.6 there exist $c_{ij}\in A$ and $z_j\in N$ such that

$$\sum_i a_ic_{ij} = 0 \quad \text{for all } j, \quad \text{and} \quad u(y_i) = \sum_j c_{ij}z_j \quad \text{for all } i.$$

By choice of $a_1,\ldots,a_r$, all the $c_{ij}\in\mathfrak{m}$, and hence $u(y_i)\in\mathfrak{m}N$ and $\bar{u}(\bar{y}_i)=0$, so that $\bar{y}_i=0$, and $y_i\in\mathfrak{m}M$. Therefore $x\in\mathfrak{m}^{n+1}M$. We have proved that $x\in\bigcap_n\mathfrak{m}^nM=0$, and hence u is injective. Now from $0\to M\longrightarrow N\longrightarrow N/u(M)\to 0$ we get $\mathrm{Tor}_1^A(k,N/u(M))=0$, so that by Theorem 3, $N/u(M)$ is flat over A. ∎

Corollary. Let A, B and $A\longrightarrow B$ be as above, and M a finite B-module; set $\bar{B}=B\otimes_A k=B/\mathfrak{m}B$, and for $x_1,\ldots,x_n\in\mathfrak{n}$ write $\bar{x}_i$ for the images in $\bar{B}$ of x_i. Then the following conditions are equivalent:

(1) $x_1,\ldots x_n$ is an M-sequence and $M_n=M/\sum_1^n x_iM$ is flat over A;

(2) $\bar{x}_1,\ldots,\bar{x}_n$ is an $M\otimes k$-sequence and M is flat over A.

Proof. (2)⇒(1) follows at once from the theorem. For (1)⇒(2) we must prove that $M_i=M/(x_1M+\cdots+x_iM)$ is flat for $i=1,\ldots,n$; but if M_i is flat over A then by Theorem 2, so is M_{i-1}. ∎

Theorem 22.6. Let A be a Noetherian ring, B a Noetherian A-algebra, M a finite B-module, and $b\in B$ a given element. Suppose that M is flat over A and that b is $M/(P\cap A)$-regular for every maximal ideal P of B; then b is M-regular and M/bM is flat over A.

Proof. Write K for the kernel of $M\xrightarrow{b}M$; then $K=0\Leftrightarrow K_P=0$ for all P. Hence b is M-regular if and only if b is M_P-regular for all P. Moreover, according to Theorem 7.1, A-flatness is also a local property in both A and B, so that we can replace B by B_P (for a maximal ideal P of B), A by $A_{(P\cap A)}$ and M by M_P, and this case reduces to Theorem 5. ∎

Corollary. Let A be a Noetherian ring, $B=A[X_1,\ldots,X_n]$ the polynomial ring over A, and let $f(X)\in B$. If the ideal of A generated by the coefficients of f contains 1 then f is a non-zero-divisor of B, and B/fB is flat over A. The same thing holds for the formal power series ring $B=A[\![X_1,\ldots,X_n]\!]$.

Proof. The polynomial ring is a free A-module, and therefore flat; the formal power series ring is flat by Ex. 7.4. Furthermore, for $\mathfrak{p}\in\mathrm{Spec}\,A$, if $B=A[X_1,\ldots,X_n]$ then $B/\mathfrak{p}B=(A/\mathfrak{p})[X_1,\ldots,X_n]$, and in the formal power series case we also have $B/\mathfrak{p}B=(A/\mathfrak{p})[\![X_1,\ldots,X_n]\!]$ since $\mathfrak{p}$ is finitely generated. In either case $B/\mathfrak{p}B$ is an integral domain, so that the assertion follows directly from the theorem. ∎

Remark (Flatness of a graded module). Let G be an Abelian group, $R=\bigoplus_{g\in G}R_g$ a G-graded ring and $M=\bigoplus_{g\in G}M_g$ a graded R-module, not necessarily finitely generated.

(1) The following three conditions are equivalent:

(*a*) M is R-flat;

(b) If $\mathscr{S}:\cdots\longrightarrow N\longrightarrow N'\longrightarrow N''\longrightarrow\cdots$ is an exact sequence of graded R-modules and R-linear maps preserving degrees, then $\mathscr{S}\otimes M$ is exact;

(*c*) $\operatorname{Tor}_1^R(M, R/H) = 0$ for every finitely generated homogeneous ideal H of R. The proof is left to the reader as an exercise, or can be found in Herrmann and Orbanz [3]. Using this criterion one can adapt the proof of Theorem 3 to prove the following graded version.

(2) Let I be a (not necessarily homogeneous) ideal of R. Suppose that

(i) for every finitely generated homogeneous ideal H of R, the R-module $H \otimes_R M$ is I-adically separated;

(ii) $M_0 = M/IM$ is R/I-flat;

(iii) $\operatorname{Tor}_1^R(M, R/I) = 0$.

Then M is R-flat.

As an application one can prove the following:

(3) Let $A = \bigoplus_{n \geq 0} A_n$ and $B = \bigoplus_{n \geq 0} B_n$ be graded Noetherian rings. Assume that A_0, B_0 are local rings with maximal ideals $\mathfrak{m}$, $\mathfrak{n}$ and set $M = \mathfrak{m} + A_1 + A_2 + \cdots$, $N = \mathfrak{n} + B_1 + B_2 + \cdots$; let $f: A \longrightarrow B$ be a ring homomorphism of degree 0 such that $f(\mathfrak{m}) \subset \mathfrak{n}$. Then the following are equivalent:

(*a*) B is A-flat;

(*b*) B_N is A-flat;

(*c*) B_N is A_M-flat.

Exercises to §22. Prove the following propositions.

22.1. (The Nagata flatness theorem, see [N1], p. 65). Let $(A, \mathfrak{m}, k)$ and $(B, \mathfrak{n}, k')$ be Noetherian local rings, and suppose that $A \subset B$ and that $\mathfrak{m}B$ is an $\mathfrak{n}$-primary ideal. We say that the *transition theorem* holds between A and B if $l_A(A/\mathfrak{q}) \cdot l_B(B/\mathfrak{m}B) = l_B(B/\mathfrak{q}B)$ for every $\mathfrak{m}$-primary ideal $\mathfrak{q}$ of A. This holds if and only if B is flat over A.

22.2. Let $(A, \mathfrak{m})$ be a Noetherian local ring, and $k \subset A$ a subfield. If $x_1, \ldots, x_n \in \mathfrak{m}$ is an A-sequence then $x_1, \ldots, x_n$ are algebraically independent over k, and A is flat over $C = k[x_1, \ldots, x_n]$ (Hartshorne [2]).

22.3. Let $(A, \mathfrak{m}, k)$ be a Noetherian local ring, B a Noetherian A-algebra, and M a finite B-module. Suppose that $\mathfrak{m}B \subset \operatorname{rad}(B)$. If $x \in \mathfrak{m}$ is both A-regular and M-regular, and if M/xM is flat over A/xA then M is flat over A.

22.4. Let A be a Noetherian ring and B a flat Noetherian A-algebra; if I and J are ideals of A and B such that $IB \subset J$ then the J-adic completion of B is flat over the I-adic completion of A.

23 Flatness and fibres

Let $(A, \mathfrak{m})$ and $(B, \mathfrak{n})$ be Noetherian local rings, and $\varphi: A \longrightarrow B$ a local homomorphism. We set $F = B \otimes_A k(\mathfrak{m}) = B/\mathfrak{m}B$ for the fibre ring of φ over $\mathfrak{m}$. If B is flat over A then according to Theorem 15.1, we have

$$(*) \quad \dim B = \dim A + \dim F.$$

As the following shows, under certain conditions the converse holds.

Theorem 23.1. Let A, B and F be as above. If A is a regular local ring, B is Cohen–Macaulay, and $\dim B = \dim A + \dim F$ then B is flat over A.
Proof. By induction on $\dim A$. If $\dim A = 0$ then A is a field, and we are done. If $\dim A > 0$, take $x \in \mathfrak{m} - \mathfrak{m}^2$ and set $A' = A/xA$ and $B' = B/xB$. By Theorem 15.1,

$$\dim B' \leqslant \dim A' + \dim F = \dim A - 1 + \dim F = \dim B - 1,$$

and using a system of parameters of B' one sees that $\dim B' \geqslant \dim B - 1$, so that

$$\dim B' = \dim A' + \dim F = \dim B - 1.$$

One sees easily from this that x is B-regular and B' is a CM ring. Hence by induction B' is flat over A'. Thus $\mathrm{Tor}_1^{A'}(A/\mathfrak{m}, B') = 0$; moreover, x is both A-regular and B-regular, so that $\mathrm{Tor}_1^{A'}(A/\mathfrak{m}, B') = \mathrm{Tor}_1^{A}(A/\mathfrak{m}, B)$. Therefore by Theorem 22.3, B is flat over A. ∎

We give a translation of the above theorem into algebraic geometry for ease of application. (The language is that of modern algebraic geometry, see for example [Ha], Ch. 2.)

Corollary. Let k be a field, X and Y irreducible algebraic k-schemes, and let $f: Y \longrightarrow X$ be a morphism. Set $\dim X = n$, $\dim Y = m$, and suppose that the following conditions hold: (1) X is regular; (2) Y is Cohen–Macaulay; (3) f takes closed points of Y into closed points of X (this holds for example if f is proper); (4) for every closed point $x \in X$ the fibre $f^{-1}(x)$ is $(m-n)$-dimensional (or empty). Then f is flat.
Proof. Let $y \in Y$ be a closed point, and set $x = f(y)$, $A = \mathcal{O}_{X,x}$ and $B = \mathcal{O}_{Y,y}$. We have $\dim A = n$, $\dim B = m$, and since by Theorem 15.1 $\dim B/\mathfrak{m}_x B \geqslant m - n$, we get $\dim B/\mathfrak{m}_x B = m - n$ from (4). Therefore by the above theorem B is flat over A, and this is what was required to prove. ∎

Theorem 23.2. Let $\varphi: A \longrightarrow B$ be a homomorphism of Noetherian rings, and let E be an A-module and G a B-module. Suppose that G is flat over A; then we have the following:

(i) if $\mathfrak{p} \in \mathrm{Spec}\, A$ and $G/\mathfrak{p}G \neq 0$ then

$$^a\varphi(\mathrm{Ass}_B(G/\mathfrak{p}G)) = \mathrm{Ass}_A(G/\mathfrak{p}G) = \{\mathfrak{p}\};$$

(ii) $\mathrm{Ass}_A(E \otimes_A G) = \bigcup_{\mathfrak{p} \in \mathrm{Ass}_A(E)} \mathrm{Ass}_B(G/\mathfrak{p}G)$.

Proof. (i) $G/\mathfrak{p}G = G \otimes_A (A/\mathfrak{p})$ is flat over $A/\mathfrak{p}$, and $A/\mathfrak{p}$ is an integral domain, so that any non-zero element of $A/\mathfrak{p}$ is $G/\mathfrak{p}G$-regular (see Ex. 7.5.). In other words, the elements of $A - \mathfrak{p}$ are $G/\mathfrak{p}G$-regular. This gives $\mathrm{Ass}_A(G/\mathfrak{p}G) = \{\mathfrak{p}\}$. Also, if $P \in \mathrm{Ass}_B(G/\mathfrak{p}G)$ then there exists $\xi \in G/\mathfrak{p}G$ such that $\mathrm{ann}_B(\xi) = P$, and then $P \cap A = \mathrm{ann}_A(\xi) \in \mathrm{Ass}_A(G/\mathfrak{p}G) = \{\mathfrak{p}\}$.

(ii) If $\mathfrak{p} \in \mathrm{Ass}_A(E)$ then there is an exact sequence of the form

$0 \to A/\mathfrak{p} \longrightarrow E$, and since G is flat the sequence $0 \to G/\mathfrak{p}G \longrightarrow E \otimes G$ is also exact; thus

$$\mathrm{Ass}_B(G/\mathfrak{p}G) \subset \mathrm{Ass}_B(E \otimes G).$$

Conversely, if $P \in \mathrm{Ass}(E \otimes G)$ then there is an $\eta \in E \otimes G$ such that $\mathrm{ann}_B(\eta) = P$. We write $\eta = \sum_1^n x_i \otimes y_i$ with $x_i \in E$ and $y_i \in G$, and set $E' = \sum_1^n Ax_i$; then by flatness of G, we can view $E' \otimes G$ as a submodule $E' \otimes G \subset E \otimes G$. Since $\eta \in E' \otimes G$ we have $P \in \mathrm{Ass}_B(E' \otimes G)$. Now E' is a finite A-module, so that we can choose a shortest primary decomposition of 0 in E', say $0 = Q_1 \cap \cdots \cap Q_r$. Since E' can be embedded in $\bigoplus (E'/Q_i)$, if we set $E'_i = E'/Q_i$ then

$$\mathrm{Ass}_B(E' \otimes G) \subset \bigcup_i \mathrm{Ass}_B(E'_i \otimes G),$$

and therefore $P \in \mathrm{Ass}_B(E'_i \otimes G)$ for some i. This E'_i is a finite A-module having just one associated prime, say $\mathfrak{p}$. We have $\mathfrak{p} \in \mathrm{Ass}_A(E') \subset \mathrm{Ass}_A(E)$. For large enough ν we get $\mathfrak{p}^\nu E'_i = 0$, so that $\mathfrak{p}^\nu(E'_i \otimes G) = 0$, and thus $\mathfrak{p} \subset P \cap A$. Moreover, an element of $A - \mathfrak{p}$ is E'_i-regular, and hence also $E'_i \otimes G$-regular, so that finally $\mathfrak{p} = P \cap A$. Now choose a chain of submodules of E'_i,

$$E'_i = E_0 \supset E_1 \supset \cdots \supset E_r = 0$$

such that $E_j/E_{j+1} \simeq A/\mathfrak{p}_j$ with $\mathfrak{p}_j \in \mathrm{Spec}\, A$. Then also

$$E'_i \otimes G \supset E_1 \otimes G \supset \cdots \supset E_r \otimes G = 0,$$

with

$$(E_j \otimes G)/(E_{j+1} \otimes G) \simeq (A/\mathfrak{p}_j) \otimes G = G/\mathfrak{p}_j G,$$

so that $\mathrm{Ass}_B(E'_i \otimes G) \subset \bigcup_j \mathrm{Ass}_B(G/\mathfrak{p}_j G)$. Therefore $P \in \mathrm{Ass}_B(G/\mathfrak{p}_j G)$ for some j, but by (i), $P \cap A = \mathfrak{p}_j$, so that $\mathfrak{p}_j = \mathfrak{p}$ and $P \in \mathrm{Ass}_B(G/\mathfrak{p}G)$. ■

Theorem 23.3. Let $(A, \mathfrak{m}, k)$ and $(B, \mathfrak{n}, k')$ be Noetherian local rings, and $\varphi: A \longrightarrow B$ a local homomorphism. Let M be a finite A-module, N a finite B-module, and assume that N is flat over A. Then

$$\mathrm{depth}_B(M \otimes_A N) = \mathrm{depth}_A M + \mathrm{depth}_B(N/\mathfrak{m}N).$$

Proof. Let $x_1, \ldots, x_r \in \mathfrak{m}$ be a maximal M-sequence, and $y_1, \ldots, y_s \in \mathfrak{n}$ a maximal $N/\mathfrak{m}N$-sequence. Writing x'_i for the images of x_i in B, let us prove that $x'_1, \ldots, x'_r, y_1, \ldots, y_s$ is a maximal $M \otimes N$-sequence. Now $x'_1, \ldots, x'_r$ is an $M \otimes N$-sequence, and if we set $M_r = M/\sum x_i M$ then

$$\mathfrak{m} \in \mathrm{Ass}_A(M_r), \quad \text{and} \quad (M \otimes N)/\sum_{i=1}^r x'_i(M \otimes N) = M_r \otimes N.$$

Moreover, by the corollary of Theorem 22.5, y_1 is N-regular, and

$N_1 = N/y_1N$ is flat over A, so that from the exact sequence $0 \to N \xrightarrow{y_1} N \longrightarrow N_1 \to 0$ we get the exact sequence $0 \to M_r \otimes N \longrightarrow M_r \otimes N \longrightarrow M_r \otimes N_1 \to 0$. Proceeding in the same way we see that $y_1, \ldots, y_s$ is an $M_r \otimes N$-sequence. After this we need only prove that the B-module.

$$(M \otimes N)/(\sum x_i'(M \otimes N) + \sum y_j(M \otimes N)) = M_r \otimes N_s$$

has depth 0, that is $\mathfrak{n} \in \mathrm{Ass}_B(M_r \otimes N_s)$; however, $\mathfrak{m} \in \mathrm{Ass}_A(M_r)$ and $\mathfrak{n} \in \mathrm{Ass}_B(N_s/\mathfrak{m}N_s)$, so that this follows at once from the previous theorem.

Corollary. Let $A \longrightarrow B$ be a local homomorphism of Noetherian rings as in the theorem, and set $F = B/\mathfrak{m}B$. Assume that B is flat over A. Then

(i) $\operatorname{depth} B = \operatorname{depth} A + \operatorname{depth} F$;

(ii) B is CM $\Leftrightarrow A$ and F are both CM.

Proof. (i) is the case $M = A$, $N = B$ of the theorem. From (i) and (*) we have

$$\dim B - \operatorname{depth} B = (\dim A - \operatorname{depth} A) + (\dim F - \operatorname{depth} F)$$

and in view of $\dim A \geqslant \operatorname{depth} A$ and $\dim F \geqslant \operatorname{depth} F$, (ii) is clear. ■

Theorem 23.4. Let $A \longrightarrow B$ be a local homomorphism of Noetherian local rings, set $\mathfrak{m} = \mathrm{rad}(A)$ and $F = B/\mathfrak{m}B$. We assume that B is flat over A; then

$$B \text{ is Gorenstein} \Leftrightarrow A \text{ and } F \text{ are both Gorenstein.}$$

Proof (K. Watanabe [1]). By the corollary just proved, we can assume that A, B and F are CM. Set $\dim A = r$ and $\dim F = s$, and let $\{x_1, \ldots, x_r\}$ be a system of parameters of A, and $\{y_1, \ldots, y_s\}$ a subset of B which reduces to a system of parameters of F modulo $\mathfrak{m}B$. Then as we have seen in the proof of Theorem 3, $\{x_1, \ldots, x_r, y_1, \ldots, y_s\}$ is a B-sequence, and therefore a system of parameters of B, and $\bar{B} = B/(\underline{x}, \underline{y})B$ is flat over $\bar{A} = A/(\underline{x})A$. Thus replacing A and B by $\bar{A}$ and $\bar{B}$, we can reduce to the case $\dim A = \dim B = 0$. Now in general, a zero-dimensional local ring (R, M) is Gorenstein if and only if $\mathrm{Hom}_R(R/M, R) = (0:M)_R$ is isomorphic to R/M. Now set

$$\mathrm{rad}(B) = \mathfrak{n}, \quad \mathrm{rad}(F) = \mathfrak{n}/\mathfrak{m}B = \bar{\mathfrak{n}} \quad \text{and} \quad (0:\mathfrak{m})_A = I.$$

Then I is of the form $I \simeq (A/\mathfrak{m})^t$ for some t, and $(0:\mathfrak{m}B)_B = IB \simeq (A/\mathfrak{m})^t \otimes B = F^t$. Furthermore, we have $(0:\mathfrak{n})_B = (0:\mathfrak{n})_{IB} \simeq ((0:\bar{\mathfrak{n}})_F)^t$, and hence if we set $(0:\bar{\mathfrak{n}})_F \simeq (F/\bar{\mathfrak{n}})^u = (B/\mathfrak{n})^u$ then $(0:\mathfrak{n})_B \simeq (B/\mathfrak{n})^{tu}$. Therefore

$$B \text{ is Gorenstein} \Leftrightarrow tu = 1 \Leftrightarrow t = u = 1 \Leftrightarrow A \text{ are } F \text{ are Gorenstein.} \quad ■$$

Theorem 23.5. If A is Gorenstein then so are $A[X]$ and $A[\![X]\!]$.

Proof. We write B for either of $A[X]$ or $A[\![X]\!]$, so that B is flat over A.

For any maximal ideal M of B we set $M \cap A = \mathfrak{p}$ and $A_\mathfrak{p}/\mathfrak{p}A_\mathfrak{p} = \kappa(\mathfrak{p})$. In case $B = A[X]$, the local ring B_M is a localisation of $B \otimes_A A_\mathfrak{p} = A_\mathfrak{p}[X]$, and the fibre ring of $A_\mathfrak{p} \longrightarrow B_M$ is a localisation of $\kappa(\mathfrak{p})[X]$, hence regular. In case $B = A[\![X]\!]$ then $X \in M$, and $\mathfrak{p}$ a maximal ideal of A, so that $\kappa(\mathfrak{p}) = A/\mathfrak{p}$ and

$$B \otimes_A \kappa(\mathfrak{p}) = (A/\mathfrak{p})[\![X]\!] = \kappa(\mathfrak{p})[\![X]\!].$$

This is a regular local ring, and is the fibre ring of $A_\mathfrak{p} \longrightarrow B_M$. Thus in either case B_M is Gorenstein by the previous theorem. ■

Theorem 23.6. Let A be a Gorenstein ring containing a field k; then for any finitely generated field extension K of k, the ring $A \otimes_k K$ is Gorenstein.

Proof. We need only consider the case that K is generated over k by one element x. If x is transcendental over k then $A \otimes_k K$ is isomorphic to a localisation of $A \otimes_k k[X] = A[X]$, and since $A[X]$ is Gorenstein, so is $A \otimes K$. If x is algebraic over k then since $K \simeq k[X]/(f(X))$ with $f(X) \in k[X]$ a monic polynomial, we have

$$A \otimes K = A[X]/(f(X));$$

now $A[X]$ is Gorenstein and $f(X)$ is a non-zero-divisor of $A[X]$, so that we see that $A \otimes K$ is also Gorenstein. ■

Remark. Theorems 5 and 6 also hold on replacing Gorenstein by Cohen–Macaulay; the proofs are exactly the same. For complete intersection rings the counterpart of Theorem 4 also holds, so that the analogs of Theorems 5 and 6 follow; the proof involves André homology (Avramov [1]). As we see in the next theorem, a slightly weaker form of the same result holds for regular rings.

Theorem 23.7. Let $(A, \mathfrak{m}, k)$ and $(B, \mathfrak{n}, k')$ be Noetherian local rings, and $A \longrightarrow B$ a local homomorphism; set $F = B/\mathfrak{m}B$. We assume that B is flat over A.

(i) If B is regular then so is A.

(ii) If A and F are regular then so is B.

Proof. (i) We have $\mathrm{Tor}_i^A(k, k) \otimes_A B = \mathrm{Tor}_i^B(B \otimes k, B \otimes k)$, and the right-hand side is zero for $i > \dim B$. Since B is faithfully flat over A, we have $\mathrm{Tor}_i^A(k, k) = 0$ for $i \gg 0$, so that by §19, Lemma 1, (i), $\mathrm{proj\,dim}_A k < \infty$, and since $\mathrm{proj\,dim}\, k = \mathrm{gl\,dim}\, A$, by Theorem 19.2, A is regular.

(ii) Set $r = \dim A$ and $s = \dim F$. Let $\{x_1, \dots, x_r\}$ be a regular system of parameters of A, and $\{y_1, \dots, y_s\}$ a subset of $\mathfrak{n}$ which maps to a regular system of parameters of F. Since $A \longrightarrow B$ is injective, we can view A as a subring $A \subset B$. Then $\{x_1, \dots, x_r, y_1, \dots, y_s\}$ generates $\mathfrak{n}$, but $\dim B = r + s$, so that B is regular. ■

Remark. In Theorem 7, even if B is regular, F need not be. For example, let k be a field, x an indeterminate over k, and $B = k[x]_{(x)}$, $A = k[x^2]_{(x^2)} \subset B$; then $F = B/x^2B = k[x]/(x^2)$ has a nilpotent element. By Theorem 1, or directly, we see that B is flat over A. (From a geometrical point of view, this example corresponds to the projection of the plane curve $y = x^2$ onto the y-axis, and, not surprisingly, the fibre over the origin is singular.)

Consider the following conditions (R_i) and (S_i) for $i = 0, 1, 2, \dots$ on a Noetherian ring A:

(R_i) A_P is regular for all $P \in \operatorname{Spec} A$ with $\operatorname{ht} P \leqslant i$;

(S_i) depth $A_P \geqslant \min(\operatorname{ht} P, i)$ for all $P \in \operatorname{Spec} A$.

(S_0) always holds. (S_1) says that all the associated primes of A are minimal, that is A does not have embedded associated primes. $(R_0) + (S_1)$ is the necessary and sufficient condition for A to be reduced. (S_i) for all $i \geqslant 0$ is just the definition of a CM ring.

For an integral domain A, (S_2) is equivalent to the condition that every prime divisor of a non-zero principal ideal has height 1. The characterisation of normal integral domain given in the corollary to Theorem 11.5 can be somewhat generalised as follows.

Theorem 23.8 (Serre). $(R_1) + (S_2)$ are necessary and sufficient conditions for a Noetherian ring A to be normal.

Proof. We defined a normal ring (see §9), by the condition that the localisation at every prime is an integrally closed domain. The conditions (R_i) and (S_i) are also conditions on localisations, so that we can assume that A is local.

Necessity. This follows from Theorems 11.2 and 11.5.

Sufficiency. Since A satisfies (R_0) and (S_1) it is reduced, and the shortest primary decomposition of (0) is $(0) = P_1 \cap \cdots \cap P_r$, where P_i are the minimal primes of A. Thus if we set K for the total ring of fractions of A, we have

$$K = K_1 \times \cdots \times K_r, \text{ with } K_i \text{ the field of fractions of } A/P_i.$$

First of all we show that A is integrally closed in K. Suppose that we have a relation in K of the form

$$(a/b)^n + c_1(a/b)^{n-1} + \cdots + c_n = 0,$$

with $a, b, c_1, \dots, c_n \in A$ and b an A-regular element. This is equivalent to a relation

$$a^n + \sum_1^n c_i a^{n-i} b^i = 0$$

in A. Let $P \in \operatorname{Spec} A$ be such that $\operatorname{ht} P = 1$; then by (R_1), A_P is regular, and therefore normal, so that $a_P \in b_P A_P$, where we write a_P, b_P for the images in A_P of a, b. Now b is A-regular, so that by (S_2), all the prime divisors of the principal ideal bA have height 1; thus if $bA = \mathfrak{q}_1 \cap \cdots \cap \mathfrak{q}_m$ is a shortest primary decomposition and we set $\mathfrak{p}_i$ for the prime divisor of $\mathfrak{q}_i$, then $a \in bA_{\mathfrak{p}_i} \cap A = \mathfrak{q}_i$ for all i, and hence $a \in bA$, so that $a/b \in A$. Therefore A is integrally closed in K; in particular, the idempotents e_i of K, which satisfy $e_i^2 - e_i = 0$, must belong to A, so that from $1 = \sum e_i$ and $e_i e_j = 0$ for $i \neq j$ we get

$$A = Ae_1 \times \cdots \times Ae_r.$$

Now since A is supposed to be local, we must have $r = 1$, so that A is an integrally closed domain. ∎

Theorem 23.9. Let $(A, \mathfrak{m})$ and $(B, \mathfrak{n})$ be Noetherian local rings and $A \longrightarrow B$ a local homomorphism. Suppose that B is flat over A, and that $i \geqslant 0$ is a given integer. Then

(i) if B satisfies (R_i), so does A;

(ii) if both A and the fibre ring $B \otimes_A k(\mathfrak{p})$ over every prime ideal $\mathfrak{p}$ of A satisfy (R_i), so does B.

(iii) The above two statements also hold with (S_i) in place of (R_i).

Proof. (i) For $\mathfrak{p} \in \operatorname{Spec} A$, since B is faithfully flat over A, there is a prime ideal of B lying over $\mathfrak{p}$; if we let P be a minimal element among these then $\operatorname{ht}(P/\mathfrak{p}B) = 0$, so that $\operatorname{ht} P = \operatorname{ht} \mathfrak{p}$. Hence $\operatorname{ht} \mathfrak{p} \leqslant i \Rightarrow B_P$ is regular, so that by Theorem 7, $A_\mathfrak{p}$ is regular. Also, by the corollary to Theorem 3, $\operatorname{depth} B_P = \operatorname{depth} A_\mathfrak{p}$, so that one sees easily that (S_i) for B implies (S_i) for A.

(ii) Let $P \in \operatorname{Spec} B$ and set $P \cap A = \mathfrak{p}$. If $\operatorname{ht} P \leqslant i$ then we have $\operatorname{ht} \mathfrak{p} \leqslant i$ and $\operatorname{ht}(P/\mathfrak{p}B) \leqslant i$, hence $A_\mathfrak{p}$ and $B_P/\mathfrak{p}B_P$ are both regular, so by Theorem 7, (ii), B_P is regular. Hence B satisfies (R_i). Moreover, for (S_i) we have

$$\begin{aligned}\operatorname{depth} B_P &= \operatorname{depth} A_\mathfrak{p} + \operatorname{depth} B_P/\mathfrak{p}B_P \\ &\geqslant \min(\operatorname{ht}\mathfrak{p}, i) + \min(\operatorname{ht} P/\mathfrak{p}B, i) \\ &\geqslant \min(\operatorname{ht}\mathfrak{p} + \operatorname{ht} P/\mathfrak{p}B, i) = \min(\operatorname{ht} P, i).\end{aligned}$$ ∎

Corollary. Under the same assumptions as Theorem 9, we have

(i) if B is normal (or reduced) then so is A;

(ii) if both A and the fibre rings of $A \longrightarrow B$ are normal (or reduced) then so is B.

Remark. If A and the closed fibre ring $F = B/\mathfrak{m}B$ only are normal, then B does not have to be; for instance, there are known examples of normal Noetherian rings for which the completion is not normal.

Finally, we would like to draw the reader's attention to the following obvious, but useful, fact concerning the fibre ring. Let $\varphi' : A' \longrightarrow B'$ be a ring homomorphism and I an ideal of A'; we set $A = A'/I$, $B = B'/IB'$, and write $\varphi : A \longrightarrow B$ for the map induced by φ'. If $\mathfrak{p}' \in \operatorname{Spec} A'$ is such that

$I \subset \mathfrak{p}'$, we set $\mathfrak{p} = \mathfrak{p}'/I$; then the fibre of φ' over $\mathfrak{p}'$ coincides with the fibre of φ over $\mathfrak{p}$. To see this,

$$B' \otimes_{A'} \kappa(\mathfrak{p}') = B' \otimes_{A'} (A'/\mathfrak{p}')_{\mathfrak{p}'} = B \otimes_A (A/\mathfrak{p})_{\mathfrak{p}} = B \otimes_A \kappa(\mathfrak{p}).$$

It follows from this that if all the fibre rings of φ' have a good property, the same is true of φ. For an example of this, see Ex. 23.2.

Exercises to §23. Prove the following propositions.

23.1. If A is a Gorenstein local ring then all the fibre rings of $A \longrightarrow \hat{A}$ are again Gorenstein; the same thing holds for Cohen–Macaulay.

23.2. If A is a quotient of a CM local ring, and satisfies (S_i), then the completion $\hat{A}$ also satisfies (S_i). In particular, if A does not have embedded associated primes then neither does $\hat{A}$.

23.3. Give another proof of Theorem 4 along the following lines: (1) Using $\mathrm{Ext}^i_A(A/\mathfrak{m}, A) \otimes_A B = \mathrm{Ext}^i_B(F, B)$, show that B Gorenstein implies A Gorenstein. (2) Assuming that A is Gorenstein, prove that F is Gorenstein if and only if B is. Firstly reduce to the case $\dim A = 0$. Then prove that $\mathrm{Ext}^i_B(F, B) = 0$ for $i > 0$ and $\simeq F$ for $i = 0$, and deduce that if $0 \to B \longrightarrow I^{\boldsymbol{\cdot}}$ is an injective resolution of B as a B-module then $0 \to F \longrightarrow \mathrm{Hom}_B(F, I^{\boldsymbol{\cdot}})$ is an injective resolution of F as an F-module, so that, writing k for the residue field of B, we have $\mathrm{Ext}^i_B(k, B) = \mathrm{Ext}^i_F(k, F)$ for all i.

24 Generic freeness and open loci results

Let A be a Noetherian integral domain, and M a finite A-module. Then there exists $0 \neq a \in A$ such that M_a is a free A_a-module. This follows from Theorem 4.10, or can be proved as follows: choose a filtration

$$M = M_0 \supset M_1 \supset \cdots \supset M_r = 0$$

such that $M_{i-1}/M_i \simeq A/\mathfrak{p}_i$, with $\mathfrak{p}_i \in \mathrm{Spec}\, A$; then if we take $a \neq 0$ contained in every non-zero $\mathfrak{p}_i$ we see that every $(M_{i-1}/M_i)_a$ is either zero or isomorphic to A_a, so that M_a is a free A_a-module.

For applications, we require a more general version of this, which does not assume M to be finite. We give below a theorem due to Hochster and Roberts [1]. First we give the following lemma.

Lemma. Let B be a Noetherian ring, and C a B-algebra generated over B by a single element x; let E be a finite C-module, and $F \subset E$ a finite B-module such that $CF = E$. Then $D = E/F$ has a filtration

$$0 = G_0 \subset G_1 \subset \cdots \subset G_i \subset G_{i+1} \subset \cdots \subset D \quad \text{with} \quad D = \bigcup_{i=0}^{\infty} G_i$$

such that the successive quotients G_{i+1}/G_i are isomorphic to a finite number of finite B-modules.

Proof. Set

$$G_i' = F + xF + \cdots + x^iF \subset E, \quad G_i = G_i'/F,$$

and

$$F_i = \{f \in F \mid x^{i+1} f \in G_i\} \subset F.$$

Then $0 \subset G_1 \subset \cdots \subset G_k \subset G_{k+1} \subset \cdots$ is a filtration of D, and $G_{i+1}/G_i \simeq F/F_i$; on the other hand, $F_0 \subset F_1 \subset \cdots \subset F_i \subset \cdots$ is an increasing chain of B-submodules of F, so must terminate. ■

Theorem 24.1. Let A be a Noetherian integral domain, R a finitely generated A-algebra, and S a finitely generated R-algebra; we let E be a finite S-module, $M \subset E$ an R-submodule which is finite over R, and $N \subset E$ an A-submodule which is finite over A, and set $D = E/(M + N)$. Then there exists $0 \neq a \in A$ such that D_a is a free A_a-module.

Proof. Write A' for the image of A in R, and suppose that $R = A'[u_1, \ldots, u_h]$; similarly, write R' for the image of R in S, and suppose that $S = R'[v_1, \ldots, v_k]$. We work by induction on $h + k$; if $h = k = 0$ then D is a finite A-module, and we have already dealt with this case.

Write $R_j = A'[u_1, \ldots, u_j]$ for $0 \leqslant j \leqslant h$, and $S_j = R'[v_1, \ldots, v_j]$ for $0 \leqslant j \leqslant k$.

Suppose first that $k > 0$; set $M + N = M' \subset E$. We have a filtration

$$S_0M' \subset S_1M' \subset \cdots \subset S_kM' = SM' \subset E,$$

the successive quotients of which are S_0M', $S_1M'/S_0M', \ldots, S_{k-1}M'/S_{k-2}M'$, $S_kM'/S_{k-1}M'$, E/SM'. We can apply the induction hypothesis to each of these except the last two. By virtue of the lemma, $S_kM'/S_{k-1}M'$ has a filtration with (up to isomorphism) just a finite number of finite S_{k-1}-modules appearing as quotients, and so we can apply the induction hypothesis again. For the final term, write $E' = E/SM'$, and let $e_1, \ldots, e_n$ be a set of generators of E' over S; write $E_{k-1} = S_{k-1}e_1 + \cdots + S_{k-1}e_n$. Then $SE_{k-1} = E'$, so that the lemma again gives a filtration of E' with essentially finitely many finite S_{k-1}-modules appearing as quotients, and we can apply the induction hypothesis to this term also.

If $k = 0$ then E is a finite R-module, and replacing E by E/M we can assume that $M = 0$. The preceding proof then applies almost verbatim to this case, with R_j instead of S_j. ■

Theorem 24.2 (topological Nagata criterion). Let A be a Noetherian ring, and $U \subset \operatorname{Spec} A$ a subset. Then the following two conditions are necessary and sufficient for $U \subset \operatorname{Spec} A$ to be open.

(1) for $P, Q \in \operatorname{Spec} A$, $P \in U$ and $P \supset Q \Rightarrow Q \in U$;

(2) if $P \in U$ then U contains a non-empty open subset of $V(P)$.

Proof. Necessity is obvious, and we prove sufficiency. Let $V_1, \ldots, V_r$ be the irreducible components of the closure of $U^c = \operatorname{Spec} A - U$, and let P_i

be their generic points. If $P_i \in U$ then by (2) there is a proper closed subset W of V_i such that $U^c \cap V_i \subset W$, and so $U^c \subset W \cup (\bigcup_{j \neq 1} V_j)$, which contradicts the definition of V_i. Thus $P_i \notin U$, so that by (1), $V_i \subset U^c$ for all i and therefore U^c is closed. ∎

Theorem 24.3. Let A be a Noetherian ring, B a finitely generated A-algebra, and M a finite B-module. Set $U = \{P \in \operatorname{Spec} B \mid M_P \text{ is flat over } A\}$; then U is open in $\operatorname{Spec} B$.

Proof. We verify the conditions (1) and (2) of Theorem 2.

(1) If $P \supset Q$ are prime ideals of B then for an A-module N we have $N \otimes_A M_Q = (N \otimes_A M_P) \otimes_{B_P} B_Q$, so that if M_P is flat over A then so is M_Q.

(2) Let $P \in U$ and $\mathfrak{p} = P \cap A$; set $\bar{A} = A/\mathfrak{p}$. Now if $Q \in V(P)$, we have $\mathfrak{p}B_Q \subset \operatorname{rad}(B_Q)$, and hence by Theorem 22.3, M_Q is flat over A if and only if $M_Q/\mathfrak{p}M_Q$ is flat over $\bar{A}$ and $\operatorname{Tor}_1^A(M_Q, \bar{A}) = 0$. Now $\operatorname{Tor}_1^A(M_P, \bar{A}) = 0$, and the left-hand side is equal to $\operatorname{Tor}_1^A(M, \bar{A}) \otimes_B B_P$. By computing the Tor by means of a finite free resolution of $\bar{A}$ over A, we see that $\operatorname{Tor}_1^A(M, \bar{A})$ is a finite B-module, so that there is a neighbourhood W of P in $\operatorname{Spec} B$ such that $\operatorname{Tor}_1^A(M_Q, \bar{A}) = 0$ for $Q \in W$. Moreover, by Theorem 1, there exists $a \in A - \mathfrak{p}$ such that $M_a/\mathfrak{p}M_a$ is a free $\bar{A}_a$-module, so that if $Q \notin V(aB)$, then $M_Q/\mathfrak{p}M_Q$ is flat over $\bar{A}$. Thus the open set $(W \cap V(P)) - V(aB)$ of $V(P)$ is contained in U. ∎

Remark. If A is Noetherian and B is a finitely generated A-algebra which is flat over A then it is also known that the map $\operatorname{Spec} B \longrightarrow \operatorname{Spec} A$ is open; see [M], p. 48 or [G2], (2.4.6).

Let A be a ring, and **P** a property of local rings; we define a subset $\mathbf{P}(A) \subset \operatorname{Spec} A$ by $\mathbf{P}(A) = \{\mathfrak{p} \in \operatorname{Spec} A \mid \mathbf{P} \text{ holds for } A_\mathfrak{p}\}$. For example, if $\mathbf{P}$ = regular, complete intersection, Gorenstein or CM we write Reg(A), CI(A), Gor(A) or CM(A) for these loci. The question as to whether $\mathbf{P}(A)$ is open is an interesting and important question. For Reg(A) this is a classical question, but for the other properties the systematic study was initiated by Grothendieck.

The following proposition is called the (ring-theoretic) *Nagata criterion* for the property **P**, and we abbreviate this to (NC).

(NC): Let A be a Noetherian ring. If $\mathbf{P}(A/\mathfrak{p})$ contains a non-empty open subset of $\operatorname{Spec}(A/\mathfrak{p})$ for every $\mathfrak{p} \in \operatorname{Spec} A$, then $\mathbf{P}(A)$ is open in $\operatorname{Spec} A$.

The truth or otherwise of this proposition depends on **P**; in the remainder of this section we discuss some **P** for which (NC) holds. In Ex. 24.2 and Ex. 24.3 we illustrate how (NC) can be applied to prove openness results.

Theorem 24.4 (Nagata). (NC) holds for $\mathbf{P}$ = regular.

Proof. Let $U = \operatorname{Reg}(A)$. A localisation of a regular local ring is again regular, so that U satisfies condition (1) of Theorem 2. We now check condition (2). If $P \in U$ then A_P is regular, so that we can take $x_1, \dots, x_n \in P$ to form a regular system of parameters of A_P (where $n = \operatorname{ht} P$). Then there exists a neighbourhood W of P in $\operatorname{Spec} A$ such that

$$PA_Q = (x_1, \dots, x_n)A_Q$$

for all $Q \in W$. (In fact, if $a \in A$ is an element not contained in P, but contained in every other prime divisor of $(x_1, \dots, x_n)$ then $PA_a = (x_1, \dots, x_n)A_a$.) Moreover, by the hypothesis in (NC) there exists a neighbourhood W' of P in $V(P)$ such that A_Q/PA_Q is regular for $Q \in W'$. Then A_Q is regular for $Q \in W' \cap W$, so that $W' \cap W \subset U$. ■

Theorem 24.5. (NC) also holds for $\mathbf{P} = \mathrm{CM}$.

Proof. As with the previous proof, we reduce to checking condition (2) of Theorem 2. Let $P \in \mathrm{CM}(A)$. If we take $a \in A - P$ and replace A by A_a then we are considering a neighbourhood of P in $\operatorname{Spec} A$, so that we will refer to this procedure as 'passing to a smaller neighbourhood of P'. Since A_P is CM, if $\operatorname{ht} P = n$ we can choose an A_P-sequence $y_1, \dots, y_n \in P$. One sees easily that after passing to a smaller neighbourhood of P, we can assume that

(*a*) $y_1, \dots, y_n$ is an A-sequence; and

(*b*) $I = (y_1, \dots, y_n)A$ is a P-primary ideal.

Then for $Q \in V(P)$, it is equivalent to say that A_Q is CM or that A_Q/IA_Q is CM. Thus replacing A by A/I we can assume that 0 is a P-primary ideal. Then $P^r = 0$ for some $r > 0$. Now consider the filtration $0 \subset P^{r-1} \subset \cdots \subset P \subset A$ of A. Each P^i/P^{i+1} is a finite A/P-module, but A/P is an integral domain, so that passing to a smaller neighbourhood of P we can assume that P^i/P^{i+1} is a free A/P-module for $0 \leqslant i < r$. It is then easy to see that if $x_1, \dots, x_m \in A$ is an A/P-sequence, it is also an A-sequence. However, according to the hypothesis in (NC), passing to a smaller neighbourhood of P, we can assume that A/P is a CM ring. Then for $Q \in V(P)$ the ring A_Q/PA_Q is CM, so that from what we have said above,

$$\operatorname{depth} A_Q \geqslant \operatorname{depth} A_Q/PA_Q = \dim A_Q/PA_Q = \dim A_Q,$$

and A_Q is CM. ■

Let A be a Noetherian ring and I an ideal of A; we set $B = A/I$ and write Y for the closed subset $V(I) \subset \operatorname{Spec} A$. Let M be a finite A-module. We say that M is *normally flat* along Y if the B-module $\operatorname{gr}_I(M) = \bigoplus_{i=0}^{\infty} I^iM/I^{i+1}M$ is flat over B. If B is a local ring, this is the same as saying that each $I^iM/I^{i+1}M$ is a free B-module. Normal flatness is an important notion introduced by Hironaka, and it plays a leading role in the problem of resolution of singularities; we have used it in the above proof in the statement that if P is nilpotent and A is normally flat along

$V(P)$ then an A/P-sequence is an A-sequence. However, in this book we do not have space to discuss the theory of normal flatness any further, and we refer to Hironaka [1] and [G2], (6.10).

Theorem 24.6. (NC) holds for $\mathbf{P}$ = Gorenstein.

Proof. Once more we reduce to verifying condition (2) of Theorem 2. Suppose that $P \in \mathrm{Gor}(A)$; if $\mathrm{ht}\, P = n$ then since A_P is CM, we can take $x_1, \ldots, x_n \in P$ forming an A_P-sequence. Passing to a smaller neighbourhood of P, we can assume that $x_1, \ldots, x_n$ is an A-sequence. Moreover, replacing A by $A/(x_1, \ldots, x_n)$ we can assume that $\mathrm{ht}\, P = 0$. In addition, we can assume that P is the unique minimal prime ideal of A. Since A_P is a zero-dimensional Gorenstein ring, we have

$$\mathrm{Ext}^1_A(A/P, A) \otimes_A A_P = \mathrm{Ext}^1_{A_P}(\kappa(P), A_P) = 0$$

and

$$\mathrm{Hom}_A(A/P, A) \otimes_A A_P = \mathrm{Hom}_{A_P}(\kappa(P), A_P) = \kappa(P).$$

Thus passing to a smaller neighbourhood, we can assume that $\mathrm{Ext}^1_A(A/P, A) = 0$ and $\mathrm{Hom}_A(A/P, A) \simeq A/P$. In addition, as in the proof of the previous theorem, we can assume that P^i/P^{i+1} is a free A/P-module for $i = 0, \ldots, r-1$, where $P^r = 0$. Then using

$$0 \to P^i/P^{i+1} \longrightarrow P/P^{i+1} \longrightarrow P/P^i \to 0,$$

we get by induction that $\mathrm{Ext}^1_A(P, A) = 0$; from this it follows that $\mathrm{Ext}^2_A(A/P, A) = 0$, and in turn by induction that $\mathrm{Ext}^2_A(P, A) = 0$, so that $\mathrm{Ext}^3_A(A/P, A) = 0$. Proceeding in the same way we see that $\mathrm{Ext}^i_A(A/P, A) = 0$ for every $i > 0$. If we take an injective resolution $0 \to A \longrightarrow I^{\cdot}$ of A as an A-module, and consider the complex obtained by applying $\mathrm{Hom}_A(A/P, -)$ to it, from what we have just said we obtain an exact sequence $0 \to A/P \longrightarrow \mathrm{Hom}_A(A/P, I^{\cdot})$, and this is an injective resolution of A/P as an A/P-module. The same thing holds on replacing A by A_Q for $Q \in V(P)$, and then setting $k = \kappa(Q)$, we get $\mathrm{Ext}^i_{A_Q/PA_Q}(k, A_Q/PA_Q) = \mathrm{Ext}^i_{A_Q}(k, A_Q)$. Thus it is equivalent to say that A_Q is Gorenstein or that A_Q/PA_Q is Gorenstein. Therefore from the hypothesis in (NC) we have that $\mathrm{Gor}(A) \cap V(P)$ contains a neighbourhood of P in $V(P)$. ∎

The above proof is due to Greco and Marinari [1]. Their paper also proves that (NC) also holds for $\mathbf{P}$ = complete intersection.

Exercises to §24. Prove the following propositions.

24.1. Let A be a Noetherian ring, and I an ideal of A; assume that $I^r = 0$, and that I^i/I^{i+1} is a free A/I-module for $1 \leqslant i < r$. Then for $x_1, \ldots, x_s \in A$, it is equivalent for $(x_1, \ldots, x_s)$ to be an A-sequence or an A/I-sequence.

24.2. If A is a quotient of a CM ring R then $\mathrm{CM}(A)$ is open in $\mathrm{Spec}\, A$.

24.3. If A is a quotient of a Gorenstein ring then $\mathrm{Gor}(A)$ is open in $\mathrm{Spec}\, A$.

9

Derivations

This chapter can be read independently of the preceding ones; the main themes are derivations of rings and modules of differentials. The results of this chapter will be applied in the proof of the structure theorem for complete local rings in the next chapter, but in addition derivations and modules of differentials have an important influence on properties of rings, for example via the connection with regularity.

In §25 we discuss the general theory of modules of differentials, and also prove the Hochschild formula for derivations of rings in characteristic p. §26 is pure field theory; Theorem 26.8, which states that a p-basis of a separable extension is algebraically independent, is taken from Matsumura [3]. The terminology 0-etale is due to André, and corresponds to 'formally etale for the discrete topology' in EGA. In §27 we treat the higher derivations of Hasse and F. K. Schmidt, concentrating on the extension problem which they did not treat, in a version due to author.

25 Derivations and differentials

Let A be a ring and M an A-module. A *derivation* from A to M is a map $D: A \longrightarrow M$ satisfying $D(a+b) = Da + Db$ and $D(ab) = bDa + aDb$; the set of all these is written $\mathrm{Der}(A, M)$. It becomes an A-module in a natural way, with $D + D'$ and aD defined by $(D+D')a = Da + D'a$ and $(aD)b = a(Db)$.

If A is a k-algebra via a ring homomorphism $f: k \longrightarrow A$, we say that D is a *k-derivation*, or a derivation over k, if $D \circ f = 0$; the set of all k-derivations of A into M is written $\mathrm{Der}_k(A, M)$. It is an A-submodule of $\mathrm{Der}(A, M)$. Since $1 \cdot 1 = 1$, for any $D \in \mathrm{Der}(A, M)$ we have $D(1) = D(1) + D(1)$, so that $D(1) = 0$, and so viewing A as $\mathbb{Z}$-algebra we have $\mathrm{Der}(A, M) = \mathrm{Der}_{\mathbb{Z}}(A, M)$.

In the particular case $M = A$, we write $\mathrm{Der}_k(A)$ for $\mathrm{Der}_k(A, A)$. If D, $D' \in D_k(A)$, we can compose D and D' as maps $A \longrightarrow A$, and it is easy to see that the bracket $[D, D'] = DD' - D'D$ is again an element of $\mathrm{Der}_k(A)$, and that $\mathrm{Der}_k(A)$ becomes a Lie algebra with this bracket.

Quite generally, for $D \in \mathrm{Der}(A, M)$ and $a \in A$ one sees at once that $D(a^n) = na^{n-1}Da$. Hence if A is a ring of characteristic p we have $D(a^p) = 0$.

Also, in general we have a Leibnitz formula for powers of D,

$$D^n(ab) = \sum_{i=0}^{n} \binom{n}{i} D^i a \cdot D^{n-i} b;$$

if A has characteristic p then this reduces to $D^p(ab) = D^p a \cdot b + a \cdot D^p b$, so that also $D^p \in \operatorname{Der}(A)$.

Let k be a ring, B a k-algebra, and N an ideal of B with $N^2 = 0$; set $A = B/N$. The B-module N can in fact be viewed as an A-module. In this situation, we say that B is an *extension* of the k-algebra A by the A-module N; (note that B does not contain A, so that this is a different usage of extension). We write this extension as usual in the form of an exact sequence

$$0 \to N \xrightarrow{i} B \xrightarrow{f} A \to 0.$$

We say that this extension is *split*, or is the trivial extension, if there exists a k-algebra homomorphism $\varphi : A \longrightarrow B$ such that $f \circ \varphi = 1_A$ (the identity map of A). Then we can identify A and $\varphi(A)$, and we have $B = A \oplus N$ as a k-module. Conversely, starting from any k-algebra A and an A-module N, we can make the direct sum $A \oplus N$ of k-modules into a trivial extension of A by N, by defining the product

$$(a, x)(a', x') = (aa', ax' + a'x) \quad \text{for } a, a' \in A \text{ and } x, x' \in N.$$

In this book, we will write $A * N$ for this algebra.

In general, given a commutative diagram in the category of k-algebras

$$\begin{array}{ccc} B & \xrightarrow{f} & A \\ & \nwarrow^{h} & \uparrow{\scriptstyle g} \\ & & C \end{array}$$

where we think of f as being fixed, we say that h is a *lifting* of g to B. Write N for the ideal $\operatorname{Ker} f$ of B. If $h' : C \longrightarrow B$ is another lifting of g, then $h - h'$ is a map from C to N. If $N^2 = 0$ then N is an $f(B)$-module, and moreover, by means of $g : C \longrightarrow f(B) \subset A$, we can consider N as a C-module. Then it is easy to see that $h - h' : C \longrightarrow N$ is a k-derivation of C to the C-module N. Conversely, if $D \in \operatorname{Der}_k(C, N)$ then $h + D$ is another lifting of g to B.

Let k be a ring and A a k-algebra, and write $\mathcal{M}_A$ for the category of A-modules. We have a covariant functor $M \mapsto \operatorname{Der}_k(A, M)$ from $\mathcal{M}_A$ to itself, which turns out to be a representable functor. In other words, there exists an A-module M_0 and a derivation $\mathrm{d} \in \operatorname{Der}_k(A, M_0)$ with the following universal property: for any A-module M and any $D \in \operatorname{Der}_k(A, M)$, there exists a unique A-linear map $f : M_0 \longrightarrow M$ such that $D = f \circ \mathrm{d}$. We are now going to prove this. Firstly, define $\mu : A \otimes_k A \longrightarrow k$ by

$$\mu(x \otimes y) = xy;$$

then μ is a homomorphism of k-algebras. Set

$$I = \operatorname{Ker}\mu, \quad \Omega_{A/k} = I/I^2 \quad \text{and} \quad B = (A \otimes_k A)/I^2;$$

then μ induces $\mu': B \longrightarrow A$, and

$$0 \to \Omega_{A/k} \longrightarrow B \xrightarrow{\mu'} A \to 0$$

is an extension of the k-algebra A by $\Omega_{A/k}$; this extension splits, and in fact defining $\lambda_i: A \longrightarrow B$ for $i = 1, 2$ by

$$\lambda_1(a) = a \otimes 1 \bmod I^2 \quad \text{and} \quad \lambda_2(a) = 1 \otimes a \bmod I^2,$$

we get two liftings of $1_A: A \longrightarrow A$. Hence $\mathrm{d} = \lambda_2 - \lambda_1$ is a derivation of A to $\Omega_{A/k}$. Now we prove that the pair $(\Omega_{A/k}, \mathrm{d})$ satisfies the conditions for the above (M_0, d). If $D \in \operatorname{Der}_k(A, M)$ and we define $\varphi: A \otimes_k A \longrightarrow A * M$ by $\varphi(x \otimes y) = (xy, xDy)$ then φ is a homomorphism of k-algebras, and

$$\mu\left(\sum x_i \otimes y_i\right) = \sum x_i y_i = 0 \Rightarrow \varphi\left(\sum x_i \otimes y_i\right) = \left(0, \sum x_i Dy_i\right);$$

hence φ maps I into M. Now $M^2 = 0$, so that we finally get $f: I/I^2 = \Omega_{A/k} \longrightarrow M$. For $a \in A$ we have

$$\begin{aligned} f(\mathrm{d}a) &= f(1 \otimes a - a \otimes 1 \bmod I^2) = \varphi(1 \otimes a) - \varphi(a \otimes 1) \\ &= Da - a \cdot D(1) = Da, \end{aligned}$$

so that $D = f \circ \mathrm{d}$. Moreover, $\Omega_{A/k}$ has the A-module structure induced by multiplication by $a \otimes 1$ in $A \otimes A$ (or multiplication by $1 \otimes a$; since $a \otimes 1 - 1 \otimes a \in I$, they both come to the same thing); thus if $\xi = \sum x_i \otimes y_i \bmod I^2 \in \Omega_{A/k}$ then $a\xi = \sum a x_i \otimes y_i \bmod I^2$, and $f(a\xi) = \sum a x_i D y_i = a f(\xi)$, so that f is A-linear. We have

$$a \otimes a' = (a \otimes 1)(1 \otimes a' - a' \otimes 1) + aa' \otimes 1,$$

so that if $\omega = \sum x_i \otimes y_i \in I$ then $\omega \bmod I^2 = \sum x_i \, \mathrm{d}y_i$. Hence $\Omega_{A/k}$ is generated as an A-module by $\{\mathrm{d}a \mid a \in A\}$, so that the uniqueness of a linear map $f: \Omega_{A/k} \longrightarrow M$ satisfying $D = f \circ \mathrm{d}$ is obvious.

The A-module $\Omega_{A/k}$ which we have just obtained is called the *module of differentials* of A over k, or the module of Kähler differentials, and for $a \in A$ the element $\mathrm{d}a \in \Omega_{A/k}$ is called the differential of a. We can write $\mathrm{d}_{A/k}$ for d to be more specific. From the definition, we see that

$$\operatorname{Der}_k(A, M) \simeq \operatorname{Hom}_A(\Omega_{A/k}, M).$$

Example. If A is generated as a k-algebra by a subset $U \subset A$ then $\Omega_{A/k}$ is generated as an A-module by $\{\mathrm{d}a \mid a \in U\}$. Indeed, if $a \in A$ then there exist $a_i \in U$ and a polynomial $f(X) \in k[X_1, \ldots, X_n]$ such that $a = f(a_1, \ldots, a_n)$, and then from the definition of derivation we have

$$\mathrm{d}a = \sum_1^n f_i(a_1, \ldots, a_n)\mathrm{d}a_i, \quad \text{where } f_i = \partial f / \partial X_i.$$

In particular if $A = k[X_1, \ldots, X_n]$ then $\Omega_{A/k} = A\mathrm{d}X_1 + \cdots + A\mathrm{d}X_n$, and

$dX_1, \ldots, dX_n$ are linearly independent over A; this follows at once from the fact that there are $D_i \in \mathrm{Der}_k(A)$ such that $D_i X_j = \delta_{ij}$.

We say that a k-algebra A is 0-*smooth* (over k) if it has the following property: for any k-algebra C, any ideal N of C satisfying $N^2 = 0$, and any $k =$ algebra homomorphism $u: A \longrightarrow C/N$, there exists a lifting $v: A \longrightarrow C$ of u to C, as a k-algebra homomorphism. In terms of diagrams, given an commutative diagram

$$\begin{array}{ccc} A & \xrightarrow{u} & C/N \\ \uparrow & & \uparrow \\ k & \longrightarrow & C, \end{array}$$

there exists v such that

$$\begin{array}{ccc} A & \xrightarrow{u} & C/N \\ \uparrow & \searrow^{v} & \uparrow \\ k & \longrightarrow & C, \end{array}$$

is commutative. Moreover, we say that A is 0-*unramified* over k (or 0-neat) if there exists at most one such v. When A is both 0-smooth and 0-unramified, that is when for given u there exists a unique v, we say that A is 0-*etale*. The condition for A to be 0-unramified over k is that $\Omega_{A/k} = 0$: sufficiency is obvious, and if we recall that in the construction of $\Omega_{A/k}$ we had $d = \lambda_2 - \lambda_1$, necessity is clear.

If A is a ring and $S \subset A$ is a multiplicative set then the localisation A_S is 0-etale over A. This follows from the fact (Ex. 1.1) that if $x \in C$ is a unit modulo a nilpotent ideal, then it is itself a unit. We leave the details to the reader.

Theorem 25.1 (First fundamental exact sequence). A composite $k \xrightarrow{f} A \xrightarrow{g} B$ of ring homomorphisms leads to an exact sequence of B-modules

$$(1) \quad \Omega_{A/k} \otimes_A B \xrightarrow{\alpha} \Omega_{B/k} \xrightarrow{\beta} \Omega_{B/A} \to 0,$$

where the maps are given by $\alpha(d_{A/k}a \otimes b) = b d_{B/k} g(a)$ and $\beta(d_{B/k}b) = d_{B/A}b$ for $a \in A$ and $b \in B$. If moreover B is 0-smooth over A then the sequence

$$(2) \ 0 \to \Omega_{A/k} \otimes B \longrightarrow \Omega_{B/k} \longrightarrow \Omega_{B/A} \to 0,$$

obtained from (1) by adding $0 \to$ at the left, is a split exact sequence.

Proof. In order for a sequence $N' \xrightarrow{\alpha} N \xrightarrow{\beta} N''$ of B-modules to be exact, it is sufficient that for every B-module T, the induced sequence

$$\mathrm{Hom}_B(N', T) \xleftarrow{\alpha^*} \mathrm{Hom}_B(N, T) \xleftarrow{\beta^*} \mathrm{Hom}_B(N'', T)$$

is exact. Indeed, taking $T = N''$, we get $\alpha^* \beta^*(1_T) = 0$, and therefore $\beta\alpha = 0$; and taking $T = N/\mathrm{Im}\,\alpha$, we see easily that $\mathrm{Ker}\,\beta = \mathrm{Im}\,\alpha$. From this, to prove that (1) above is exact, it is enough to show that for any B-module T,

(3) $\mathrm{Der}_k(A, T) \longleftarrow \mathrm{Der}_k(B, T) \longleftarrow \mathrm{Der}_A(B, T) \leftarrow 0$

is exact, but this is obvious.

Now suppose that B is 0-smooth over A. Choose $D \in \mathrm{Der}_k(A, T)$ and consider the commutative diagram

$$\begin{array}{ccc} B & \xrightarrow{1_B} & B \\ {\scriptstyle g}\uparrow & & \uparrow \\ A & \xrightarrow{\varphi} & B * T \end{array} \qquad \text{with } \varphi(a) = (ga, Da).$$

Then by assumption, there exists $h: B \longrightarrow B * T$ which can be added to the diagram, leaving it commutative. If we write $h(b) = (b, D'b)$ then $D': B \longrightarrow T$ is a derivation of B such that $D = D' \circ g$, and D' corresponds to a B-linear map $\alpha': \Omega_{B/k} \longrightarrow T$. Now take T to be $\Omega_{A/k} \otimes B$, and define D by $D(a) = d_{A/k}(a) \otimes 1$, so that $D = D' \circ g$ implies that $\alpha'\alpha = 1_T$. Thus (2) is split. ■

Now consider the case $k \xrightarrow{f} A \xrightarrow{g} B$ when g is surjective; set $\mathrm{Ker}\, g = \mathfrak{m}$, so $B = A/\mathfrak{m}$. Then in (1) of the previous theorem we of course have $\Omega_{B/A} = 0$, and we want to determine $\mathrm{Ker}\,\alpha$.

Theorem 25.2 (Second fundamental exact sequence). In the above notation, we have an exact sequence

(4) $\quad \mathfrak{m}/\mathfrak{m}^2 \xrightarrow{\delta} \Omega_{A/k} \otimes_A B \xrightarrow{\alpha} \Omega_{B/k} \to 0$

where δ is the B-linear map defined by $\delta(x \bmod \mathfrak{m}^2) = \mathrm{d}_{A/k} x \otimes 1$. If B is 0-smooth over k then

(5) $\quad 0 \to \mathfrak{m}/\mathfrak{m}^2 \longrightarrow \Omega_{A/k} \otimes B \longrightarrow \Omega_{B/k} \to 0$

is a split exact sequence.

Proof. We once more take an arbitrary B-module T and consider

(6) $\mathrm{Hom}_B(\mathfrak{m}/\mathfrak{m}^2, T) \xleftarrow{\delta^*} \mathrm{Der}_k(A, T) \xleftarrow{\alpha^*} \mathrm{Der}_k(B, T)$.

For $D \in \mathrm{Der}_k(A, T)$, to say that $\delta^*(D) = 0$ is just to say that $D(\mathfrak{m}) = 0$, so that D can be considered as a derivation from $B = A/\mathfrak{m}$; hence (6) is exact. If B is 0-smooth over k then the extension

$$0 \to \mathfrak{m}/\mathfrak{m}^2 \longrightarrow A/\mathfrak{m}^2 \xrightarrow{g} B \to 0$$

of the k-algebra B by $\mathfrak{m}/\mathfrak{m}^2$ splits, that is there exists a homomorphism of k-algebras $s: B \longrightarrow A/\mathfrak{m}^2$ such that $gs = 1_B$. Now $sg: A/\mathfrak{m}^2 \longrightarrow A/\mathfrak{m}^2$ is a homomorphism vanishing on $\mathfrak{m}/\mathfrak{m}^2$, and $g(1 - sg) = 0$, so that if we set

$D = 1 - sg$ then $D: A/\mathfrak{m}^2 \longrightarrow \mathfrak{m}/\mathfrak{m}^2$ is a derivation. If $\psi \in \mathrm{Hom}_B(\mathfrak{m}/\mathfrak{m}^2, T)$ then the composite D' of

$$A \longrightarrow A/\mathfrak{m}^2 \xrightarrow{D} \mathfrak{m}/\mathfrak{m}^2 \xrightarrow{\psi} T$$

is an element of $\mathrm{Der}_k(A, T)$ satisfying $\delta^*(D') = \psi$. Indeed, for $x \in \mathfrak{m}$, if we let $\bar{x} = x \bmod \mathfrak{m}^2$ then

$$D'(x) = \psi(D(\bar{x})) = \psi(\bar{x} - sg(\bar{x})) = \psi(\bar{x}).$$

Therefore δ^* is surjective. If we set $T = \mathfrak{m}/\mathfrak{m}^2$ then we see that (5) is a split exact sequence. ■

Example. Suppose that $B = k[X_1, \ldots, X_n]/(f_1, \ldots, f_m) = k[x_1, \ldots, x_n]$; then setting $A = k[X_1, \ldots, X_n]$ and using the above theorem, we have

$$\Omega_{B/k} = (\Omega_{A/k} \otimes B)/\sum B df_i = F/R,$$

where F is the free B-module with basis $dX_1, \ldots, dX_n$, and R is the submodule of F generated by $df_i = \sum_j (\partial f_i/\partial X_j) dX_j$ for $1 \leqslant i \leqslant m$. For example, if k is a field of characteristic $\neq 2$ and

$$B = k[X, Y]/(X^2 + Y^2) = k[x, y],$$

then $\Omega_{B/k} = B dx + B dy$, where the only relation between dx and dy is $x dx + y dy = 0$. If k has characteristic 2 then $\Omega_{B/k}$ is the free B-module of rank 2 with basis dx, dy.

Theorem 25.3. Suppose that a field L is a separable algebraic extension of a subfield K; then L is 0-etale over K. Moreover, for any subfield $k \subset K$ we have $\Omega_{L/k} = \Omega_{K/k} \otimes_K L$.

Proof. Suppose that $0 \to N \longrightarrow C \longrightarrow C/N \to 0$ is an extension of K-algebras with $N^2 = 0$, and that $u: L \longrightarrow C/N$ is a given K-algebra homomorphism. If L' is an intermediate field $K \subset L' \subset L$ with L' finite over K, then, as is well-known in field theory, we can write $L' = K(\alpha)$; let $f(X)$ be the minimal polynomial of α over K so that $L' \simeq K[X]/(f)$, and $f'(\alpha) \neq 0$. Thus to lift $u_{|L'}: L' \longrightarrow C/N$ to C, we need only find an element $y \in C$ satisfying $f(y) = 0$ and $y \bmod N = u(\alpha)$. Now choose some inverse image $y \in C$ of $u(\alpha)$; then $f(y) \bmod N = u(f(\alpha)) = 0$, so that $f(y) \in N$. Moreover, $N^2 = 0$, so that for $\eta \in N$ we get

$$f(y + \eta) = f(y) + f'(y) \cdot \eta;$$

but $f'(\alpha)$ is a unit of L, so that $u(f'(\alpha)) = f'(y) \bmod N$ is a unit of C/N, and hence $f'(y)$ is a unit of C by Ex. 1.1. Thus if we set $\eta = -f(y)/f'(y)$ we have $\eta \in N$ and $f(y + \eta) = 0$. The K-algebra homomorphism $v: L' \longrightarrow C$ obtained by taking α to $v(\alpha) = y + \eta$ is a lifting of $u_{|L'}$, and one can see by the construction that v is unique. Thus for every $\alpha \in L$ there is a uniquely determined lifting $v_\alpha: K(\alpha) \longrightarrow C$ of $u_{|L(\alpha)}$, and we can define $v: L \longrightarrow C$

by $v(\alpha) = v_\alpha(\alpha)$. In fact, for $\alpha, \beta \in L$ there exists $\gamma \in L$ such that $K(\gamma)$ contains both α and β, and then by uniqueness we have

$$v_{\gamma|K(\alpha)} = v_\alpha \quad \text{and} \quad v_{\gamma|K(\beta)} = v_\beta.$$

The second half comes from $\Omega_{L/K} = 0$ and Theorem 1. ■

We turn now to derivations. As we have seen, if A is a ring of characteristic p then for $D \in \operatorname{Der}(A)$ we have $D^p \in \operatorname{Der}(A)$. What can we say if $i < p$?

Theorem 25.4. Let K be a field of characteristic p, and let $0 \neq D \in \operatorname{Der}(K)$.

(i) $1, D, D^2, \ldots, D^{p-1}$ are linearly independent over K;

(ii) the only way in which $c_0 + c_1 D + \cdots + c_{p-1} D^{p-1}$ with $c_i \in K$ can be a derivation is if $c_0 = c_2 = \cdots = c_{p-1} = 0$.

Proof. For $a \in K$, write a_L for the operation of multiplying by a; then the property $D(ax) = D(a) \cdot x + a \cdot Dx$ of a derivation means that $D \circ a_L = D(a)_L + aD$. We can write the Leibnitz formula as

$$D^i \circ a_L = aD^i + i \cdot D(a) D^{i-1} + \binom{i}{2} D^2(a) D^{i-2} + \cdots + D^i(a)_L;$$

our proof exploits this formula.

(i) For some $i < p$ suppose that $1, D, \ldots, D^{i-1}$ are linearly independent over K, but that $1, D, \ldots, D^i$ are not. Then we can write $D^i = c_{i-1} D^{i-1} + \cdots + c_0$, with $c_\nu \in K$. If we choose some $a \in K$ such that $D(a) \neq 0$, then in view of $D^i \circ a_L = c_{i-1} D^{i-1} \circ a_L + \cdots$, we get

$$aD^i + i \cdot D(a) D^{i-1} + \cdots = c_{i-1} a D^{i-1} + \cdots,$$

where ... indicates a linear combination of $1, D, \ldots, D^{i-2}$. Subtracting a times our original relation from this gives a relation of the form

$$i \cdot D(a) D^{i-1} = \cdots,$$

and this contradicts the assumption that $1, D, \ldots, D^{i-1}$ are linearly independent.

(ii) Suppose that $E = c_i D^i + \cdots + c_1 D + c_0$ is a derivation of K, with $i < p$ and $c_i \neq 0$. Then $E(1) = c_0$, so that $c_0 = 0$. Now if $i > 1$ then take $a \in K$ such that $D(a) \neq 0$, and substitute both sides of $E \circ a_L = c_i D^i \circ a_L + \cdots$ in the Leibnitz formula: we get

$$aE + E(a)_L = ac_i D^i + [i \cdot c_i \cdot D(a) + ac_{i-1}] D^{i-1} + \cdots,$$

but then in view of the linear independence of $1, D, \ldots, D^{p-1}$, the coefficients of D^{i-1} on both sides must be equal; therefore $i \cdot c_i \cdot D(a) = 0$, which is a contradiction. ■

Remarks. (i) The theorem also holds if $\operatorname{char} K = 0$.

(ii) If K is not a field, this result does not necessarily hold. For example, let k be a field of characteristic p, and set $A = k[X]/(X^p) = k[x]$, with

$x^p = 0$; then every derivation of $k[X]$ will take the ideal (X^p) into itself, and therefore induces a derivation of A. In particular, the derivation $X^{p-1}\cdot\partial/\partial X$ of $k[X]$ induces $D \in \mathrm{Der}_k(A)$ such that $D(x) = x^{p-1}$, but $D(x^i) = i\cdot x^{i-1}x^{p-1} = 0$ if $i > 1$, and therefore for $p > 2$ we have $D^2 = 0$.

Theorem 25.5 (the Hochschild formula). Let A be a ring of characteristic p; then for $a \in A$ and $D \in \mathrm{Der}(A)$ we have

$$(aD)^p = a^p D^p + (aD)^{p-1}(a)\cdot D.$$

Proof. Set $E = aD$. Then $E^2 = E \circ a_L \circ D = (aE + E(a))D = a^2D^2 + E(a)D$, and proceeding by induction, we get a relation of the form

$$E^k = a^k D^k + \sum_{i=2}^{k-1} b_{k,i} D^i + E^{k-1}(a)D,$$

where $b_{k,i}$ are elements of A given by a purely formal computation, so that

$$b_{k,i} = f_{k,i}(a, D(a), D^2(a), \ldots, D^{k-i}(a)),$$

where the $f_{k,i}$ are polynomials with coefficients in $\mathbb{Z}/(p)$ not depending on A, on a or on D. Now to prove our theorem we need only show that $f_{p,i} = 0$ for $1 < i < p$. Let k be a field of characteristic p, and let $x_1, x_2, \ldots$ be a countable number of indeterminates over k; set $K = k(x_1, x_2, \ldots)$. Define a k-derivation D of K by $Dx_i = x_{i+1}$. (Since $\Omega_{A/k}$ is the free K-module with basis $dx_1, dx_2, \ldots$, given any $f_i \in K$ there exists a unique $D \in \mathrm{Der}_k(K)$ such that $Dx_i = f_i$.) For this D, we set $E = x_1D$; then since $E^p - x_1^pD^p = b_{p,p-1}D^{p-1} + \cdots + b_{p,2}D^2 + E^{p-1}(a)\cdot D$ is a derivation, by the previous theorem we must have $b_{p,i} = 0$ for $1 < i < p$. Therefore

$$b_{p,i} = f_{p,i}(x_1, x_2, \ldots, x_{p-i+1}) = 0,$$

and this proves that $f_{p,i} = 0$. ■

This formula is known as the Hochschild formula, although it is also reported to have been first proved by Serre. Be that as it may, it is an important fact that $(aD)^p$ is a linear combination of D^p and D.

Exercises to §25. Prove the following propositions.

25.1. Let A be a ring, $a, b \in A$ and $D, D' \in \mathrm{Der}(A)$; then

$$[aD, bD'] = ab[D, D'] + aD(b)D' - bD'(a)D.$$

Hence in order for an A-submodule $\mathfrak{g} \subset \mathrm{Der}(A)$ to be closed under [,], it is enough to have $\mathfrak{g} = \sum_{i\in I} AD_i$ with $[D_i, D_j] \in \mathfrak{g}$ for all $i, j \in I$.

25.2. Let A be a ring containing the rational field $\mathbb{Q}$. Suppose that $x \in A$ and $D \in \mathrm{Der}(A)$ are such that $Dx = 1$ and $\bigcap_{n=1}^{\infty} x^nA = (0)$; then x is a non-zero-divisor of A.

25.3. Let A be a ring, and I an ideal of A; set $\hat{A}$ for the I-adic completion of A. Then for $D \in \mathrm{Der}(A)$ we have $D(I^n) \subset I^{n-1}$ for all $n > 0$, so that D is I-

adically continuous, and hence induces a derivation of $\hat{A}$. Also for a multiplicative set $S \subset A$, a derivation D induces a derivation of A_S by means of $D(a/s) = (D(a)\cdot s - a\cdot D(s))/s^2$.

25.4. Let k be a ring, k' and A two k-algebras, and set $A' = k' \otimes_k A$; let $S \subset A$ be a multiplicative set. Then $\Omega_{A'/k'} = \Omega_{A/k} \otimes_k k' = \Omega_{A/k} \otimes_A A'$, and $\Omega_{A_S/k} = \Omega_{A/k} \otimes_A A_S$.

25.5. Let A be a ring of characteristic p, and $x \in A$, $D \in \mathrm{Der}(A)$ elements such that $D^p = 0$ and $Dx = 1$; set $A_0 = \{a \in A \mid Da = 0\}$. Then A_0 is a subring of A, and $A = A_0[x] = A_0 + A_0 x + \cdots + A_0 x^{p-1}$, with $1, x, \ldots, x^{p-1}$ linearly independent over A_0.

26 Separability

Let k be a field and A a k-algebra. We say that A is *separable* over k if for every extension field k' of k, the ring $A' = A \otimes_k k'$ is reduced, that is does not contain nilpotents. From the definition, one sees at once the following:

(1) a subalgebra of a separable k-algebra is separable;

(2) A is separable over k if and only if every finitely generated k-subalgebra of A is separable over k;

(3) for A to be separable over k it is sufficient that $A \otimes_k k'$ is reduced for every finitely generated extension field k' of k;

(4) if A is separable over k and k' is an extension field of k then $A \otimes_k k'$ is separable over k'.

Remark. When A is a finite k-algebra, the separability condition can be checked using the discriminant. The *trace* of an element α of A, denoted by $\mathrm{tr}_{A/k}(\alpha)$, is defined to be the trace of the k-linear mapping $A \longrightarrow A$ induced by multiplication by α. Let $\omega_1, \ldots, \omega_n$ be a linear basis of A over k. Then $d = \det(\mathrm{tr}_{A/k}(\omega_i\omega_j))$ is called a discriminant of A over k. If we use another basis $\omega'_1, \ldots, \omega'_n$, and if $\omega'_i = \sum c_{ij}\omega_j$, then the discriminant with respect to this basis is $\det(c_{ij})^2\cdot d$. Thus $d = 0$ or $d \neq 0$ is a property of A independent of the choice of basis. Now we claim that A is separable if and only if $d \neq 0$.
Proof. If k' is an extension field of k and $A' = A \otimes_k k'$, then $\omega_1, \ldots, \omega_n$ is also a linear basis of A' over k', and so d is also a discriminant of A' over k'. If A' is not reduced, let $N = \mathrm{nil}(A')$. Take a basis $\omega'_1, \ldots, \omega'_n$ of A' such that $\omega'_1, \ldots, \omega'_r$ span N. Then $\omega'_i\omega'_j$ is nilpotent for $i \leqslant r$, hence its trace is zero. It follows that $\det(\mathrm{tr}(\omega'_i\omega'_j)) = 0$, and so $d = 0$. Conversely, if A is separable over k, take an algebraic closure K of k. Then $A \otimes_k K$ is reduced. Therefore we need only prove that if k is algebraically closed and A is reduced then $d \neq 0$. Now A is an Artinian reduced ring, hence is a finite product of fields, each of which (as a finite extension of k) is

isomorphic to k. Thus $A = ke_1 + \cdots + ke_n$ with $e_ie_j = 0$ for $i \neq j$ and $e_i^2 = e_i$. Hence $\operatorname{tr}(e_i) = 1$ and $d \neq 0$. ∎

In what follows we consider mainly the case when A is a field. If K is an algebraic extension field of k, and is separable in the usual sense (that is every element of K is a root of a polynomial with coefficients in k having no multiple roots), then K is separable over k in our sense. To see this, by (2) above we can assume that K is finitely generated over k, and then according to the well-known primitive element theorem of field theory, $K \simeq k[X]/(f(X))$, where $f \in k[X]$ is irreducible and with no multiple roots. Then if k' is an extension field of k we have

$$K \otimes_k k' \simeq k'[x]/(f(X)),$$

and when we factorise f into primes in $k'[X]$ we get $f = f_1 \dots f_r$ with $(f_i, f_j) = 1$ for $i \neq j$, so that by Theorem 1.4,

$$k'[X]/(f) \simeq k'[X]/(f_1) \times \cdots \times k'[X]/(f_r);$$

since this is a direct product of fields, it is reduced. ∎

We say that an extension field K of k is *separably generated* over k if K has a separating transcendence basis over k, that is a transcendence basis Γ such that K is a separable algebraic extension of $k(\Gamma)$.

Theorem 26.1. A separably generated extension field is separable.

Proof. Let k be a field and K a separably generated extension of k, with Γ a separating transcendence basis of K. If k' is any extension field of k then $k(\Gamma) \otimes_k k'$ is a ring of fractions of $k[\Gamma] \otimes_k k' = k'[\Gamma]$, so that it is an integral domain with field of fractions $k'(\Gamma)$. Thus $K \otimes_k k' = K \otimes_{k(\Gamma)} (k(\Gamma) \otimes_k k')$ is a subring of $K \otimes_{k(\Gamma)} k'(\Gamma)$. Now K is a separable algebraic extension of $k(\Gamma)$, so that as we have seen above $K \otimes_{k(\Gamma)} k'(\Gamma)$ is reduced. ∎

Theorem 26.2. Let k be a field of characteristic p, and K a finitely generated extension field of k; then the following conditions are equivalent:

(1) K is separable over k;

(2) $K \otimes_k k^{1/p}$ is reduced;

(3) K is separably generated over k.

Proof. (1)⇒(2) is trivial and (3)⇒(1) has just been proved.

(2)⇒(3) Let $K = k(x_1, \dots, x_n)$; we can assume that $x_1, \dots, x_r$ is a transcendence basis for K over k. Assume furthermore that $x_{r+1}, \dots, x_q$ are separable algebraic over $k(x_1, \dots, x_r)$, and that x_{q+1} is not; set $y = x_{q+1}$ and let $f(Y^p)$ be the minimal polynomial of y over $k(x_1, \dots, x_r)$. The coefficients of $f(Y^p)$ are rational functions of $x_1, \dots, x_r$, so that clearing denominators we get an irreducible polynomial $F(X_1, \dots X_r, Y^p) \in k[X_1, \dots, X_r, Y]$, with $F(x, y^p) = 0$. Now if $\partial F/\partial X_i = 0$ for $1 \leqslant i \leqslant r$ then $F(X, Y^p)$ is the pth power of a polynomial $G(X, Y)$ with coefficients in $k^{1/p}$, but then we would have

$$k[x_1,\ldots,x_r,y]\otimes_k k^{1/p}=(k[X,Y]/(F(X,Y))\otimes_k k^{1/p}$$
$$=k^{1/p}[X,Y]/(G(X,Y)^p);$$

this is a subring of $K\otimes_k k^{1/p}$ containing nilpotent elements, and this contradicts (2). Hence we can assume that $\partial F/\partial X_1\neq 0$. Then x_1 is separable algebraic over $k(x_2,\ldots,x_r,y)$, and hence so are $x_{r+1},\ldots,x_q$. Therefore exchanging x_1 and $y=x_{q+1}$, we find that $x_{r+1},\ldots,x_{q+1}$ are separably algebraic over $k(x_1,\ldots,x_r)$, so that by induction on q, we have (3). ■

Remark. As we have seen in the proof, if $K=k(x_1,\ldots,x_n)$ is separable over k then we can choose a separating transcendence basis from among $x_1,\ldots,x_n$.

Theorem 26.3. If k is a perfect field then every extension field K of k is separable over k, and a k-algebra A is separable if and only if it is reduced.

Proof. Recall that a field k is perfect if every algebraic extension of k is separable. If k has characteristic 0 then every extension field K is separably generated, and therefore separable. In characteristic p, perfect implies $k=k^{1/p}$, so that by the previous theorem, all subfields of K finitely generated over k are separable, so that K itself is separable over k. (Caution: K may fail to be separably generated over k; for a counter-example, let x be an indeterminate over k, and $K=k(x,x^{p^{-1}},x^{p^{-2}},\ldots)$.) Now we show that if A is a reduced k-algebra then A is separable. We can assume that A is finitely generated over k. Then A is a Noetherian ring, and the total ring of fractions K of A is of the form $K=K_1\times\cdots\times K_r$ by Ex. 6.5. Each K_i is separable over k, so that K is also separable, and hence so is its subring A. ■

In general two subfields K, K' of a given field L are said to be *linearly disjoint* over a common subfield k if the following three equivalent conditions are satisfied:

(*a*) if $\alpha_1,\ldots,\alpha_n\in K$ are linearly independent over k they are also linearly independent over K';

(*b*) the same with K and K' interchanged;

(*c*) if we write $K[K']$ for the subring of L generated by K and K', the natural map $K\otimes_k K'\longrightarrow K[K']$ is an isomorphism.

Proof of equivalence. (*a*)⇒(*c*) Let $\xi=\sum_1^m x_i\otimes y_i$ be an element of the kernel of $K\otimes_k K'\longrightarrow K[K']$. Suppose that $x_1,\ldots,x_r$ are linearly independent over k, and that the remainder $x_{r+1},\ldots,x_m$ are linear combinations of them, and rewrite $\xi=\sum_1^r x_i\otimes y_i'$. The image in $K[K']$ of ξ is $\sum x_iy_i'$, but if this is 0 then by (*a*) we have $y_i'=0$ for all i, so that $\xi=0$. This proves (*c*).

(*c*)⇒(*a*) is also easy; finally, since (*c*) is symmetric in K and K', we of course also get (*a*)⇔(*b*).

Let k be a field of characteristic p, and K an extension field of k. Inside an algebraic closure $\bar{K}$ of K, consider the subfields $k^{p^{-n}} = \{\alpha \in \bar{K} \mid \alpha^{p^n} \in k\}$ and $k^{p^{-\infty}} = \bigcup_{n>0} k^{p^{-n}}$. These are purely inseparable extension fields of k, and $k^{p^{-\infty}}$ is the smallest perfect field containing k.

Theorem 26.4 (S. MacLane). Let k and K be as above. We have

(i) if K is separable over k then K and $k^{p^{-\infty}}$ are linearly disjoint over k;

(ii) if K and $k^{p^{-n}}$ are linearly disjoint over k for some $n > 0$ then K is separable over k.

Proof. (i) Suppose that $\alpha_1, \ldots, \alpha_n \in K$ are linearly independent over k. If $\sum \alpha_i \xi_i = 0$ with $\xi_i \in k^{p^{-\infty}}$, we set $k_1 = k(\xi_1, \ldots, \xi_n)$, so that k_1 is a finite extension of k; for some sufficiently large n we have $k_1^{p^n} \subset k$, and if we set $A = K \otimes_k k_1$, then A is a reduced ring. However, A is finite as a K-module, so is a zero-dimensional ring, but the p^nth power of any element of A is in K, so that we see that A has only one prime ideal. Hence A is a field, and $A \simeq K[k_1]$. From this we get $\sum \alpha_i \otimes \xi_i = 0$, that is $\xi_i = 0$ for all i.

(ii) If K and $k^{p^{-n}}$ are linearly disjoint, then $k^{p^{-1}} \subset k^{p^{-n}}$, and hence K and $k^{p^{-1}}$ are also linearly disjoint over k, so that $K \otimes_k k^{p^{-1}}$ is a field. If K' is a subfield of K which is finitely generated over k then by Theorem 2, K' is separable over k. Hence K is also separable over k. ■

Differential bases

Let K be an extension field of a field k; then $\Omega_{K/k}$ is a vector space over K, and is generated by $\{dx \mid x \in K\}$, so that there exists a subset $B \subset K$ such that $\{dx \mid x \in B\}$ forms a basis of the vector space $\Omega_{K/k}$. A subset $B \subset K$ with this property is called a *differential basis* of K over k. The following condition (*) is necessary and sufficient for a subset $\{x_\lambda\}_{\lambda \in \Lambda} \subset K$ to form a differential basis for K over k.

(*) if $y_\lambda \in K$ are specified for every $\lambda \in \Lambda$ in an arbitrary way, then there exists a unique $D \in \mathrm{Der}_k(K)$ such that $D(x_\lambda) = y_\lambda$ for all λ.

For $x_1, \ldots, x_n \in K$, let us study the condition for $dx_1, \ldots, dx_n \in \Omega_{K/k}$ to be linearly independent over k. If k has characteristic 0 then this is equivalent to $x_1, \ldots, x_n$ being algebraically independent over k. Indeed, if there exists $0 \neq f(X) \in k[X_1, \ldots, X_n]$ such that $f(x_1, \ldots, x_n) = 0$, then choose such a relation of smallest degree; suppose for instance that X_1 actually appears in f, so that $f_1 = \partial f / \partial X_1$ is non-zero, but of smaller degree than f, and hence $f_1(x) \neq 0$. Then $f(x) = 0$ gives

$$0 = df = \sum f_i(x) dx_i,$$

so that $dx_1, \ldots, dx_n$ are linearly dependent. Conversely, if $x_1, \ldots, x_n$ are algebraically independent over k then there exists a transcendence basis B of K/k containing these, so that there exists k-derivations D_i of the purely transcendental extension $k(B)$ satisfying

$$D_i(x_i) = 1 \quad \text{and} \quad D_i(y) = 0 \quad \text{for} \quad x_i \neq y \in B,$$

(namely, $\partial/\partial x_i$). Moreover, K is a separable algebraic extension of $k(B)$, and so by Theorem 25.3 is 0-etale, so that the derivations D_i extend to derivations from K to K. Then since $D_i(x_j) = \delta_{ij}$, the differentials $dx_1, \ldots, dx_n \in \Omega_{K/k}$ are clearly linearly independent. Thus in this case a differential basis is the same thing as a transcendence basis.

Now consider a field k of characteristic p. We say that elements $x_1, \ldots, x_n \in K$ of an extension field K are *p-independent* over k if $[K^p(k, x_1, \ldots, x_n): K^p(k)] = p^n$, and a subset $B \subset K$ is p-independent if any finite subset of B is p-independent. This condition means precisely that the set

$$\Gamma_B = \left\{ x_1^{\alpha_1} \ldots x_n^{\alpha_n} \,\middle|\, \begin{array}{l} x_1, \ldots, x_n \text{ are distinct} \\ \text{elements of } B \text{ and } 0 \leqslant \alpha_i < p \end{array} \right\}$$

is linearly independent over $K^p(k)$; the elements of Γ_B are called the p-monomials of B. If B is not p-independent, we say it is p-dependent. The condition of p-independence is not just a property of B and k, but also depends on K. If $B \subset K$ is p-independent over k and $K = K^p(k, B)$, we say that B is a *p-basis* of K/k. If $C \subset K$ is p-independent over k then one can easily show by Zorn's lemma that there exists a p-basis of K/k containing C.

One sees easily that B is a p-basis of K/k is equivalent to Γ_B being a basis of K over $K^p(k)$ in the sense of linear algebra. If this holds then any map $D\colon B \longrightarrow K$ has a unique extension to an element $D \in \mathrm{Der}_k(K)$. Indeed, for a p-monomial of B we set

$$D(x_1^{\alpha_1} \ldots x_n^{\alpha_n}) = \textstyle\sum_{i=1}^n \alpha_i x_1^{\alpha_1} \ldots x_i^{\alpha_i - 1} \ldots x_n^{\alpha_n} D(x_i),$$

and extend D to K as a $K^p(k)$-linear map; then D is a k-derivation. Thus a p-basis B is a differential basis of K/k. Conversely, if B' is a differential basis of K/k then B' is p-independent over k; for if $x_1, \ldots, x_n \in B'$ are p-dependent, we can assume that $x_1 \in K^p(k, x_2, \ldots, x_n)$, so that we can write $x_1 = f(x_2, \ldots, x_n)$, where f is a polynomial with coefficients in $K^p(k)$. Then in $\Omega_{K/k}$ we get $dx_1 = \sum_2^n (\partial f/\partial x_i) dx_i$, which contradicts the linear independence of $dx_1, \ldots, dx_n$. Now if we take a p-basis B of K/k containing B', then since both B and B' are differential bases, we have $B = B'$. We summarise the above as follows:

Theorem 26.5. The notion of differential basis coincides with transcendence basis in characteristic 0, and with p-basis in characteristic p.

Now we look at the relation between separability and differential bases. Letting $\Pi \subset k$ denote the prime subfield of k, we write Ω_k for $\Omega_{k/\Pi}$.

Theorem 26.6. For a field extension K/k, the following conditions are equivalent:

(1) K/k is separable;

(2) for any subfield $k' \subset k$ the standard map $\Omega_{k/k'} \otimes_k K \longrightarrow \Omega_{K/k'}$ is injective;

(2′) for any subfield $k' \subset k$ and any differential basis B of k/k', there exists a differential basis of K/k' containing B;

(3) $\Omega_k \otimes_k K \longrightarrow \Omega_K$ is injective;

(4) any derivation of k to an arbitrary K-module M extends to a derivation from K to M.

Proof. (2) and (2′) are clearly equivalent, and (2)$\Rightarrow$(3)$\Leftrightarrow$(4) are trivial. In characteristic 0, both (1) and (2′) hold, so that we need only consider the case of characteristic p.

(1)$\Rightarrow$(2′) Since K and $k^{1/p}$ are linearly disjoint over k, we can apply the isomorphism $x \mapsto x^p$ to all three of these to get K^p and k linearly disjoint over k^p. Hence $K^p(k^p, k') = K^p(k')$ and k are linearly disjoint over $k^p(k')$ (see Ex. 26.1 below). If we choose a p-basis B of k over k' then the set Γ_B of p-monomials of B is linearly independent over $k^p(k')$, hence also linearly independent over $K^p(k')$, and B as a subset of K is also p-independent over k'. Therefore B can be extended to a p-basis of K/k'.

(3)$\Rightarrow$(1) If we take a p-basis B of k over Π then the set Γ_B of p-monomials of B is a basis of k over k^p. $\{dx \mid x \in B\}$ is a basis of Ω_k over k, and by assumption is linearly independent in Ω_K over K, so that Γ_B is also linearly independent over K^p. Therefore

$$k \otimes_{k^p} K^p \simeq k(K^p),$$

and k and K^p are linearly disjoint over k^p, so that by Theorem 4, K/k is separable. ∎

Let k be a field of characteristic p, and $\Pi \subset k$ the prime subfield; a p-basis of k/Π is called an *absolute p-basis* of k. If $k_0 \subset k$ is any perfect field contained in k then $k^p(k_0) = k^p = k^p(\Pi)$, so that an absolute p-basis of k is also a p-basis of k/k_0.

Theorem 26.7. Let k be a field of characteristic p, and K an extension field of k. If an absolute p-basis of k is also an absolute p-basis of K, then K is 0-etale over k, and conversely.

Proof. Consider a commutative diagram of ring homomorphisms

$$\begin{array}{ccc} K & \xrightarrow{\;u\;} & \bar{C} \\ {\scriptstyle i}\uparrow & & \uparrow{\scriptstyle g} \\ k & \xrightarrow{\;j\;} & C, \end{array}$$

where $\bar{C} = C/N$, with N an ideal of C satisfying $N^2 = 0$, and g the natural map. For $\alpha \in K$, if we choose $a \in C$ such that $u(\alpha) = g(a)$, then a^p is independent of the choice of a. For if $g(a) = g(a')$ then we can write $a' = a + x$ with $x \in N$, and since C is a k-algebra via j, it is a ring of characteristic p, so that

$$a'^p = a^p + x^p = a^p.$$

Now we define a map $v_0: K^p \longrightarrow C$ by $v_0(\alpha^p) = a^p$ for $\alpha \in K$, where $a \in C$ is such that $u(\alpha) = g(a)$; one checks easily that v_0 is a homomorphism, and coincides with j on k^p. So far we have not used the assumption on K/k. Now since by assumption K is separable over k, and $K = K^p[k]$, we can think of K as

$$K = K^p \otimes_{k^p} k,$$

and thus we can define $v: K \longrightarrow C$ by letting v be equal to v_0 on K^p, and equal to j on k; this is a lifting of u to C. Uniqueness of the lifting is clear from the fact that $K^p(k) = K$, so that $\Omega_{K/k} = 0$.

Conversely, if we assume that K/k is 0-etale, then first of all, from 0-unramified we have $\Omega_{K/k} = 0$, so that by 0-smoothness and Theorem 25.1 we have $\Omega_K = \Omega_k \otimes_k K$. Thus an absolute p-basis of k is also an absolute p-basis of K. ∎

Theorem 26.8. Let K/k be a separable extension of fields of characteristic p, and let B be a p-basis of K/k. Then B is algebraically independent over k, and K is 0-etale over $k(B)$.

Proof. Suppose by contradiction that $b_1, \ldots, b_n \in B$ are algebraically dependent over k. Suppose that $0 \neq f \in k[X_1, \ldots, X_n]$ is a polynomial of minimal degree among all those with $f(b) = 0$, and set $\deg f = d$. Then write

$$f(X_1, \ldots, X_n) = \sum_{0 \leqslant i_1, \ldots, i_n < p} g_{i_1 \ldots i_n}(X_1^p, \ldots, X_n^p) X_1^{i_1} \ldots X_n^{i_n}$$

Then since $b_1, \ldots, b_n$ are p-independent over k and $f(b) = 0$, we have $g_{i_1 \ldots i_n}(b^p) = 0$ for all values of $i_1, \ldots, i_n$. However, since

$$d \geqslant \deg g_{i_1 \ldots i_n}(X^p) + i_1 + \cdots + i_n,$$

by choice of f we must have $f(X) = g_{0 \ldots 0}(X^p)$. Hence we can write f in the form $f(X) = h(X)^p$, with $h \in k^{1/p}[X_1, \ldots, X_n]$. However, since K and $k^{1/p}$ are linearly disjoint over k, the monomials of degree $< d$ in $b_1, \ldots, b_n$, being linearly independent over k, must also be linearly independent over $k^{1/p}$. Thus $h(b) \neq 0$, but this contradicts $h(b)^p = f(b) = 0$. For the second half, see the proof of the following theorem. ∎

Remark. Although $k(B)$ is purely transcendental over k, it does not follow that K is algebraic over $k(B)$; for a counter-example, let $K = k(x, x^{p^{-1}}, x^{p^{-2}}, \ldots)$, with x an indeterminate over k. In this case $B = \varnothing$.

Theorem 26.9. If K is a separable field extension of a field k, then K is 0-smooth, and conversely.

Proof. Let B be a differential basis of K/k. If K/k is separable then by Theorem 5 and Theorem 8, $k(B)$ is purely transcendental over k. Then, as one sees at once from the definition, $k(B)$ is 0-smooth over k. Moreover, $K/k(B)$ is 0-etale. In characteristic 0 this follows from Theorem 25.3; in characteristic p, by Theorem 6 we have an exact sequence $0 \to \Omega_k \otimes K \longrightarrow \Omega_K \longrightarrow \Omega_{K/k} \to 0$, so that putting together an absolute p-basis of k with B we have an absolute p-basis of K, and this is clearly also an absolute p-basis of $k(B)$. Hence $K/k(B)$ is 0-etale by Theorem 7. Therefore K/k is 0-smooth.

Conversely, if K/k is 0-smooth then by Theorem 25.1, $\Omega_k \otimes K \longrightarrow \Omega_K$ is injective, so that by Theorem 6, K/k is separable. ■

Imperfection modules and the Cartier equality

Quite generally, if $k \longrightarrow A \longrightarrow B$ are ring homomorphisms, we write $\Gamma_{B/A/k}$ for the kernel of $\Omega_{A/k} \otimes_A B \longrightarrow \Omega_{B/k}$, and call it the *imperfection module* of the A-algebra B over k. If $k = \mathbb{Z}$ or $k = \mathbb{Z}/p$ (the prime field of characteristic p) we omit k, and write $\Gamma_{B/A}$.

Lemma 1. Let $k \longrightarrow K \longrightarrow L \longrightarrow L'$ be field homomorphisms. Then there exists a natural exact sequence

$$\begin{aligned} 0 \to \Gamma_{L/K/k} \otimes_L L' &\longrightarrow \Gamma_{L'/K/k} \longrightarrow \Gamma_{L'/L/k} \\ &\longrightarrow \Omega_{L/K} \otimes_L L' \longrightarrow \Omega_{L'/K} \longrightarrow \Omega_{L'/L} \to 0. \end{aligned}$$

Proof. We have a commutative diagram with exact rows.

$$\begin{array}{ccccccccc} 0 \to & \Gamma_{L/K/k} \otimes_L L' & \longrightarrow & \Omega_{K/k} \otimes_K L' & \longrightarrow & \Omega_{L/k} \otimes_L L' & \longrightarrow & \Omega_{L/K} \otimes_L L' & \to 0 \\ & {\scriptstyle f_1}\downarrow & & \| & & {\scriptstyle f_2}\downarrow & & {\scriptstyle f_3}\downarrow & \\ 0 \to & \Gamma_{L'/K/k} & \longrightarrow & \Omega_{K/k} \otimes L' & \longrightarrow & \Omega_{L'/k} & \longrightarrow & \Omega_{L'/K} & \to 0. \end{array}$$

We abbreviate this as

$$\begin{array}{ccccccccc} 0 \to & X & \longrightarrow & A & \longrightarrow & B & \longrightarrow & P & \to 0 \\ & \downarrow & & \| & & \downarrow & & \downarrow & \\ 0 \to & Y & \longrightarrow & A & \longrightarrow & C & \longrightarrow & Q & \to 0. \end{array}$$

and from it construct

$$\begin{array}{ccccccc} 0 \to & A/X & \longrightarrow & B & \longrightarrow & P & \to 0 \\ & \downarrow & & {\scriptstyle f_2}\downarrow & & {\scriptstyle f_3}\downarrow & \\ 0 \to & A/Y & \longrightarrow & C & \longrightarrow & Q & \to 0; \end{array}$$

applying the snake lemma gives the exact sequence

$$0 \to Y/X \longrightarrow \operatorname{Ker} f_2 \longrightarrow \operatorname{Ker} f_3 \to 0,$$

from which we easily get

$$0 \to X \longrightarrow Y \longrightarrow \operatorname{Ker} f_2 \longrightarrow P \longrightarrow Q \longrightarrow \operatorname{Coker} f_3 \to 0.$$

This is just what we wanted to prove. ■

Theorem 26.10 (the Cartier equality). Let k be a perfect field, K an extension of k and L a finitely generated extension field of K; then

(*) $\quad \operatorname{rk}_L \Omega_{L/K} = \operatorname{tr.deg}_K L + \operatorname{rk}_L \Gamma_{L/K/k}$,

(where rk_L denotes the dimension of a vector space over L).

Proof. Suppose that $k \subset K \subset L \subset L'$, with L finitely generated over K and L' finitely generated over L. If the theorem holds for $k \subset L \subset L'$ and for $k \subset K \subset L$ then both $\Gamma_{L'/L/k}$ and $\Gamma_{L/K/k} \otimes L'$ are finite-dimensional over L', so that by the lemma, $\Gamma_{L'/K/k}$ is also finite-dimensional, and

$$\begin{aligned}&\mathrm{rk}_{L'}\Omega_{L'/K} - \mathrm{rk}_{L'/K/k}\\ &= (\mathrm{rk}_{L'}\Omega_{L'/L} - \mathrm{rk}_{L'}\Gamma_{L'/L/k}) + (\mathrm{rk}_L\Omega_{L/K} - \mathrm{rk}_L\Gamma_{L/K/k})\\ &= \mathrm{tr.deg}_L L' + \mathrm{tr.deg}_K L = \mathrm{tr.deg}_K L';\end{aligned}$$

thus the theorem also holds for $k \subset K \subset L'$. Now every finitely generated field extension can be obtained by a succession of the following three kinds of simple extensions:

(1) $L = K(\alpha)$ where α is transcendental over K;

(2) $L = K(\alpha)$ where α is separable algebraic over K;

(3) $L = K(\alpha)$ where $\mathrm{char}\, K = p$, and $\alpha^p = a \in K$, but $\alpha \notin K$.

Hence we need only prove (*) in each of these special cases. (1) and (2) are easy. For (3), if we write $L = K[X]/(X^p - a)$ we see that

$$\begin{aligned}\Omega_{L/k} &= (\Omega_{K[X]/k} \otimes L)/L\mathrm{d}a\\ &= (\Omega_{K/k}/K\mathrm{d}a) \otimes L \oplus L\mathrm{d}\alpha,\end{aligned}$$

and $\mathrm{d}\alpha \neq 0$. Furthermore, since k is a perfect field, we have $a \notin K^p = kK^p$, so that in $\Omega_{K/k} (= \Omega_K)$ we have $\mathrm{d}a \neq 0$, $rk\,\Omega_{L/K} = 1$ and $rk\,\Gamma_{L/K/k} = 1$, so that (*) also holds in this case. ■

Remark (Harper's theorem). An ideal I of a ring R is called a *differential ideal* if it maps into itself under every derivation of R to R. A ring R is said to be *differentiably simple* if it has no non-trivial differential ideals. The following beautiful theorem is due to L. Harper, Jr.:

Theorem. *A Noetherian ring R of characteristic p is differentiably simple if and only if it has the form $R = k[T_1, \ldots, T_n]/(T_1^p, \ldots, T_n^p)$, where k is a field of characteristic p.*

The 'if' part is easy. The proof of the 'only if' part is not so easy and we refer the reader to Harper [1] and Yuan [1]. Recently this theorem was used by Kimura–Niitsuma [1] to prove the following theorem which had been known as Kunz' Conjecture:

Theorem. *Let R be a regular local ring of characteristic p, and let S be a local subring of R containing R^p. Assume that R is a finite S-module. Then R has a p-basis over S if and only if S is regular.*

Exercises to §26. Prove the following propositions.

26.1. Let L be a field and k, k', K, K' subfields of L; assume that $k \subset k' \subset K$ and $k \subset K'$, and that K and K' are linearly disjoint over k. Then we have (i) $K \cap K' = k$, and (ii) K and $k'(K')$ are linearly disjoint over k'.

26.2. Let L be a separable extension of a field K; then $L((T_1,\ldots,T_n))$ is separable over $K((T_1,\ldots,T_n))$. Here $L((T_1,\ldots,T_n))$ denotes the field of fractions of $L[\![T_1,\ldots,T_n]\!]$.

27 Higher derivations

Let $k \xrightarrow{f} A \xrightarrow{g} B$ be ring homomorphisms. Let t be an indeterminate over B, and set $B_m = B[t]/(t^{m+1})$ for $m = 0, 1, \ldots$, and $B_\infty = B[\![t]\!]$. We can view B_m as a k-algebra in a natural way (for $m \leqslant \infty$).

For $m \leqslant \infty$ we define a *higher derivation* (over k) of length m from A to B to be a sequence $\underline{\mathrm{D}} = (\mathrm{D}_0, \mathrm{D}_1, \ldots, \mathrm{D}_m)$ of k-linear maps D_i: $A \longrightarrow B$, satisfying the conditions

$$(*)\quad \mathrm{D}_0 = g \quad\text{and}\quad \mathrm{D}_i(xy) = \sum_{r+s=i} \mathrm{D}_r(x)\mathrm{D}_s(y)$$

for $1 \leqslant i \leqslant m$ and $x, y \in A$. These conditions are equivalent to saying that the map $E_t\colon A \longrightarrow B_m$ defined by

$$E_t(x) = \sum_{i=0}^{m} \mathrm{D}_i(x)t^i$$

is a k-algebra homomorphism with $E_t(x) \equiv g(x) \bmod t$.

When $A = B$ and $g = 1$ (the identity map of A) then we speak simply of a higher k-derivation of A. In what follows we consider mainly this case.

If $\underline{\mathrm{D}} = (\mathrm{D}_0, \mathrm{D}_1, \mathrm{D}_2, \ldots)$ is a higher derivation then $\mathrm{D}_1 \in \mathrm{Der}_k(A, B)$. Furthermore, D_i is 0 on $f(k)$ for $i > 0$. In general we say that $\underline{\mathrm{D}}$ is trivial on $a \in A$ if $\mathrm{D}_i(a) = 0$ for $i > 0$; this means precisely that a goes over into a constant under the homomorphism E_t corresponding to $\underline{\mathrm{D}}$.

The theory of higher derivations was initiated by Hasse and F. K. Schmidt [1]. In view of this, in this book we write $\mathrm{HS}_k(A, m)$ for the set of all higher k-derivations of length m of A, and we also write $\mathrm{HS}_k(A)$ for $\mathrm{HS}_k(A, \infty)$. When we are not concerned with k, we simplify this to $\mathrm{HS}(A, m)$ and $\mathrm{HS}(A)$. These sets do not have a module structure like that of $\mathrm{Der}_k(A)$, but they do have a group structure (generally non-Abelian), which we now explain. The homomorphism $E_t\colon A \longrightarrow A_m$ corresponding to $\underline{\mathrm{D}} \in \mathrm{HS}_k(A, m)$ can be extended to an endomorphism of A_m by setting

$$E_t\left(\sum_\nu a_\nu t^\nu\right) = \sum_\nu E_t(a_\nu)t^\nu.$$

Now E_t is injective, since if $\xi = a_r t^r + a_{r+1}t^{r+1} + \cdots \in A_m$ with $a_r \neq 0$ then $E_t(\xi) \equiv a_r t^r \bmod t^{r+1}$; also, by setting

$$\xi - E_t(a_r t^r) = b_{r+1}t^{r+1} + \cdots,$$
$$\xi - E_t(a_r t^r + b_{r+1}t^{r+1}) = c_{r+2}t^{r+2} + \cdots$$

and proceeding in the same way, we see easily that E_t is surjective. In other words, E_t is an automorphism of A_m. Conversely, an automorphism E of the k-algebra A_m satisfying $E(a) \equiv a \bmod t$ corresponds to a higher derivation. Thus if we identify $\mathrm{HS}_k(A, m)$ with a subgroup of the automorphism group $\mathrm{Aut}_k(A_m)$ of A_m, it acquires the group structures which we are looking for. Let us start computing this structure. For $\underline{D} = (D_0, D_1, \ldots)$, $\underline{D}' = (D'_0, D'_1, \ldots)$, set

$$\underline{D} \cdot \underline{D}' = (D''_0, D''_1, \ldots), \quad \text{and} \quad \underline{D}^{-1} = (D^*_0, D^*_1, \ldots);$$

then

$$\begin{aligned} E_t(E'_t(a)) &= E_t(a + D'_1(a)t + D'_2(a)t^2 + \cdots) \\ &= (a + D_1(a)t + D_2(a)t^2 + \cdots) \\ &\quad + (D'_1(a) + D_1(D'_1(a))t + D_2(D'_1(a))t^2 + \cdots)t \\ &\quad + (D'_2(a) + D_1(D'_2(a)t + \cdots)t^2 + \cdots \\ &= a + (D_1 + D'_1)(a)t + (D_2 + D_1 D'_1 + D'_2)(a)t^2 + \cdots, \end{aligned}$$

so that

$$D''_i = \sum_{p+q=i} D_p D'_q \quad \text{for all} \quad i;$$

and the D^*_i are obtained by solving $\sum_{p+q=i} D_p D^*_q = 0$ for $i > 0$, that is

$$D^*_0 = D_0 = 1, \; D^*_1 = -D_1, \; D^*_2 = D_1^2 - D_2,$$
$$D^*_3 = -D_1^3 + D_1 D_2 + D_2 D_1 - D_3, \ldots.$$

If $S \subset A$ and $T \subset B$ are multiplicative sets such that $g(S) \subset T$, then the given homomorphism $g: A \longrightarrow B$ induces a homomorphism $A_S \longrightarrow B_T$. Now we show that in a similar way, a higher derivation has a unique extension to the localisations. To see this, let $\underline{D} = (D_0, D_1, \ldots, D_m)$ be a higher derivation of length m from A to B; if we compose the homomorphism $E_t: A \longrightarrow B_m$ corresponding to $\underline{D}$ with the localisation $B_m \longrightarrow (B_T)_m$, then an element $x \in S$ maps to $g(x)_T + D_1(x)_T t + \cdots$, and this is a unit of $(B_T)_m$, since the constant term $g(x)_T$ is a unit of B_T. This E_t induces a homomorphism $A_S \longrightarrow (B_T)_m$, which provides a higher derivation $A_S \longrightarrow B_T$.

Let $\underline{D} = (D_0, D_1, \ldots, D_m)$ be a higher k-derivation of length $m < \infty$ from A to B. Consider the problem of extending this to a higher derivation of length $m + 1$. If $E_{t,m}: A \longrightarrow B_m$ is the homomorphism corresponding to $\underline{D}$, the problem of extending $\underline{D}$ is equivalent to that lifting $E_{t,m}$ to a homomorphism $A \longrightarrow B_{m+1}$. The following theorem is then clear from this.

Theorem 27.1. If the ring A is 0-smooth over a ring k, then a higher derivation of length $m < \infty$ over k from A to an A-algebra B can be extended to a derivation of length ∞.

This theorem can be applied for example to the case of a field k and a separable extension field A of k.

If A is a ring of characteristic p then an ordinary derivation of A is zero on the subring A^p, but higher derivations need not vanish on A^p. If $\underline{D} = (D_0, D_1, D_2, \dots)$ is a higher derivation of length $m \geqslant p$ then from $E_t(a^p) = E_t(a)^p = a^p + D_1(a)^p \cdot t^p + \cdots$, we get

$$D_p(a^p) = D_1(a)^p, \quad \text{and in general} \quad D_{p^r}(a^{p^r}) = D_1(a)^{p^r}.$$

For example, it follows from this that if k is a field of characteristic p, and $K = k(\alpha)$ with $\alpha^p \in k$ but $\alpha \notin k$, then although there exists $D \in \mathrm{Der}_k(K)$ such that $D(\alpha) = 1$, this D cannot be extended to a higher k-derivation of length $\geqslant p$. Since K is separable over the prime subfield, D can be extended to a higher derivation of length ∞ (over the prime subfield), but this extension cannot be trivial on k.

We say that a higher derivation $\underline{D} = (D_0, D_1, \dots) \in \mathrm{HS}_k(A)$ is *iterative* if it satisfies the following conditions:

$$D_i \circ D_j = \binom{i+j}{i} D_{i+j} \quad \text{for all} \quad i, j.$$

This condition is equivalent to asking that the following diagram is commutative:

$$\begin{array}{ccc} A[\![t]\!] & \xrightarrow{E_u} & A[\![t,u]\!] \\ {\scriptstyle E_t}\uparrow & & \uparrow \\ A & \xrightarrow{E_{t+u}} & A[\![t+u]\!], \end{array}$$

where $E_t(a) = \sum t^\nu D_\nu(a)$ and $E_u(\sum t^\nu a_\nu) = \sum t^\nu E_u(a_\nu)$, and the right-hand vertical arrow is the inclusion map. Indeed,

$$E_u(E_t(a)) = E_u(\textstyle\sum t^\nu D_\nu(a)) = \sum\limits_\nu t^\nu \sum\limits_\mu u^\mu D_\mu D_\nu(a),$$

and

$$E_{t+u}(a) = \sum_\lambda (t+u)^\lambda D_\lambda(a) = \sum_\nu t^\nu \sum_\mu u^\mu \binom{\nu+\mu}{\nu} D_{\nu+\mu}(a).$$

If A contains the rational field $\mathbb{Q}$, then one sees by induction on n that an iterative higher derivation satisfies $D_n = D_1^n/n!$, and is hence determined by D_1 only. Conversely, for $D \in \mathrm{Der}_k(A)$, we see that $(1, D, D^2/2!, D^3/3!, \dots)$ is an iterative higher derivation. If A has characteristic p then for an iterative $\underline{D} = (D_0, D_1, \dots)$ we have $D_i = D_1^i/i!$ for $i < p$, and $D_1^p = 0$. Thus one cannot hope to extend any derivation to an iterative higher derivation, even if A is a field.

Theorem 27.2. Let $k \xrightarrow{f} A \xrightarrow{g} B$ be ring homomorphisms, and suppose that B is 0-etale over A. Then given $\underline{D} = (D_0, D_1, \dots) \in \mathrm{HS}_k(A, B)$, there exists a unique $\underline{D}' = (D'_0, D'_1, \dots) \in \mathrm{HS}_k(B)$ such that $D'_i(g(a)) =$

$D_i(a)$ for all i. Moreover, if $\underline{D}^* = (D_0^*, D_1^*, \ldots) \in HS_k(A)$ is such that $D_i = g \circ D_i^*$ for all i, and if $\underline{D}^*$ is iterative, then $\underline{D}'$ is also iterative.

Proof. There is no problem about $D_0' = 1_B$. Now assume that $(D_0', \ldots, D_m') \in HS_k(B, m)$ has been constructed so that $D_i' \circ g = D_i$ for $i \leqslant m$; then if we define

$$h: A \longrightarrow B_{m+1} = B[t]/(t^{m+2}) \text{ by } h(a) = \sum_0^{m+1} t^\nu D_\nu(a),$$

and $u: B \longrightarrow B_m$ by $u(b) = \sum_0^m t^\nu D_\nu'(b)$, we obtain the left-hand commutative diagram.

$$\begin{array}{ccc} B & \xrightarrow{u} & B_m \\ {\scriptstyle g}\uparrow & & \uparrow \\ A & \xrightarrow{h} & B_{m+1} \end{array} \qquad\qquad \begin{array}{ccc} B & \xrightarrow{u} & B_m \\ {\scriptstyle g}\uparrow & \searrow^{v} & \uparrow \\ A & \xrightarrow{h} & B_{m+1} \end{array}$$

Hence by the 0-etale assumption, there exists a unique $v: B \longrightarrow B_{m+1}$ which makes the right-hand diagram commutative. Repeating this we see that $\underline{D}'$ exists and is unique. If $\underline{D}^*$ is iterative, and we consider the homomorphisms $E_t: A \longrightarrow A[\![t]\!]$ and $E_t': B \longrightarrow B[\![t]\!]$ corresponding respectively to $\underline{D}^*$ and $\underline{D}'$, then we know $E_u \circ E_t = E_{t+u}$, and we need only prove that $E_u' \circ E_t' = E_{t+u}'$. By induction on m, assume that

$$E_u'(E_t'(b)) \equiv E_{t+u}'(b) \bmod (t, u)^{m+1} \text{ for all } b \in B;$$

then from the commutativity of

$$\begin{array}{ccccc} B & & \xrightarrow{E_{t+u}'} & & B[\![t,u]\!]/(t,u)^{m+1} \\ \uparrow & & & & \uparrow \\ A & \xrightarrow{E_{t+u}} & A[\![t,u]\!] & \longrightarrow & B[\![t,u]\!]/(t,u)^{m+2}, \end{array}$$

from $E_{t+u} = E_u \circ E_t$ and from the assumption that B is 0-etale over A, we get

$$E_u'(E_t'(b)) \equiv E_{t+u}'(b) \bmod (t, u)^{m+2} \quad \text{for all} \quad b \in B.$$

This proves that $E_u' \circ E_t' = E_{t+u}'$. ■

Remark. The above $\underline{D}'$ will be called an extension of $\underline{D}$ (or of $\underline{D}^*$) to B, (even if A is not a subring of B).

Theorem 27.3. (i) Let A be a ring of characteristic p, and suppose that $x \in A$, $D \in \mathrm{Der}(A)$ satisfy $Dx = 1$ and $D^p = 0$; set $A_0 = \{a \in A \mid Da = 0\}$. Then A is a free module over A_0 with basis $1, x, \ldots, x^{p-1}$.

(ii) Let k be a field of characteristic p, and K a separable extension of k;

let $D \in \mathrm{Der}_k(K)$ be such that $D \neq 0$, $D^p = 0$. Set $K_0 = \{a \in K \mid Da = 0\}$. Then there exists an $x \in K$ such that $Dx = 1$, and a subset $B_0 \subset K_0$ such that $B = \{x\} \cup B_0$ is a p-basis for K over k.

Proof. (i) Suppose that $\alpha_0 + \alpha_1 x + \cdots + \alpha_i x^i = 0$ for some $i < p$ and $\alpha_j \in A_0$. Then applying D^i we get $i!\alpha_i = 0$, hence $\alpha_i = 0$. Hence by downward induction on i we see that $1, x, \ldots, x^{p-1}$ are linearly independent over A_0. Now in view of $D^p = 0$, for every $a \in A$ we have $D^{i+1}a = 0$ for some $0 \leqslant i < p$. If $i = 0$ we have $a \in A_0$. If $i > 0$ then $D^i(a - x^i D^i a/i!) = 0$, so that we see by induction that

$$D^{i+1}a = 0 \Rightarrow a \in A_0 + A_0 x + \cdots + A_0 x^i;$$

Setting $i = p - 1$ in this gives $A = A_0 + A_0 x + \cdots + A_0 x^{p-1}$.

(ii) Since $D \neq 0$ we can find $z \in K$ such that $Dz \neq 0$. Now in view of $D^p z = 0$, there is an i such that $D^i z \neq 0$ but $D^{i+1} z = 0$. If we set $y = D^i z$ and $x = (D^{i-1}z)/y$ then $Dx = 1$, so that by (i) we have $K = K_0(x)$ and $[K:K_0] = p$. Now if we had $x^p \in K_0^p k$, then $x \in K_0 k^{1/p}$, and we could write $x = \sum_1^n \omega_i \alpha_i$ with $\omega_1, \ldots, \omega_n \in K_0$ linearly independent over k, and $\alpha \in k^{1/p}$. Now $k \subset K_0$ and $x \notin K_0$, so that $x, \omega_1, \ldots, \omega_n$ are linearly independent over k, and hence by the assumption that K is separable over k, they are also linearly independent over $k^{1/p}$, which contradicts $x = \sum \omega_i \alpha_i$. Thus $x^p \notin K_0^p k$. Hence we can choose a p-basis C of K_0 over k such that $x^p \in C$; set $B_0 = C - \{x^p\}$. Then if $y_1, \ldots, y_n$ are distinct elements of B_0, we have $[K_0^p k(x^p, y_1, \ldots, y_n) : K_0^p k] = p^{n+1}$, and together with $K = K_0(x)$ this gives $[K^p k(y_1, \ldots, y_n) : K^p k] = p^n$. Thus B_0 as a subset of K is p-independent over k. Since also $K_0 = K_0^p k(x^p, B_0)$, we have $K = K_0(x) = K^p k(x, B_0)$, so that setting $B = B_0 \cup \{x\}$ we get a p-basis of K over k. ■

Theorem 27.4. Let K be a field of characteristic p, and $k \subset K$ a subfield such that K is separable over k. Then a necessary and sufficient condition for $D \in \mathrm{Der}_k(K)$ to extend to an iterative element of $\mathrm{HS}_k(K)$ is that $D^p = 0$.

Proof. We have already seen necessity, and we prove sufficiency. We can assume that $D \neq 0$; if $D^p = 0$ then we can choose K_0, x and B_0 as in Theorem 3, (ii). We set $K' = k(B_0)$; then D is a K'-derivation, K is 0-etale over $K'(x)$, and $K'(x)$ is a purely transcendental extension of K'. We define a homomorphism $E_t : K'(x) \longrightarrow K'(x)[\![t]\!]$ by setting $E_t(\alpha) = \alpha$ for $\alpha \in K'$ and $E_t(x) = x + t$; then

$$E_u(E_t(x)) = x + u + t = E_{t+u}(x),$$

so that $E_u \circ E_t = E_{t+u}$ holds over the whole of $K'(x)$. Thus E_t defines a iterative higher derivation $\underline{D}$ of $K'(x)$ over K'. Since K is 0-etale over $K'(x)$, by Theorem 2 there is an extension of $\underline{D}$ to an iterative higher derivation of

K over K'; the term of degree 1 in $\underline{D}$ is D, so that $\underline{D}$ is an extension of D (or more precisely, of $(1, D)$). ■

Exercises to §27.

27.1. Let k be a ring, A a k-algebra and $D \in \mathrm{Der}_k(A)$. Say that D is *integrable* over k if there exists an extension $\underline{D} \in \mathrm{HS}_k(A)$ with $\underline{D} = (D_0, D_1, \ldots)$ and $D = D_1$ (then $\underline{D}$ is an *integral* of D); set $\mathrm{Ider}_k(A) = \{D \in \mathrm{Der}_k(A) \mid D$ is integrable over $k\}$. Then prove that $\mathrm{Ider}_k(A) \subset \mathrm{Der}_k(A)$ is an A-submodule.

27.2. In the notation of this section, consider the construction of $E_t: A \longrightarrow A[\![t]\!]$ corresponding to $\underline{D}$; then if $t' \in A[\![t]\!]$ is any power series with no constant term, we have $A[\![t']\!] \subset A[\![t]\!]$, so that $E_{t'}: A \longrightarrow A[\![t']\!]$ can be composed to a homomorphism $E_{t'}: A \longrightarrow A[\![t]\!]$, to give a different higher derivation. Thus for $\underline{D} = (D_0, D_1, \ldots)$, the homomorphism E_{t^2} corresponds to the higher derivation $\underline{D}' = (D_0, 0, D_1, 0, D_2, \ldots)$; taking the product $\underline{D} \cdot \underline{D}'$ we get an integral of D_1 different from $\underline{D}$. Thus for given $D \in \mathrm{Ider}_k(A)$ there will in general exist many integrals of D; verify that if $D^p = 0$ and we impose the condition that the integral should be iterative, it is still not uniquely determined.

10

I-smoothness

I-smoothness is a notion which Grothendieck obtained by reformulating the theory of simple (non-singular) points in algebraic geometry in terms of an algebraic 'infinitesimal analysis', which makes effective use of nilpotent elements. The definition looks complicated at first sight, but it has various alternative formulations, and is a natural and useful notion. In §28, along with the general theory of I-smoothness following [G1] we prove the existence of a coefficient field for a complete local ring of equal characteristic, relating this to the author's idea of quasi-coefficient field (Theorem 28.3), and discuss Faltings' very simple proof of the equivalence of $\mathfrak{m}$-smoothness and geometric regularity for local rings. In §29 we deduce the existence of a coefficient ring for a complete local ring of unequal characteristic from Theorem 28.10, and prove some classical theorems of Cohen on complete local rings; these results are of decisive significance for the usefulness of taking completions. §30 is something of a jumble of various theories, but is for the most part occupied with the so-called Jacobian criterion for regularity. On this subject, we treat the simple and powerful method obtained by the author's 1972 seminar in the case of a ring containing a field of characteristic 0; in the most difficult case of power series rings in characteristic p, the only method currently available is that of Nagata, and we explain this as simply as possible.

28 I-smoothness

Let A be a ring, B an A-algebra, and I an ideal of B; we consider B with the I-adic topology. We say that B is I-smooth over A if given an A-algebra C, an ideal N of C satisfying $N^2=0$, and an A-algebra homomorphism $u:B\longrightarrow C/N$ which is continuous for the discrete topology of C/N (that is, such that $u(I^\nu)=0$ for some ν), then there exists a lifting $v:B\longrightarrow C$ of u to C.

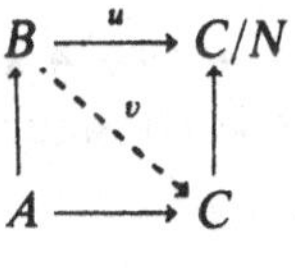

If $I = (0)$, that is if no continuity condition is imposed on u, then this is the definition of 0-smooth given in §25. Write $f: C \longrightarrow C/N$ for the natural map; then from $fv(I^\nu) = u(I^\nu) = 0$ we have $v(I^\nu) \subset N$, hence $v(I^{2\nu}) \subset N^2 = 0$, so that $v: B \longrightarrow C$ (assuming it exist) is continuous for the discrete topology of C. From this, one sees that if B is I-smooth over A, and instead of the condition $N^2 = 0$ we assume that C is an N-adically complete ring, then a continuous homomorphism $u: B \longrightarrow C/N$ has a lifting $v: B \longrightarrow C$, and v is continuous with respect to the N-adic topology of C; this is because we can lift u successively to $B \longrightarrow C/N^i$ of $i = 1, 2, \ldots$, and then v is given by $B \longrightarrow \varprojlim C/N^i = C$.

We now return to the original assumption $N^2 = 0$; we say that B is *I-unramified* (or I-neat) over A if given C, N and a continuous homomorphism $u: B \longrightarrow C/N$, there exists at most one lifting of u to C. If B is both I-smooth and I-unramified over A, we say that B is *I-etale*. These conditions become weaker if we replace I by a larger ideal.

Theorem 28.1 (Transitivity). Let $A \xrightarrow{g} B \xrightarrow{g'} B'$ be ring homomorphisms, and suppose that g' is continuous for the I-adic topology of B and the I'-adic topology of B'; if B is I-smooth over A, and B' is I'-smooth over B then B' is I' smooth over A. The same thing holds with I-unramified in place of I-smooth.

Proof. Suppose that u is given in the diagram;

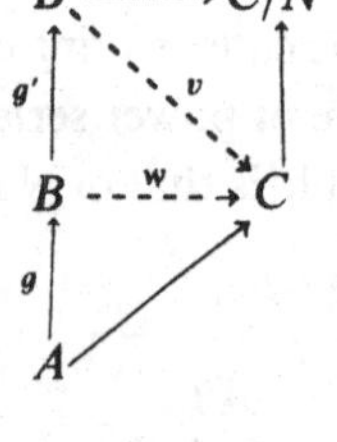

then since $ug': B \longrightarrow C/N$ is continuous, by the I-smoothness of B, there exists a lifting $w: B \longrightarrow C$. Next by the I'-smoothness of B' over B, we can lift u to $v: B' \longrightarrow C$. Also if B is I-unramified over A, and the map v in the diagram exists, then $w = vg'$ is unique, and if in addition B' is I'-unramified over B then v is unique. ∎

Theorem 28.2 (Base-change). Let A be a ring, B and A' two A-algebras, and set $B' = B \otimes_A A'$. If B is I-smooth over A then B' is IB'-smooth over A'. The same thing holds for I-unramified.

Proof. We have the diagram

$$\begin{array}{ccccc} B & \xrightarrow{p} & B' & \xrightarrow{u} & C/N \\ \uparrow & & \uparrow{\scriptstyle q} & & \uparrow \\ A & \longrightarrow & A' & \xrightarrow{\lambda} & C, \end{array}$$

where p, q are the natural homomorphisms. Then if u satisfies $u(I^\nu B') = 0$, there is a lifting $v: B \longrightarrow C$ of up. Then if we define $v': B' = B \otimes_A A' \longrightarrow C$, by $u' = v \otimes \lambda$, this is a lifting of u to C. For unramified, this is clear from the fact that v' is uniquely determined by v. ∎

Example 1. Let k be a ring, $(A, \mathfrak{m})$ a local ring, $(\hat{A}, \hat{\mathfrak{m}})$ its completion, and $k \longrightarrow A$ a homomorphism. Then

(i) $\hat{A}$ is $\hat{\mathfrak{m}}$-etale over A;

(ii) A is $\mathfrak{m}$-smooth (or $\mathfrak{m}$-unramified) over $k \Leftrightarrow \hat{A}$ is $\hat{\mathfrak{m}}$-smooth (or $\hat{\mathfrak{m}}$-unramified) over k.

We get a proof at once from the fact that $\hat{A}/\hat{\mathfrak{m}}^\nu \simeq A/\mathfrak{m}^\nu$ for all ν.

Example 2. Let A be any ring, and set $B = A[\![X_1, \ldots, X_n]\!]$ and $I = \sum_1^n X_i B$; we give B the I-adic topology. Then B is I-smooth over A.

Remark. The gap between I-smoothness and 0-smoothness has been studied by Tanimoto [1], [2]. For instance, if k is a field, then $k[\![X_1, \ldots, X_n]\!]$ is 0-smooth over k only when $\operatorname{char} k = p$ and $[k:k^p] < \infty$.

Let $(A, \mathfrak{m}, K)$ be a local ring. If A is of characteristic p, then also $\operatorname{char} K = p$; moreover, if $\operatorname{char} K = 0$, then $\operatorname{char} A = 0$, and A contains the rational number field $\mathbb{Q}$. In either of these cases A is said to be *equicharacteristic*, or a local ring *of equal characteristic*; this is equivalent to saying that A contains a field. If A is not of equal characteristic, then either

$$\operatorname{char} A = 0 \quad \text{and} \quad \operatorname{char} K = p,$$

or

$$\operatorname{char} A = p^n \quad \text{for some} \quad n > 1 \quad \text{and} \quad \operatorname{char} K = p.$$

In this case we say that A is a local ring of unequal characteristic.

Let A be an equicharacteristic local ring and let K' be a subfield of A. We say that K' is a *coefficient field* of A if K' maps isomorphically to K under the natural map $A \longrightarrow A/\mathfrak{m} = K$, or equivalently, if $A = K' + \mathfrak{m}$. Moreover, we say that K' is a *quasi-coefficient field* of A if K is 0-etale over K' (or rather, over the image of K' in K).

Theorem 28.3. Let $(A, \mathfrak{m}, K)$ be an equicharacteristic local ring. Then

(i) A has a quasi-coefficient field;

(ii) if A is complete, it has a coefficient field;

(iii) if the residue field K of A is separable over a subfield $k \subset A$ then A has a quasi-coefficient field K' containing k;

(iv) if K' is a quasi-coefficient field of A, then there exists a unique coefficient field K'' of the completion $\hat{A}$ containing K'.

Proof. (iii) Suppose that $B = \{\xi_1, \xi_2, \ldots\}$ is a differential basis of K/k, and for each ξ_i, choose an inverse image $x_i \in A$. Then by Theorem 26.8, $\xi_1, \xi_2, \ldots$ are algebraically independent over k, so that the subring $k[x_1, x_2, \ldots]$ of A meets $\mathfrak{m}$ in $\{0\}$, and hence A contains the field $K' = k(x_1, x_2, \ldots)$. We identify K' with its image $k(B)$ in K, so that K is clearly 0-etale over K', and K' is a quasi-coefficient field, as required.

(i) By assumption A contains a field, so that it contains a perfect field (for example, the prime subfield). We need only apply (iii) to this.

$$\begin{array}{ccc} K & = & \hat{A}/\hat{\mathfrak{m}} \\ \uparrow & & \uparrow \\ K' & \longrightarrow & \hat{A} \end{array}$$

(iv) In the diagram above, there exists a unique lifting of the identity map $K \longrightarrow \hat{A}/\hat{\mathfrak{m}}$ to $K \longrightarrow \hat{A}$, and its image is the required coefficient field.

(ii) follows from (i) and (iv). ∎

The next lemma will be made more precise in Theorem 28.7.

Lemma 1. Suppose that $(A, \mathfrak{m}, K)$ is a Noetherian local ring containing a field k. If A is $\mathfrak{m}$-smooth over k then A is regular. The converse holds if the residue field K is separable over k.

Proof. Take a perfect subfield $k_0 \subset k$; then k is 0-smooth over k_0, so that by transitivity, A is also $\mathfrak{m}$-smooth over k_0, so that we can assume that k is a perfect field. Also, replacing A by $\hat{A}$, we can assume that A is complete. Then A has a coefficient field containing k; for ease of notation we write K for this, and identify it with the residue field. If $\{x_1, \ldots, x_n\}$ is a minimal basis of $\mathfrak{m}$ then as K-algebras we have

$$A/\mathfrak{m}^2 \simeq K[X_1, \ldots, X_n]/(X_1, \ldots, X_n)^2.$$

The composite

$$A \longrightarrow A/\mathfrak{m}^2 \xrightarrow{\sim} K[X_1, \ldots, X_n]/(X)^2 \xrightarrow{\sim} K[\![X_1, \ldots, X_n]\!]/(X)^2$$

lifts to $A \longrightarrow K[\![X_1, \ldots, X_n]\!]$, and by Theorem 8.4, this is surjective. Thus $\dim A \geqslant \dim K[\![X_1, \ldots, X_n]\!] = n$, and together with $\operatorname{emb dim} A = n$ this gives the regularity of A.

Conversely, if A is regular and K is separable over k, then $\hat{A}$ has a coefficient field K containing k. Let $\{x_1, \ldots, x_n\}$ be a regular system of parameters of $\hat{A}$, and define a homomorphism of K-algebras $\psi: K[\![X_1, \ldots, X_n]\!] \longrightarrow \hat{A}$ by $\psi(X_i) = x_i$; then once more by Theorem 8.4, ψ is surjective, and comparing dimensions, we see that

$$K[\![X_1, \ldots, X_n]\!] \xrightarrow{\sim} \hat{A}.$$

Therefore $\hat{A}$ is $\hat{\mathfrak{m}}$-smooth over K, and since K is 0-smooth over k, we see that $\hat{A}$ is $\hat{\mathfrak{m}}$-smooth over k, and therefore A is $\mathfrak{m}$-smooth over k. ■

Let $k \longrightarrow A \longrightarrow B$ be ring homomorphisms, and let I be an ideal of B; we consider B in the I-adic topology. We say that B is *I-smooth* over A *relative to* k if the following condition holds: for any A-algebra C, and an ideal N of C such that $N^2 = 0$, given an A-algebra homomorphism $u: B \longrightarrow C/N$ satisfying $u(I^\nu) = 0$ for sufficiently large ν, if u has lifting $v': B \longrightarrow C$ as a k-algebra homomorphism, it also has a lifting $v': B \longrightarrow C$ as an A-algebra homomorphism:

$$\begin{array}{ccccc} & & B & \xrightarrow{u} & C/N \\ & & \uparrow & \searrow & \uparrow \\ k & \longrightarrow & A & \longrightarrow & C. \end{array}$$

Theorem 28.4 Let $k \xrightarrow{f} A \xrightarrow{g} B$ and $I \subset B$ be as above; then the following three conditions are equivalent:

(1) B is I-smooth over A relative to k;

(2) if N is a B-module such that $I^\nu N = 0$ for sufficiently large ν, then $\mathrm{Der}_k(B, N) \longrightarrow \mathrm{Der}_k(A, N)$ is surjective;

(3) for every $\nu > 0$, the map $\Omega_{A/k} \otimes_A (B/I^\nu) \longrightarrow \Omega_{B/k} \otimes_B (B/I^\nu)$ has a left inverse (that is, it maps injectively onto a direct summand).

Proof. (1)⇒(2) If $I^\nu N = 0$, set $C = (B/I^\nu) * N$, and let $u: B \longrightarrow B/I^\nu = C/N$ be the natural map. Given $D \in \mathrm{Der}_k(A, N)$, define $\lambda: A \longrightarrow C$ by $\lambda(a) = (ug(a), D(a))$. If we consider C as an A-algebra via λ, then $b \mapsto (u(b), 0) \in C$ is a k-algebra homomorphism from B to C lifting u, so that by assumption there exists a lifting $v': B \longrightarrow C$ of u as an A-algebra homomorphism; then writing v' in the form

$$v'(b) = (u(b), D'(b)), \quad \text{with} \quad D' \in \mathrm{Der}_k(B, N),$$

we have $v'g = \lambda$, so that $D'g = D$.

(2)⇒(1) Suppose given a commutative diagram

$$\begin{array}{ccccc} & & B & \xrightarrow{u} & C/N \\ & & {\scriptstyle g}\uparrow & & \uparrow{\scriptstyle j} \\ k & \xrightarrow{f} & A & \xrightarrow{\lambda} & C \end{array}$$

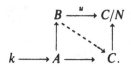

with j the natural map, and $u(I^\nu)=0$; if $v:B \longrightarrow C$ is a k-algebra homomorphism satisfying $jv = u$ and $vgf = \lambda f$, then setting $D = \lambda - vg$, we can view D as an element of $\mathrm{Der}_k(A, N)$. By assumption there exists $D' \in \mathrm{Der}_k(B, N)$ such that $D = D'g$. Using this, we set $v' = v + D'$; then

$$v'g = vg + D'g = \lambda - D + D = \lambda, \quad \text{and} \quad jv' = u.$$

(2)$\Leftrightarrow$(3) comes from observing the general fact that for a ring R, a map $\varphi: M \longrightarrow M'$ of R-modules has a left inverse if and only if for every R-module N the induced map

$$\mathrm{Hom}_R(M', N) \longrightarrow \mathrm{Hom}_R(M, N)$$

is surjective. ∎

Theorem 28.5. Let A be a ring, B an A-algebra, and I an ideal of B; set $\bar{B} = B/I$, and assume that B is I-smooth over A. Then $\Omega_{B/A} \otimes_B \bar{B}$ is projective as a $\bar{B}$-module.

Proof. It is enough to show that for an exact sequence $L \xrightarrow{\varphi} M \to 0$ of $\bar{B}$-modules, the sequence

$$\mathrm{Hom}_{\bar{B}}(\Omega_{B/A} \otimes \bar{B}, L) \longrightarrow \mathrm{Hom}_{\bar{B}}(\Omega_{B/A} \otimes \bar{B}, M) \to 0$$

is exact, that is that

$$\mathrm{Der}_A(B, L) \longrightarrow \mathrm{Der}_A(B, M) \to 0$$

is exact. Set $C = \bar{B} * L$ and $N = \mathrm{Ker}\,\varphi$. If we view both L and N as ideals of C, we have $L^2 = N^2 = 0$ and $C/N \simeq \bar{B} * M$. Now for any $D \in \mathrm{Der}_A(B, M)$ we have an A-algebra homomorphism $B \longrightarrow C/N$ given by

$$b \mapsto (\bar{b}, D(b)) \in \bar{B} * M,$$

and lifting this to $B \longrightarrow C$ is equivalent to lifting D to an element of $\mathrm{Der}_A(B, L)$. ∎

Lemma 2. Let B be a ring and I an ideal of B, and let $u: L \longrightarrow M$ be a map of B-modules; assume that M is projective. Suppose also that one of the following two conditions hold:

(α) I is nilpotent;

or (β) L is a finite B-module and $I \subset \mathrm{rad}(B)$.

Then

$$u \text{ has a left inverse} \Leftrightarrow \bar{u}: L/IL \longrightarrow M/IM \text{ has a left inverse.}$$

Proof. ($\Rightarrow$) is trivial. To prove ($\Leftarrow$), suppose that $\bar{v}: M/IM \longrightarrow L/IL$ is a left inverse of $\bar{u}$. Since M is projective, there is a map $v: M \longrightarrow L$ such that the diagram

$$\begin{array}{ccc} M & \xrightarrow{v} & L \\ \downarrow & & \downarrow \\ M/IM & \xrightarrow{\bar{v}} & L/IL \end{array}$$

commutes. Set $w = vu$. Then w induces the identity on L/IL, so that $L = w(L) + IL$, so that by NAK, $L = w(L)$. Hence if L is a finite B-module, then by Theorem 2.4, w is also injective. Furthermore, if $I^\nu = 0$ we do the following: if $x \in \operatorname{Ker} w$ then $0 = w(x) \equiv x \bmod IL$, so that $x \in IL$, and we can write $x = \sum a_i y_i$ with $a_i \in I$ and $y_i \in L$. Then

$$0 = w(x) = \sum a_i w(y_i) \equiv \sum a_i y_i = x \bmod I^2 L,$$

so that $x \in I^2 L$, and proceeding in the same way we arrive at $x \in I^\nu L = 0$. Hence also in this case w is an automorphism of L, and $w^{-1}v$ is the required left inverse of u. ■

Theorem 28.6. Let $k \longrightarrow A \longrightarrow B$ be ring homomorphisms, I an ideal of B, and suppose that B is I-smooth over k. Set $B_1 = B/I$. Then the following conditions are equivalent:

(1) B is I-smooth over A;

(2) $\Omega_{A/k} \otimes_A B_1 \longrightarrow \Omega_{B/k} \otimes_B B_1$ has a B_1-linear left inverse.

Proof. (1)⇒(2) is contained in Theorem 4. Conversely, suppose that (2) holds. For any $\nu > 0$, set $B_\nu = B/I^\nu$; then since I-smoothness and I^ν-smoothness are the same, by Theorem 5, $\Omega_{B/k} \otimes B_\nu$ is a projective B_ν-module. Now set $I_\nu = I/I^\nu$; then $B_\nu/I_\nu = B_1$, and $(I_\nu)^\nu = 0$, so that applying Lemma 2, we see that $\Omega_{A/k} \otimes_A B_\nu \longrightarrow \Omega_{B/k} \otimes_B B_\nu$ has a left inverse. By Theorem 4, B is I-smooth over A relative to k, but since it is also I-smooth over k, it is also I-smooth over A. ■

Corollary. Let $(A, \mathfrak{m}, K)$ be a regular local ring containing a field k; then

$$A \text{ is } \mathfrak{m}\text{-smooth over } k \Leftrightarrow \Omega_k \otimes_k K \longrightarrow \Omega_A \otimes_A K \text{ is injective.}$$

Proof. Let $k_0 \subset k$ be the prime subfield. Then by Lemma 1, A is $\mathfrak{m}$-smooth over k_0, so that we need only apply the theorem to $k_0 \longrightarrow k \longrightarrow A$. ■

Let A be a Noetherian local ring, and $k \subset A$ a subfield. We say that A is *geometrically regular* over k if $A \otimes_k k'$ is a regular ring for every finite extension field k' of k.

Theorem 28.7. Let $(A, \mathfrak{m}, K)$ be a Noetherian local ring, and $k \subset A$ a subfield; then

$$A \text{ is } \mathfrak{m}\text{-smooth over } k \Leftrightarrow A \text{ is geometrically regular over } k.$$

Proof. (⇒) Let k' be a finite extension field of k. Then $A \otimes_k k' = A'$ is $\mathfrak{m}A'$-smooth over k' by base-change. Let $\mathfrak{n}$ be any maximal ideal of A'; then A' is a finite A-module, hence integral over A, so that $\mathfrak{n} \supset \mathfrak{m}A'$. Thus if we set $A'' = A'_\mathfrak{n}$ and $\mathfrak{m}'' = \mathfrak{n}A''$ then $A' \longrightarrow A''$ is continuous for the $\mathfrak{m}A'$-adic topology of A' and the $\mathfrak{m}''$-adic topology of A'', but as a localisation A'' is 0-etale over A', so that by Theorem 1, A'' is $\mathfrak{m}''$-smooth over k', and hence by Lemma 1, $A'' = A'_\mathfrak{n}$ is a regular local ring. This is what was required to prove.

($\Leftarrow$) According to Lemma 1, there is only a problem if k is of characteristic p. The proof we now give was discovered in 1977 by G. Faltings [1] while he was still a student.

By the corollary of the previous theorem, we need only prove that $\Omega_k \otimes_k K \longrightarrow \Omega_A \otimes_A K$ is injective. For this, we let $x_1, \ldots, x_r \in k$ be p-independent elements, and prove that $dx_1, \ldots, dx_r \in \Omega_A \otimes K$ are linearly independent over K. Write α_i for pth roots of the x_i, and set $k' = k(\alpha_1, \ldots, \alpha_r)$. Then

$$B = A \otimes_k k' = A[T_1, \ldots, T_r]/(T_1^p - x_1, \ldots, T_r^p - x_r)$$

is a Noetherian local ring; write $\mathfrak{n}$ for its maximal ideal, and L for the residue field $L = B/\mathfrak{n}$. By Theorem 25.2 the sequence

$$0 \to \mathfrak{n}/\mathfrak{n}^2 \longrightarrow \Omega_B \otimes_B L \longrightarrow \Omega_L \to 0$$

is exact. Similarly,

$$0 \to \mathfrak{m}/\mathfrak{m}^2 \longrightarrow \Omega_A \otimes_A K \longrightarrow \Omega_K \to 0$$

is exact. Now consider the commutative diagram:

$$\begin{array}{ccccc} 0 \to \mathfrak{n}/\mathfrak{n}^2 & \longrightarrow & \Omega_B \otimes_B L & \longrightarrow & \Omega_L \to 0 \\ \uparrow \varphi_1 & & \uparrow \varphi_2 & & \uparrow \varphi_3 \\ 0 \to (\mathfrak{m}/\mathfrak{m}^2) \otimes_K L & \longrightarrow & \Omega_A \otimes_A L & \longrightarrow & \Omega_K \otimes_K L \to 0 \end{array}$$

where φ_1, φ_2 and φ_3 are the natural maps. Then by the snake lemma, we get a long exact sequence of L-modules

$$\begin{aligned} 0 \to \operatorname{Ker} \varphi_1 &\longrightarrow \operatorname{Ker} \varphi_2 \longrightarrow \operatorname{Ker} \varphi_3 \\ &\longrightarrow \operatorname{Coker} \varphi_1 \longrightarrow \operatorname{Coker} \varphi_2 \longrightarrow \operatorname{Coker} \varphi_3 \to 0. \end{aligned}$$

By assumption A and B are regular local rings of the same dimension, so that $\operatorname{rank} \mathfrak{m}/\mathfrak{m}^2 = \dim A = \operatorname{rank} \mathfrak{n}/\mathfrak{n}^2$, so $\operatorname{rank} \operatorname{Ker} \varphi_1$ and $\operatorname{rank} \operatorname{Coker} \varphi_1$ are finite and equal; moreover, since L is a finite extension field of K, both $\operatorname{rank} \operatorname{Ker} \varphi_3$ and $\operatorname{rank} \operatorname{Coker} \varphi_3$ are finite and equal (the Cartier equality). Therefore by the above exact sequence, we get

$$\operatorname{rank} \operatorname{Ker} \varphi_2 = \operatorname{rank} \operatorname{Coker} \varphi_2.$$

However, $\operatorname{Coker} \varphi_2 = \Omega_{B/A} \otimes_B L$, and by Theorem 25.2,

$$\Omega_{B/A} = B dT_1 + \cdots + B dT_r \simeq B^r,$$

so that both of $\operatorname{Ker} \varphi_2$ and $\operatorname{Coker} \varphi_2$ have rank equal to r. Now if we set $J = (T_1^p - x_1, \ldots, T_r^p - x_r)$ we have an exact sequence

$$J/J^2 \xrightarrow{\ \delta\ } \Omega_{A[T_1, \ldots, T_r]} \otimes B = \Omega_A \otimes B \oplus \sum B dT_i \longrightarrow \Omega_B \to 0,$$

and since this remains exact after performing $\otimes_B L$, we see that $\operatorname{Ker} \varphi_2$ is generated by $dx_1, \ldots, dx_r$. Therefore $dx_1, \ldots, dx_r \in \Omega_A \otimes L$ are linearly independent over L, so that they must also be linearly independent over K as elements of $\Omega_A \otimes K$. This is what we had to prove. ■

Let B be an A-algebra, I an ideal of B, and consider B with the I-adic topology. Let N be a B-module such that $I^\nu N = 0$ for some $\nu > 0$; in what follows we will say that a B-module with this property is *discrete*. An A-bilinear map $f: B \times B \longrightarrow N$ will be called a *continuous symmetric 2-cocycle* if it satisfies the three conditions.

(α) $xf(y,z) - f(xy,z) + f(x,yz) - f(x,y)z = 0$ for all $x, y, z \in B$,

(β) $f(x,y) = f(y,x)$,

(γ) there exists $\mu \geqslant \nu$ such that $f(x,y) = 0$ if either $x \in I^\mu$ or $y \in I^\mu$.

If this holds, we set $f(1,1) = \tau$; then substituting $y = z = 1$ in (α) gives $x\tau = f(x,1)$.

Define a product on the A-module $C = (B/I^\mu) \oplus N$ by

$$(\bar{x}, \xi)(\bar{y}, \eta) = (\overline{xy}, -f(x,y) + x\eta + y\xi)$$

for $x, y \in B$; then C is a commutative ring with unit $(1, \tau)$, and N is an ideal of C satisfying $N^2 = 0$. If we define a map $A \longrightarrow C$ by $a \mapsto (\bar{a}, a\tau)$ then this is a ring homomorphism, and the diagram

$$\begin{array}{ccc} B & \xrightarrow{u} & B/I^\mu = C/N \\ \uparrow & & \uparrow \\ A & \longrightarrow & C \end{array}$$

is commutative; then a necessary and sufficient condition for u to have a lifting to $B \longrightarrow C$ is that there should exist an A-linear map $g: B \longrightarrow N$ such that

(α') $f(x,y) = xg(y) - g(xy) + g(x)y$ for all $x, y \in B$.

For if g exists, then defining $v: B \longrightarrow C$ by $v(x) = (\bar{x}, g(x))$ we find that v is a lifting of u, and conversely, if there is a lifting v of u one checks easily that one can find a g as above.

We say that the 2-cocycle f *splits* if there exists g satisfying (α'). For any A-linear map $g: B \longrightarrow N$, we write δg for the bilinear map $B \times B \longrightarrow N$ given by the right-hand side of (α'); this satisfies (α) and (β), and if g is continuous (that is, $\exists \mu$ such that $g(I^\mu) = 0$), then it also satisfies (γ).

Theorem 28.8. Let A be a ring and B an A-algebra with an I-adic topology.

(i) If B is I-smooth over A then every continuous symmetric 2-cocycle $f: B \times B \longrightarrow N$ with values in a discrete B-module N splits.

(ii) If B/I^n is projective as an A-module for infinitely many n, and if every continuous symmetric 2-cocycle with values in a discrete B-module splits, then B is I-smooth over A.

Proof. (i) is what we have just said.

(ii) Suppose that we are given a commutative diagram

$$\begin{array}{ccc} B & \xrightarrow{u} & C/N \\ \uparrow & & \uparrow \\ A & \longrightarrow & C \end{array}$$

with $N^2=0$ and $u(I^\nu)=0$; then in view of $N^2=0$, the C-module N can be viewed as a C/N-module, and by means of u as a B-module; but then $I^\nu N=0$, so that N is a discrete B-module. Take an integer $n>\nu$ such that B/I^n is projective as an A-module. Then u can be lifted as a map of A-modules to $\lambda: B \longrightarrow C$ such that $\lambda(I^n)=0$. For $x, y \in B$ we set

$$f(x, y)=\lambda(xy)-\lambda(x)\lambda(y),$$

and since λ is a ring homomorphism modulo N, we have $f(x, y) \in N$. Now for $\xi \in N$ and $x \in B$, by definition we have $\lambda(x)\cdot\xi = x\cdot\xi$ (both sides are products evaluated in C), and using this one computes the left-hand side of (α) to be zero. The symmetry (β) is obvious. Also $\lambda(I^n)=0$, so that we also get (γ). Thus f is a continuous symmetric 2-cocycle, and hence by assumption there is an A-linear map $g: B \longrightarrow N$ satisfying

$$f(x, y)=xg(y)-g(xy)+g(x)y.$$

Now if we set $v=\lambda+g$, we have

$$\begin{aligned} v(xy) &= \lambda(xy)+g(xy) \\ &= \lambda(x)\lambda(y)+f(x, y)+g(xy) \\ &= \lambda(x)\lambda(y)+\lambda(x)g(y)+g(x)\lambda(y) \\ &= v(x)v(y), \end{aligned}$$

so that v is an A-algebra homomorphism, and is a lifting of u. ■

Theorem 28.9 ([G1], 19.7.1). Let $(A, \mathfrak{m}, k)$ and $(B, \mathfrak{n}, k')$ be Noetherian local rings, and $\varphi: A \longrightarrow B$ a local homomorphism; set $B_0 = B \otimes_A k = B/\mathfrak{m}B$ and $\mathfrak{n}_0=\mathfrak{n}/\mathfrak{m}B$. Then the following conditions are equivalent:

(1) B is $\mathfrak{n}$-smooth over A,

(2) B is flat over A and B_0 is $\mathfrak{n}_0$-smooth over k.

This is an extremely important theorem, but the proof is long and difficult, and we refer to [G1] for it. We content ourselves with proving the following analogous theorem, which is all that we will need in what follows.

Theorem 28.10. Let $(A, \mathfrak{m}, k)$ be a local ring, and B a flat A-algebra; suppose that $B_0=B\otimes_A k$ is 0-smooth over k. Then B is $\mathfrak{m}B$-smooth over A.

Proof. As one sees from the definition of $\mathfrak{m}B$-smoothness, it is enough to prove that $B/\mathfrak{m}^\nu B$ is 0-smooth over $A/\mathfrak{m}^\nu$ for every $\nu>0$. Since $B/\mathfrak{m}^\nu B$ is flat over $A/\mathfrak{m}^\nu$, we can assume that $\mathfrak{m}$ is nilpotent. Then a flat A-module is free (by Theorem 7.10), so that B is a projective A-module, and hence by Theorem 8, we need only show that every symmetric 2-cocycle $f: B\times B \longrightarrow N$ with values in a B-module N splits. First of all, in the case that N satisfies $\mathfrak{m}N=0$, then since f is A-bilinear f is essentially a 2-cocycle over B_0, that is there is a map $f_0: B_0 \times B_0 \longrightarrow N$ such that $f(x, y)=f_0(\bar{x}, \bar{y})$.

Now B_0 is 0-smooth over k, so that by Theorem 8, f_0 splits, that is there is a map $g_0: B_0 \longrightarrow N$ such that $f_0 = \delta g_0$. Thus setting $g(x) = g_0(\bar{x})$ we get

$$f = \delta g.$$

In the general case, write φ for the natural map $N \longrightarrow N/\mathfrak{m}N$, and consider $\varphi \circ f$; then this splits, that is there exists $\bar{g}: B \longrightarrow N/\mathfrak{m}N$ such that

$$\varphi \circ f = \delta \bar{g}.$$

Now since B is projective over A, we can lift $\bar{g}$ to an A-linear map $g: B \longrightarrow N$, and then $f - \delta g$ is a 2-cocycle with values in $\mathfrak{m}N$. Doing the same thing once more, we find $h: B \longrightarrow \mathfrak{m}N$ such that $f - \delta(g + h)$ is a 2-cocycle with values in $\mathfrak{m}^2 N$. Proceeding in the same way, since $\mathfrak{m}$ is nilpotent we finally see that f splits. ■

Exercises to §28. Prove the following propositions.

28.1. Theorem 28.10 also holds on replacing smooth by unramified or etale.

28.2. Let k be a non-perfect field of characteristic p, and $a \in k - k^p$; set $A = k[X]_{(X^p - a)}$. Then the residue field $k(a^{1/p})$ of A is inseparable over k. This ring A does not have a coefficient field containing k, but is 0-smooth over k.

29 The structure theorems for complete local rings

By Theorem 28.3, a complete local ring of equal characteristic A has a coefficient field. If K is a coefficient field of A, and $x_1, \ldots, x_n$ are generators of the maximal ideal, then any element of A can be expanded as a power series in $x_1, \ldots, x_n$ with coefficients in K, and therefore A is a quotient of the regular local ring $K[\![X_1, \ldots, X_n]\!]$. We now want to extend all of this to the case of unequal characteristic.

We say that a DVR of characteristic 0 is a *p-ring* if its maximal ideal is generated by the prime number p. If K is a given field of characteristic p, then there exists a p-ring having K as its residue field. This follows by applying the next theorem to $A = \mathbb{Z}_{p\mathbb{Z}}$.

Theorem 29.1. Let (A, tA, k) be a DVR and K an extension field of k; then there exists a DVR (B, tB, K) containing A.

Proof. Let $\{x_\lambda\}_{\lambda \in \Lambda}$ be a transcendence basis of K over k, and set $k_1 = k(\{x_\lambda\})$. We take indeterminates $\{X_\lambda\}_{\lambda \in \Lambda}$ over A in bijection with the $\{x_\lambda\}$, and set $A[\{X_\lambda\}] = A'$ and $A_1 = (A')_{tA'}$. Now A' is a free A-module, and hence separated for the t-adic topology; therefore, so is A_1, and A_1 is a DVR with $A_1/tA_1 \simeq k_1$. Hence, replacing A and k by A_1 and k_1, we can assume that K is algebraic over k. Let L be an algebraic closure of the field of fractions of A, and let $\mathscr{F}$ be the set of all pairs (B, φ), where B is an intermediate ring $A \subset B \subset L$ and $\varphi: B \longrightarrow K$ is an A-algebra homomorphism, satisfying the conditions

(*) B is a DVR and $\operatorname{Ker}\varphi = \operatorname{rad}(B)$ is generated by t.

We introduce an order on $\mathscr{F}$ by defining $(B, \varphi) < (C, \psi)$ if $B \subset C$ and $\psi|_B = \varphi$. Let us show that $\mathscr{F}$ has a maximal element. Suppose that

$$\mathscr{G} = \{(B_i, \varphi_i)\}_{i \in I}$$

is a totally ordered subset of $\mathscr{F}$, and set

$$B_0 = \bigcup B_i;$$

then one sees easily that B_0 is a local ring with maximal ideal tB_0. If $0 \neq x \in B_0$ then $x \in B_i$ for some i, and since B_i is a DVR we can write $x = t^n u$ with u an unit of B_i and some n. From this we get $x \notin t^{n+1} B_0$, so that B_0 is t-adically separated. Hence B_0 is a DVR, and if $\varphi_0: B_0 \longrightarrow K$ is defined to be equal to φ_i on B_i then $(B_0, \varphi_0) \in \mathscr{F}$. Hence by Zorn's lemma $\mathscr{F}$ has a maximal element; suppose that (B, φ) is one. Then if $\varphi(B) \neq K$, take $a \in K - \varphi(B)$, let $\bar{f}(X)$ be the minimal polynomial of a over $\varphi(B)$, and take a monic $f(X) \in B[X]$ which is an inverse image of $\bar{f}(X)$. Then $f(X)$ is irreducible in $B[X]$, and hence (by Ex. 9.6) also irreducible over the field of fractions of B. Let α be a root of $f(X)$ in L, and set $B' = B[\alpha]$; then $B' = B[X]/(f)$. Therefore

$$B'/tB' = B[X]/(t, f) = \varphi(B)[X]/(\bar{f}) = \varphi(B)(a)$$

is a field; since B' is integral over B, every maximal ideal of B' contains tB', so that B' is a local integral domain with maximal ideal tB', and is Noetherian because it is finite over B, therefore a DVR. This contradicts the maximality of B, so that we must have $\varphi(B) = K$.

Remark. Since B is an integral domain containing A, it is flat over A by Ex. 10.2. In EGA 0_{III}, (10.3.1), the following more general fact is proved: let $(A, \mathfrak{m}, k)$ be a Noetherian local ring, and K an extension field of k, then there exists a Noetherian ring B containing A satisfying the three conditions (1) $\operatorname{rad}(B) = \mathfrak{m}B$, (2) $B/\mathfrak{m}B$ is isomorphic over k to K, and (3) B is flat over A.

Theorem 29.2. Let $(A, \mathfrak{m}, K)$ be a complete local ring, (R, pR, k) a p-ring, and $\varphi_0: k \longrightarrow K$ a field homomorphism; then there exists a local homomorphism $\varphi: R \longrightarrow A$ which induces φ_0 on the residue fields.

Proof. Set $S = \mathbb{Z}_{p\mathbb{Z}}$, and let $k_0 \subset k$ be the prime subfield. Since $\varphi_0(k_0) \subset K$. the prime number p, viewed as an element of A, belongs to $\mathfrak{m}$. Hence the standard homomorphism $\mathbb{Z} \longrightarrow A$ extends to a local homomorphism $S \longrightarrow A$. Now $R \otimes_S k_0 = R/pR = k$ is a separable extension of k_0, and hence 0-smooth over k_0; also R is a torsion-free S-module, hence flat over

S, so that by Theorem 28.10, R is pR-smooth over S.

$$\begin{array}{ccc} R \to k & \xrightarrow{\varphi_0} & K \\ \uparrow & \searrow^{\varphi} & \uparrow \\ S & \longrightarrow & A \end{array} \qquad\qquad \begin{array}{ccc} R & \longrightarrow & A/\mathfrak{m}^i \\ \uparrow & \searrow & \uparrow \\ S & \longrightarrow & A/\mathfrak{m}^{i+1} \end{array}$$

Therefore, as we discussed at the beginning of §28, we can lift $R \longrightarrow A/\mathfrak{m} = K$ successively to $R \longrightarrow A/\mathfrak{m}^i$, and using the fact that $A = \varprojlim A/\mathfrak{m}^i$, we get $\varphi: R \longrightarrow A$ making the left-hand diagram commute. ∎

Corollary. A complete p-ring is uniquely determined up to isomorphism by its residue field.

Proof. Suppose that R and R' are both complete p-rings with residue field k; then by the theorem there exists a local homomorphism $\varphi: R \longrightarrow R'$ which induces the identity map on the residue field. We have $R' = \varphi(R) + pR'$, and of course $\varphi(p) = p$, so that by the completeness of R we see that φ is surjective, and is also injective, since $p^n R$ is not contained in $\operatorname{Ker} \varphi$ for any n. Therefore $R \simeq R'$. ∎

Let $(A, \mathfrak{m}, k)$ be a complete local ring of unequal characteristic, and let $p = \operatorname{char} k$. We say that a subring $A_0 \subset A$ is a *coefficient ring* of A if A_0 is a complete Noetherian local ring with maximal ideal pA_0 and

$$A = A_0 + \mathfrak{m}, \quad \text{that is,} \quad k = A/\mathfrak{m} \simeq A_0/pA_0.$$

By Theorem 1 applied to the residue field k of A, there exists a p-ring S such that $S/pS = k$; write R for the completion of S, so that R is a complete p-ring with residue field k. By Theorem 2, there exists a local homomorphism $\varphi: R \longrightarrow A$ inducing an isomorphism on the residue fields. If we set $\varphi(R) = A_0$ then this is clearly a coefficient ring of A. If A has characteristic 0 then φ is injective and $A_0 \simeq R$. If A has characteristic p^n then $A_0 \simeq R/p^n R$. We summarise the above discussion in the following theorem.

Theorem 29.3. If $(A, \mathfrak{m}, k)$ is a complete local ring and $p = \operatorname{char} k$ then A has a coefficient ring A_0. If A has characteristic 0 then A_0 is a complete DVR.

In what follows, in order to include the case of equal characteristic in our discussion, we also consider coefficient fields as being coefficient rings. From the previous theorem and Theorem 28.3, we get the following important result.

Theorem 29.4. (i) If $(A, \mathfrak{m})$ is a complete local ring and $\mathfrak{m}$ is finitely generated, then A is Noetherian.

(ii) A Noetherian complete local ring is a quotient of a regular local ring; in particular it is universally catenary.

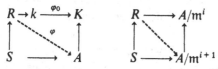

(iii) If A is a Noetherian complete local ring (and in the case of unequal characteristic, A is an integral domain), then there exists a subring $A' \subset A$ with the following properties: A' is a complete regular local ring with the same residue field as A, and A is finitely generated as an A'-module.

Proof. We choose a coefficient ring A_0 of A. If $\mathfrak{m} = (x_1, \ldots, x_n)$ then every element of A can be expanded as a power series in $x_1, \ldots, x_n$ with coefficients in A_0, so that A is a quotient of $A_0[\![X_1, \ldots, X_n]\!]$, and hence Noetherian. Now A_0 is a quotient of a p-ring R, so that A is a quotient of $R[\![X_1, \ldots, X_n]\!]$, which is a regular local ring, hence a CM ring, and therefore according to Theorem 17.9, A is universally catenary. To prove (iii), set $\dim A = n$, and in the case of equal characteristic let $\{y_1, \ldots, y_n\}$ be any system of parameters of A; if A is an integral domain of characteristic 0 and $\operatorname{char} k = p$, we can choose a system of parameters $\{y_1 = p, y_2, \ldots, y_n\}$ of A starting with p. In either case $R = A_0$, so that we set $A' = R[\![y]\!]$. Then A' is the image of $\varphi: R[\![Y]\!] \longrightarrow A$, where $R[\![Y]\!]$ is the regular local ring

$$R[\![Y]\!] = R[\![Y_1, \ldots, Y_n]\!], \quad \text{or} \quad R[\![Y_2, \ldots, Y_n]\!] \quad \text{if} \quad y_1 = p,$$

and φ is the R-algebra homomorphism defined by $\varphi(Y_i) = y_i$. Set $\mathfrak{m}' = \sum_1^n y_i A'$. Since $A/\mathfrak{m} = A'/\mathfrak{m}'$, every A-module of finite length has the same length when viewed as an A'-module. In particular $A/\mathfrak{m}'A$ is a finite module over $A'/\mathfrak{m}'$ and A is $\mathfrak{m}'$-adically separated, so that by Theorem 8.4 A is a finite A'-module. Therefore, we have

$$\dim A' = \dim A = n.$$

$R[\![Y]\!]$ is an n-dimensional integral domain, and if $\operatorname{Ker} \varphi \neq 0$ we would have $\dim A' < n$, which is a contradiction. Therefore φ is injective and $A' \simeq R[\![Y]\!]$.

Remark. In the case of unequal characteristic when A is not an integral domain, (iii) can fail to hold. If A is of characteristic p^m with $m > 1$, then every subring of A has the same characteristic p^m, so cannot be regular. Even if A has characteristic 0, the following is a counter-example: let R be a complete p-ring and $A = R[\![X]\!]/(pX)$; then A is a complete one-dimensional Noetherian local ring, but if a subring A' as in (iii) were to exist, A' would be a one-dimensional regular local ring, hence a DVR, and since A' has characteristic 0 and its residue field characteristic p, A'/pA' would be an Artinian ring, and hence also A/pA would be Artinian. But $A/pA \simeq k[\![X]\!]$ is one-dimensional, and this is a contradiction.

The proof of Theorem 4 shows that it is sufficient to assume that p is not in any minimal prime ideal of A.

Corresponding to the definition of quasi-coefficient field of an equi-characteristic local ring, let us define quasi-coefficient rings in the case of unequal characteristic. Let $(A, \mathfrak{m}, K)$ be a possibly non-complete local ring, and suppose $\operatorname{char} K = p$. A subring $S \subset A$ is said to be a *quasi-coefficient ring* if it satisfies the following two conditions:

(1) S is a Noetherian local ring with maximal ideal pS;

(2) $K = A/\mathfrak{m}$ is 0-etale over S/pS.

In view of (1), if A has characteristic 0 then S is a DVR, and if A has characteristic p^m then S is an Artinian ring.

Theorem 29.5. Let $(A, \mathfrak{m}, K)$ be a local ring, and suppose $\operatorname{char} K = p$. Let $C \subset A$ be a subring, and assume that C is a Noetherian local ring with maximal ideal pC, and that $K = A/\mathfrak{m}$ is separable over C/pC. Then there exists a quasi-coefficient ring S of A containing C; moreover, if A is flat over C, then it is also flat over S.

Proof. Let $\{\beta_\lambda\}_{\lambda\in\Lambda}$ be a p-basis of K over C/pC, and choose an inverse image $b_\lambda \in A$ for each β_λ. Setting $C[\{b_\lambda\}] = C'$, by Theorem 26.8 we see that $C'/(\mathfrak{m}\cap C') = (C/pC)[\{\beta_\lambda\}]$ is a polynomial ring over C/pC. Hence if $f(\dots X_\lambda \dots)$ is a non-zero polynomial with coefficients in C which satisfies $f(b_\lambda) \in \mathfrak{m}$, then setting p^r for the highest common factor of the coefficients of f, we have

$$f(X) = p^r f_0(X) \quad \text{and} \quad \bar{f}_0(\beta_\lambda) \neq 0.$$

Thus $f_0(b_\lambda) \notin \mathfrak{m}$, and we must have $r > 0$, so that we have shown that $\mathfrak{m}\cap C' = pC'$. Setting $S = (C')_{pC'}$ we have $S \subset A$, $\mathfrak{m}\cap S = pS$ and $S/pS = (C/pC)(\{\beta_\lambda\})$. Since C' is p-adically separated, so is S, and hence all the ideals of S are of the form (0) or (p^n). Thus S is Noetherian, and it satisfies all the conditions for a quasi-coefficient ring of A. If A is flat over C, then for any n we have

$$p^n C \otimes_C A \simeq p^n A.$$

and hence the composite

$$p^n C \otimes_C A = (p^n C \otimes_C S) \otimes_S A \longrightarrow p^n S \otimes_S A \longrightarrow p^n A$$

is injective; but the first arrow is surjective, so that the second arrow $p^n S \otimes A \longrightarrow p^n A$ is injective. By Theorem 7.7, this proves that A is flat over S. ■

All the assumptions of this theorem are satisfied by taking C to be the image of $\mathbb{Z}_{p\mathbb{Z}} \longrightarrow A$, so that this proves that every local ring has a quasi-coefficient ring (including the quasi-coefficient field of a local ring in the equal characteristic case).

Theorem 29.6. Let $(A, \mathfrak{m}, K)$ be a local ring, and $\hat{A}$ its completion, and suppose $\operatorname{char} K = p$. Let S be a quasi-coefficient ring of A, and write S' for its image in $\hat{A}$; then there exists a unique coefficient ring A_0 of $\hat{A}$ containing S'.

Proof. Since S' is a quasi-coefficient ring of $\hat{A}$, we can assume that A is a complete local ring. If A has characteristic 0 then S is a DVR, and by Theorem 1 there exists a complete p-ring R containing S and with residue field K. Now K is 0-etale over S/pS, and R is flat over S, so that R is pR-etale over S (by Ex. 28.1). Hence there exists a unique S-algebra homomorphism $R \longrightarrow A$ which induces the identity map on the residue fields; we write A_0 for the image of R. Then $R \simeq A_0$ and A_0 is a coefficient ring of A. If A had another coefficient ring B containing S then B would also be a complete p-ring, and for the same reason as above, there exist unique S-algebra homomorphisms $B \longrightarrow A_0$ and $B \longrightarrow A$, so that we must have $B = A_0$.

If A has characteristic p^n then S is an Artinian ring, and therefore complete, so that applying Theorem 3 to S itself, we can write $S = R_0/(p^n)$ with R_0 a complete p-ring. Let R be a complete p-ring containing R_0 and with residue field K; then by Ex. 28.1, R is pR-etale over R_0, so that there exists a unique R_0-algebra homomorphism $R \longrightarrow A$ inducing the identity map on the residue field K. The image is a coefficient ring of A containing S; the uniqueness is proved as in the case of characteristic 0. ■

Next we study the structure of complete regular local rings. Let $(A, \mathfrak{m}, K)$ be a local ring of unequal characteristic, and suppose that $\operatorname{char} K = p$; then A is said to be *ramified* if $p \in \mathfrak{m}^2$ and *unramified* if $p \notin \mathfrak{m}^2$. We will also say that A is unramified in the case of equal characteristic.

Theorem 29.7. An unramified complete regular local ring is a formal power series ring over a field or over a complete p-ring.

Proof. Let R be a coefficient ring of A. In the case of equal characteristic, R is a field, and if $x_1, \dots, x_n$ is a regular system of parameters of A then $A = R[\![x_1, \dots, x_n]\!] \simeq R[\![X_1, \dots, X_n]\!]$ (see the proof of Theorem 4). In the case of unequal characteristic, R is a complete p-ring, and since $p \in \mathfrak{m} - \mathfrak{m}^2$, we can choose a regular system of parameters $\{p, x_2, \dots, x_n\}$ of A containing p. Then $A = R[\![x_2, \dots, x_n]\!] \simeq R[\![X_2, \dots, X_n]\!]$. ■

In the ramified case, it is not necessarily the case that A can be expressed as a formal power series ring over a DVR. To give the structure theorem in this case we need the notion of an Eisenstein extension.

Lemma 1 (Eisenstein's irreducibility criterion). Let A be a ring, and $f(X) = X^n + a_1 X^{n-1} + \cdots + a_n$, with $a_i \in A$. If there exists a prime ideal $\mathfrak{p}$ of A such that $a_1, \dots, a_n \in \mathfrak{p}$ but $a_n \notin \mathfrak{p}^2$ then f is irreducible in $A[X]$. If in addition A is an integrally closed domain then the principal ideal (f) is a prime ideal of $A[X]$.

Proof. If f is reducible, we can write $f = (X^r + b_1 X^{r-1} + \cdots + b_r)(X^s + c_1 X^{s-1} + \cdots + c_s)$ with $0 < r < n$, $s = n - r$ and b_i, $c_j \in A$. Reducing the coefficients on either side modulo $\mathfrak{p}$ we have

$$X^n = (X^r + \bar{b}_1 X^{r-1} + \cdots + \bar{b}_r)(X^s + \bar{c}_1 X^{s-1} + \cdots + \bar{c}_s)$$

in $(A/\mathfrak{p})[X]$, so that we must have $b_i, c_j \in \mathfrak{p}$ for all i, j, but then $a_n = b_r c_s \in \mathfrak{p}^2$, which contradicts the assumption. If A is an integrally closed domain and K is the field of fractions of A, then by Ex. 9.6, f remains irreducible in $K[X]$. Also, f is monic, so that we have $f \cdot A[X] = f \cdot K[X] \cap A[X]$, and this is a prime ideal of $A[X]$. ■

Let $(A, \mathfrak{m})$ be a normal local ring; then an extension ring

$$B = A[X]/(f) = A[x]$$

defined by an Eisenstein polynomial

$$f = X^n + a_1 X^{n-1} + \cdots + a_n \quad \text{with } a_i \in \mathfrak{m} \text{ for all } i, \quad \text{and } a_n \notin \mathfrak{m}^2$$

is called an *Eisenstein extension* of A. By the lemma, B is an integral domain, and is integral over A. We have $B/\mathfrak{m}B = (A/\mathfrak{m})[X]/(X^n)$, so that B has just one maximal ideal $\mathfrak{n} = \mathfrak{m}B + xB$. Hence B is a local ring, and its residue field coincides with that of A.

Theorem 29.8. (i) If $(A, \mathfrak{m})$ is a regular local ring, then an Eisenstein extension of A is again a regular local ring.

(ii) If A is a ramified complete regular local ring and R is a coefficient ring of A then there is a subring $A_0 \subset A$ with the following properties:

(1) A_0 is an unramified complete regular local ring containing R, and hence can be expressed as a formal power series ring over R;

(2) A is an Eisenstein extension of A_0.

Proof. (i) Let $B = A[x]$, and $x^n + a_1 x^{n-1} + \cdots + a_n = 0$, with $a_i \in \mathfrak{m}$ and $a_n \notin \mathfrak{m}^2$. Then there exists a regular system of parameters $\{y_1, \ldots, y_d = a_n\}$ of A with a_n as an element. As we have seen above, the maximal ideal of B is $\mathfrak{m}B + xB$, but $a_n \in xB$, so that $\{y_1, \ldots, y_{d-1}, x\}$ is a regular system of parameters of B.

(ii) Since $\operatorname{ht} pA = 1$, by a skilful choice of a regular system of parameters $\{x_1, \ldots, x_d\}$ of A, we can arrange that $\{p, x_2, \ldots, x_d\}$ is a system of parameters of A. If we set $A_0 = R[\![x_2, \ldots, x_d]\!]$ then A_0 is a complete unramified regular local ring, and A is a finite module over A_0 (see the proof of Theorem 4). We set $\mathfrak{m}_0$ for the maximal ideal of A_0. Now $A = \mathfrak{m}_0 A + A_0[x_1]$, so that by Theorem 8.4 (or by NAK), $A = A_0[x_1]$. Let

$$f(X) = X^n + a_1 X^{n-1} + \cdots + a_n \quad \text{with} \quad a_i \in A_0$$

be the minimal polynomial of x_1 over A_0. Then $a_n \in x_1 A \subset \mathfrak{m}$, so that $a_n \in \mathfrak{m}_0$. Therefore by Hensel's lemma (Theorem 8.3), all the $a_i \in \mathfrak{m}_0$. We are left to prove that $a_n \notin \mathfrak{m}_0^2$. Write $p = \sum_1^d b_i x_i$ with $b_i \in A$, and express the b_i in the form $b_i = \varphi_i(x_1)$, with $\varphi_i(X) \in A_0[X]$; then x_1 is a root of

$$F(X) = \varphi_1(X)X + \textstyle\sum_2^d \varphi_i(X)x_i - p,$$

so that $F(X)$ is divisible by $f(X)$. Hence the constant term $F(0)$ of F is divisible by a_n. However, $F(0) = \sum_2^d \varphi_i(0)x_i - p$, and $p, x_2, \ldots, x_d$ is a regular system of parameters, so that $F(0) \notin \mathfrak{m}_0^2$, hence also $a_n \notin \mathfrak{m}_0^2$. ∎

Exercises to §29. Prove the following propositions.

29.1. Let A be a complete p-ring, y an indeterminate over A, and $B = A[\![y]\!]$; let $C = B[x]$ be the Eisenstein extension of B given by $x^2 + yx + p = 0$. Then C is a two-dimensional complete regular local ring, but is not a formal power series ring over a DVR of characteristic 0.

29.2. In Theorem 29.2, if k is a perfect field then φ is uniquely determined by φ_0.

30 Connections with derivations

Theorem 30.1 (Nagata–Zariski–Lipman). Let $(A, \mathfrak{m})$ be a complete Noetherian local ring with $\mathbb{Q} \subset A$. Suppose that $x_1, \ldots, x_r \in \mathfrak{m}$ and $D_1, \ldots, D_r \in \operatorname{Der}(A)$ are elements satisfying $\det(D_i x_j) \notin \mathfrak{m}$. Then

(i) There is a subring $C \subset A$ such that

$$A = C[\![x_1, \ldots, x_r]\!] \simeq C[\![X_1, \ldots, X_r]\!]$$

Therefore $x_1, \ldots, x_r$ are analytically independent over C, and A is I-smooth over C, where $I = \sum_1^r Ax_i$, and therefore also $\mathfrak{m}$-smooth over C.

(ii) If $\mathfrak{g} = \sum_1^r AD_i$ is a Lie algebra, (that is if $[D_i, D_j] \in \mathfrak{g}$ for all i, j) then we can take C to be $\{a \in A \mid D_1 a = \cdots = D_r a = 0\}$.

Proof. Letting (c_{ij}) be the inverse matrix of $(D_i x_j)$, and setting $D_i' = \sum c_{ij} D_j$, we have $D_i' x_j = \delta_{ij}$, so that we can assume that $D_i x_j = \delta_{ij}$. Quite generally, for an element $t \in \mathfrak{m}$ and a derivation $D \in \operatorname{Der}(A)$, we define a map $E(D, t): A \longrightarrow A$ by

$$E(D, t) = \sum_{n=0}^{\infty} \frac{t^n}{n!} D^n;$$

by our assumptions, $E(D, t)(a) = \sum (t^n/n!) D^n(a)$ is meaningful, and one sees easily that $E(D, t)$ is a ring homomorphism. Now set

$$E_1 = E(D_1, -x_1) \quad \text{and} \quad C_1 = \operatorname{Im}(E_1);$$

then C_1 is a subring of A, and by computation we see that

$$D_1\left(\sum_0^{\infty}\left(\frac{(-x_1)^n}{n!}\right) D_1^n\right) = \sum_1^{\infty} -\frac{(-x_1)^{n-1}}{(n-1)!} D_1^n + \sum_0^{\infty} \frac{(-x_1)^n}{n!} D_1^{n+1} = 0,$$

so that $C_1 \subset \{a \in A \mid D_1 a = 0\}$. Conversely, if $D_1 a = 0$ then $E_1(a) = a$, so that $C_1 = \{a \in A \mid D_1 a = 0\}$. Also, since for any $a \in A$ we have $E_1(a) \equiv a \bmod x_1 A$, we see that elements of A can be expanded in power series in x_1 with coefficients in C_1, so that $A = C_1[\![x_1]\!]$. Now $E_1(x_1) = x_1 - x_1 = 0$, so that $x_1 A \subset \operatorname{Ker} E_1$, and conversely, if $E_1(a) = a - x_1 Da + \cdots = 0$ then $a \in x_1 A$, and therefore $\operatorname{Ker} E_1 = x_1 A$. Also, $c \in C_1 \Leftrightarrow D_1 c = 0 \Leftrightarrow E_1(c) = c$, so that $C_1 \cap x_1 A = 0$. Now we prove that x_1 is analytically independent over C_1; by contradiction, suppose that

$$c_r x_1^r + c_{r+1} x_1^{r+1} + \cdots = 0 \quad \text{with} \quad c_i \in C_1 \quad \text{and} \quad c_r \neq 0.$$

Then by Ex. 25.2, since x_1 is not a zero-divisor in A, we have $c_r \in x_1 A$, which is a contradiction. Thus if $0 \neq \varphi(X) \in C_1[\![X]\!]$ then $\varphi(x_1) \neq 0$, as required.

If $r > 1$, then write D_i' for the restriction to C_1 of $E_1 \circ D_i$; then $D_i' \in \operatorname{Der}(C_1)$, and $x_j \in C_1$ with $D_i' x_j = \delta_{ij}$ for $2 \leqslant i, j \leqslant r$, so that by induction we have

$$C_1 = C[\![x_2, \ldots, x_r]\!] \simeq C[\![X_2, \ldots, X_r]\!].$$

(i) follows from this.

If $\mathfrak{g}$ is a Lie algebra, then we first arrange as before that $D_i x_j = \delta_{ij}$, and then set $[D_i, D_j] = \sum_\nu a_{ij\nu} D_\nu$ with $a_{ij\nu} \in A$; then $[D_i, D_j] x_\nu = D_i(\delta_{j\nu}) - D_j(\delta_{i\nu}) = 0$, so that $a_{ij\nu} = 0$, hence $[D_i, D_j] = 0$, and $D_1(D_i(C_1)) = D_i(D_1(C_1)) = D_i(0) = 0$. Therefore $D_i(C_i) \subset C_1$ for $i > 1$, and then in the above notation $D_i' = D_i$ for $i > 1$. Thus by induction we have $C = \{a \in A \mid D_1 a = \cdots = D_r a = 0\}$. ■

Corollary. Let $(A, \mathfrak{m})$ be a reduced n-dimensional local ring containing $\mathbb{Q}$, and suppose that the completion $\hat{A}$ of A is also reduced. If there exist elements $D_1, \ldots, D_n \in \operatorname{Der}(A)$ and $x_1, \ldots, x_n \in \mathfrak{m}$ such that $\det(D_i x_j) \notin \mathfrak{m}$ then A is a regular local ring and $x_1, \ldots, x_n$ is a regular system of parameters of A. Suppose in addition that $\mathfrak{g} = \sum_1^n A D_i$ is a Lie algebra; then $k = \{a \in \hat{A} \mid D_1 a = \cdots = D_n a = 0\}$ is a coefficient field of $\hat{A}$.

Proof. Consider $\hat{A}$, with each of the D_i extended to $\hat{A}$. If $[D_i, D_j] = \sum a_{ij\nu} D_\nu$ holds in A then it also holds in $\hat{A}$, so that if $\mathfrak{g}$ is a Lie algebra, so is $\sum \hat{A} D_\nu$. By the theorem, $\hat{A} = C[\![x_1, \ldots, x_n]\!]$, and C is isomorphic to $\hat{A}/\sum x_i \hat{A}$, so is a zero-dimensional local ring. Now by assumption C is also reduced, so that C must be a field. Therefore $\hat{A}$ is a formal power series ring over a field, and hence is regular, so that A is also regular. The other assertion is also clear. ■

Remark. If we view this corollary as a criterion for regularity, then the condition that $\hat{A}$ should be reduced is rather a nuisance; however, as we will see later, for a very wide variety of local rings, we have A is reduced $\Leftrightarrow$ $\hat{A}$ is reduced. This is the case (corollary of Theorem 32.6) if A is a localisation of a ring B which is finitely generated over a field (such an

A is said to be *essentially of finite type* over K). Note also that if we start off with a regular local ring A, then the corollary gives a concrete method of constructing a coefficient field of $\hat{A}$.

Next we consider rings which are finitely generated over a field, which are important in algebraic geometry.

Theorem 30.2. Let k be a field, and $A = k[x_1, \ldots, x_n]$ a finitely generated ring over k. If $A_\mathfrak{p}$ is 0-smooth over k for every $\mathfrak{p} \in \text{m-Spec}\,A$, then A is 0-smooth over k.

Proof. Write $k[X]$ for $k[X_1, \ldots, X_n]$, and let $I = \{f(X) \in k[X] \mid f(x) = 0\}$, so that $A = k[X]/I$. Suppose that $I = (f_1, \ldots, f_s)$. Consider a commutative diagram

$$\begin{array}{ccc} A & \xrightarrow{\psi} & C/N \\ \uparrow & & \uparrow \varphi \\ k & \longrightarrow & C, \end{array}$$

where C is a ring, and $N \subset C$ is an ideal satisfying $N^2 = 0$. To lift ψ to $A \longrightarrow C$, we first of all choose $u_i \in C$ such that $\psi(x_i) = \varphi(u_i)$. If $f \in I$ then $f(u_1, \ldots, u_n) \in N$. Now if we can choose $y_i \in N$ for $1 \leqslant i \leqslant n$ such that $f_j(u + y) = 0$ for all j, the homomorphism $A \longrightarrow C$ defined by $x_i \mapsto u_i + y_i$ is a lifting of ψ. We have $f_j(u + y) = f_j(u) + \sum_{i=1}^n (\partial f_j/\partial X_i)(u) \cdot y_i$, so that we are looking for solutions in N of the system of linear equations in $y_1, \ldots, y_n$:

$$(*) \quad f_j(u) + \sum_{i=1}^{n} \left(\frac{\partial f_j}{\partial X_i}\right)(u) \cdot y_i = 0 \quad \text{for} \quad j = 1, \ldots, s.$$

For each maximal ideal $\mathfrak{p}$, the local ring $A_\mathfrak{p}$ is 0-smooth over k, so that if we set $\bar{S} = \psi(A - \mathfrak{p})$ and $S = \varphi^{-1}(\bar{S})$ then in the diagram

$$\begin{array}{ccc} A_\mathfrak{p} & \xrightarrow{\psi_\mathfrak{p}} & (C/N)_{\bar{S}} = C_S/N_S \\ \uparrow & & \uparrow \\ k & \longrightarrow & C_S \end{array}$$

there exists a $\Psi_\mathfrak{p} : A_\mathfrak{p} \longrightarrow C_S$ lifting $\psi_\mathfrak{p}$. From this we see that (*), as a system of equations in N_S, has a solution in N_S. If we view N as a C/N-module then $N_S = N_{\bar{S}}$. Thus the theorem reduces to the following lemma.

Lemma 1. Let A and B be rings, $\psi: A \longrightarrow B$ a ring homomorphism, and N a B-module; suppose that $b_{ij} \in B$ and $\beta_i \in N$. If the system of linear equations

$$\sum_{j=1}^{n} b_{ij} Y_j = \beta_i \quad (\text{for } i = 1, \ldots, s)$$

has a solution in $N_{\psi(A-\mathfrak{p})}$ for every $\mathfrak{p} \in \text{m-Spec}\,A$, then it has a solution in N.

Proof. The assumption that there is a solution in $N_{\psi(A-\mathfrak{p})}$ means that there exist $\eta_{j\mathfrak{p}} \in N$ for $1 \leqslant j \leqslant n$, and $t_\mathfrak{p} \in A - \mathfrak{p}$ such that

$$\sum_j b_{ij} \eta_{j\mathfrak{p}} - \psi(t_\mathfrak{p}) \beta_i = 0 \quad \text{for} \quad 1 \leqslant i \leqslant s.$$

Now since $\sum_{\mathfrak{p}} t_{\mathfrak{p}} A = A$, there is a finite set $\{\mathfrak{p}_1, \ldots, \mathfrak{p}_r\} \subset \text{m-Spec}\, A$ and $a_\nu \in A$ such that

$$\sum_{\nu=1}^{r} a_\nu t_{\mathfrak{p}_\nu} = 1.$$

Hence if we set

$$\eta_j = \sum_{\nu=1}^{r} \eta_{j\mathfrak{p}_\nu} \psi(a_\nu),$$

we get $\sum_j b_{ij}\eta_j = \beta_i$ for $1 \geqslant i \geqslant s$. ∎

Theorem 30.3. Let k be a field, $S = k[X_1, \ldots, X_n]$, and I, P ideals of S such that $I \subset P \in \operatorname{Spec} S$. Set

$$S_P = R, \quad \operatorname{rad}(R) = PR = M, \quad \text{and} \quad R/IR = A, \quad \operatorname{rad}(A) = \mathfrak{m},$$
$$R/M = A/\mathfrak{m} = K,$$

and suppose that $\operatorname{ht} IR = r$ and $I = (f_1(X), \ldots, f_t(X))$. Then the following conditions are equivalent.

(1) $\operatorname{rank}(\partial(f_1, \ldots, f_t)/\partial(X_1, \ldots, X_n) \bmod P) = r$;

(2) A is 0-smooth over k;

(3) A is $\mathfrak{m}$-smooth over k;

(4) $\Omega_{A/k}$ is a free A-module of rank $n - r$;

(5) A is an integral domain, its field of fractions is separable over k, and $\Omega_{A/k}$ is a free A-module.

Proof. Note that R is a regular local ring. (1)⇒(2) By assumption, for a suitable choice of r elements $D_1, \ldots, D_r$ from $\partial/\partial X_1, \ldots, \partial/\partial X_n$ and of r elements $g_1, \ldots, g_r$ from $f_1, \ldots, f_t$, we have $\det(D_i g_j) \notin M$. We observe that taking $f \in M$ into $(D_1 f \bmod M, \ldots, D_r f \bmod M) \in K^r$ induces a linear map $M/M^2 \longrightarrow K^r$, so that the images of $g_1, \ldots, g_r$ in M/M^2 are linearly independent. Therefore $\sum_1^r g_i R$ is a height r prime ideal contained in IR, and hence $\sum_1^r g_i R = IR$. Given a commutative diagram

$$\begin{array}{ccc} A & \xrightarrow{\ \psi\ } & C/N \\ \uparrow & & \uparrow \\ R & \varphi & | \\ \uparrow & & | \\ k & \longrightarrow & C \end{array}$$

with $N^2 = 0$, write $x_i \in A$ for the image of X_i, and choose $u_i \in C$ such that $\varphi(u_i) = \psi(x_i) \in C/N$. Then there is a homomorphism $R \longrightarrow C$ defined by $X_i \mapsto u_i$, and this induces a lifting of ψ to C if and only if $g_i(u) = 0$ for $1 \leqslant i \leqslant r$. Therefore, as in the proof of the previous theorem, we need only solve the system of equations in unknowns $y_1, \ldots, y_n \in N$:

$$(*) \quad g_i(u) + \sum_{j=1}^{n} \left(\frac{\partial g_i}{\partial X_j}\right)(u) \cdot y_j = 0 \quad \text{for} \quad 1 \leqslant i \leqslant r.$$

However, we view N as a C/N-module, then via ψ as an A-module, so that we can replace the coefficients $(\partial g_i/\partial X_j)(u)$ of this system by $(\partial g_i/\partial X_j)(x) \in A$. Now by assumption, one $r \times r$ minor of this $r \times n$ matrix of coefficients is a unit of A, so that (*) can always be solved.

(2)⇒(3) is trivial.

(3)⇒(1) By §28, Lemma 1, A is regular, so that IR is generated by r elements forming a subset of a regular system of parameters of R, and the image of the natural map $IR \longrightarrow M \longrightarrow M/M^2$ is an r-dimensional K-vector space. Let $k_0 \subset k$ be the prime subfield; then by Theorem 26.9 $K = R/M$ is 0-smooth over k_0, so that by Theorem 25.2, the sequence

$$0 \to M/M^2 \longrightarrow \Omega_R \otimes K \longrightarrow \Omega_K \to 0$$

is exact. We can write $\Omega_S = (\Omega_k \otimes S) \oplus F$, where F is the free S-module with basis $dX_1, \ldots, dX_n$ (for example, by Theorem 25.1), and localising we get

$$\Omega_R = (\Omega_k \otimes_k R) \oplus (F \otimes R);$$

hence $\Omega_R \otimes K = (\Omega_k \otimes_k K) \oplus (F \otimes K)$. However, from

$$IR/I^2R \longrightarrow \Omega_R \otimes A \longrightarrow \Omega_A \to 0,$$

we get the exact sequence $(I/I^2) \otimes_S K \longrightarrow \Omega_R \otimes_R K \longrightarrow \Omega_A \otimes_A K \to 0$, and if A is $\mathfrak{m}$-smooth over k then by the corollary to Theorem 28.6, $\Omega_k \otimes K \longrightarrow \Omega_A \otimes K$ is injective, so that $V = \operatorname{Im}\{(I/I^2) \otimes K \longrightarrow \Omega_R \otimes K\}$ maps isomorphically to its projection $W \subset F \otimes K$ in the second factor of the decomposition $\Omega_R \otimes K = (\Omega_k \otimes K) \oplus (F \otimes K)$. Now factor $I/I^2 \otimes K \longrightarrow \Omega_R \otimes K$ as the composite $I/I^2 \otimes K \longrightarrow M/M^2 \longrightarrow \Omega_R \otimes K$; as we have seen above, the first arrow has rank r, and the second is injective, so that $\operatorname{rank} V = r$. Now $F \otimes K = K\,dX_1 + \cdots + K\,dX_n$, and if we write $(\partial f_i/\partial X_j) \bmod P = \alpha_{ij}$ then W is spanned by $\sum_1^n \alpha_{ij} dX_j$ for $1 \leqslant i \leqslant t$. Therefore $\operatorname{rank}(\alpha_{ij}) = r$, and this proves (1).

(2)⇒(5) By Theorem 25.2,

$$0 \to IR/I^2R \longrightarrow \Omega_{R/k} \otimes A \longrightarrow \Omega_{A/k} \to 0$$

is a split exact sequence, and since $\Omega_{R/k} \otimes A$ is a free A-module with basis $dX_1, \ldots, dX_n$, the A-module $\Omega_{A/k}$ is projective; but A is a local ring, so that $\Omega_{A/k}$ is free. Also A is a regular local ring, therefore an integral domain, and if L is its field of fractions then L is 0-smooth over A, hence also 0-smooth over k, so that L is separable over k.

(5)⇒(4) By Theorem 26.2, the field of fractions L of A is separably generated over k, so that a separating transcendence basis of L over k is a differential basis, and $\operatorname{rank} \Omega_{L/k} = \operatorname{tr.deg}_k L = n - r$; but $\Omega_{L/k} = \Omega_{A/k} \otimes_A L$, and hence $\operatorname{rank}_A \Omega_{A/k} = \operatorname{rank}_L \Omega_{L/k} = n - r$.

(4)⇒(1) In the exact sequence

$$IR/I^2R \longrightarrow \Omega_{R/k} \otimes A \longrightarrow \Omega_{A/k} \to 0,$$

set $E=\mathrm{Im}\{IR/I^2R \longrightarrow \Omega_{R/k}\otimes A\}$. Then $\Omega_{R/k}\otimes A \simeq E\oplus A^{n-r}$, so that $E\simeq A^r$, and therefore $E\otimes K\simeq K^r$. This gives (1). ∎

Remark 1. In the above proof, the equivalence of (1), (2), (4) and (5) was comparatively easy. The proof of (3)⇒(1) used the corollary to Theorem 28.6, and so is not very elementary.

Remark 2. If k is a perfect field, or more generally if the residue field K is separable over k, then $\mathfrak{m}$-smoothness is equivalent to A being a regular local ring, so that Theorem 3 gives a criterion for regularity. In the case of an imperfect field k, if A is 0-smooth over k then so is the field of fractions L of A, but the residue field K is not necessarily separable over k; for example, $A=k[X]_{(X^p-a)}$ with $a\in k-k^p$. For the case of an imperfect field k, we give a regularity criterion for A in Theorem 5.

Quite generally, let A be a ring and P a prime ideal of A, and let $D_1,\ldots,D_s\in\mathrm{Der}(A)$ and $f_1,\ldots f_t\in A$; then we write $J(f_1,\ldots,f_t;D_1,\ldots,D_s)(P) = (D_if_j \bmod P)$. This is an $s\times t$ matrix with entries in the integral domain A/P.

Theorem 30.4. Let R be a regular ring, $P\in\mathrm{Spec}\,R$, and let $I\subset P$ be an ideal of R; suppose that $\mathrm{ht}\,IR_P=r$.

(i) for any $D_1,\ldots,D_s\in\mathrm{Der}(R)$ and $f_1,\ldots f_t\in I$ we have

$$\mathrm{rank}\,J(f_1,\ldots,f_t;D_1,\ldots,D_s)(P)\leqslant r;$$

(ii) if $D_1,\ldots,D_r\in\mathrm{Der}(R)$ and $f_1,\ldots f_r\in I$ are such that $\det(D_if_j)\notin P$ then $IR_P=(f_1,\ldots,f_r)R_P$ and R_P/IR_P is regular.

Proof. (i) If Q is a prime ideal of R with $I\subset Q\subset P$ and $\mathrm{ht}\,Q=r$ then

$$\mathrm{rank}\,J(f_1,\ldots;D_1,\ldots)(P)\leqslant\mathrm{rank}\,J(f_1,\ldots;D_1,\ldots)(Q),$$

and if we set $QR_Q=\mathfrak{m}$ then R_Q is an r-dimensional regular local ring, so that $\mathfrak{m}$ is generated by r elements, $\mathfrak{m}=(g_1,\ldots,g_r)$. Working in R_Q, we can write

$$f_j=\sum_1^r g_\nu\alpha_{\nu j}\quad\text{with}\quad \alpha_{\nu j}\in R_Q$$

so that

$$D_if_j\equiv\sum_{\nu=1}^r (D_ig_\nu)\cdot\alpha_{\nu j}\mod Q,$$

and therefore

$$\mathrm{rank}\,J(f_1,\ldots,f_t;D_1,\ldots,D_s)(Q)$$
$$\leqslant\mathrm{rank}\,J(g_1,\ldots,g_r;D_1,\ldots,D_s)(Q)\leqslant r.$$

(ii) Set $M=PR_p$; then if $\det(D_if_j)\notin M$ one sees easily that the images in M/M^2 of $f_1,\ldots,f_r$ are linearly independent over $R_P/M=\kappa(M)$, so that $\sum_1^r f_iR_P$ is a prime ideal of height r, and therefore coincides with IR_P. Also, R_P/IR_P is regular. ∎

Theorem 30.5 (Zariski). Let k be a field of characteristic p, $S = k[X_1,\ldots,X_n]$, and I and P ideals of S with $I \subset P \in \operatorname{Spec} S$. Set $S_P = R$ and $R/IR = A$, and suppose that $\operatorname{ht} IR = r$ and $I = (f_1,\ldots,f_t)$. Then the following three conditions are equivalent.

(1) A is a regular local ring.

(2) For any p-basis $\{u_\gamma\}_{\gamma\in\Gamma}$ of k, define $D_\gamma \in \operatorname{Der}(S)$ by $D_\gamma(u_{\gamma'}) = \delta_{\gamma\gamma'}$ (the Kronecker δ) and $D_\gamma(X_i) = 0$; then there are a finite number of elements $\alpha, \beta, \ldots, \gamma \in \Gamma$ such that

$$\operatorname{rank} J(f_1,\ldots,f_t; D_\alpha, D_\beta,\ldots, D_\gamma, \partial/\partial X_1,\ldots,\partial/\partial X_n)(P) = r.$$

(3) There exists a subfield $k' \subset k$ with the following properties: $k^p \subset k'$, $[k:k'] < \infty$, and $\Omega_{A/k'}$ is a free A-module, with

$$\operatorname{rank} \Omega_{A/k'} = n - r + \operatorname{rank} \Omega_{k/k'}.$$

Proof. (2)$\Rightarrow$(1) comes from Theorem 4, (ii).

(1)$\Rightarrow$(2), (3) If A is regular then IR is generated by r elements, and these form an R-sequence, so that IR/I^2R is a free A-module of rank r. Set $M = PR$, $\mathfrak{m} = M/IR$ and $K = R/M = A/\mathfrak{m}$; then the image of $IR \longrightarrow M/M^2$ is an r-dimensional K-vector space, so that the natural map $(IR/I^2R)\otimes_A K \longrightarrow M/M^2$ is injective. From the exact sequence

$$IR/I^2R \longrightarrow \Omega_R \otimes_R A \longrightarrow \Omega_A \to 0$$

we get the exact sequence

$$(*) \quad (IR/I^2R)\otimes_A K \longrightarrow \Omega_R \otimes_R K \longrightarrow \Omega_A \otimes_A K \to 0.$$

Now by Theorem 25.2,

$$0 \to M/M^2 \longrightarrow \Omega_R \otimes_R K \longrightarrow \Omega_K \to 0$$

is an exact sequence. The first arrow in (*) is the composite $(IR/I^2R) \longrightarrow M/M^2 \longrightarrow \Omega_R \otimes K$, and so is injective. Thus

$$(**) \quad 0 \to (IR/I^2R)\otimes K \longrightarrow \Omega_R \otimes_R K \longrightarrow \Omega_A \otimes_A K \to 0$$

is exact. Let $\{du_\gamma \mid \gamma \in \Gamma\}$ be a basis of Ω_k over k; then Ω_S is the free S-module with basis $\{du_\gamma \mid \gamma\in\Gamma\} \cup \{dX_1,\ldots,dX_n\}$, and $\Omega_R = \Omega_S \otimes_S R$, $\Omega_R \otimes_R K = \Omega_S \otimes_S K$. Now reorder $f_1,\ldots,f_t$ so that $f_1,\ldots,f_r$ are generators of IR; then $df_1,\ldots,df_r \in \Omega_R \otimes_R K$ can be expressed using $dX_1,\ldots,dX_n$ together with finitely many elements $du_\alpha,\ldots,du_\gamma$, and this gives (2). Now if we let k' be the field obtained by adjoining to k^p all the elements u_σ for $\sigma \in \Gamma$ other than the $du_\alpha,\ldots,du_\gamma$ just used, then from (**) we see that

$$0 \to (IR/I^2R)\otimes K \longrightarrow \Omega_{R/k'} \otimes K \longrightarrow \Omega_{A/k'} \otimes K \to 0$$

is also exact. In the exact sequence

$$(\dagger) \quad IR/I^2R \longrightarrow \Omega_{R/k'} \otimes A \longrightarrow \Omega_{A/k'} \to 0,$$

the middle term is the free A-module with basis $du_\alpha,\ldots,du_\gamma, dX_1,\ldots,dX_n$;

now the generators $f_1,\ldots,f_r$ of IR map to $\mathrm{d}f_1,\ldots,\mathrm{d}f_r \in \Omega_{R/k'} \otimes A$, and since these are linearly independent over K, they base a direct summand of $\Omega_{R/k'} \otimes A$. Thus $\Omega_{A/k'}$ is also a free A-module, and rank $\Omega_{A/k'} + r =$ rank $\Omega_{R/k'} = \mathrm{rank}\,\Omega_{k/k'} + n$.

(3)$\Rightarrow$(1) In the exact sequence (†) the two terms $\Omega_{R/k'} \otimes A$ and $\Omega_{A/k'}$ are both free modules, and the difference between their ranks is r, so that IR/I^2R maps onto a direct summand of $\Omega_{R/k'} \otimes A$, which is a free module of rank r. Thus we can choose $f_1,\ldots,f_r \in IR$ such that $\mathrm{d}f_1 \otimes 1,\ldots,\mathrm{d}f_r \otimes 1$ base this direct summand, and then by NAK we see that $R\mathrm{d}f_1 + \cdots + R\mathrm{d}f_r$ is a direct summand of $\Omega_{R/k'}$. Hence there exist $D_1,\ldots,D_r \in \mathrm{Der}_{k'}(R)$ such that $\det(D_i f_j) \notin M$. Therefore, by Theorem 4, A is regular. ■

Corollary. Let k be a field and $S = k[X_1,\ldots,X_n]$; let I be an ideal of S, and set $B = S/I$. Define $U = \{\mathfrak{p} \in \mathrm{Spec}\,B \mid B_\mathfrak{p}$ is 0-smooth over $k\}$ and $\mathrm{Reg}(B) = \{\mathfrak{p} \in \mathrm{Spec}\,B \mid B_\mathfrak{p}$ is regular$\}$. Then both U and $\mathrm{Reg}(B)$ are open subsets of Spec B.

Proof. Set $V = \mathrm{Spec}\,B$, and let $V_1,\ldots,V_h$ be the irreducible components of V. To say that $\mathfrak{p} \in V_i \cap V_j$ for $i \neq j$ means just that $B_\mathfrak{p}$ has at least two minimal prime ideals, and such points cannot belong to $\mathrm{Reg}(B)$, (nor *a fortiori* to U). Thus first of all we can throw out the closed subset $W = \bigcup_{i\neq j}(V_i \cap V_j)$, and therefore we need only prove that $V_i \cap U$ and $V_i \cap \mathrm{Reg}(B)$ are open in V_i for each i; we fix i, and set $\dim V_i = n - r$. Then by Theorems 3 and 4, if we let $\Delta_1,\ldots,\Delta_\lambda$ denote the images in B of the $r \times r$ minors of the Jacobian matrix $(\partial f_i/\partial X_j)$, where $I = (f_1,\ldots,f_t)$, then $V_i - U$ is the intersection of V_i with the closed subset of V defined by the ideal $(\Delta_1,\ldots,\Delta_\lambda)B$ of B, and is thus closed in V_i. Using Theorem 5, we can argue similarly for $\mathrm{Reg}(B)$; the only difference is that, instead of one Jacobian matrix, we have to consider the closed subsets of V defined by the ideal of B generated by all the $r \times r$ minors of the infinitely many Jacobian matrices $J(f_1,\ldots,f_t;\ D_\alpha,\ D_\beta,\ldots,D_\gamma, \partial/\partial X_1,\ldots,\partial/\partial X_n)$, where $\{D_\alpha,\ldots,D_\gamma\}$ runs through all finite subsets of the set of derivations $\{D_\gamma\}_{\gamma\in\Gamma}$ appearing in Theorem 5, (2). ■

Theorems 3 and 5 contain the result known as the *Jacobian criterion for regularity* in polynomial rings. There is some purpose in trying to extend this to more general rings. In cases when the module of differentials is not finitely generated, then the above method cannot be used as it stands, so we approach the problem using modules of derivations.

Quite generally, let A be an integral domain with field of fractions L; then for an A-module M, we write $\mathrm{rank}_A M$ for the dimension over L of the vector space $M \otimes_A L$.

Theorem 30.6 (M. Nomura). Let $(R, \mathfrak{m})$ be an equicharacteristic n-dimensional regular local ring, and R^* the completion of R; suppose that k is a quasi-coefficient field of R and K a coefficient field of R^* containing k. Let $x_1,\ldots,x_n$ be a regular system of parameters of R.

(i) $R^* = K[\![x_1,\ldots,x_n]\!]$, and if we write $\partial/\partial x_i$ for the partial derivatives in this representation, then $\mathrm{Der}_k(R^*) = \mathrm{Der}_K(R^*)$ is the free R^*-module with basis $\partial/\partial x_i$ for $1 \leqslant i \leqslant n$.

(ii) The following conditions are equivalent:

(1) $\partial/\partial x_i$ maps R into R for $1 \leqslant i \leqslant n$, so that they can be considered as elements of $\mathrm{Der}_k(R)$;

(2) there exist $D_1,\ldots,D_n \in \mathrm{Der}_k(R)$ and $a_1,\ldots,a_n \in R$ such that $D_i a_j = \delta_{ij}$;

(3) there exist $D_1,\ldots,D_n \in \mathrm{Der}_k(R)$ and $a_1,\ldots,a_n \in R$ such that $\det(D_i a_j) \notin \mathfrak{m}$;

(4) $\mathrm{Der}_k(R)$ is a free R-module of rank n;

(5) $\mathrm{rank}\ \mathrm{Der}_k(R) = n$.

Proof. (i) Since K is 0-etale over k, any derivation of R^* which vanishes on k also vanishes on K. If $D \in \mathrm{Der}_K(R^*)$, set $Dx_i = y_i$; then for any $f(x) \in R^* = K[\![x_1,\ldots,x_n]\!]$ we have $D(f) = \sum_{i=1}^n (\partial f/\partial x_i)\cdot y_i$, and hence $D = \sum y_i \partial/\partial x_i$. Conversely, for any given y_i we can construct a derivation by this formula, so that $\mathrm{Der}_K(R^*)$ is the free R^*-module with basis $\partial/\partial x_1,\ldots,\partial/\partial x_n$.

(ii) (1)$\Rightarrow$(2)$\Rightarrow$(3) and (4)$\Rightarrow$(5) are trivial. If (3) holds then $D_1,\ldots,D_n$ are linearly independent over R and over R^*, so that by (i), any $D \in \mathrm{Der}_k(R)$ can be written as a combination $D = \sum c_i D_i$ of the D_i with coefficients in the field of fractions of R^*, but since $Da_j = \sum c_i D_i(a_j)$ for $j = 1,\ldots,n$, we have $c_i \in R$; therefore $D_1,\ldots,D_n$ form a basis of $\mathrm{Der}_k(R)$, which proves (4).

(5)$\Rightarrow$(1) If $D_1,\ldots,D_n$ are linearly independent over R then there exist $a_1,\ldots,a_n \in R$ such that $\det(D_i a_j) \neq 0$. Therefore $D_1,\ldots,D_n$ are also linearly independent over R^*. Thus writing L' for the field of fractions of R^* we can write $\partial/\partial x_i = \sum c_{ij} D_j$, with $c_{ij} \in L'$. From this we get $\delta_{ih} = \sum c_{ij} D_j x_h$, so that the matrix (c_{ij}) is the inverse of $(D_j x_h)$, and $c_{ij} \in L$, where L is the field of fractions of R; therefore $(\partial/\partial x_i)(R) \subset L \cap R^* = R$. ■

Lemma 2. Let R be a regular ring and $P \in \mathrm{Spec}\, R$ with $\mathrm{ht}\, P = r$; then the following two conditions are equivalent:

(1) there exist $D_1,\ldots,D_r \in \mathrm{Der}(R)$ and $f_1,\ldots,f_r \in P$ such that $\det(D_i f_j) \notin P$;

(2) for all $Q \in \mathrm{Spec}\, R$ with $\mathrm{ht}\, Q = s \leqslant r$ such that $Q \subset P$ and R_P/QR_P is regular, there exist $D_1,\ldots,D_s \in \mathrm{Der}(R)$ and $g_1,\ldots,g_s \in Q$ such that $\det(D_i g_j) \notin P$.

Proof. (2) contains (1) as the special case $P = Q$; conversely, suppose that (1) holds. By Theorem 4 we see that $f_1,\ldots,f_r$ is a regular system of parameters of R_P. If R_P/QR_P is regular then we can take $g_1,\ldots,g_s\in Q$ forming a minimal basis of QR_P, and then take $g_{s+1},\ldots,g_r\in P$ such that $g_1,\ldots,g_r$ is a regular system of parameters of R_P. Then from $\det(D_if_j)\notin P$ we deduce that $\det(D_ig_j)\notin P$, or in other words $\operatorname{rank} J(g_1,\ldots,g_r;D_1,\ldots,D_r)(P)=r$. Therefore

$$\operatorname{rank} J(g_1,\ldots,g_s;D_1,\ldots,D_r)(P)=s. \quad \blacksquare$$

If the above condition (1) holds, we say that the *weak Jacobian condition* (WJ) holds at P. If (WJ) holds at every $P\in\operatorname{Spec} R$ then we say that (WJ) holds in R. If this holds, then for any $P,Q\in\operatorname{Spec} R$ with $Q\subset P$, setting $\operatorname{ht} Q=s$ we have

$$R_P/QR_P \text{ is regular} \Leftrightarrow \begin{cases}\text{there exist } D_1,\ldots,D_s\in\operatorname{Der}(R) \text{ and} \\ f_1,\ldots,f_s\in Q \text{ such that } \det(D_i f_j)\notin P,\end{cases}$$

(the implication ($\Leftarrow$) is given by Theorem 4). This statement is the Jacobian criterion for regularity. We can use $\operatorname{Der}_k(R)$ in place of $\operatorname{Der}(R)$, and we then write $(\mathrm{WJ})_k$.

Theorem 30.7. Let $(A,\mathfrak{m})$ be an n-dimensional Noetherian local integral domain containing $\mathbb{Q}$, and let $k\subset A$ be a subfield such that $\operatorname{tr.deg}_k(A/\mathfrak{m})=r<\infty$. Then $\operatorname{Der}_k(A)$ is isomorphic to a submodule of A^{n+r}, and is therefore a finite A-module, with

$$\operatorname{rank}\operatorname{Der}_k(A)\leqslant \dim A+\operatorname{tr.deg}_k(A/\mathfrak{m}).$$

Proof. We write A^* for the completion of A; choose a quasi-coefficient field k' of A containing k, and let K be a coefficient field of A^* containing k'. Let $u_1,\ldots,u_r$ be a transcendence basis of k' over k, and $x_1,\ldots,x_n$ a system of parameters of A. We define $\varphi:\operatorname{Der}_k(A)\longrightarrow A^{n+r}$ by $\varphi(D)=(Du_1,\ldots,Du_r,Dx_1,\ldots,Dx_n)$; then φ is A-linear, so that we need only prove that it is injective. Suppose then that $Du_i=Dx_j=0$ for all i and j; then D has a continuous extension to A^*, and this vanishes on $B=K[\![x_1,\ldots,x_n]\!]$. Now we do not know whether A^* is an integral domain, but it is finite as a B-module, so that any $a\in A$ is integral over B, and if $f(X)\in B[X]$ has a as a root, and has minimal degree, then $f(a)=0$, $f'(a)\neq 0$. Then $0=D(f(a))=f'(a)\cdot Da$, and $Da\in A$, so that, since a non-zero element of A cannot be a zero-divisor in A^*, we must have $Da=0$. Hence $D=0$. $\blacksquare$

Remark. If k is an imperfect field then there are counter-examples even if $k=A/\mathfrak{m}$: suppose $\operatorname{char} k=p$ and $a\in k-k^p$, and set $A=k[X,Y]_{(X,Y)}/(X^p+aY^p)$; then $\dim A=1$ but $\operatorname{rank}\operatorname{Der}_k(A)=2$.

Theorem 30.8. Let $(R,\mathfrak{m})$ be a regular local ring containing $\mathbb{Q}$, and k a quasi-coefficient field of R. Then the following three conditions are equivalent:

(1) $(\mathrm{WJ})_k$ holds at $\mathfrak{m}$;

(2) $\operatorname{rank}\operatorname{Der}_k(R) = \dim R$;

(3) $(\mathrm{WJ})_k$ holds at every $P \in \operatorname{Spec} R$.

Furthermore, if these conditions hold, then for any $P \in \operatorname{Spec} R$, every element of $\operatorname{Der}_k(R/P)$ is induced by an element of $\operatorname{Der}_k(R)$ and

$$\operatorname{rank}\operatorname{Der}_k(R/P) = \dim R/P.$$

Proof. (1)$\Leftrightarrow$(2) is known from Theorem 6, and (1) is contained in (3).

(1)$\Rightarrow$(3) Write R^* for the completion of R, and let K be the coefficient field of R^* containing k; then by Theorem 6, if $x_1, \ldots, x_n$ is a regular system of parameters of R then the derivations $\partial/\partial x_i$ of $R^* = K[\![x_1, \ldots, x_n]\!]$ belong to $\operatorname{Der}_k(R)$ for $1 \leqslant i \leqslant n$, and form a basis of it. Now let $P \in \operatorname{Spec} R$, and write $\varphi: R \longrightarrow R/P$ for the natural homomorphism; to say that $D' \in \operatorname{Der}_k(R/P)$ is induced by $D \in \operatorname{Der}_k(R)$ means that there is a commutative diagram

$$\begin{array}{ccc} R & \xrightarrow{D} & R \\ {\scriptstyle\varphi}\downarrow & & \downarrow{\scriptstyle\varphi} \\ R/P & \xrightarrow{D'} & R/P. \end{array}$$

Suppose then that D' is given; then $D' \circ \varphi \in \operatorname{Der}_k(R, R/P)$, and this has a unique extension to an element of $\operatorname{Der}_k(R^*, R^*/PR^*)$, so that it is uniquely determined by its values on $x_1, \ldots, x_n$. Therefore if we choose $b_1, \ldots, b_n \in R$ such that $D'(\varphi(x_i)) = \varphi(b_i)$, and set $D = \sum b_i \partial/\partial x_i$, then D' is induced by D.

Now $\operatorname{Der}_k(R, R/P)$ is a free R/P-module with basis $\varphi \circ \partial/\partial x_i$ for $1 \leqslant i \leqslant n$, and $\operatorname{Der}_k(R/P)$ can be identified with the submodule

$$N = \{\delta \in \operatorname{Der}_k(R, R/P) \mid \delta(f) = 0 \text{ for all } f \in P\}.$$

Therefore, if $P = (f_1, \ldots, f_t)$ and $\operatorname{ht} P = r$ then

$$\operatorname{rank}\operatorname{Der}_k(R/P) = n - \operatorname{rank} J(f_1, \ldots, f_t; \partial/\partial x_1, \ldots, \partial/\partial x_n)(P);$$

according to Theorem 4, the right-hand side is $\geqslant n - r$, and by Theorem 7 the left-hand side is $\leqslant \dim R/P = n - r$, so we see that

$$\operatorname{rank} J(f_1, \ldots, f_t; \partial/\partial x_1, \ldots, \partial/\partial x_n)(P) = r,$$
$$\operatorname{rank}\operatorname{Der}_k(R/P) = \dim R/P. \quad \blacksquare$$

This theorem has many applications. For example, in the ring R of convergent power series in n variables over $k = \mathbb{R}$ or $\mathbb{C}$ (R is denoted by $k\langle\!\langle X_1, \ldots, X_n \rangle\!\rangle$ in [N1], and elsewhere by $k\{X_1, \ldots, X_n\}$) we have $\partial X_i/\partial X_j = \delta_{ij}$, so that the Jacobian criterion for regularity holds in R. In characteristic 0, the assumptions of the theorem are satisfied by the formal power series ring over any field, or more generally by the formal power series ring $R[\![Y_1,\ldots,Y_m]\!]$ over a regular local ring R which satisfies the assumptions of the theorem. Even if R is not local, but is a regular ring containing a field k of characteristic 0, and such that the residue field at every maximal ideal is algebraic over k, then if $(\mathrm{WJ})_k$ holds at every maximal ideal of R, it in fact holds at every prime ideal of R, so that the Jacobian criterion for regularity holds for example in rings such as $k[X_1,\ldots,X_n][\![Y_1,\ldots,Y_m]\!]$. A further extension of this theorem can be found in Matsumura [2].

Now we are going to prove the results analogous to Theorem 5 for formal power series rings over a field of characteristic p. This is a difficult theorem obtained by Nagata [6]. First of all we have to do some preparatory work.

Let k be a field and $k' \subset k$ a subfield; we say that k' is *cofinite* in k if $[k:k'] < \infty$. We say that a family $\mathscr{F} = \{k_\alpha\}_{\alpha\in I}$ of subfields of k is a *directed family* if for any $\alpha, \beta \in I$ there exists $\gamma \in I$ such that $k_\gamma \subset k_\alpha \cap k_\beta$.

Let K be a field of characteristic p, and $k \subset K$ a subfield. Then there exists a directed family $\mathscr{F} = \{k_\alpha\}_{\alpha\in I}$ of intermediate fields $k \subset k_\alpha \subset K$ cofinite in K such that $\bigcap k_\alpha = k(K^p)$. To construct this, we let B be a fixed p-basis of K over k, and let I be the set of all finite subsets of B; then we only need take $k_\alpha = k(K^p, B - \alpha)$ for $\alpha \in I$.

Lemma 3. Let K be a field, $\{k_\alpha\}_{\alpha\in I}$ a directed family of subfields of K, and set $k = \bigcap k_\alpha$. Then if V is a vector space over K, and $v_1,\ldots,v_n \in V$ are linearly independent over k, there exists $\alpha \in I$ such that $v_1,\ldots,v_n$ are also linearly independent over k_α.

Proof. For each $\alpha \in I$, write $q(\alpha)$ for the number of linearly independent elements over k_α among $v_1,\ldots,v_n$; let α be such that $q(\alpha)$ is maximal, and set $q = q(\alpha)$. Now if $q < n$, and we assume that $v_1,\ldots,v_q$ are linearly independent over k_α, we have $v_n = \sum_1^q c_i v_i$ with $c_i \in k_\alpha$. Since the v_i are linearly independent over k, at least one of the c_i does not belong to k, so that we can assume $c_1 \notin k$. Hence there exists $\beta \in I$ such that $c_1 \notin k_\beta$, and also $\gamma \in I$ such that $k_\gamma \subset k_\alpha \cap k_\beta$, so that $v_1,\ldots,v_q$ and v_n are linearly independent over k_γ; this contradicts the maximality of q, so that $q = n$. ■

Lemma 4. Let $k \subset K$ be fields of characteristic p, and let $\mathscr{F} = \{k_\alpha\}_{\alpha\in I}$ be a directed family of intermediate fields $k \subset k_\alpha \subset K$; then the following conditions are equivalent:

(1) $\bigcap_\alpha k_\alpha(K^p) = k(K^p)$;

(2) the natural map $\Omega_{K/k} \longrightarrow \varprojlim \Omega_{K/k_\alpha}$ is injective;

(3) if a finite subset $\{u_1,\ldots,u_n\} \subset K$ is p-independent over k, then it is also p-independent over k_α for some α;

(4) there exists a p-basis B of K over k such that every finite subset of B is p-independent over k_α for some α.

Proof. (1)$\Rightarrow$(3) If $\{u_1,\ldots,u_n\}$ is p-independent over k then the p^n monomials $u_1^{\nu_1}\cdots u_n^{\nu_n}$ for $0\leqslant \nu_i<p$ are linearly independent over $k(K^p)$, so that by the previous lemma they are also linearly independent over $k_\alpha(K^p)$ for some α.

(3)$\Rightarrow$(4) is trivial; any p-basis will do.

(4)$\Rightarrow$(2) If $0\neq\omega\in\Omega_{K/k}$ and B is a p-basis then there is a unique expression $\omega=\sum c_i\mathrm{d}_{K/k}b_i$, with $\{b_1,\ldots,b_n\}\subset B$ a finite set and $c_i\in K$. If we take α such that $b_1,\ldots,b_n$ are p-independent over k_α then $\mathrm{d}_{K/k_\alpha}b_i$ for $1\leqslant i\leqslant n$ are linearly independent as elements of Ω_{K/k_α}, so that ω has non-zero image in Ω_{K/k_α}.

(2)$\Rightarrow$(1) Let $a\in K$ be such that $a\notin k(K^p)$; then $\mathrm{d}_{K/k}a\neq 0$, so that $\mathrm{d}_{K/k_\alpha}a\neq 0$ for some k_α, in other words $a\notin k_\alpha(K^p)$. ■

Lemma 5. Let $k\subset K$ and $\mathscr{F}=\{k_\alpha\}_{\alpha\in I}$ be as in the previous lemma, and assume that $\bigcap_\alpha k_\alpha(K^p)=k(K^p)$. Then if L is an extension field of K which is either separable over K or finitely generated over K, we again have $\bigcap_\alpha k_\alpha(L^p)=k(L^p)$.

Proof. (i) If L/K is separable, choose a p-basis B of K/k and a p-basis C of L/K; then in view of the exact sequence (Theorem 25.1)

$$0\to\Omega_{K/k}\otimes L\longrightarrow\Omega_{L/k}\longrightarrow\Omega_{L/K}\to 0,$$

$B\cup C$ is a p-basis of L/k. If $b_1,\ldots,b_m\in B$ and $c_1,\ldots,c_n\in C$ are finitely many distinct elements, then by the previous lemma, $b_1,\ldots,b_m$ are p-independent over some k_α, and from this (replacing k by k_α in the above exact sequence) we see that $b_1,\ldots,b_m,c_1,\ldots,c_n$ are p-independent over k_α.

(ii) If L/K is finitely generated, then since any finitely generated extension can be obtained as a succession of elementary extensions of type

(*a*) $L=K(x)$ with x separable over K,

or

(*b*) $L=K(x)$ with $x^p=a\in K$,

we need only consider case (*b*). We further divide this into two subcases: in the first, $\mathrm{d}_{K/k}a=0$, and then from Theorem 25.2 and the fact that $L\simeq K[X]/(X^p-a)$ we get $\Omega_{L/k}=(\Omega_{K/k}\otimes L)\oplus L\mathrm{d}x$; from this one sees easily that $\Omega_{L/k}\longrightarrow\varprojlim\Omega_{L/k_\alpha}$ is injective. In the second subcase, $\mathrm{d}_{K/k}a\neq 0$ and now we have $\Omega_{L/k}\simeq((\Omega_{K/k}\otimes L)/L\cdot\mathrm{d}_{K/k}a)\oplus L\mathrm{d}x$. Hence if we take a p-basis of K/k of the form $\{a\}\cup B'$ with $a\notin B'$, then $\{x\}\cup B'$ is a p-basis of L/k. Now L/k satisfies condition (4) of the previous lemma, since for $b_1,\ldots,b_m\in B'$, if we choose α such that $\{a,b_1,\ldots,b_m\}\subset K$ is p-independent over k_α, then $\{x,b_1,\ldots,b_m\}\subset L$ is p-independent over k_α. ■

Lemma 6. Let K be a field of characteristic p, and let $\{K_\alpha\}$ be a directed family of cofinite subfields $K_\alpha\subset K$ such that $\bigcap_\alpha K_\alpha=K^p$. Then if L is a finite field extension of K, there exists α such that

$$\operatorname{rank}_L\Omega_{L/K'}=\operatorname{rank}_K\Omega_{K/K'}$$

for all subfields $K'\subset K_\alpha$ with $[K:K']<\infty$.

Proof. Let $K = K_0 \subset K_1 \subset \cdots \subset K_t = L$ be a chain of intermediate fields such that $K_i = K_{i-1}(x_i)$, with x_i either separable algebraic over K_{i-1} or $x_i^p \in K_{i-1}$. Then by the previous lemma we have $\bigcap_\alpha K_\alpha(K_i^p) = K_i^p$, so that we need only prove the lemma for $t = 1$; hence suppose that $L = K(x)$. If L is separable over K then in view of $\Omega_{L/K'} = \Omega_{K/K'} \otimes_K L$, the assertion is clear (any α will do). Thus suppose that $x^p = a \in K$, but $a \notin K^p$; then there exists α such that $a \notin K_\alpha$. If $K' \subset K_\alpha$ then computing by means of $L = K[X]/(X^p - a)$ we see that

$$\Omega_{L/K'} = (\Omega_{K/K'} \otimes_K L \oplus L\,\mathrm{d}x)/L \cdot \mathrm{d}_{K/K'}a,$$

and if $\operatorname{rank} \Omega_{K/K'} < \infty$ then $\operatorname{rank} \Omega_{L/K'} = \operatorname{rank} \Omega_{K/K'}$. ■

Theorem 30.9 (Nagata). Let k be a field of characteristic p, $S = k[\![Y_1, \ldots, Y_m]\!]$ and $P \in \operatorname{Spec} S$; suppose that $\mathscr{F} = \{k_\alpha\}_{\alpha \in I}$ is a directed family of cofinite subfields of k such that $\bigcap_\alpha k_\alpha = k^p$. Then there exists $\alpha \in I$ such that for every intermediate field $k^p \subset k' \subset k_\alpha$ with $[k:k'] < \infty$ the following formula holds:

$$\operatorname{rank} \operatorname{Der}_{k'}(S/P) = \dim(S/P) + \operatorname{rank} \operatorname{Der}_{k'}(k).$$

Proof. Set $A = S/P$, let L be the field of fractions of A, and $\dim A = n$. Choose a system of parameters $x_1, \ldots, x_n$ of A, set $B = k[\![x_1, \ldots, x_n]\!]$, let K be the field of fractions of B and $\mathfrak{m}_B$ its maximal ideal. Then A is a finite B-module, and hence $[L:K] < \infty$. If k' is an intermediate field $k^p \subset k' \subset k$ with $[k:k'] = p^r < \infty$, and if $u_1, \ldots, u_r$ is a p-basis of k over k', then $\{u_1, \ldots, u_r, x_1, \ldots, x_n\}$ is a p-basis of B over $C' = k'[\![x_1^p, \ldots, x_n^p]\!]$, in the sense that B is the free C'-module with basis the set of p-monomials in $u_1, \ldots, u_r, x_1, \ldots, x_n$. A derivation from B to B is continuous in the $\mathfrak{m}_B$-adic topology, and any element of $\operatorname{Der}_{k'}(B)$ is 0 on C'. Therefore $\operatorname{Der}_{k'}(B) = \operatorname{Der}_{C'}(B) = \operatorname{Hom}_B(\Omega_{B/C'}, B)$, and $\Omega_{B/C'}$ is the free B-module of rank $n + r$ with basis $\mathrm{d}u_1, \ldots, \mathrm{d}u_r$, $\mathrm{d}x_1, \ldots, \mathrm{d}x_n$. Therefore $\operatorname{Der}_{k'}(B)$ is also a free B-module of rank $n + r$, and the theorem holds in case $A = B$.

We write F' for the field of fractions of C', and for $k_\alpha \in \mathscr{F}$ we set $C_\alpha = k_\alpha[\![x_1^p, \ldots, x_n^p]\!]$ and write K_α for the field of fractions of C_α. As above we have

$$\operatorname{Der}_{k'}(A) = \operatorname{Der}_{C'}(A) = \operatorname{Hom}_A(\Omega_{A/C'}, A),$$

and since A is a finite C'-module, $\Omega_{A/C'}$ is a finite A-module. Hence

$$\operatorname{Der}_{k'}(A) \otimes L = \operatorname{Hom}_L(\Omega_{A/C'} \otimes L, L) = \operatorname{Hom}_L(\Omega_{L/F'}, L),$$

so that $\operatorname{rank}_A \operatorname{Der}_{k'}(A) = \operatorname{rank}_L \Omega_{L/F'}$. Moreover,

$$n + \operatorname{rank} \operatorname{Der}_{k'}(k) = \operatorname{rank} \Omega_{B/C'} = \operatorname{rank} \Omega_{K/F'},$$

so that the conclusion of the theorem can be rewritten

$$\operatorname{rank} \Omega_{L/F'} = \operatorname{rank} \Omega_{K/F'}.$$

Any element of K_α can be written with its denominator in

$$k^p[\![x_1^p,\ldots,x_n^p]\!]=B^p.$$

However, B is faithfully flat over C_α, so that $K_\alpha\cap B=C_\alpha$. From this one deduces easily that $\bigcap_\alpha K_\alpha=K^p$. Thus the theorem follows from the previous lemma.

Theorem 30.10 (Nagata's Jacobian criterion). Let k be a field of characteristic p, $S=k[\![X_1,\ldots,X_n]\!]$, and let I, P be ideals of S such that $I\subset P\in\operatorname{Spec}S$; set $S_P=R$, $R/IR=A$, and suppose that $\operatorname{ht}IR=r$ and $I=(f_1,\ldots,f_t)$. Choose a p-basis $\{u_\gamma\}_{\gamma\in\Gamma}$ of k, and define $D_\gamma\in\operatorname{Der}(k)$ by $D_\gamma(u_{\gamma'})=\delta_{\gamma\gamma'}$. We make any element $D\in\operatorname{Der}(k)$ act on the coefficients of power series, thus extending D to an element of $\operatorname{Der}(S)$.

Then the following conditions are equivalent:

(1) A is a regular local ring;

(2) there exists a finite number of elements $\alpha,\beta,\ldots,\gamma\in\Gamma$ such that

$$\operatorname{rank}J(f_1,\ldots,f_t;D_\alpha,\ldots,D_\gamma,\partial/\partial X_1,\ldots,\partial/\partial X_n)(P)=r.$$

Proof. (2)⇒(1) follows from Theorem 4.

(1)⇒(2) As we see from the proof of Lemma 2, we need only prove (2) in the case $I=P$. We let $\mathscr{F}$ be the family of subfields of k obtained by adjoining all but a finite number of $\{u_\gamma\}_{\gamma\in\Gamma}$ to k^p; then the conditions of Theorem 9 are satisfied, and there exists $k'\in\mathscr{F}$ such that

$$\operatorname{rank}\operatorname{Der}_{k'}(S/P)=n-\operatorname{ht}P+\operatorname{rank}\operatorname{Der}_{k'}(k).$$

If we set $[k:k']=p^s$ then by construction there exist $\gamma_1,\ldots,\gamma_s\in\Gamma$ such that $k=k'(u_{\gamma_1},\ldots,u_{\gamma_s})$; now set $C=k'[\![X_1^p,\ldots,X_n^p]\!]$, so that $\Omega_{S/C}$ is the free S-module with basis $du_{\gamma_1},\ldots,du_{\gamma_s},dX_1,\ldots,dX_n$, and arguing as in the proof of Theorem 8, we see that the derivations of S/P over k' are all induced by derivations of S over k', and that if $g_1,\ldots,g_m$ are generators of P, we have

$$\operatorname{rank}J(g_1,\ldots,g_m;D_{\gamma_1},\ldots,D_{\gamma_s},\partial/\partial X_1,\ldots,\partial/\partial X_n)(P)=\operatorname{ht}P.\quad\blacksquare$$

Remark. Conditions (1) and (2) in Theorem 10 are of the same form as the corresponding conditions in Theorem 5. Condition (3) of Theorem 5 is not applicable as it stands to the present situation, since $\Omega_{A/k'}$ is in general not a finite A-module. However, in general for a module M over a local ring $(R,\mathfrak{m})$, if we write (in temporary notation) $\bar{M}=M/\bigcap_n\mathfrak{m}^nM$ for the associated separated module, then for any R-module N which is separated ($N=\bar{N}$) we have $\operatorname{Hom}_R(M,N)=\operatorname{Hom}_R(\bar{M},N)$. Therefore in the situation of the above theorem we have $\operatorname{Der}_{k'}(S)=\operatorname{Hom}_S(\bar{\Omega}_{S/k'},S)$ and $\operatorname{Der}_{k'}(A)=\operatorname{Hom}_A(\bar{\Omega}_{A/k'},A)$, and moreover $\bar{\Omega}_{S/k'}$ is a free S-module of rank $n+\operatorname{rank}\Omega_{k/k'}$. From this, using the same argument as in the proof of Theorem 5, we see that replacing $\Omega_{A/k'}$ by $\bar{\Omega}_{A/k'}$ in (3) of Theorem 5, this condition is equivalent to (1) and (2) of Theorem 10. Verifying this is a suitable exercise for the reader.

Corollary. Let A be a complete Noetherian local ring; then $\operatorname{Reg}(A)$ is an open subset of $\operatorname{Spec} A$.

Proof. If A is equicharacteristic, and k is a coefficient field of A, then A is of the form $A = S/I$ with $S = k[\![X_1, \ldots, X_n]\!]$, so that by Theorems 8 and 10, we see as in the corollary of Theorem 5 that $\operatorname{Reg}(A)$ is open.

If A is of unequal characteristic, then by Theorem 24.4, it is enough to prove, under the assumption that A is an integral domain, that $\operatorname{Reg}(A)$ contains a non-empty open subset of $\operatorname{Spec} A$. Now by Theorem 29.4, A contains a regular local ring B, and is a finite module over B, so that if we let L and K be the fields of fractions of A and B, then L is a finite extension of K, and is separable since $\operatorname{char} K = 0$. Replacing A by A_b and B by B_b for some suitable $0 \neq b \in B$ we can assume that A is a free B-module (although A and B are no longer local rings, B remains regular). Suppose that $\omega_1, \ldots, \omega_r$ are a basis of A as a B-module, and consider the discriminant

$$d = \det(\operatorname{tr}_{L/K}(\omega_i \omega_j)).$$

Let us prove that if $P \in \operatorname{Spec} A$ is such that $d \notin P$ then $P \in \operatorname{Reg}(A)$. Set $\mathfrak{p} = P \cap B$; then A_P is flat over $B_\mathfrak{p}$ and $B_\mathfrak{p}$ is regular, so that we need only prove that the fibre $A_P \otimes_B \kappa(\mathfrak{p})$ is regular. Now $A = \sum B\omega_i$, so that $A \otimes \kappa(\mathfrak{p}) = \sum \kappa(\mathfrak{p})\bar{\omega}_i$, and in $\kappa(\mathfrak{p})$ we have $\det(\operatorname{tr}(\bar{\omega}_i \bar{\omega}_j)) = \bar{d} \neq 0$; hence $A \otimes \kappa(\mathfrak{p})$ is reduced, and is therefore a direct product of fields. Therefore $A_P \otimes \kappa(\mathfrak{p})$ is a field; this proves that $\operatorname{Reg}(A)$ contains a non-empty open subset of $\operatorname{Spec} A$. ■

Exercises to §30.

30.1. Let $(A, \mathfrak{m})$ be a complete Noetherian local ring, and $\underline{D} = (D_0, D_1, \ldots) \in HS(A)$. Suppose that $x \in \mathfrak{m}$ satisfies $D_1 x = 1$ and $D_i x = 0$ for $i > 0$, that is $E_t(x) = x + t$, and define $\varphi = E_{-x}$ by $\varphi(a) = \sum_0^\infty (-x)^n D_n a$. Then φ is an endomorphism of A to A, with $\operatorname{Ker} \varphi = xA$; and if we set $C = \operatorname{Im} \varphi$ then $A = C[\![x]\!] \simeq C[\![X]\!]$,

30.2. If the conditions of Theorem 5 hold, is it true that A is 0-smooth over the k' appearing in (3)?

30.3. Are conditions (1) and (3) of Theorem 3 still equivalent if $S = k[\![X_1, \ldots, X_n]\!]$?

30.4. Let R be a regular ring containing a field of characteristic 0; if (WJ) holds in R then it also holds in the polynomial ring $R[X_1, \ldots, X_n]$.

11

Applications of complete local rings

It has become clear in the previous chapter that the completion of a local ring has a number of good properties. In this chapter we give some applications of this. §31 centres on the work of Ratliff, giving characterisations of catenary and universally catenary rings; Ratliff is practically the only current practitioner of the Krull and Nagata tradition, obtaining deep results by a fluent command of the methods of classical ideal theory, and there is something about his proofs which is to be savoured. In §32 we discuss Grothendieck's theory of the formal fibre; this book is already long enough, and we have only covered a part of the theory of G-rings, referring to [G2] and [M] for more details. In §33 we discuss some further important applications, again sending the reader to appropriate references for the details.

31 Chains of prime ideals

Theorem 31.1. Let A be a Noetherian ring and $P \in \operatorname{Spec}(A)$. Then there are at most finitely many prime ideals P' of A satisfying $P \subset P'$, $\operatorname{ht}(P'/P) = 1$ and $\operatorname{ht} P' > \operatorname{ht} P + 1$. (Ratliff [3] in the semi-local case, and McAdam [3] in the general case.)

Proof. Let $\operatorname{ht} P = n$ and take $a_1, \dots, a_n \in P$ such that $\operatorname{ht}(a_1, \dots, a_n) = n$. Set $I = (a_1, \dots, a_n)$ and let $P_1 = P, P_2, \dots, P_r$ be the minimal prime divisors of I. In general, if $\{Q_\lambda\}_\lambda$ is an *infinite* set of prime ideals such that $Q_\lambda \supset P$ and $\operatorname{ht}(Q_\lambda/P) = 1$, then $\bigcap_\lambda Q_\lambda = P$. This is because $\bigcap_\lambda Q_\lambda$ is equal to its own radical, hence is a finite intersection of prime ideals containing P, hence either

$$\bigcap Q_\lambda = P$$

or

$$\bigcap Q_\lambda = Q_{\lambda_1} \cap \dots \cap Q_{\lambda_t};$$

but the second case cannot occur since $Q_\lambda \not\supset Q_{\lambda_1} \cap \dots \cap Q_{\lambda_t}$ for $\lambda \notin \{\lambda_1, \dots, \lambda_t\}$. Therefore if there were an infinite number of prime ideals Q_λ such that $Q_\lambda \supset P$, $\operatorname{ht}(Q_\lambda/P) = 1$ and $\operatorname{ht} Q_\lambda \neq n + 1$, then $\bigcap_\lambda Q_\lambda = P$. Hence there would exist a Q_λ, say Q_0, which does not contain $P_2, \dots, P_r$. Then P would be the only prime ideal lying between Q_0 and I. Let $b \in Q_0 - P$. Then Q_0 would be a minimal prime divisor of $I + bA = (a_1, \ldots, a_n, b)$, so that ht $Q_0 \leqslant n + 1$, a contradiction. ■

Theorem 31.2 (Ratliff's weak existence theorem). Let A be a Noetherian ring, and $\mathfrak{p}$, $P \in \operatorname{Spec} A$ be such that $\mathfrak{p} \subset P$, $\operatorname{ht} \mathfrak{p} = h$ and $\operatorname{ht}(P/\mathfrak{p}) = d > 1$; then there exist infinitely many $\mathfrak{p}' \in \operatorname{Spec} A$ with the properties

$$\mathfrak{p} \subset \mathfrak{p}' \subset P, \quad \operatorname{ht} \mathfrak{p}' = h + 1 \quad \text{and} \quad \operatorname{ht}(P/\mathfrak{p}') = d - 1.$$

Proof. We first observe that if $P \supset \mathfrak{p}_1 \supset \mathfrak{p}_2 \supset \cdots \supset \mathfrak{p}_d = \mathfrak{p}$ is a strictly decreasing chain of prime ideals, and if $\mathfrak{p}_{d-2} \supset \mathfrak{p}' \supset \mathfrak{p}$ with $\operatorname{ht} \mathfrak{p}' = h + 1$ then $\operatorname{ht}(P/\mathfrak{p}') = d - 1$. Now there exist infinitely many $\mathfrak{p}' \in \operatorname{Spec} A$ such that $\mathfrak{p}_{d-2} \supset \mathfrak{p}' \supset \mathfrak{p}$ and $\operatorname{ht}(\mathfrak{p}'/\mathfrak{p}) = 1$: for if $\mathfrak{p}'_1, \ldots, \mathfrak{p}'_m$ are a finite set of these, let $a \in \mathfrak{p}_{d-2} - \bigcup \mathfrak{p}'_i$, and let $\mathfrak{p}'_{m+1}$ be a minimal prime divisor of $\mathfrak{p} + aA$ contained in $\mathfrak{p}_{d-2}$; then $\operatorname{ht}(\mathfrak{p}'_{m+1}/\mathfrak{p}) = 1$. By the previous theorem, all but finitely many of these satisfy $\operatorname{ht} \mathfrak{p}' = \operatorname{ht} \mathfrak{p} + 1$. ■

Lemma 1. Let A be a Noetherian ring, and $P \in \operatorname{Spec} A$ with $\operatorname{ht} P = h > 1$; suppose that $u \in P$ is such that $\operatorname{ht}(uA) = 1$. Then there exist infinitely many prime ideals $Q \subset P$ such that

$$u \notin Q \quad \text{and} \quad \operatorname{ht} Q = h - 1.$$

Proof. Suppose that $\mathfrak{p}_1, \ldots, \mathfrak{p}_t$ are the minimal prime ideals of A, and let $P_1, \ldots, P_r$ be finitely many given height 1 prime ideals not containing u. Let $Q_1, \ldots, Q_s$ be the minimal prime divisors of uA, so that these are also height 1 prime ideals of A. Since $h > 1$, there exists $v \in P$ not contained in any $\mathfrak{p}_i, P_j$ or Q_k. Then

$$\operatorname{ht}(u, v) = 2 \quad \text{and} \quad \operatorname{ht}(v) = 1.$$

Now let $P_{r+1}, \ldots, P_{r+n}$ be the minimal prime divisors of (v); continuing in the same way we can find infinitely many height 1 prime ideals not containing u, so that if $h = 2$ we are done. If $h > 2$ we set $\bar{A} = A/(v)$ and $\bar{P} = P/(v)$, so that $\operatorname{ht} \bar{P} = h - 1$, and since the image $\bar{u}$ of u in $\bar{A}$ satisfies $\operatorname{ht}(\bar{u}) = 1$, by induction on h we can find infinitely many prime ideals $\bar{P}_\alpha$ of $\bar{A}$ satisfying

$$\bar{u} \notin \bar{P}_\alpha \subset \bar{P} \quad \text{and} \quad \operatorname{ht} \bar{P}_\alpha = h - 2.$$

The inverse image P_α of $\bar{P}_\alpha$ in A_α does not contain u, and from $P_\alpha/(v) = \bar{P}_\alpha$ we have $\operatorname{ht} P_\alpha = h - 1$. ■

Theorem 31.3 (Ratliff's strong existence theorem). Let A be a Noetherian integral domain, $\mathfrak{p}$, $P \in \operatorname{Spec} A$, and suppose that $\operatorname{ht} \mathfrak{p} = h > 0$ and $\operatorname{ht}(P/\mathfrak{p}) = d$. Then for each i with $0 \leqslant i < d$ the set

$$\{\mathfrak{p}' \in \operatorname{Spec} A \mid \mathfrak{p}' \subset P, \operatorname{ht}(P/\mathfrak{p}') = d - i \text{ and } \operatorname{ht} \mathfrak{p}' = h + i\}$$

is infinite.

Proof. For $i>0$ this follows at once from the weak existence theorem, so that we consider the case $i=0$.

Step 1. Replacing A by A_P we can assume that (A, P) is a local integral domain. Choose $a_1,\dots,a_h\in\mathfrak{p}$ such that $\operatorname{ht}(a_1,\dots,a_j)=j$ for $j=1,\dots,h$, and set

$$\mathfrak{a}=(a_1,\dots,a_h) \quad\text{and}\quad \mathfrak{b}=(a_1,\dots,a_{h-1}).$$

Then $\mathfrak{p}$ is a minimal prime divisor of $\mathfrak{a}$. Let

$$\mathfrak{a}=\mathfrak{a}_1\cap\cdots\cap\mathfrak{a}_r \quad\text{and}\quad \mathfrak{b}=\mathfrak{b}_1\cap\cdots\cap\mathfrak{b}_s$$

be shortest primary decompositions of $\mathfrak{a}$ and $\mathfrak{b}$, and let $\mathfrak{p}_i, \mathfrak{p}'_j$ be the prime divisors of $\mathfrak{a}_i$ and $\mathfrak{b}_j$, respectively; we can assume that $\mathfrak{p}=\mathfrak{p}_1$. Suppose that $\mathfrak{b}_{t+1},\dots,\mathfrak{b}_s$ are all the $\mathfrak{b}_j$ not contained in $\mathfrak{p}$. Then since

$$\mathfrak{a}_2\cap\cdots\cap\mathfrak{a}_r\cap\mathfrak{b}_{t+1}\cap\cdots\cap\mathfrak{b}_s\not\subset\mathfrak{p},$$

we can choose an element $y\in P$ contained in the left-hand side and not contained in $\mathfrak{p}$. Now $\mathfrak{p}'_j\subset\mathfrak{p}$ for $1\leqslant j\leqslant t$, so that $y\notin\mathfrak{p}'_j$, and hence

$$\mathfrak{a}:yA=\mathfrak{a}_1 \quad\text{and}\quad \mathfrak{b}:yA=\mathfrak{b}_1\cap\cdots\cap\mathfrak{b}_t.$$

Now set

$$B=A[x_1,\dots,x_h], \quad\text{where}\quad x_i=a_i/y,$$
$$I=(x_1,\dots,x_h)B \quad\text{and}\quad Q=PB+I.$$

Step 2. We prove that

$$B/I\simeq A/\mathfrak{a}_1,\quad Q\in\operatorname{Spec}B \quad\text{and}\quad \operatorname{ht}Q=h+d.$$

A general element of B can be written in the form

$$a/y^\nu,\quad\text{with}\quad a\in(\mathfrak{a}+yA)^\nu,\quad\text{for some}\quad \nu\geqslant 0.$$

Now $B=A+I$, so that $B/I\simeq A/(I\cap A)$. Now if $\alpha\in I\cap A$, then there exists $\nu>0$ such that $y^\nu\alpha\in\mathfrak{a}(\mathfrak{a}+yA)^{\nu-1}$. Since $\mathfrak{a}:y^\nu=\mathfrak{a}_1$ we have $\alpha\in\mathfrak{a}_1$. Conversely, $y\mathfrak{a}_1\subset\mathfrak{a}$, giving $\mathfrak{a}_1\subset I\cap A$, and hence $I\cap A=\mathfrak{a}_1$. Therefore

$$B/I\simeq A/\mathfrak{a}_1.$$

Under this isomorphism, the prime ideals $\mathfrak{p}/\mathfrak{a}_1$ and $P/\mathfrak{a}_1$ of $A/\mathfrak{a}_1$ correspond to $(\mathfrak{p}B+I)/I$ and $(PB+I)/I$ respectively in B/I; hence, setting

$$\mathfrak{q}=\mathfrak{p}B+I=\mathfrak{p}+I \quad\text{and}\quad Q=PB+I=P+I,$$

we have $\mathfrak{q}, Q\in\operatorname{Spec}B$, with $B/\mathfrak{q}=A/\mathfrak{p}$ and $B/Q=A/P$, and Q is a maximal ideal of B. Also, $B/I=A/\mathfrak{a}_1$ and since I is generated by h elements and $\mathfrak{a}_1$ is a $\mathfrak{p}$-primary ideal, we get

$$\operatorname{ht}Q=\dim B_Q\leqslant h+\operatorname{ht}(P/\mathfrak{a}_1)=h+\operatorname{ht}(P/\mathfrak{p})=h+d.$$

On the other hand, from $y\notin\mathfrak{p}$ we get $\mathfrak{q}=\mathfrak{p}A[y^{-1}]\cap B$: indeed, if $\alpha\in\mathfrak{p}A[y^{-1}]\cap B$, then we can write $\alpha=c/y^\nu$ with $c\in\mathfrak{p}\cap(\mathfrak{a}+yA)^\nu$; since $\mathfrak{a}\subset\mathfrak{p}$ and $y\notin\mathfrak{p}$, we have

$$\mathfrak{p}\cap(\mathfrak{a}+yA)^\nu=\mathfrak{a}(\mathfrak{a}+yA)^{\nu-1}+y^\nu\mathfrak{p},$$

and therefore $\alpha \in I + \mathfrak{p} = \mathfrak{q}$. Also, $A[y^{-1}] = B[y^{-1}]$, so that

$$\operatorname{ht}\mathfrak{q} = \operatorname{ht}\mathfrak{q}A[y^{-1}] = \operatorname{ht}\mathfrak{p}A[y^{-1}] = \operatorname{ht}\mathfrak{p} = h.$$

However, $Q/\mathfrak{q} = P/\mathfrak{p}$, so that

$$\operatorname{ht}(Q/\mathfrak{q}) = \operatorname{ht}(P/\mathfrak{p}) = d,$$

and therefore $\operatorname{ht}Q \geqslant d + h$. Putting this together with the previous inequality, we see that $\operatorname{ht}Q = d + h$.

Step 3. For $\nu = 1, 2, \ldots$, set

$$J_\nu = (x_1, \ldots, x_{h-1}, x_h - y^\nu)B,$$

let Q_ν be a minimal prime divisor of J_ν satisfying

$$Q_\nu \subset Q \quad \text{and} \quad \operatorname{ht}(Q/Q_\nu) = \operatorname{ht}(Q/J_\nu),$$

and set $P'_\nu = Q_\nu \cap A$. We will complete the proof by showing that $P'_1, P'_2, \ldots$ are all distinct, and that

$$\operatorname{ht}P'_\nu = h, \quad \operatorname{ht}(P/P'_\nu) = d \quad \text{for all } \nu.$$

Now $B = A + J_\nu$, so that $B/Q_\nu \simeq A/P'_\nu$ and $Q/Q_\nu \simeq P/P'_\nu$, hence $\operatorname{ht}(P/P'_\nu) = \operatorname{ht}(Q/Q_\nu) = \operatorname{ht}(Q/J_\nu)$; moreover, $\operatorname{ht}Q = d + h$, and since J_ν is generated by h elements,

$$\operatorname{ht}(P/P'_\nu) \geqslant d + h - h = d.$$

Therefore we have

$$d \leqslant \operatorname{ht}(P/P'_\nu) = \operatorname{ht}(Q/Q_\nu) \leqslant \operatorname{ht}Q - \operatorname{ht}Q_\nu = d + h - \operatorname{ht}Q_\nu,$$

and so if we can prove that

(*) $\operatorname{ht}Q_\nu = \operatorname{ht}P'_\nu \geqslant h$

then this will show simultaneously that $\operatorname{ht}(P/P'_\nu) = d$ and $\operatorname{ht}P'_\nu = h$.

We have already seen that $\mathfrak{q} = \mathfrak{p}A[y^{-1}] \cap B$, so that $\mathfrak{q} \cap A = \mathfrak{p}$, and hence $y \notin \mathfrak{q}$. Also, $\mathfrak{a}_1$ is a $\mathfrak{p}$-primary ideal with $B/I \simeq A/\mathfrak{a}_1$, so that I is a $\mathfrak{q}$-primary ideal, and from $\operatorname{ht}\mathfrak{q} = h$ we get

$$\operatorname{ht}(I + yB) = \operatorname{ht}(x_1, \ldots, x_h, y)B = h + 1.$$

In addition, we have $I + yB = J_\nu + yB$, and since Q_ν is a minimal prime divisor of the ideal J_ν, which is generated by h elements, $\operatorname{ht}Q_\nu \leqslant h$, and hence $y \notin Q_\nu$, and

$$Q_\nu = Q_\nu A[y^{-1}] \cap B \quad \text{and} \quad P'_\nu = Q_\nu A[y^{-1}] \cap A,$$

so that, by p.20, Example 1,

$$\operatorname{ht}Q_\nu = \operatorname{ht}P'_\nu.$$

Furthermore

$$(\mathfrak{b}:yA)B \subset (x_1, \ldots, x_{h-1})B,$$
$$(\mathfrak{b}:yA) + (a_h - y^{\nu+1})A \subset J_\nu \cap A \subset Q_\nu \cap A = P'_\nu,$$

and since all prime divisors of $(\mathfrak{b}:yA)$ are also prime divisors of $\mathfrak{b}$, we have $\operatorname{ht}(\mathfrak{b}:yA) = h - 1$. Now $a_h \in \mathfrak{p}$ and $y \notin \mathfrak{p}$, so that $a_h - y^{\nu+1} \notin \mathfrak{p}$, and since all minimal prime divisors of $(\mathfrak{b}:yA)$ are contained in $\mathfrak{p}$ they do not contain $a_h - y^{\nu+1}$. Therefore $\operatorname{ht} P'_\nu \geqslant h$. This completes the proof of (*).

Step 4. If $\nu < \mu$ then $(Q_\nu + Q_\mu)B_Q$ contains $y^\nu = (y^\nu - y^\mu)/(1 - y^{\nu-\mu})$, and therefore contains $(J_\nu + y^\nu B)B_Q = (I + y^\nu B)B_Q$, so that

$$\operatorname{ht}(Q_\nu + Q_\mu)B_Q \geqslant h+1, \quad \text{but} \quad \operatorname{ht} Q_\nu \leqslant h,$$

and therefore $Q_\nu \neq Q_\mu$. From $Q_\nu A[y^{-1}] = P'_\nu A[y^{-1}]$ we see that $P'_1, P'_2, \ldots$ must also all be distinct. ∎

Theorem 31.4. A Noetherian local integral domain $(A, \mathfrak{m})$ is catenary if and only if

$$\operatorname{ht} \mathfrak{p} + \operatorname{coht} \mathfrak{p} = \dim A \quad \text{for all} \quad \mathfrak{p} \in \operatorname{Spec} A.$$

Proof. 'Only if' is trivial, and we prove 'if'. Let $\dim A = n$; if A is not catenary, then there exist $\mathfrak{p}, P \in \operatorname{Spec} A$ such that

$$\mathfrak{p} \subset P, \quad \operatorname{ht}(P/\mathfrak{p}) = 1 \quad \text{but} \quad \operatorname{ht} P > \operatorname{ht} \mathfrak{p} + 1.$$

Set $\operatorname{ht}(\mathfrak{m}/P) = d$. Applying the strong existence theorem to $A/\mathfrak{p}$ we see that there exist infinitely many $P_\lambda \in \operatorname{Spec} A$ such that

$$\mathfrak{p} \subset P_\lambda, \quad \operatorname{ht}(P_\lambda/\mathfrak{p}) = 1 \quad \text{and} \quad \operatorname{ht}(\mathfrak{m}/P_\lambda) = d.$$

However, by assumption, $\operatorname{ht}(\mathfrak{m}/P_\lambda) + \operatorname{ht} P_\lambda = n$, so that

$$\operatorname{ht} P_\lambda = n - d = \operatorname{ht} P > \operatorname{ht} \mathfrak{p} + 1.$$

But according to Theorem 1, there are only finitely many such P_λ, and we have a contradiction. ∎

If A is a ring of finite Krull dimension, we say that A is *equidimensional* if $\dim A/\mathfrak{p} = \dim A$ for every minimal prime $\mathfrak{p}$ of A.

Lemma 2. If an equidimensional local ring $(A, \mathfrak{m})$ is catenary then

$$\operatorname{ht} \mathfrak{p}_2 = \operatorname{ht} \mathfrak{p}_1 + \operatorname{ht}(\mathfrak{p}_2/\mathfrak{p}_1) \quad \text{for all} \quad \mathfrak{p}_1, \mathfrak{p}_2 \in \operatorname{Spec} A \text{ with } \mathfrak{p}_1 \subset \mathfrak{p}_2.$$

Proof. If we choose a minimal prime ideal $\mathfrak{p} \subset \mathfrak{p}_1$ then $\operatorname{ht}(\mathfrak{p}_1/\mathfrak{p}) = \operatorname{ht}(\mathfrak{m}/\mathfrak{p}) - \operatorname{ht}(\mathfrak{m}/\mathfrak{p}_1) = \dim A - \operatorname{ht}(\mathfrak{m}/\mathfrak{p}_1)$, and this is independent of the choice of $\mathfrak{p}$, so that $\operatorname{ht} \mathfrak{p}_1 = \operatorname{ht}(\mathfrak{p}_1/\mathfrak{p})$. Similarly, $\operatorname{ht} \mathfrak{p}_2 = \operatorname{ht}(\mathfrak{p}_2/\mathfrak{p})$, so that $\operatorname{ht} \mathfrak{p}_2 = \operatorname{ht}(\mathfrak{p}_2/\mathfrak{p}_1) + \operatorname{ht} \mathfrak{p}_1$. ∎

Theorem 31.5. Let A, B be Noetherian local rings, and $A \longrightarrow B$ a local homomorphism. If B is equidimensional and catenary and is flat over A then A is also equidimensional and catenary, and $B/\mathfrak{p}B$ is equidimensional for every $\mathfrak{p} \in \operatorname{Spec} A$.

Proof. Write $\mathfrak{m}$ and $\mathfrak{M}$ for the maximal ideals of A and B. For any minimal prime ideal $\mathfrak{p}_0$ of A there exists a minimal prime ideal P_0 of B lying over $\mathfrak{p}_0$; then $\dim B/P_0 = \dim B$, so that $\dim B/\mathfrak{p}_0 B = \dim B$, and then by Theorem 15.1 we have

$$\operatorname{ht}(\mathfrak{m}/\mathfrak{p}_0) = \operatorname{ht}(\mathfrak{M}/\mathfrak{p}_0 B) - \operatorname{ht}(\mathfrak{M}/\mathfrak{m}B) = \dim B - \operatorname{ht}(\mathfrak{M}/\mathfrak{m}B).$$

This is independent of the choice of $\mathfrak{p}_0$, so that A is equidimensional. If $\mathfrak{p} \in \operatorname{Spec} A$ and P is a minimal prime divisor of $\mathfrak{p}B$ then from the going-down theorem (Theorem 9.5) we see that $P \cap A = \mathfrak{p}$, so that by Theorem 15.1, $\operatorname{ht} P = \operatorname{ht} \mathfrak{p}$, and therefore $\operatorname{ht}(\mathfrak{M}/P) = \operatorname{ht} \mathfrak{M} - \operatorname{ht} P = \operatorname{ht} \mathfrak{M} - \operatorname{ht} \mathfrak{p}$ is determined by $\mathfrak{p}$ only. That is, $B/\mathfrak{p}B$ is equidimensional. Also, if $\mathfrak{p}' \in \operatorname{Spec} A$ is such that $\mathfrak{p}' \subset \mathfrak{p}$ and $\operatorname{ht}(\mathfrak{p}/\mathfrak{p}') = 1$, we let P' be a minimal prime divisor of $\mathfrak{p}'B$ contained in P; then $B/\mathfrak{p}'B$ is also equidimensional and flat over $A/\mathfrak{p}'$, so that $\operatorname{ht}(P/P') = \operatorname{ht}(P/\mathfrak{p}'B) = \operatorname{ht}(\mathfrak{p}/\mathfrak{p}') = 1$. However, B is equidimensional and catenary, so that $\operatorname{ht}(P/P') = \operatorname{ht} P - \operatorname{ht} P' = \operatorname{ht} \mathfrak{p} - \operatorname{ht} \mathfrak{p}'$, and therefore $\operatorname{ht} \mathfrak{p} = \operatorname{ht} \mathfrak{p}' + 1$, and so A is catenary. ■

Corollary. Let A be a quotient of a regular local ring R. If A is equidimensional then so is its completion A^*.

Proof. Let P_0 be a minimal prime ideal of A^* and $\mathfrak{p}_0 = P_0 \cap A$; then writing $\mathfrak{p} \subset R$ for the inverse image of $\mathfrak{p}_0$, we have $R^*/\mathfrak{p}R^* = A^*/\mathfrak{p}_0 A^*$. R^* is an integral domain, and therefore equidimensional, so we can apply the theorem to $R \longrightarrow R^*$ and see that $R^*/\mathfrak{p}R^*$ is equidimensional. Hence

$$\dim A^*/P_0 = \dim A^*/\mathfrak{p}_0 A^* = \dim A/\mathfrak{p}_0 = \dim A. \quad ■$$

Definition. We say that a Noetherian local ring A is *formally equidimensional* (or *quasi-unmixed*) if its completion A^* is equidimensional.

Theorem 31.6. Let $(A, \mathfrak{m})$ be a formally equidimensional Noetherian local ring.

(i) $A_\mathfrak{p}$ is formally equidimensional for every $\mathfrak{p} \in \operatorname{Spec} A$.

(ii) If I is an ideal of A, then

$$A/I \text{ is equidimensional} \Leftrightarrow A/I \text{ is formally equidimensional.}$$

(iii) If B is a local ring which is essentially of finite type over A (see p. 232), and if B is equidimensional then it is also formally equidimensional.

(iv) A is universally catenary.

Proof. (i) Let $P \in \operatorname{Spec}(A^*)$ be such that $P \cap A = \mathfrak{p}$, and set $B = (A^*)_P$. Since B is flat over $A_\mathfrak{p}$, by Theorem 22.4, B^* is flat over $(A_\mathfrak{p})^*$. Now B is a quotient of a regular local ring, and is equidimensional, so that by the above corollary, B^* is also equidimensional. Hence by Theorem 5, $(A_\mathfrak{p})^*$ is also equidimensional.

(ii) follows easily from Theorem 5.

(iii) B is a localisation of a quotient of $A[X_1, \ldots, X_n]$ for some n, so that by (ii) we need only show that a localisation B of $A[X_1, \ldots, X_n]$ is formally equidimensional. Now $A^*[X_1, \ldots, X_n]$ is faithfully flat over $A[X_1, \ldots, X_n]$, so that there is a local ring C which is a localisation of $A^*[X_1, \ldots, X_n]$, and a local homomorphism $B \longrightarrow C$ such that C is flat over B, and hence C^* is flat over B^*. By the remark after Theorem 15.5, C is equidimensional, and is a quotient of a regular local ring, so that by the corollary of Theorem 5, C^* is also equidimensional; hence by Theorem 5, B^* is also equidimensional.

(iv) Any local integral domain essentially of finite type over A is formally equidimensional, and hence catenary, so that any integral domain which is finitely generated over A is catenary. ■

We say that a Noetherian local ring A is *formally catenary* ([G2], (7.1.9)) if $A/\mathfrak{p}$ is formally equidimensional for every $\mathfrak{p} \in \operatorname{Spec} A$. One sees easily from the above theorem that formally catenary implies universally catenary. The converse of this was proved by Ratliff [2]. Universally catenary is a property of finitely generated A-algebras, and we have to deduce from this a property of the completion, so that the proof is difficult. Before giving Ratliff's proof we make the following observation.

Let $(R, \mathfrak{m})$ be a Noetherian local integral domain, K the field of fractions of R, and R' the integral closure of R in K; let S be an intermediate ring $R \subset S \subset R'$ such that S is a finite R-module. S is a semilocal ring, and its completion S^* (with respect to the $\mathfrak{m}$-adic topology, which coincides with the $\operatorname{rad}(S)$-adic topology) can be identified with $R^* \otimes_R S$. Now R^* is flat over R, so that $R \subset S \subset R' \subset K$ gives $R^* \subset S^* \subset R^* \otimes_R R' \subset R^* \otimes_R K$. The ring $R^* \otimes_R K$ is the localisation of R^* with respect to $R - \{0\}$, so that writing T for the total ring of fractions of R^*, we can consider $R^* \otimes_R K \subset T$, and hence $R^* \subset S^* \subset T$. This leads to the possibility that properties of R^* will be reflected in some S.

Theorem 31.7. For a Noetherian local ring A, the following conditions are equivalent:

(1) A is formally catenary,

(2) A is universally catenary.

(3) $A[X]$ is catenary.

Proof. (1)⇒(2) was proved in Theorem 6 and (2)⇒(3) is trivial. We prove (3)⇒(1). Suppose then that $A[X]$ is catenary; we will prove that A^* is equidimensional by assuming the contrary and deriving a contradiction. The proof breaks up into several lemmas.

Lemma 3. Let $(R, \mathfrak{m})$ be a catenary Noetherian local integral domain, and let R^* be its completion. Let $\dim R = n$, and suppose that there exists a minimal prime Q of R^* such that

$$1 < \dim(R^*/Q) = d < n.$$

For $i = 1, 2, \ldots, d-1$, write Φ_i for the set of $\mathfrak{p} \in \operatorname{Spec} R$ satisfying the conditions

(1) $\operatorname{ht} \mathfrak{p} = i$,

and (2) there exists a minimal prime divisor P of $\mathfrak{p}R^*$ such that $Q \subset P$ and $\dim(R^*/P) = d - i$.

Then Φ_i is non-empty for each i.

Proof. We work by induction on i. If $0 \neq a \in \mathfrak{m}$ then any minimal prime divisor P of $aR^* + Q$ satisfies $\mathrm{ht}(P/Q) = 1$, $P \cap R \neq 0$ and contains a. Hence if we set $M = \{P \in \mathrm{Spec}(R^*) | Q \subset P,\ \mathrm{ht}(P/Q) = 1 \text{ and } P \cap R \neq 0\}$, then $\mathfrak{m} = \bigcup_{P \in M}(P \cap R)$. Now $\mathrm{ht}(\mathfrak{m}R^*/Q) = d > 1$, so that $\mathfrak{m}R^* \notin M$, and hence $\mathfrak{m}$ itself is not of the form $P \cap R$ for $P \in M$; therefore both M and $\{P \cap R | P \in M\}$ are infinite sets. By Theorem 1, $M' = \{P \in M | \mathrm{ht}\, P = 1\}$ is also infinite; choose any $P \in M'$, and set $\mathfrak{p} = P \cap R$. Then $0 < \mathrm{ht}\,\mathfrak{p} = \mathrm{ht}\,\mathfrak{p}R^* \leqslant \mathrm{ht}\, P = 1$, so that $\mathrm{ht}\,\mathfrak{p} = 1$, and P is a minimal prime divisor of $\mathfrak{p}R^*$. Since R^* is catenary, $\dim(R^*/P) = \dim(R^*/Q) - \mathrm{ht}(P/Q) = d - 1$; hence $\mathfrak{p} \in \Phi_1$ and the assertion is true for $i = 1$.

If $i > 1$, take $\mathfrak{p}$ as above, and set $\bar{R} = R/\mathfrak{p}$ and $\bar{P} = P/\mathfrak{p}R^*$; then since R is catenary, $\dim \bar{R} = n - 1$, and $\bar{P}$ is a minimal prime divisor of $\bar{R}^* = R^*/\mathfrak{p}R^*$, with $\dim(\bar{R}^*/\bar{P}) = \dim(R^*/P) = d - 1 < n - 1$. Hence by induction there exists a prime ideal $\bar{\mathfrak{p}}_i = \mathfrak{p}_i/\mathfrak{p}$ of $\bar{R}$ of height $i - 1$, and a minimal prime divisor $\bar{P}_i$ of $\bar{\mathfrak{p}}_i\bar{R}^*$ such that $\bar{P} \subset \bar{P}_i$ and $\dim(\bar{R}^*/\bar{P}_i) = (d-1) - (i-1) = d - i$. If $\bar{P}_i = P_i/\mathfrak{p}R^*$ then P_i is a minimal prime divisor of $\mathfrak{p}_iR^*$ containing P, and hence Q, and $R^*/P_i = \bar{R}^*/\bar{P}_i$ is $(d-i)$-dimensional, so that from the fact that R is catenary we get $\mathrm{ht}\,\mathfrak{p}_i = \mathrm{ht}\,\bar{\mathfrak{p}}_i + \mathrm{ht}\,\mathfrak{p} = i$, and $\mathfrak{p}_i \in \Phi_i$. ■

Lemma 4. Let $(R, \mathfrak{m})$ be a Noetherian local integral domain, R^* its completion, and let $0 = \mathfrak{q}_1 \cap \cdots \cap \mathfrak{q}_r$ be a shortest primary decomposition of 0 in R^*, with $P_i = \sqrt{\mathfrak{q}_i}$ for $1 \leqslant i \leqslant r$. Suppose that P_1 satisfies

$$\mathrm{ht}\, P_1 = 0, \quad \mathrm{coht}\, P_1 = 1 < \dim R.$$

Then there exist $b, c \in \mathfrak{m}$ and $\delta \in (\mathfrak{q}_2 \cap \cdots \cap \mathfrak{q}_r) - P_1$ with the following properties:

(1) $b - \delta \in \mathfrak{q}_1$,

(2) $(b, \delta)R^* = (b, c)R^*$,

and (3) $c/b \notin R$ but is integral over R.

Proof.

Step 1. P_1 is a minimal prime ideal and $r > 1$, so that $\mathfrak{q}_2 \ldots \mathfrak{q}_r \not\subset P_1$, and we can choose $\delta' \in (\mathfrak{q}_2 \cap \cdots \cap \mathfrak{q}_r) - P_1$. Since $\mathrm{coht}\, P_1 = 1$, it follows that $\mathfrak{q}_1 + \delta'R^*$ is $(\mathfrak{m}R^*)$-primary, and hence if we set

$$\mathfrak{a} = (\mathfrak{q}_1 + \delta'R^*) \cap R,$$

then this is an $\mathfrak{m}$-primary ideal. Now if $0 \neq b \in \mathfrak{a}$ is any element, then in R^* we can write

$$b = \zeta + \beta\delta' \quad \text{with} \quad \zeta \in \mathfrak{q}_1 \quad \text{and} \quad \beta \in R^*.$$

Since b is not a zero-divisor in R^* we have $\beta\delta' \notin P_1$, so that setting $\delta = \beta\delta'$, we get (1).

Step 2. We prove that $\delta \notin bR^*$. By contradiction, suppose that $\delta = b\xi$ with $\xi \in R^*$; then $b - \delta = b(1-\xi) \in \mathfrak{q}_1$, and since $b \notin P_1$ we have $1 - \xi \in \mathfrak{q}_1 \subset \mathfrak{m}R^*$, so that ξ is a unit of R^*, and $b \in \delta R^* \subset \mathfrak{q}_2$. This contradicts the fact that b is not a zero-divisor of R^*.

Step 3. If $\mathfrak{b}$ is a non-zero ideal of R such that $\mathfrak{m}$ is not a prime divisor of $\mathfrak{b}$ then $\delta \in \mathfrak{b}R^*$. Indeed, $\mathfrak{b}:\mathfrak{m} = \mathfrak{b}$, so that $\mathfrak{b}R^*:\mathfrak{m}R^* = (\mathfrak{b}:\mathfrak{m})R^* = \mathfrak{b}R^*$, and so $\mathfrak{m}R^*$ is not a prime divisor of $\mathfrak{b}R^*$; if P is any prime divisor of $\mathfrak{b}R^*$, then $P \neq \mathfrak{m}R^*$ and $P \neq P_1$ (in view of $P_1 \cap R = 0$), so that $\operatorname{coht} P_1 = 1$ implies $P \not\supset P_1$, hence $P \not\supset \mathfrak{q}_1$, and there exists $\alpha \in \mathfrak{q}_1 - P$. If we write Q for the P-primary component of $\mathfrak{b}R^*$ then $\alpha\delta = 0 \in Q$, so that $\delta \in Q$. Thus finally $\delta \in \mathfrak{b}R^*$.

Step 4. By the previous two steps $\mathfrak{m}$ is a prime divisor of bR. Hence we can write

$$bR = I \cap J \quad \text{with} \quad I \text{ an } \mathfrak{m}\text{-primary ideal and} \quad J:\mathfrak{m} = J,$$

and then $bR^* = IR^* \cap JR^*$ with $\delta \in JR^*$ and $\delta \notin IR^*$. Moreover, $(IR^* + JR^*)/IR^* \simeq (I+J)/I$, so we can choose $c \in J$ such that $\delta - c \in IR^*$. Then

$$\delta - c \in IR^* \cap JR^* = bR^*,$$

so that $(b,c)R^* = (b,\delta)R^*$, and (2) is proved. If $c \in bR$ we would have $(b,\delta)R^* = bR^*$, contradicting Step 2, so that $c/b \notin R$. On the other hand, we have $b - \delta \in \mathfrak{q}_1$ so that $\delta(b-\delta) = 0$, that is $b\delta = \delta^2$, and $c - \delta \in bR^*$. Set $c = \delta + b\gamma$; then $c^2 = \delta^2 + 2b\delta\gamma + b^2\gamma^2 \in b(\delta, b)R^* = b(c,b)R^*$, so that

$$c^2 \in (bc, b^2)R^* \cap R = (bc, b^2)R.$$

From this, we get $c^2 = bcu + b^2v$ with $u, v \in R$, which proves that c/b is integral over R. Thus we have proved (3). ■

Lemma 5. In the notation and assumptions of Lemma 4, set $S = R[c/b]$; then S has a maximal ideal of height 1.

Proof. Write T for the total ring of fractions of R^*; then we can view S^* as an intermediate ring $R^* \subset S^* = R^*[c/b] \subset T$, and T is the total ring of fractions of S^*. In Lemma 4 we had $\operatorname{Ass}(R^*) = \{P_1, \ldots, P_r\}$, so that setting $Q_i = P_iT \cap S^*$, we get

$$\operatorname{Ass}(S^*) = \{Q_1, \ldots, Q_r\} \quad \text{with} \quad \operatorname{ht} Q_i = \operatorname{ht} P_i.$$

Moreover, S^* is integral over R^*, so that S^*/Q_i is integral over R^*/P_i, and hence also $\operatorname{coht} Q_i = \operatorname{coht} P_i$. Let P^* be any maximal ideal of S^* containing Q_1. Then from $(b,c)R^* = (b,\delta)R^*$ we get $S^* = R^*[c/b] = R^*[\delta/b]$, and since $\delta \in \mathfrak{q}_2 \cap \cdots \cap \mathfrak{q}_r$ and $\delta - b \in \mathfrak{q}_1$ we have

$$\delta/b \in Q_2 \cap \cdots \cap Q_r \quad \text{and} \quad \delta/b - 1 \in Q_1,$$

so that $Q_1 + Q_i = S^*$ for all $i > 1$. Therefore Q_1 is the only minimal prime ideal contained in P^*. However, coht $Q_1 = 1$ and $P^* \cap R^* = \mathfrak{m}R^*$, so that ht $P^* = 1$. Setting $P = P^* \cap S$, we have ht P = ht $P^* = 1$, and P is a maximal ideal of S, since P^* is a maximal ideal of S^*. ■

Now we return to the proof of Theorem 7. Let A be a Noetherian local integral domain, and suppose that A^* is not equidimensional; then by Lemma 3, there exists a prime ideal $\mathfrak{p}$ of A such that $(A/\mathfrak{p})^* = A^*/\mathfrak{p}A^*$ has dimension > 1, and has a minimal prime ideal of coheight 1. Set $R = A/\mathfrak{p}$; then by Lemma 4 and Lemma 5 applied to R, there exists a subring S of the integral closure R' of R generated by one element, and having a maximal ideal P with ht $P = 1 < \dim R$. Let $f: R[X] \longrightarrow S$ be a surjective homomorphism of R-algebras and let $\mathfrak{q}$ be its kernel. Set $Q = f^{-1}(P)$. Then $P = Q/\mathfrak{q}$ and $Q \cap R = \mathfrak{m}$. Since $R \subset S$ we have $\mathfrak{q} \cap R = (0)$, hence ht Q = ht $\mathfrak{m} + 1$ and ht $\mathfrak{q} = 1$ by the remark after Theorem 15.5. Thus ht Q − ht $\mathfrak{q}$ = ht $\mathfrak{m}$ = dim $A/\mathfrak{p} > 1$ = ht$(Q/\mathfrak{q})$, so that $R[X]$ (and hence also $A[X]$) is not catenary. ■

Corollary 1. A Noetherian ring A is universally catenary if and only if $A[X]$ is catenary.
Proof. Suppose $A[X]$ is catenary and set $B = A[X_1, \ldots, X_n]$. In order to prove that B is catenary it suffices to prove that B_P is catenary for every $P \in \operatorname{Spec} B$. Let $\mathfrak{p} = P \cap A$. Then B_P is a localisation of $A_\mathfrak{p}[X_1, \ldots, X_n]$. Since $A_\mathfrak{p}[X_1]$ is catenary, $A_\mathfrak{p}[X_1, \ldots, X_n]$ is catenary by the theorem. ■

Corollary 2. A Noetherian ring of dimension d is catenary if $d \leqslant 2$ and is universally catenary if $d \leqslant 1$.
Proof. The first assertion is obvious from the definitions and the second assertion follows from the first because $\dim A[X] = d + 1$. ■

32 The formal fibre

Let $(A, \mathfrak{m})$ be a Noetherian local ring and A^* its completion. The fibre ring of the natural homomorphism $A \longrightarrow A^*$ over any $\mathfrak{p} \in \operatorname{Spec} A$ is called a *formal fibre* of A (although strictly speaking we should distinguish between the fibre and the fibre ring, we will not do so in what follows). If I is an ideal of A then $(A/I)^* = A^*/IA^*$, so that a formal fibre of A/I is also a formal fibre of A.

Let A be a Noetherian ring and $k \subset A$ a subfield. We say that A is *geometrically regular* over k if $A \otimes_k k'$ is a regular ring for every finite extension k' of k (see §28). This is equivalent to saying that $A_\mathfrak{p}$ is geometrically regular over k for every maximal ideal $\mathfrak{p}$ of A.

We say that a homomorphism $\varphi: A \longrightarrow B$ of Noetherian rings is *regular* if φ is flat, and for every $\mathfrak{p} \in \operatorname{Spec} A$, the fibre $B \otimes_A \kappa(\mathfrak{p})$ of φ over $\mathfrak{p}$ is geometrically regular over the field $\kappa(\mathfrak{p})$.

A Noetherian ring A is said to be a *G-ring* (here G stands for Grothendieck) if $A_\mathfrak{p} \longrightarrow (A_\mathfrak{p})^*$ is regular for every prime ideal $\mathfrak{p}$ of A; this means that all the formal fibres of all the local rings of A are geometrically regular.

Theorem 32.1. Let $A \xrightarrow{\varphi} B \xrightarrow{\psi} C$ be homomorphisms of Noetherian rings; then

(i) if φ and ψ are regular then so is $\psi\varphi$;

(ii) if $\psi\varphi$ is regular and ψ is faithfully flat then φ is also regular.

Proof. (i) Clearly $\psi\varphi$ is flat. For $\mathfrak{p} \in \operatorname{Spec} A$, write $K = \kappa(\mathfrak{p})$, and let L be a finite extension field of K. Set $B \otimes_A L = B_L$ and $C \otimes_A L = C_L$; then the homomorphism $\psi_L : B_L \longrightarrow C_L$ induced by ψ is also regular. Indeed, if P is a prime ideal of B_L then $C \otimes_B B_L = C \otimes_B (B \otimes_A L) = C \otimes_A L = C_L$, and hence if F is a finite extension of $\kappa(P)$ then $C_L \otimes_{B_L} F = C \otimes_B F$; setting $P \cap B = Q$, since B_L is a finite B-module we have $[\kappa(P) : \kappa(Q)] < \infty$, and hence $[F : \kappa(Q)] < \infty$, so that $C \otimes_B F$ is a regular ring. Now φ is regular, so that B_L is a regular ring, and hence by Theorem 23.7, (ii), C_L is a regular ring.

(ii) The flatness of φ is obvious. If we let $\mathfrak{p}$, K and L be as above, then C_L is a regular ring and is flat over B_L, so that by Theorem 23.7, (i), B_L is also regular. ∎

Theorem 32.2. Let $\varphi : A \longrightarrow B$ be a homomorphism of Noetherian rings, and assume that φ is faithfully flat and regular.

(i) A is regular (or normal, reduced, CM, or Gorenstein) if and only if B has the same property.

(ii) If B is a G-ring then so is A (the converse is not true).

Proof. (i) follows from Theorem 23.7, the corollaries to Theorems 23.9 and 23.3, and Theorem 23.4.

(ii) Let $\mathfrak{p} \in \operatorname{Spec} A$, choose $P \in \operatorname{Spec} B$ lying over $\mathfrak{p}$, and consider the commutative diagram

$$\begin{array}{ccc} (A_\mathfrak{p})^* & \xrightarrow{f^*} & (B_P)^* \\ {\scriptstyle\alpha}\uparrow & & \uparrow{\scriptstyle\beta} \\ A_\mathfrak{p} & \xrightarrow{f} & B_P. \end{array}$$

Here f is the map induced by φ, and f^* is the map induced by f, and the vertical arrows are the natural maps. Now f and β are both regular, and f^* is faithfully flat, so that according to the previous theorem, α is also regular.

To construct an example where A is a G-ring and B is not, we let $A = k$ be a perfect field, and B a regular local ring containing k. Then $k \longrightarrow B$ is certainly faithfully flat and regular, and k is a field, and so trivially a G-ring. However, there are known examples in which B is not a G-ring. (See the appendix to [N1]; a counter-example is provided by the ring R in (E3.1) if char $k = p$, and by R in Example 7 if char $k = 0$. In (E3.1) the field k is not perfect, but R is geometrically regular over k.)

Theorem 32.3. A complete Noetherian local ring is a G-ring.

Proof. Let A be a complete Noetherian local ring and $\mathfrak{p} \in \operatorname{Spec} A$; set $B = A_\mathfrak{p}$, and let B^* be the completion of B. We prove that $B \longrightarrow B^*$ is a regular homomorphism; that is, for any prime ideal $\mathfrak{p}'$ of B, we need to show that $B^* \otimes_B \kappa(\mathfrak{p}')$ is geometrically regular over $\kappa(\mathfrak{p}')$. However, $A/\mathfrak{p}' \cap A$ is also a complete local ring, so that we can replace A by $A/\mathfrak{p}' \cap A$ and reduce to the case $\mathfrak{p}' = (0)$. Thus assume that A is an integral domain, and let L be the common field of fractions of A and B; we must show that $B^* \otimes_B L$ is geometrically regular over L.

The problem can be further reduced to the case when A is a regular local ring. In fact, by Theorem 29.4, A contains a complete regular local ring R and is a finite module over R. Set $\mathfrak{p} \cap R = \mathfrak{q}$, $R_\mathfrak{q} = S$ and $B' = A_\mathfrak{q} = A \otimes_R S$; then B' is a semilocal ring, and B is a localisation of B' at a maximal ideal, so that B^* is one direct factor of $B'^* = B' \otimes_S S^*$. Write K for the common field of fractions of R and S.

$$\begin{array}{ccccccc} & & B'^* = B' \otimes_S S^* & \longrightarrow & B^* & & \\ & & \uparrow & & \uparrow & & \\ A & \longrightarrow & B' = A \otimes_R S & \longrightarrow & B = A_\mathfrak{p} & \longrightarrow & L \\ \uparrow & & \uparrow & & & & \uparrow \\ R & \longrightarrow & S = R_\mathfrak{q} & & \longrightarrow & & K \end{array}$$

Now $B^* \otimes_B L$ can be written as $B^* \otimes_{B'} L$, and it is hence a direct factor of $B'^* \otimes_{B'} L = S^* \otimes_S L = (S^* \otimes_S K) \otimes_K L$, so that we need only show that $S^* \otimes_S K$ is geometrically regular over K.

Now R, S and S^* are regular local rings, and $S^* \otimes_S K$ is a localisation of S^*, so is a regular ring. Hence if char $K = 0$, there is nothing to prove. We assume that char $K = p$ in what follows. Then R has a coefficient field k, and can be written $R = k[\![X_1, \ldots, X_n]\!]$. Choose a directed family $\{k_\alpha\}$ of cofinite subfields $k_\alpha \subset k$ such that $\bigcap_\alpha k_\alpha = k^p$, and set $R_\alpha = k_\alpha[\![X_1^p, \ldots, X_n^p]\!]$; write K_α for the field of fractions of R_α. Then one sees easily (compare the proof of Theorem 30.9) that $\bigcap_\alpha K_\alpha = K^p$.

We set $\mathfrak{q}_\alpha = \mathfrak{q} \cap R_\alpha$; then since $R_\alpha \supset R^p$, we see that $\mathfrak{q}$ is the unique prime ideal of R lying over $\mathfrak{q}_\alpha$. Hence if we let $S_\alpha = (R_\alpha)_{\mathfrak{q}_\alpha}$ then $S = R_\mathfrak{q} = R \otimes_{R_\alpha} S_\alpha$, and S is a finite module over S_α. Hence $S^* = S \otimes_{S_\alpha} S_\alpha^*$; let us prove that S^* is 0-smooth over S relative to S_α (see §28). Suppose we are given a commutative diagram of the form

$$\begin{array}{ccccc} S_\alpha^* & \longrightarrow & S^* & \xrightarrow{v} & C/N \\ \uparrow & & \uparrow & & \uparrow \\ S_\alpha & \longrightarrow & S & \xrightarrow{u} & C, \end{array}$$

where C is a ring and N is an ideal of C with $N^2 = 0$. If there is a lifting $v': S^* \longrightarrow C$ of v as a homomorphism of S_α-algebras, set $w = v'|_{S_\alpha^*}$, and let $v'' = u \otimes w: S^* = S \otimes_{S_\alpha} S_\alpha^* \longrightarrow C$; then one sees easily that v'' is a lifting of v over S. Hence S^* is 0-smooth over S relative to S_α. Now for $Q \in \mathrm{Spec}(S^*)$, let $Q \cap S = (0)$; then $(S^*)_Q$ is a local ring of $S^* \otimes_S K$, and conversely, every local ring of $S^* \otimes_S K$ is of this form. From the diagram

$$\begin{array}{ccccc} S^* & \longrightarrow & (S^*)_Q & \longrightarrow & C/N \\ \uparrow & & \uparrow & & \uparrow \\ S & \longrightarrow & K & \longrightarrow & C \\ \uparrow & & \uparrow & & \\ S_\alpha & \longrightarrow & K_\alpha, & & \end{array}$$

one sees that $(S^*)_Q$ is 0-smooth over K relative to K_α. Set $E = (S^*)_Q$ and $\mathfrak{m} = \mathrm{rad}(E)$; then E is $\mathfrak{m}$-smooth over K relative to K_α, so that according to Theorem 28.4,

$$\Omega_{K/K_\alpha} \otimes_K (E/\mathfrak{m}) \longrightarrow \Omega_{E/K_\alpha} \otimes_E (E/\mathfrak{m})$$

is injective for every α. Moreover, since $\bigcap_\alpha K_\alpha = K^p$, by §30, Lemma 4,

$$\Omega_K \longrightarrow \varprojlim \Omega_{K/K_\alpha}$$

is injective, and hence

$$\Omega_K \otimes (E/\mathfrak{m}) \longrightarrow \varprojlim (\Omega_{K/K_\alpha} \otimes (E/\mathfrak{m}))$$

is also injective. Therefore, from the commutative diagram

$$\begin{array}{ccc} \Omega_K \otimes_K (E/\mathfrak{m}) & \longrightarrow & \Omega_E \otimes (E/\mathfrak{m}) \\ \downarrow & & \downarrow \\ \varprojlim (\Omega_{K/K_\alpha} \otimes (E/\mathfrak{m})) & \longrightarrow & \varprojlim (\Omega_{E/K_\alpha} \otimes (E/\mathfrak{m})), \end{array}$$

we finally see that $\Omega_K \otimes (E/\mathfrak{m}) \longrightarrow \Omega_E \otimes (E/\mathfrak{m})$ is injective. Hence it follows from the corollary to Theorem 28.6 that E is $\mathfrak{m}$-smooth over K, and thus is geometrically regular. Since E is an arbitrary local ring of $S^* \otimes_S K$, we see that $S^* \otimes_S K$ is geometrically regular over K. ■

Theorem 32.4. Let A be a Noetherian ring; if $A_\mathfrak{m} \longrightarrow (A_\mathfrak{m})^*$ is regular for every maximal ideal $\mathfrak{m}$ of A, then A is a G-ring.

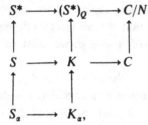

Proof. Since $(A_\mathfrak{m})^*$ is a G-ring, by Theorem 2, $A_\mathfrak{m}$ is also a G-ring. For any $\mathfrak{p}\in\operatorname{Spec}A$, if we let $\mathfrak{m}$ be a maximal ideal of A containing $\mathfrak{p}$ then $A_{\mathfrak{p}}$ is a localisation of the G-ring $A_{\mathfrak{m}}$, and hence $A_{\mathfrak{p}} \longrightarrow (A_{\mathfrak{p}})^*$ is regular. ∎

Theorem 4 makes it much easier to distinguish G-rings. For example, the next theorem is based on Theorem 4.

Theorem 32.5. Let A be a Noetherian semilocal ring; then a sufficient condition for A to be a G-ring is that if C is a finite A-algebra which is an integral domain, $\mathfrak{m}$ is a maximal ideal of C and we write $B = C_{\mathfrak{m}}$, then $(B^*)_Q$ is a regular local ring for every $Q\in\operatorname{Spec}(B^*)$ such that $Q\cap B = (0)$.

Remark. It is easy to see that the condition is also necessary.

Proof. By the previous theorem, we need only show that under the given condition, $A \longrightarrow A^*$ is regular. Let $\mathfrak{p}\in\operatorname{Spec}A$, and let L be a finite extension of $\kappa(\mathfrak{p})$; we prove that $A^*\otimes_A L$ is regular. Suppose that $L = \kappa(\mathfrak{p})(t_1,\dots,t_n)$; then multiplying each t_i by an element of $A/\mathfrak{p}$ we can assume that t_i is integral over $A/\mathfrak{p}$, so that if we set $C = (A/\mathfrak{p})[t_1,\dots,t_n]$, then C is a finite A-module, and the field of fractions of C is L. Now $C^* = A^*\otimes_A C$, and if we write $\mathfrak{m}_1,\dots,\mathfrak{m}_r$ for the maximal ideals of C and set $B_i = C_{\mathfrak{m}_i}$, then $C^* = B_1^*\times\dots\times B_r^*$. We can identify any local ring of $A^*\otimes_A L = C^*\otimes_C L$ with the localisation $(B_i^*)_Q$ of one of the factors B_i^* at some prime ideal Q of B_i^* with $Q\cap B_i = (0)$, and by assumption this is regular. Hence $A^*\otimes_A L$ is a regular ring. ∎

Theorem 32.6 (H. Mizutani). Let R be a regular ring. If the weak Jacobian condition (WJ) of §30 holds for $R[X_1,\dots,X_n]$ for every $n\geqslant 0$, then R is a G-ring.

Proof. Since (*WJ*) is inherited by any localisation we can assume that R is local. We prove that the condition of Theorem 5 holds. Set $R_n = R[X_1,\dots,X_n]$. Any integral domain C which is finite as an R-module can be expressed as $C = R_n/Q$ with $Q\in\operatorname{Spec}(R_n)$ for some n. Let $\mathfrak{m}$ be a maximal ideal of C, M the maximal ideal of R_n corresponding to $\mathfrak{m}$, and $S = (R_n)_M$; then it is enough to show that $(S^*)_P/Q(S^*)_P$ is regular for every $P\in\operatorname{Spec}(S^*)$ with $P\cap S = QS$. If $\operatorname{ht}Q = r$ then $\operatorname{ht}Q(S^*)_P = r$, and by assumption there exist $D_1,\dots,D_r\in\operatorname{Der}(R_n)$ and $f_1,\dots,f_r\in Q$ such that $\det(D_if_j)\notin Q$. Now D_i has a natural extension to S, and then to S^*, and since $P\cap R_n = Q$, we have $\det(D_if_j)\notin P$, so that by Theorem 30.4, $(S^*)_P/Q(S^*)_P$ is regular. ∎

Corollary. A ring which is finitely generated over a field, or a localisation of such a ring, is a G-ring.

Proof. It follows from the definition that a quotient or localisation of a G-ring is again a G-ring, so that we need only show that for a field k, the ring $k[X_1,\dots,X_n]$ is a G-ring; but by Theorems 30.3 and 30.5, (WJ) holds in $k[X_1,\dots,X_{n+m}]$, so that $k[X_1,\dots,X_n]$ is a G-ring by the theorem. ∎

Remark 1. The local rings which appear in algebraic geometry are essentially of finite type over a field, and therefore G-rings. Hence for these rings, properties such as reduced and normal pass to the completion.

Remark 2. If R is a regular ring containing a field of characteristic 0, and (WJ) holds in R, then by Ex. 30.4, it also holds in $R[X_1, \ldots, X_n]$. Hence R and $R[X_1, \ldots, X_n]$ are G-rings. In particular by Theorem 30.8, rings of convergent power series over $\mathbb{R}$ and $\mathbb{C}$ are G-rings.

A theorem proved by Grothendieck asserts that if A is a G-ring, then so is $A[X]$; the proof is very hard, and we omit it, referring only to [M], Theorem 77. The analogous statement for $A[\![X]\!]$ remained unsolved for a long time, but was recently proved for a semilocal ring A by C. Rotthaus [3]. In the non-semilocal case she proved in [4] that, if A is a finite-dimensional excellent ring containing the rational numbers, then $A[\![X]\!]$ is excellent. On the other hand, Nishimura [3] showed that there exists a *G*-ring A such that $A[\![X]\!]$ is not a *G*-ring.

Nagata [8] studied the condition that Reg(A) is open in Spec A; putting together Nagata's work with his own theory of G-rings, Grothendieck gave the definition of excellent ring in [G2].

Definition. A Noetherian ring A is *excellent* if it satisfies the following three conditions:

(1) A is universally catenary;

(2) A is a G-ring;

(3) $\mathrm{Reg}(B) \subset \mathrm{Spec}\, B$ is open for every finitely generated A-algebra B.

A Noetherian ring satisfying (2) and (3) is said to be quasi-excellent.

One can prove that the classes of rings satisfying each of (1), (2) and (3) are closed under localisation, finitely generated extensions and passing to quotients. It can also be proved that (2) implies (3) for semilocal Noetherian rings. A complete Noetherian local ring is excellent, as are practically all Noetherian rings in applications. For more information on excellent rings, see [M], Ch. 13, or [G2].

R.Y. Sharp [5] defined the notion of an *acceptable ring*, replacing condition (2) by the condition that all formal fibres of all localisations of A are Gorenstein, and replacing Reg(B) by Gor(B) in (3), and showed that the resulting theory is analogous to the theory of excellent rings (see also Greco–Marinari [1], Sharp [6]).

Using his cohomology theory, M. André [1] proved the following theorem. Let A, B be Noetherian local rings, and $\varphi: A \longrightarrow B$ a local homomorphism; suppose that A is quasi-excellent and B is $\mathfrak{m}_B$-smooth over A, where $\mathfrak{m}_B = \mathrm{rad}(B)$; then φ is regular. This is an extremely strong theorem; the result of Rotthaus mentioned above also makes use of this.

33 Some other applications

Dimension of intersection

Let k be a field and V and W irreducible algebraic varieties in affine n-space over k. (Here one may either assume that k is algebraically closed and identify the varieties with the corresponding subsets of k^n, or take the scheme-theoretic viewpoint.) Then it is well known that every irreducible component of $V \cap W$ has dimension $\geqslant \dim V + \dim W - n$. Algebraically, this is equivalent to the following theorem.

Let P and P' be two prime ideals in the polynomial ring $R = k[X_1, \ldots, X_n]$ over a field k, and let Q be a minimal prime divisor of $P + P'$. Then

$$\dim R/Q \geqslant \dim(R/P) + \dim(R/P') - n,$$

or equivalently,

$$(*) \quad \operatorname{ht} Q \leqslant \operatorname{ht} P + \operatorname{ht} P'.$$

The idea of the proof consists of transforming the intersection $V \cap W$ in k^n into the intersection $\Delta \cap (V \times W)$ in k^{2n}, where Δ is the diagonal, and availing oneself of the fact that Δ is defined by n equations $x_i - y_i = 0$, for $i = 1, \ldots, n$ (see [M], p. 93).

Now in algebraic geometry, the theorem remains true if one replaces affine n-space by a non-singular (smooth) variety; namely, if V and W are irreducible subvarieties of a non-singular variety U, then every irreducible component of $V \cap W$ has dimension $\geqslant \dim V + \dim W - \dim U$. Algebraically, the inequality (*) still holds if R is an arbitrary regular local ring containing a field. One can easily reduce to the case where R is complete, and then by I.S. Cohen's structure theorem R is isomorphic to $k[\![X_1, \ldots, X_n]\!]$, and so one can apply the same diagonal trick. Thus it is clear that (*) holds in an arbitrary regular local ring of equal characteristic. How about the unequal characteristic case? If R is unramified, then its completion R^* is a formal power series ring over a DVR by Theorem 29.7, and a slight modification of the argument used in the case of $k[\![X_1, \ldots, X_n]\!]$ suffices. When R is ramified, by Theorem 29.8, R^* is of the form $D[\![X_1, \ldots, X_n]\!]/(f)$, where D is a complete DVR and f is an Eisenstein polynomial. Using this, and applying his deep results on intersection multiplicity, J.-P. Serre proved the inequality (*) for general regular local rings R in Chapter V of his book [Se]. We recommend this excellent book to the reader.

Integral closure of a Noetherian integral domain

Let A be a Noetherian integral domain with field of fractions K, and let A' denote the integral closure of A in K (the so-called derived normal ring of A). Is A' a finite module over A? This is a difficult question, and the answer is *no* in general. When A is finitely generated over a field k (the case encountered in algebraic geometry) it is easy to prove finiteness. We need the following two lemmas.

Lemma 1. Let A be a Noetherian normal integral domain with field of fractions K; suppose that L is a finite separable extension of K, and let A' be the integral closure of A in L. Then A' is a finite A-module.

Proof. By enlarging L if necessary we can assume that L is a Galois extension of K. Write $G = \{\sigma_i | 1 \leqslant i \leqslant n\}$ for the Galois group of L/K, where $n = [L:K]$, and let $y_1, \ldots, y_n$ be elements of A' which form a basis of L over K. If $z \in A'$ and $z = \sum_1^n c_j Y_j$ with $c_j \in K$, then $\sigma_i z = \sum_j c_j \sigma_i y_j$ for $i = 1, \ldots, n$, and hence $c_j = C_j/D$, where $D = \det(\sigma_i y_j)$ and $C_j \in A'$. Putting $d = D^2$ we see $d \in K$. In fact it is easy to see that $d = (\mathrm{tr}_{L/K}(y_i y_j))$ is the discriminant of the separable K-algebra L (see p. 198). It follows that $d \neq 0$ and $dc_j \in A' \cap K = A$ for all j. Therefore A' is contained in the finite A-module $\sum_j Ad^{-1} y_j$, so that A' itself is finite over A.

Lemma 2 (Normalisation theorem of E. Noether). Let $A = k[x_1, \ldots, x_n]$ be a finitely generated algebra over a field k; then there exist $y_1, \ldots, y_r \in A$ which are algebraically independent over k such that A is integral over $k[y_1, \ldots, y_r]$.

Proof. Here we assume that k is an infinite field, referring the reader to [M], (14.G) or [N1], (14.4) for the general case. Suppose $x_1, \ldots, x_n$ are algebraically dependent over k, and let $f(x_1, \ldots, x_n) = 0$ be a relation, where $f(X_1, \ldots, X_n)$ is a non-zero polynomial with coefficients in k. Write d for the degree of f and let $f_d(X_1, \ldots, X_n)$ be the homogeneous part of f of degree d. Take $c_1, \ldots, c_{n-1} \in k$ such that $f_d(c_1, \ldots, c_{n-1}, 1) \neq 0$, and set $y_i = x_i - c_i x_n$ for $i = 1, \ldots, n-1$. Then

$$\begin{aligned} 0 = f(x_1, \ldots, x_n) &= f(y_1 + c_1 x_n, \ldots, y_{n-1} + c_{n-1} x_n, x_n) \\ &= f_d(c_1, \ldots, c_{n-1}, 1) x_n^d + g_1 x_n^{d-1} + \cdots + g_d, \end{aligned}$$

with $g_i \in k[y_1, \ldots, y_{n-1}]$, so that x_n is integral over $k[y_1, \ldots, y_{n-1}]$. Then $x_i = y_i + c_i x_n$, for $i = 1, \ldots, n-1$, are also integral over $k[y_1, \ldots, y_{n-1}]$, hence A is integral over $k[y_1, \ldots, y_{n-1}]$. Thus the assertion is proved by induction on n. ■

Now let A be a finitely generated integral domain over k, with field of fractions K and derived normal ring A'. Take $y_1, \ldots, y_r \in A$ as in Lemma 2, so that A' is also the integral closure of $k[y_1, \ldots, y_r]$ in K. Set $K' = k(y_1, \ldots, y_r)$. Then K is a finite algebraic extension of K'. If this extension is separable then A' is finite over $k[y_1, \ldots, y_r]$ by Lemma 1, hence is also finite over A. If K is inseparable over K', let $p = \mathrm{char}\, K$. Then there is a purely inseparable finite extension K'' of K' such that $K(K'')$ is separable over K''. Therefore it suffices to prove that the integral closure of $k[y_1, \ldots, y_r]$ in K'' is finite over it. But K'' is contained in a field L which is obtained by adjoining to K' the qth roots of a finite number of elements $a_1, \ldots, a_s$ of k and also the qth roots of $y_1, \ldots, y_r$, where q is a sufficiently high power of p. Then the integral closure of $k[y_1, \ldots, y_r]$ in L is $k'[y_1^{1/q}, \ldots, y_r^{1/q}]$, where $k' = k(a_1^{1/q}, \ldots, a_s^{1/q})$, and it is clear that $k'[y_1^{1/q}, \ldots, y_r^{1/q}]$ is finite over $k[y_1, \ldots, y_r]$. This completes the proof of finiteness of A' over A.

When A is a complete Noetherian local domain one can prove the finiteness of A' along the same line as above. Using Theorem 29.4, (iii), instead of Lemma 2, one reduces to proving the finiteness of the integral closure of a complete regular local ring A in a finite extension L of the field of fractions K of A. If char $K = 0$ this is proved by Lemma 1, so that we can assume char $K = p > 0$. Then $A = k[\![X_1, \ldots, X_n]\!]$ is a formal power series ring over a field k. We can also assume, as in the above proof, that L is purely inseparable over K, so that there is a power $q = p^m$ of p such that $L \subset K^{1/q}$. But there is one problem. Since a formal power series has infinitely many coefficients, it may not be possible to find a finite extension k_0 of k such that $L \subset k_0((\underline{Y}))$, where $\underline{Y} = (Y_1, \ldots, Y_n)$, $Y_i = X_i^{1/q}$ and $k_0((\underline{Y}))$ denotes the field of fractions of $k_0[\![\underline{Y}]\!]$. One can overcome this difficulty either by an argument of Nagata in [N1, p.113], or (following J. Tate) by induction on n as follows:

We may assume that $Y_i = X_i^{1/q} \in L$ $(1 \leqslant i \leqslant n)$. Since A is normal we have $A' = \{f \in L \mid f^q \in A\}$. Set $P = X_1 A$, $Q = Y_1 A'$; then $Q = \{f \in L \mid f^q \in P\}$, so that Q is the only prime ideal lying over P. Now A_P and A'_Q are DVRs by Theorem 11.2 (3), and their fields of fractions are L and K respectively. Let κ' and κ be their residue fields; then $[\kappa' : \kappa] \leqslant [L:K]$ by Ex.10.8. Since A'/Q is contained in the integral closure of $A/P = k[\![X_2, \ldots, X_n]\!]$ in κ', by the induction hypothesis, A'/Q is finite over A/P. Since

$$Q^i/Q^{i+1} = Y_1{}^i A'/Y_1{}^{i+1} A' \simeq A'/Q \text{ and } Q^q = PA'$$

we see that A'/PA' is finite over A/P. Moreover, A' is separated in the P-adic topology (which is the same as the Y_1-adic topology), because $A' \subset k^{1/q}[\![Y_1, \ldots, Y_n]\!]$. Since A is P-adically complete, A' is finite over A by Theorem 8.4. ■

From this result it is easy to derive the following theorem: *If A is a Noetherian local ring whose completion A^* is reduced, then the integral closure A' of A in its total ring of fractions is finite over A.* See [M] p. 237.

On the other hand, if A is not reduced and if the maximal ideal $\mathfrak{m}$ contains a regular element (that is, a non-zero-divisor), then A' is not finite over A (Krull [2]). In fact, if $x \neq 0$ is nilpotent, take a regular element s such that $x \notin sA$; (it is possible to find such an s since $\bigcap \mathfrak{m}^i = (0)$). Then the elements x/s^j (for $j = 1, 2, 3, \ldots$) belong to A'. If A' is finite over A then there must be some integer n such that $s^n(x/s^j) \in A$ for all j. But then $x \in \bigcap_{r>0} s^r A = (0)$, a contradiction. ■

Suppose $(A, \mathfrak{m})$ is a one-dimensional Noetherian local integral domain. Then A^* is reduced if and only if A' is finite over A (Krull [2]). In fact, if A' is finite over A then it is a semilocal ring, and for each maximal ideal P of A' the local ring$(A')_P$ is a DVR. The rad(A')-adic topology of A' coincides with the $\mathfrak{m}$-adic topology, and the completion A'^* of A' with respect to this topology is a direct product of complete DVRs by Theorem 8.15, hence is reduced. On the other hand it coincides with $A^* \otimes_A A'$ by Theorem 8.7. Since $0 \to A \longrightarrow A'$ is exact, $0 \to A^* \longrightarrow A^* \otimes A'$ is also exact. Therefore A^* is reduced. The converse holds, as already mentioned, without the restriction on dimension. ■

Rees [9] proved that a reduced Noetherian local ring A has reduced completion A^* if and only if for every finite subset Γ of the total ring of fractions K of A, the integral closure of $A[\Gamma]$ in K is finite over $A[\Gamma]$.

Akizuki [1] constructed the first example of a one-dimensional Noetherian local integral domain with non-reduced completion (see also Larfeldt-Lech [1]). To avoid such pathology, Nagata [N1] defined and studied the class of pseudo-geometric rings, which were called 'anneaux universellement japonais' by Grothendieck ([G1], [G2]). These are now known as 'Nagata rings' ([M], [B9]). A Noetherian ring A is called a *Nagata ring* if for every prime ideal P of A and for every finite extension field L of the field of fractions $\kappa(P)$ of A/P, the integral closure of A/P in L is finite over A/P. For the basic properties of Nagata rings, see [M], §31. An alternative definition is the following: a Noetherian ring A is a Nagata ring if (1) for every maximal ideal $\mathfrak{m}$ the formal fibres of $A_\mathfrak{m}$ are geometrically reduced, and (2) for every finite A-algebra B which is an integral domain, the set $\mathrm{Nor}(B) = \{P \in \operatorname{Spec} B \mid B_P \text{ is normal}\}$ is open in $\operatorname{Spec} B$. The equivalence of these two definitions can easily be proved from [G2], (7.6.4) and (7.7.2).

Although the integral closure A' of a Noetherian integral domain A may fail to be finite over A, it is a Krull ring by the theorem of Mori–Nagata mentioned in §12. Because of its importance we quote here the theorem in full.

Mori–Nagata integral closure theorem. Let A be a Noetherian integral domain and let A' be its derived normal ring. Then (1) A' is a Krull ring, and (2) for every prime ideal P of A there are only finitely many prime ideals P' of

A' lying over P, and for each such P' the field of fractions $\kappa(P')$ of A'/P' is finite over $\kappa(P)$.

For a proof, see [N1], (33.10). This proof depends on I.S. Cohen's structure theorem. There are also more recent proofs which do not use the structure theorem (Nishimura [2], Querré [1], Kiyek [1]).

Note that A' is Noetherian if $\dim A \leqslant 2$. This follows easily from the above theorem, Theorem 11.7 (Krull–Akizuki) and Theorem 12.7 (Mori–Nishimura). When $\dim A = 3$, Nagata constructed a counter-example ([N1], p. 207).

Theorem 28.9, which is due to Grothendieck and is not proved in this book, was given a new proof by Radu [5]. This interesting proof depends heavily on I.S. Cohen's structure theorem. The same remark applies also to André's proof [1] of the theorem mentioned at the end of §32.

I.S. Cohen's structure theorem is also at the basis of the theories of canonical modules ([HK]) and of dualising complexes (Sharp [3], [5]). Here, the fact that a complete Noetherian local ring is a quotient of a Gorenstein ring is important.

Appendix A

Tensor products, direct and inverse limits

Tensor products

Let A be a ring, L, M and N three A-modules. We say that a map $\varphi: M \times N \longrightarrow L$ is bilinear if fixing either of the entries it is A-linear in the other, that is if

$$\varphi(x+x', y) = \varphi(x, y) + \varphi(x', y), \quad \varphi(ax, y) = a\varphi(x, y),$$
$$\varphi(x, y+y') = \varphi(x, y) + \varphi(x, y'), \quad \varphi(x, ay) = a\varphi(x, y).$$

Write $\mathscr{L}(M, N; L)$ or $\mathscr{L}_A(M, N; L)$ for the set of all bilinear maps from $M \times N$ to L; as with $\operatorname{Hom}(M, L)$, this has an A-module structure (since we are assuming that A is commutative).

If $g: L \longrightarrow L'$ is an A-linear map and $\varphi \in \mathscr{L}(M, N; L)$ then $g \circ \varphi \in \mathscr{L}(M, N; L')$. Bearing this in mind, for given M and N, consider a bilinear map $\otimes: M \times N \longrightarrow L_0$ having the following property, where we write $x \otimes y$ instead of $\otimes(x, y)$: for any A-module L and any $\varphi \in \mathscr{L}(M, N; L)$ there exists a unique A-linear map $g: L_0 \longrightarrow L$ satisfying

$$g(x \otimes y) = \varphi(x, y).$$

If this holds we say that L_0 is the *tensor product of M and N over A*, and write $L_0 = M \otimes_A N$; we sometimes omit A and write $M \otimes N$. As usual with this kind of definition, $M \otimes_A N$, assuming it exists, is uniquely determined up to isomorphism. To prove existence, write F for the free A-module with basis the set $M \times N$, and let $R \subset F$ be the submodule generated by all elements of the form

$$(x+x', y) - (x, y) - (x', y), \quad (ax, y) - a(x, y)$$
$$(x, y+y') - (x, y) - (x, y'), \quad (x, ay) - a(x, y).$$

Then set $L_0 = F/R$ and write $x \otimes y$ for the image in L_0 of $(x, y) \in F$. It is now easy to check that L_0 and $\otimes$ satisfy the above condition.

Note that the general element of $M \otimes_A N$ is a sum of the form $\sum x_i \otimes y_i$, and cannot necessarily be written $x \otimes y$.

For A-modules M, N and L the definition of tensor product gives:

Formula 1. $\operatorname{Hom}_A(M \otimes_A N, L) \simeq \mathscr{L}(M, N; L)$.

The canonical isomorphism is obtained by taking an element φ of the right-hand side to the element g of the left-hand side satisfying $g(x \otimes y) = \varphi(x, y)$.

We can define multilinear maps from an r-fold product of A-modules $M_1, \ldots, M_r$ to an A-module L just as in the bilinear case, and get modules $\mathscr{L}(M_1, \ldots, M_r; L)$ and $M_1 \otimes_A \cdots \otimes_A M_r$; the following 'associative law' then holds:

Formula 2. $(M \otimes_A M') \otimes_A M'' = M \otimes_A M' \otimes_A M'' = M \otimes_A (M' \otimes_A M'')$.

For example, for the first equality it is enough to check that the trilinear map $M \times M' \times M'' \longrightarrow (M \otimes M') \otimes M''$ given by $(x, y, z) \mapsto (x \otimes y) \otimes z$ has the universal property for trilinear maps, and this is easy. The following Formulas 3, 4 and 5 are also easy:

Formula 3. $M \otimes_A N \simeq N \otimes_A M$ (by $x \otimes y \leftrightarrow y \otimes x$).

Formula 4. $M \otimes_A A = M$.

Formula 5. $(\bigoplus_\lambda M_\lambda) \otimes_A N = \bigoplus_\lambda (M_\lambda \otimes_A N)$.

If $f: M \longrightarrow M'$ and $g: N \longrightarrow N'$ are both A-linear then $(x, y) \mapsto f(x) \otimes g(y)$ is a bilinear map from $M \times N$ to $M' \otimes_A N'$, and so it defines a linear map $M \otimes_A N \longrightarrow M' \otimes_A N'$, which we denote $f \otimes g$. From the definition we have:

Formula 6: $(f \otimes g)(\sum_i x_i \otimes y_i) = \sum_i f(x_i) \otimes g(y_i)$.

In particular, if both f and g are surjective then we see from this that $f \otimes g$ is surjective; its kernel is generated by $\{x \otimes y \mid f(x) = 0 \text{ or } g(y) = 0\}$. Indeed, let $T \subset M \otimes N$ be the submodule generated by this set; then $T \subset \ker(f \otimes g)$ so that $f \otimes g$ induces a linear map $\alpha: (M \otimes N)/T \longrightarrow M' \otimes N'$; furthermore, we can define a bilinear map $M' \times N' \longrightarrow (M \otimes N)/T$ by

$$(x', y') \mapsto (x \otimes y \bmod T), \quad \text{where } f(x) = x',\ g(y) = y',$$

since a different choice of inverse images x and y leads to a difference belonging to T. This defines a linear map $\beta: M' \otimes N' \longrightarrow (M \otimes N)/T$, which is obviously an inverse of α. We summarise the above (writing 1 for the identity maps):

Formula 7. Suppose given exact sequences

$$0 \to K \xrightarrow{i} M \xrightarrow{f} M' \to 0 \text{ and } 0 \to L \xrightarrow{j} N \xrightarrow{g} N' \to 0;$$

then $M' \otimes N' \simeq (M \otimes N)/T$, where

$$T = (i \otimes 1)(K \otimes N) + (1 \otimes j)(M \otimes L).$$

Formula 8 (right-exactness of the tensor product). If

$$M_1 \xrightarrow{f} M_2 \xrightarrow{g} M_3 \to 0$$

is an exact sequence then so is

$$M_1 \otimes N \xrightarrow{f \otimes 1} M_2 \otimes N \xrightarrow{g \otimes 1} M_3 \otimes N \to 0.$$

In general, even if $f: M \longrightarrow M'$ is injective, $f \otimes 1: M \otimes N \longrightarrow M' \otimes N$ need not be. (Counter-example: let $A=\mathbb{Z}$ and $N=\mathbb{Z}/n\mathbb{Z}$ for some $n>1$. Let $f:\mathbb{Z}\longrightarrow\mathbb{Z}$ be multiplication by n; then $\mathbb{Z}\otimes_{\mathbb{Z}}N\simeq N\neq 0$, but $f\otimes 1:N\xrightarrow{n}N$ is the zero map, and so is not injective.) However, if $\operatorname{Im} f$ is a direct summand of M' (in which case we say that the exact sequence $0\to M\xrightarrow{f} M'\longrightarrow M'/M\to 0$ *splits*), then there is a map $g:M'\longrightarrow M$ such that $gf=1$.

$$(g\otimes 1)(f\otimes 1)=gf\otimes 1=1\otimes 1$$

is the identity map of $M\otimes N$, and hence $f\otimes 1$ is injective, and the sequence $0\to M\otimes N\longrightarrow M'\otimes N\longrightarrow (M'/M)\otimes N\to 0$ is split. In particular if A is a field then any submodule is a direct summand, so that the operation $\otimes N$ takes exact sequences into exact sequences; in other word, $\otimes N$ is an exact functor. For an arbitrary ring A, an A-module N is said to be *flat* if $\otimes N$ is an exact functor. For more on this see §7.

Change of coefficient ring

Let A and B be rings, and P a two-sided A-B-module; that is, for $a\in A$, $b\in B$ and $x\in P$ the products ax and xb are defined, and in addition to the usual conditions for A-modules and B-modules we assume that

$$(ax)b=a(xb).$$

Then multiplication by an element $b\in B$ induces an A-linear map of P to itself, which we continue to denote by b. This determines a map $1\otimes b:M\otimes_A P\longrightarrow M\otimes_A P$ for any A-module M, and by definition we take this to be scalar multiplication by b in $M\otimes_A P$; that is, we set $(\sum y_i\otimes x_i)b=\sum y_i\otimes x_i b$ for $y_i\in M$ and $x_i\in P$.

If N is a B-module, then for $\varphi\in\operatorname{Hom}_B(P,N)$ we define the product φa of φ and $a\in A$ by

$$(\varphi a)(x)=\varphi(ax)\quad\text{for}\quad x\in P;$$

we have $\varphi a\in\operatorname{Hom}_B(P,N)$, and this makes $\operatorname{Hom}_B(P,N)$ into an A-module.

Formula 9. $\operatorname{Hom}_A(M,\operatorname{Hom}_B(P,N))\simeq\operatorname{Hom}_B(M\otimes_A P,N)$.

Formula 10. $(M\otimes_A P)\otimes_B N\simeq M\otimes_A(P\otimes_B N)$.

Both of these are easy to prove, and we leave them to the reader. Formula 10 generalises Formula 2.

Given a ring homomorphism $\lambda:A\longrightarrow B$, we can think of B as a two-sided A-B-module by setting $ab=\lambda(a)b$; then for any A-module M, $M\otimes_A B$ is a B-module, called the *extension of scalars* in M from A to B, and written $M_{(B)}$. For A-modules M and M' the following formula holds, so that tensor product commutes with change of scalars.

Formula 11. $(M\otimes_A B)\otimes_B(M'\otimes_A B)=(M\otimes_A M')\otimes_A B$.

Indeed, using Formulas 10, 4 and 2, the left-hand side is equal to $M\otimes_A(B\otimes_B(M'\otimes_A B))=M\otimes_A(M'\otimes_A B)=(M\otimes_A M')\otimes_A B$.

Tensor product of A-algebras

Given a ring homomorphism $\lambda: A \longrightarrow B$ we say that B is an A-algebra. Let B' be another A-algebra defined by $\lambda': A \longrightarrow B'$. We say that a map $f: B \longrightarrow B'$ is a homomorphism of A-algebras if it is a ring homomorphism satisfying $\lambda' = f \circ \lambda$. If B and C are A-algebras, then we can take the tensor product $B \otimes_A C$ of B and C as A-modules and this is again an A-algebra. That is, we define the product by

$$\left(\sum_i b_i \otimes c_i\right)\left(\sum_j b'_j \otimes c'_j\right) = \sum_{i,j} b_i b'_j \otimes c_i c'_j,$$

and the ring homomorphism $A \longrightarrow B \otimes C$ by $a \mapsto a \otimes 1 (= 1 \otimes a)$. The fact that the above product is well-defined can easily be seen using the bilinearity of $bb' \otimes cc'$ with respect to both (b, c) and (b', c'). The algebra $B \otimes C$ contains $B \otimes 1$ (short for the subset $\{b \otimes 1 \mid b \in B\} \subset B \otimes C$) and $1 \otimes C$ as subalgebras, and is generated by these. Note that $B \otimes 1$ is not necessarily isomorphic to B.

Example 1. If $\mathfrak{a}$ is an ideal of A and $C = A/\mathfrak{a}$, then $B \otimes_A C = B \otimes_A (A/\mathfrak{a}) = B/\mathfrak{a}B$, and the above $B \otimes 1$ is also equal to $B/\mathfrak{a}B$.

Example 2. If B is an A-algebra and $A[X]$ is the polynomial ring over A then $B \otimes_A A[X]$ can be identified with $B[X]$. Indeed, $A[X]$ is a free A-module with basis $\{X^\nu \mid \nu = 0, 1, 2, \ldots\}$, so that $B \otimes_A A[X]$ is also the free B-module with basis $\{X^\nu\}$, and is isomorphic to $B[X]$ both as an A-module and as a ring. Similarly for the polynomial ring in several variables.

Direct limits

A *directed set* is a partially ordered set Λ such that for any $\lambda, \mu \in \Lambda$ there exists $\nu \in \Lambda$ with $\lambda \leqslant \nu$ and $\mu \leqslant \nu$. For example, a totally ordered set is directed; the set of finite subsets of a set S, ordered by inclusion, is a directed set which is not totally ordered.

Suppose that for each element λ of a directed set Λ we are given a set M_λ, and whenever $\lambda \leqslant \mu$ we are given a map $f_{\mu\lambda}: M_\lambda \longrightarrow M_\mu$ satisfying the conditions

$$f_{\lambda\lambda} = 1, \quad \text{and} \quad f_{\nu\mu} \circ f_{\mu\lambda} = f_{\nu\lambda} \quad \text{for} \quad \lambda \leqslant \mu \leqslant \nu;$$

we express all this data as $\{M_\lambda; f_{\mu\lambda}\}$, and refer to it as a *direct system* over Λ (or indexed by Λ). If each M_λ is an A-module, and each $f_{\mu\lambda}$ A-linear we speak of a direct system of A-modules; if each M_λ is a ring, and each $f_{\mu\lambda}$ a ring homomorphism, a direct system of rings. More generally, we can define direct systems in any category.

Given two direct systems $\mathscr{F} = \{M_\lambda; f_{\mu\lambda}\}$ and $\mathscr{F}' = \{M'_\lambda; f'_{\mu\lambda}\}$ indexed by the same set, a morphism $\varphi: \mathscr{F} \longrightarrow \mathscr{F}'$ is a system of maps

$\{\varphi_\lambda : M_\lambda \longrightarrow M'_\lambda\}$ such that

$$f'_{\mu\lambda} \circ \varphi_\lambda = \varphi_\mu \circ f_{\mu\lambda} \quad \text{for} \quad \lambda < \mu.$$

By a map from $\mathscr{F}$ to a set X we mean a system $\{\varphi_\lambda\}$ of maps $\varphi_\lambda : M_\lambda \longrightarrow X$ satisfying $\varphi_\lambda = \varphi_\mu \circ f_{\mu\lambda}$ for $\lambda < \mu$. Now if a map $\psi : \mathscr{F} \longrightarrow M_\infty$ from $\mathscr{F}$ to a set M_∞ has the universal property for maps from $\mathscr{F}$ to sets, that is, if for any map $\varphi : \mathscr{F} \longrightarrow X$ there exists a unique map $h : M_\infty \longrightarrow X$ such that $\varphi_\lambda = h \circ \psi_\lambda$ for all $\lambda \in \Lambda$, then M_∞ is called the *direct limit* of $\mathscr{F}$, or simply the limit of $\mathscr{F}$, and we write $M_\infty = \varinjlim M_\lambda$, or $M_\infty = \lim M_\lambda$. As one sees easily from the definition, a map $\varphi : \mathscr{F} \longrightarrow \mathscr{F}'$ induces a map $\lim M_\lambda \longrightarrow \lim M'_\lambda$, which in this book we write φ_∞ or $\lim \varphi$.

The limit of a direct system $\mathscr{F} = \{M_\lambda; f_{\mu\lambda}\}$ always exists. In order to construct it we do the following: take the disjoint union $\amalg_\lambda M_\lambda$ of the M_λ, and define a relation $\equiv$ by

$$x \equiv y \Leftrightarrow \begin{cases} x \in M_\lambda,\ y \in M_\mu, \text{and there exists a } \nu \\ \text{with } \lambda \leqslant \nu,\ \mu \leqslant \nu \quad \text{and} \quad f_{\nu\lambda}(x) = f_{\nu\mu}(y). \end{cases}$$

Then it is easy to see that $\equiv$ is an equivalence relation. We write M_∞ for the quotient set $(\amalg_\lambda M_\lambda)/\equiv$, that is the set of equivalence classes under $\equiv$; then one sees easily that M_∞ satisfies the conditions for a direct limit. We write $\lim x \in M_\infty$ for the equivalence class of $x \in M_\lambda$. If $\mathscr{F}$ is a direct system of modules then M_∞ can be given a natural structure of A-module, and $x \mapsto \lim x$ is an A-linear map from M_λ to M_∞. Similarly for direct systems or rings.

The above is general theory. In this book the following two theorems are of particular importance.

Theorem A1. Let A be a ring, N an A-module, and let $\mathscr{F} = \{M_\lambda; f_{\mu\lambda}\}$ be a direct system of A-modules. Then

$$\varinjlim (M_\lambda \otimes_A N) = (\varinjlim M_\lambda) \otimes_A N.$$

(In other words, tensor product commutes with direct limits.)

Proof. Set $\lim M_\lambda = M_\infty$ and $\lim(M_\lambda \otimes N) = L_\infty$. We write $\varphi_\lambda : M_\lambda \longrightarrow M_\infty$ for the A-linear map given by $x \mapsto \lim x$, so that $\{\varphi_\lambda \otimes 1\}$ is a map from the direct system $\{M_\lambda \otimes N; f_{\mu\lambda} \otimes 1\}$ to the A-module $M_\infty \otimes N$; this determines a unique A-linear map $h : L_\infty \longrightarrow M_\infty \otimes N$. For $x \in M_\lambda$ and $y \in N$ we have $h(\lim(x \otimes y)) = (\lim x) \otimes y$. On the other hand, fixing $y \in N$ we can define $g_{\lambda,y} : M_\lambda \longrightarrow L_\infty$ by $g_{\lambda,y}(x) = \lim(x \otimes y)$, and in the limit we get

$$g_y : M_\infty \longrightarrow L_\infty.$$

If $x_\infty \in M_\infty$ we can write $x_\infty = \lim x$ for some λ and some $x \in M_\lambda$. Then $g_y(x_\infty) = g_{\lambda,y}(x) = \lim(x \otimes y)$. From this we can see that $g_y(x_\infty)$ is bilinear in x_∞ and in y, and so defines an A-linear map $g : M_\infty \otimes N \longrightarrow L_\infty$ such that

$g(x_\infty \otimes y) = g_y(x_\infty)$. Now it is easy to see that g and h are inverse maps, so that $M_\infty \otimes N \cong L_\infty$. ■

Theorem A2. Suppose that we have three direct systems of A-modules indexed by the same set Λ, $\mathscr{F}' = \{M'_\lambda; f'_{\mu\lambda}\}$, $\mathscr{F} = \{M_\lambda; f_{\mu\lambda}\}$ and $\mathscr{F}'' = \{M''_\lambda; f''_{\mu\lambda}\}$, and maps $\{\varphi_\lambda\}: \mathscr{F}' \longrightarrow \mathscr{F}$ and $\{\psi_\lambda\}: \mathscr{F} \longrightarrow \mathscr{F}''$ such that for every λ,

$$M'_\lambda \xrightarrow{\varphi_\lambda} M_\lambda \xrightarrow{\psi_\lambda} M''_\lambda$$

is an exact sequence; then the sequence obtained in the limit

$$\varinjlim M'_\lambda \xrightarrow{\varphi_\infty} \varinjlim M_\lambda \xrightarrow{\psi_\infty} \varinjlim M''_\lambda$$

is also exact. (In other words, direct limit is an exact functor.)

Proof. Write M_∞ for the limit of $\mathscr{F}$, and let $y_\infty \in M_\infty$ be such that $\psi_\infty(y_\infty) = 0$. For some λ and $y \in M_\lambda$ we can write $y_\infty = \lim y$, and then $0 = \psi_\infty(\lim y) = \lim \psi_\lambda(y)$, so that for some $\mu \geqslant \lambda$ we have $f''_{\mu\lambda}(\psi_\lambda(y)) = 0$; the left-hand side here is equal to $\psi_\mu(f_{\mu\lambda}(y))$, so that by assumption there is $x \in M'_\mu$ such that $f_{\mu\lambda}(y) = \varphi_\mu(x)$. Thus $y_\infty = \lim f_{\mu\lambda}(y) = \lim \varphi_\mu(x) = \varphi_\infty(\lim x) \in \operatorname{Im}(\varphi_\infty)$. Also, $\psi_\infty \circ \varphi_\infty = 0$ is obvious. ■

Given an A module M, write $\{M_\lambda\}_{\lambda \in \Lambda}$ for the collection of all finitely generated submodules of M. We define a partial order on Λ by letting $\lambda \leqslant \mu$ if $M_\lambda \subset M_\mu$; this makes Λ into a directed set, and we write $f_{\mu\lambda}: M_\lambda \longrightarrow M_\mu$ for the natural inclusion. Then $\{M_\lambda; f_{\mu\lambda}\}$ is a direct system of A-modules, the limit of which is the original M, that is $M = \varinjlim M_\lambda$. Hence any A-module can be expressed as a direct limit of finitely generated A-modules.

In a similar way, given any ring A and a subring $A_0 \subset A$, we can express A as the direct limit of subrings which are finitely generated over A_0 as rings. If we take A_0 to be the minimal subring of A (that is, the image in A of $\mathbb{Z}$), then a ring which is finitely generated over A_0 is Noetherian, and hence every ring is a direct limit of Noetherian rings.

Inverse limits

Inverse systems and inverse limits are defined as the dual notions to direct systems and direct limits, that is by reversing all the arrows in the definitions. That is, we take a directed set Λ as indexing set; an *inverse system* of sets is the data of a set M_λ for each $\lambda \in \Lambda$, and of a map $f_{\lambda\mu}: M_\mu \longrightarrow M_\lambda$ whenever $\lambda \leqslant \mu$, such that

$$f_{\lambda\lambda} = 1, \quad \text{and} \quad f_{\lambda\mu} \circ f_{\mu\nu} = f_{\lambda\nu} \quad \text{for} \quad \lambda \leqslant \mu \leqslant \nu;$$

we write this as $\{M_\lambda; f_{\lambda\mu}\}$. A morphism between two inverse systems with the same indexing set, and a map from a set N to an inverse system

$\mathscr{F} = \{M_\lambda; f_{\lambda\mu}\}$ are defined dually to the case of direct systems. We say that M_∞ is an *inverse limit*, or projective limit of $\mathscr{F}$, and write $M_\infty = \varprojlim M_\lambda$, if there is a map $\varphi = \{\varphi_\lambda\}\colon M_\infty \longrightarrow \mathscr{F}$ which has the property that for any set X, and any map $\psi = \{\psi_\lambda\}: X \longrightarrow \mathscr{F}$, there exists a unique map $h: X \longrightarrow M_\infty$ such that $\psi_\lambda = \varphi_\lambda \circ h$ for all λ. To prove the existence of $\varprojlim M_\lambda$ we only have to let M_∞ be the following subset of the direct product $\Pi_\lambda M_\lambda$:

$$M_\infty = \{(x_\lambda)_{\lambda\in\Lambda} \mid \lambda \leqslant \mu \Rightarrow x_\lambda = \varphi_{\lambda\mu}(x_\mu)\}.$$

If each M_λ is a module and each $\varphi_{\lambda\mu}$ is a linear map then this M_∞ is a submodule of the direct product module, and is the inverse limit of modules. In a similar way, the inverse limit of an inverse system of rings is again a ring.

Example. Let $\Lambda = \{1, 2, 3, \ldots\}$ and let p be a prime number. Consider the inverse system of rings

$$\mathbb{Z}/(p) \longleftarrow \mathbb{Z}/(p^2) \longleftarrow \mathbb{Z}/(p^3) \longleftarrow \cdots,$$

where each arrow is the natural homomorphism. The inverse limit $\varprojlim \mathbb{Z}/(p^n)$ is known as the ring of p-adic integers. Its elements are of the form

$$(a_1, a_2, a_3, \ldots), \quad \text{with} \quad a_i \in \mathbb{Z}/(p^i) \quad \text{and} \quad a_i \equiv a_{i-1} \bmod p^{i-1};$$

addition and multiplication is carried out term-by-term:

$$(a_1, a_2, \ldots) + (b_1, b_2, \ldots) = (a_1 + b_1, a_2 + b_2, \ldots)$$
$$(a_1, a_2, \ldots) \cdot (b_1, b_2, \ldots) = (a_1 b_1, a_2 b_2, \ldots).$$

More generally, if A is any ring and I an ideal of A, the inverse limit $\varprojlim A/I^n$ of the inverse system of rings $A/I \longleftarrow A/I^2 \longleftarrow \cdots$ is called the I-adic completion of A (see §8).

Taking the inverse limit of an inverse system of modules is a left-exact functor, but is not an exact functor, so that the analog of Theorem A2 for inverse systems does not hold.

Example. Consider the diagram

$$\begin{array}{ccccccccc}
 & & \vdots & & \vdots & & \vdots & & \\
 & & \downarrow p & & \downarrow p & & \downarrow p & & \\
0 & \to & \mathbb{Z} & \xrightarrow{n} & \mathbb{Z} & \longrightarrow & \mathbb{Z}/(n) & \to & 0 \\
 & & \downarrow p & & \downarrow p & & \downarrow p & & \\
0 & \to & \mathbb{Z} & \xrightarrow{n} & \mathbb{Z} & \longrightarrow & \mathbb{Z}/(n) & \to & 0 \\
 & & \downarrow p & & \downarrow p & & \downarrow p & & \\
0 & \to & \mathbb{Z} & \xrightarrow{n} & \mathbb{Z} & \longrightarrow & \mathbb{Z}/(n) & \to & 0;
\end{array}$$

here p and n are coprime integers, and the arrows marked $\xrightarrow{n}$ are multiplication by n. The rows are exact sequences, and each column defines an inverse system. The left-hand and middle columns have 0 as their inverse limits, but since every arrow in the right-hand column is an isomorphism, the inverse limit is isomorphic to $\mathbb{Z}/n$. Thus going to the inverse limit, we find that $0 \longrightarrow 0 \longrightarrow \mathbb{Z}/(n)$ is exact, but the second arrow is not surjective.

Exercises to Appendix A. Prove the following propositions.

A.1. Let M and N be A-modules. If the natural map $M' \otimes_A N \longrightarrow M \otimes_A N$ is injective for every finitely generated submodule $M' \subset M$ then the same thing holds for every submodule $M' \subset M$.

A.2. Let A be a ring, and B, C, D (commutative) A-algebras. Then to give a homomorphism of A-algebras from $B \otimes_A C$ to D is the same thing as to give a pair of homomorphisms of A-algebras $B \longrightarrow D$ and $C \longrightarrow D$; in other words, $B \otimes_A C$ is the category-theoretical direct product of B and C in the category of A-algebras.

Appendix B

Some homological algebra

Let A be a ring; by a map from an A-module into another we mean an A-linear map.

Complexes

By a *complex* we mean a sequence

$$\cdots \longrightarrow K_n \xrightarrow{d_n} K_{n-1} \xrightarrow{d_{n-1}} K_{n-2} \longrightarrow \cdots$$

of A-modules and A-linear maps such that $d_{n-1} \circ d_n = 0$ for every n. This complex is written $K_.$. Since $\mathrm{Im}(d_{n+1}) \subset \mathrm{Ker}(d_n)$ we can define $H_n(K_.) = \mathrm{Ker}(d_n)/\mathrm{Im}(d_{n+1})$ to be the homology of $K_.$ in dimension n. To say that $H_n(K_.) = 0$ for all n is to say that $K_.$ is exact. We also consider complexes in which the indices go the other way, $\cdots \longrightarrow K^n \xrightarrow{d_n} K^{n+1} \longrightarrow \cdots$, and for these we write $K^\cdot$ for the complex, and $H^n(K^\cdot) = \mathrm{Ker}(d_n)/\mathrm{Im}(d_{n-1})$ for the cohomology of $K^\cdot$ in dimension n. From now on we often omit the indices, writing d for d_n. We call d the *differential* of the complex $K_.$.

A morphism $f: K_. \longrightarrow K'_.$ of complexes is a family $f = (f_n)_{n \in Z}$ of A-linear maps $f_n: K_n \longrightarrow K'_n$ satisfying $d' \circ f_n = f_{n-1} \circ d$, or in other words a commutative diagram

$$\begin{array}{ccccccccc} \cdots \longrightarrow & K_n & \longrightarrow & K_{n-1} & \longrightarrow & K_{n-2} & \longrightarrow \cdots \\ & {\scriptstyle f_n}\downarrow & & {\scriptstyle f_{n-1}}\downarrow & & {\scriptstyle f_{n-2}}\downarrow & \\ \cdots \longrightarrow & K'_n & \longrightarrow & K'_{n-1} & \longrightarrow & K'_{n-2} & \longrightarrow \cdots \end{array}$$

In an obvious way, f induces a linear map $H_n(K_.) \longrightarrow H_n(K'_.)$ between the homology modules in each dimension; this is often written f_{*n}, or simply f if there is no fear of confusion. If $f, g: K_. \longrightarrow K'_.$ are two morphisms we say that f and g are homotopic (denoted $f \sim g$) if for each n there is a linear map $h_n: K_n \longrightarrow K'_{n+1}$ such that

$$f_n - g_n = d'h_n + h_{n-1}d.$$

If this happens then f and g induce the same map $H_n(K_.) \longrightarrow H_n(K')$ on homology. Two complexes $K_.$ and $K'_.$ are said to be *homotopy equivalent* if there exist morphisms $f: K_. \longrightarrow K'_.$ and $g: K'_. \longrightarrow K_.$ such that $gf \sim 1_K$

and $fg \sim 1_{K'}$, where 1_K denotes the identity map $K. \longrightarrow K.$. Homotopy equivalent complexes have the same homology.

A sequence of complexes

$$0 \to K'_{\cdot} \xrightarrow{f} K_{\cdot} \xrightarrow{g} K''_{\cdot} \to 0$$

is said to be exact if

$$0 \to K'_n \xrightarrow{f_n} K_n \xrightarrow{g_n} K''_n \to 0$$

is exact for every n. In this case, a connecting homomorphism $\delta_n: H_n(K''_{\cdot}) \longrightarrow H_{n-1}(K'_{\cdot})$ is defined as follows: for $\xi \in H_n(K''_{\cdot})$, choose $x \in \operatorname{Ker} d''_n$ representing ξ, and take $y \in K_n$ such that $g(y) = x$; then since $g(\mathrm{d}y) = 0$ there is a well-determined $z \in K'_{n-1}$ for which $f_{n-1}(z) = \mathrm{d}y$, and $\mathrm{d}z = 0$. The class $\zeta \in H_{n-1}(K'_{\cdot})$ represented by z can easily be seen to depend only on ξ, and δ_n is defined by $\delta_n(\xi) = \zeta$. The following sequence is then exact:

$$\cdots \xrightarrow{\delta} H_n(K'_{\cdot}) \xrightarrow{f} H_n(K_{\cdot}) \xrightarrow{g} H_n(K''_{\cdot}) \xrightarrow{\delta} H_{n-1}(K'_{\cdot}) \longrightarrow \cdots.$$

The proof does not require anything new, and is well-known, so that we omit it; this should be thought of as a fundamental theorem of homology theory. The above sequence is called the homology long exact sequence of the short exact sequence $0 \to K'_{\cdot} \longrightarrow K_{\cdot} \longrightarrow K''_{\cdot} \to 0$.

Double complexes

A *double complex* of A-modules is a doubly indexed family $K_{\cdot\cdot} = \{K_{p,q}\}_{p,q \in \mathbf{Z}}$ of A-modules, with two sets of A-linear maps $d'_{pq}: K_{p,q} \longrightarrow K_{p-1,q}$ and $\mathrm{d}''_{pq}: K_{p,q} \longrightarrow K_{p,q-1}$ for which $\mathrm{d}'\mathrm{d}' = 0$, $\mathrm{d}''\mathrm{d}'' = 0$ and $\mathrm{d}'\mathrm{d}'' = \mathrm{d}''\mathrm{d}'$. Given a double complex $K_{\cdot\cdot}$, if we set

$$K_n = \bigoplus_{p+q=n} K_{p,q},$$

and define $\mathrm{d}_n: K_n \longrightarrow K_{n-1}$ by

$$\mathrm{d}x = \mathrm{d}'x + (-1)^p \mathrm{d}''x \quad \text{if} \quad x \in K_{p,q},$$

then since $\mathrm{dd} = 0$, the $\{K_n\}$ form an ordinary complex with differential d. The homology of this complex is called the homology of $K_{\cdot\cdot}$, and written $H_n(K_{\cdot\cdot})$, or simply $H_n(K)$.

To treat homology and cohomology in a unified manner, we fix the following convention on raising and lowering indices: $K_{p,q} = K^{-p}{}_q = K_p{}^{-q} = K^{-p,-q}$. For example, given a double complex $\{K_p{}^q\}$ with $\mathrm{d}': K_p{}^q \longrightarrow K_{p-1}{}^q$ and $\mathrm{d}'': K_p{}^q \longrightarrow K_p{}^{q+1}$, we think of $K_p{}^q$ as $K_{p,-q}$, and set

$$K_n = \bigoplus_{p-q=n} K_p{}^q.$$

The basic technique for studying the homology of double complexes is spectral sequences, but we leave this to specialist texts, and only consider here the extreme cases which we will use later.

We can fix the first index p in $K_{..}$ getting a complex $K_{p.}$.

$$\cdots \longrightarrow K_{p,q+1} \xrightarrow{d''} K_{p,q} \xrightarrow{d''} K_{p,q-1} \longrightarrow \cdots$$

Similarly, for fixed q, d′ defines a complex $K_{.q}$.

Now suppose that $K_{..}$ satisfies the condition $K_{pq}=0$ if p or $q<0$ (a first quadrant double complex). We set $H_0(K_{p.}) = K_{p,0}/d''K_{p,1} = X_p$; then d′ induces a map $X_p \longrightarrow X_{p-1}$, making the X_p into a complex $X_.$. Similarly, d″ makes the $H_0(K_{.q}) = Y_p$ into a complex Y. In this notation we have the following theorem.

Theorem B1. Suppose that the double complex $K_{..}$ satisfies the conditions

$$K_{pq}=0 \quad \text{for } p \text{ or } q<0,$$

and

$$H_q(K_{p.})=0 \quad \text{for } q>0 \text{ and all } p.$$

Then in the above notation we have

$$H_n(K) \simeq H_n(X_.) \quad \text{for all } n.$$

If in addition we have $H_p(K_{.q})=0$ for $p>0$ and all q then

$$H_n(X_.) \simeq H_n(K) \simeq H_n(Y_.).$$

Sketch proof. Write a_{ij} to denote an element of K_{ij}. We define a map $\Phi: K_n \longrightarrow X_n$ by taking $a=\sum_{i=0}^{n} a_{n-i,i} \in K_n$ into $\varphi(a_{n,0}) \in X_n$, where $\varphi: K_{n,0} \longrightarrow X_n$ denotes the canonical map. Then Φ is a morphism of complexes, and we prove that it induces an isomorphism on homology.

Let $x \in X_n$ with $d'x=0$. We can take $a_{n,0}$ such that $x=\varphi(a_{n,0})$, and then since $\varphi(d'a_{n,0}) = d'x = 0$ there exists $a_{n-1,1}$ such that $d'a_{n,0} = d''a_{n-1,1}$. In turn, since $d''(d'a_{n-1,1}) = d'(d''a_{n-1,1}) = d'd'a_{n,0} = 0$ and since $H_1(K_{n-2.})=0$ there exists $a_{n-2,2}$ such that $d'a_{n-1,1} = d''a_{n-2,2}$; then proceeding as before, we can choose $a_{n-i,i}$ for $0 \leqslant i \leqslant n$ such that $d'a_{n-i,i} = d''a_{n-i-1,i+1}$ for $0 \leqslant i < n$. Then for a suitable choice of $\pm$ signs, $a = \sum_0^n \pm a_{n-i,i} \in K_n$ satisfies $da=0$ and $\Phi(a)=x$, and this proves that Φ induces a surjection $H_n(K) \longrightarrow H_n(X_.)$. The proof that this is also injective is similar. The second part follows by symmetry from the first. ■

Dually, we have the following theorem for cohomology.

Theorem B2. Suppose that the double complex $K^{\cdot\cdot}$ satisfies

$$K^{pq}=0 \quad \text{for } p \text{ or } q<0$$

and

$$H^q(K^{p\cdot})=0 \quad \text{for } q>0 \text{ and all } p.$$

Then making $X^p = \mathrm{Ker}(\mathrm{d}'': K^{p,0} \longrightarrow K^{p,1})$ into a complex $X^\cdot$ by means of d′, we have

$$H^n(K) \simeq H^n(X^\cdot),$$

If in addition $H^p(K^{\cdot q}) = 0$ for $p > 0$ and all q then

$$H^n(Y^\cdot) \simeq H^n(K) \simeq H^n(X^\cdot),$$

where $Y^\cdot$ is the complex made from $Y^q = \mathrm{Ker}(\mathrm{d}': K^{0,q} \longrightarrow K^{1,q})$.

We leave the proof as a suitable exercise for the reader.

Projective and injective modules

An A-module P is said to be *projective* if it satisfies the following condition: for any surjection $f: M \longrightarrow N$ and any map $g: P \longrightarrow N$, there exists a lifting $h: P \longrightarrow M$ such that $g = fh$. A free module is projective, and we can characterise projective modules as direct summands of free modules. Indeed, if we express a projective module P as a quotient $P = F/G$ of a free module F then the identity map $P \longrightarrow P$ has a lifting such that $P \longrightarrow F \longrightarrow P$ is the identity map, and then $F \simeq P \oplus G$. Reversing the arrows and replacing surjection by injection in the definition of projective module, we get the definition of *injective module*. There is no dual notion to that of a free module, so that injective modules do not have any very simple characterisation, but we can easily prove the following theorem using Zorn's lemma.

Theorem B3. A necessary and sufficient condition for an A-module I to be injective is that for any ideal $\mathfrak{a}$ of A, and any map $\varphi: \mathfrak{a} \longrightarrow I$, it is possible to extend φ to a map from the whole of A to I.

Any A-module can be written as a quotient of a projective module (take for example a free module). Dually to this, any module can be embedded into an injective module; the proof of this is a little tricky, and we leave it to more specialist textbooks. Given a module M, consider a surjection $P_0 \overset{\varepsilon}{\longrightarrow} M$ from a projective module P_0 to M; letting K_0 be the kernel, we get an exact sequence $0 \to K_0 \longrightarrow P_0 \longrightarrow M \to 0$. In the same way, we construct for K_0 an exact sequence $0 \to K_1 \longrightarrow P_1 \longrightarrow K_0 \to 0$ with P_1 projective, and proceeding as before we get exact sequences $0 \to K_i \longrightarrow P_i \longrightarrow K_{i-1} \to 0$ for $i = 1, 2, \ldots$ with P_i projective. The resulting complex

$$P_\cdot : \cdots \longrightarrow P_n \longrightarrow P_{n-1} \longrightarrow \cdots \longrightarrow P_1 \longrightarrow P_0 \to 0$$

is called a *projective resolution* of M. Since by construction this becomes an exact sequence on replacing the final $P_0 \to 0$ by $P_0 \overset{\varepsilon}{\longrightarrow} M \to 0$, we have $H_n(P_\cdot) = 0$ for $n > 0$ and $H_0(P_\cdot) = M$. In the case that A is Noetherian and M is finite, we can take P_0 to be a free module of finite rank, and then K_0 is again finitely generated. Proceeding in the same way, we see that M has a projective resolution in which each P_n is a finite free module.

Dually, for any A-module M there exists an exact sequence of the form $0 \to M \longrightarrow Q^0 \longrightarrow Q^1 \longrightarrow \cdots$ with each Q^n an injective module. The complex $Q^{\cdot}$: $0 \to Q^0 \longrightarrow Q^1 \longrightarrow \cdots$ is called an *injective resolution* of M; it satisfies $H^0(Q^{\cdot}) = M$ and $H^n(Q^{\cdot}) = 0$ for $n > 0$.

If $f: M \longrightarrow N$ is a map of A-modules and $P_.$, $P'_.$ are projective resolutions of M and N then there exists a morphism of complexes $\varphi: P_. \longrightarrow P'_.$ for which $f\varepsilon = \varepsilon' \varphi_0$, that is, a commutative diagram

$$\begin{array}{ccccccccc} \cdots \longrightarrow & P_n & \longrightarrow & P_{n-1} & \longrightarrow \cdots \longrightarrow & P_0 & \xrightarrow{\varepsilon} & M & \to 0 \\ & \downarrow{\scriptstyle \varphi_n} & & \downarrow{\scriptstyle \varphi_{n-1}} & & \downarrow{\scriptstyle \varphi_0} & & \downarrow{\scriptstyle f} & \\ \cdots \longrightarrow & P'_n & \longrightarrow & P'_{n-1} & \longrightarrow \cdots \longrightarrow & P'_0 & \xrightarrow{\varepsilon'} & N & \to 0. \end{array}$$

The existence of $\varphi_0, \varphi_1, \ldots$ can easily be proved successively, using the fact that the P_n are projective and the exactness of the lower sequence. Up to homotopy, this φ is unique, that is if φ and ψ both have the given property then $\varphi \sim \psi$. We leave the proof of this to the reader. From this it follows that any two projective resolutions of M are homotopy equivalent. Exactly the same thing holds for injective resolutions.

The Tor *functors*

Let M and N be A-modules and $P_.$, $Q_.$ projective resolutions of M and N, respectively. We write $P_. \otimes N$ for the complex obtained by tensoring $P_.$ through with N:

$$P_. \otimes N: \cdots \longrightarrow P_n \otimes N \longrightarrow P_{n-1} \otimes N \longrightarrow \cdots \longrightarrow P_0 \otimes N \to 0.$$

The complex $M \otimes Q_.$ is constructed similarly. Moreover, we can define a double complex $K_{..}$ by $K_{p,q} = P_p \otimes_A Q_q$, with the obvious definitions of d′, d″. Each P_p is a direct summand of a free module, and is therefore flat (that is performing $\otimes P_p$ takes exact sequences into exact sequences). Thus $H_n(K_{p.}) = H_n(P_p \otimes Q_.) = 0$ for $n > 0$, and $H_0(K_{p.}) = H_0(P_p \otimes Q_.) = P_p \otimes N$. In exactly the same way, $H_n(K_{.q}) = 0$ for $n > 0$ and $H_0(K_{.q}) = M \otimes Q_q$, and therefore by Theorem B1 $H_n(P_. \otimes N) \simeq H_n(K_{..}) \simeq H_n(M \otimes Q_.)$. This module (defined uniquely up to isomorphism) is written $\mathrm{Tor}_n^A(M, N)$; it is independent of the choice of the projective resolutions of M and N chosen, since if $P_.$ and $P'_.$ are two projective resolutions, we have $P_. \sim P'_.$, and therefore $P_. \otimes N \sim P'_. \otimes N$.

The Tor functors have the following properties, (all of which can be proved directly from the definition):

(1) $\mathrm{Tor}_0^A(M, N) = M \otimes_A N$;

(2) if M is flat then $\mathrm{Tor}_n^A(M,N)=0$ for any N and $n>0$;

(3) $\mathrm{Tor}_n^A(M,N)\simeq \mathrm{Tor}_n^A(N,M)$;

(4) $\mathrm{Tor}_n^A(M,N)$ is a covariant functor in both of its entries, and each short exact sequence $0\to M'\longrightarrow M\longrightarrow M''\to 0$ leads to a long exact sequence

$$\begin{aligned}\cdots\longrightarrow \mathrm{Tor}_n^A(M',N)&\longrightarrow \mathrm{Tor}_n^A(M,N)\longrightarrow \mathrm{Tor}_n^A(M'',N)\\ \longrightarrow \mathrm{Tor}_{n-1}^A(M',N)&\longrightarrow\cdots\longrightarrow \mathrm{Tor}_1^A(M'',N)\\ \longrightarrow M'\otimes N&\longrightarrow M\otimes N\longrightarrow M''\otimes N\to 0.\end{aligned}$$

(5) If $\{N_\lambda, f_{\mu\lambda}\}$ is a direct system of A-modules then

$$\mathrm{Tor}_n^A(M, \varinjlim N_\lambda)=\varinjlim \mathrm{Tor}_n^A(M,N_\lambda).$$

The Ext *functors*

Let M and N be A-modules. The functor $\mathrm{Hom}_A(M,-)$ is left-exact, that is it takes an exact sequence $0\to N'\longrightarrow N\longrightarrow N''\to 0$ into an exact sequence $0\to\mathrm{Hom}_A(M,N')\longrightarrow\mathrm{Hom}_A(M,N)\longrightarrow\mathrm{Hom}_A(M,N'')$; and M is projective if and only if $\mathrm{Hom}_A(M,-)$ is exact. In addition, $\mathrm{Hom}_A(-,N)$ is left-exact, in the sense that it takes an exact sequence $0\to M'\longrightarrow M\longrightarrow M''\to 0$ into an exact sequence $0\to\mathrm{Hom}_A(M'',N)\longrightarrow\mathrm{Hom}_A(M,N)\longrightarrow\mathrm{Hom}_A(M',N)$, and N is injective if and only if $\mathrm{Hom}_A(-,N)$ is exact.

Choose a projective resolution $P_.$ of M and an injective resolution $Q^\cdot$ of N; we define a double complex $K^{\cdot\cdot}$ by $K^{p,q}=\mathrm{Hom}_A(P_p,Q^q)$, and construct the two complexes

$$\mathrm{Hom}_A(M,Q^\cdot)\colon 0\to\mathrm{Hom}_A(M,Q^0)\longrightarrow\mathrm{Hom}_A(M,Q^1)\longrightarrow\cdots$$

and

$$\mathrm{Hom}_A(P_.,N)\colon 0\to\mathrm{Hom}_A(P_0,N)\longrightarrow\mathrm{Hom}_A(P_1,N)\longrightarrow\cdots.$$

Then by Theorem B2 we get

$$H^n(\mathrm{Hom}_A(M,Q^\cdot))\simeq H^n(K^{\cdot\cdot})\simeq H^n(\mathrm{Hom}_A(P_.,N)).$$

Identifying these three, we write $\mathrm{Ext}_A^n(M,N)$. As with Tor, this does not depend on the choice of $P_.$ and $Q^\cdot$.

The main properties of the Ext functors are as follows:

(1) $\mathrm{Ext}_A^0(M,N)=\mathrm{Hom}_A(M,N)$,

(2) If M is projective, or if N is injective, then $\mathrm{Ext}_A^n(M,N)=0$ for $n>0$;

(3) $\mathrm{Ext}_A^n(M,N)$ is a contravariant functor in M and a covariant functor in N. A short exact sequence $0\to M'\longrightarrow M\longrightarrow M''\to 0$ gives rise to a long exact sequence

$$\begin{aligned}0\to\mathrm{Hom}_A(M'',N)&\longrightarrow\mathrm{Hom}_A(M,N)\longrightarrow\mathrm{Hom}_A(M',N)\\ \longrightarrow\mathrm{Ext}_A^1(M'',N)&\longrightarrow\mathrm{Ext}_A^1(M,N)\longrightarrow\mathrm{Ext}_A^1(M',N)\\ \longrightarrow\mathrm{Ext}_A^2(M'',N)&\longrightarrow\cdots,\end{aligned}$$

and a short exact sequence $0\to N'\longrightarrow N\longrightarrow N''\to 0$ gives rise to a long exact sequence

$$0 \to \mathrm{Hom}_A(M, N') \longrightarrow \mathrm{Hom}_A(M, N) \longrightarrow \mathrm{Hom}_A(M, N'') \longrightarrow$$
$$\longrightarrow \mathrm{Ext}_A^1(M, N') \longrightarrow \mathrm{Ext}_A^1(M, N) \longrightarrow \mathrm{Ext}_A^1(M, N'') \longrightarrow$$
$$\longrightarrow \mathrm{Ext}_A^2(M, N') \longrightarrow \cdots;$$

(4) M is projective $\Leftrightarrow \mathrm{Ext}_A^1(M, N) = 0$ for all N,

and

$$N \text{ is injective} \Leftrightarrow \mathrm{Ext}_A^1(M, N) = 0 \text{ for all } M.$$

Projective and injective dimensions

If M is an A-module for which there exists a projective resolution $P_.$ with $P_n = 0$ for $n > d$, but such that $P_d \neq 0$ for any choice of projective resolution $P_.$, then we say that M has projective dimension d, and write $\mathrm{proj\,dim}\, M = d$. If there is no such d then we write $\mathrm{proj\,dim}\, M = \infty$. The injective dimension $\mathrm{inj\,dim}\, M$ is defined in the same way using injective resolutions. Clearly $\mathrm{proj\,dim}\, M = 0$ if and only if M is projective, and $\mathrm{inj\,dim}\, M = 0$ if and only if M is injective.

For a projective resolution $P_.$ of M and some $i > 0$, let K_i denote the image of $P_i \longrightarrow P_{i-1}$; then $\cdots \longrightarrow P_n \longrightarrow P_{n-1} \longrightarrow \cdots \longrightarrow P_i \to 0$ is a projective resolution of K_i, so that for $n > i$ we have $\mathrm{Ext}_A^n(M, N) \simeq \mathrm{Ext}_A^{n-i}(K_i, N)$. Now if $\mathrm{Ext}_A^{d+1}(M, N) = 0$ for all N, we have $\mathrm{Ext}_A^1(K_d, N) = 0$ for all N, and hence K_d is projective, so that $0 \to K_d \longrightarrow P_{d-1} \longrightarrow \cdots \longrightarrow P_0 \to 0$ is also a projective resolution of P_0, and $\mathrm{proj\,dim}\, M \leqslant d$. Conversely, if $\mathrm{proj\,dim}\, M \leqslant d$ then obviously $\mathrm{Ext}_A^n(M, N) = 0$ for $n > d$.

Similarly, $\mathrm{inj\,dim}\, N \leqslant d \Leftrightarrow \mathrm{Ext}_A^{d+1}(M, N) = 0$ for all M.

Derived functors

As we have just seen, the definition of functors like Tor and Ext can be given using a resolution of just one entry. For instance, write T for the functor $\mathrm{Hom}_A(-, N)$, and let $P_.$ be a projective resolution of a given module M; construct the complex $T(P_.)\colon \cdots \longleftarrow T(P_n) \longleftarrow T(P_{n-1}) \longleftarrow \cdots \longleftarrow T(P_0) \longleftarrow 0$, and take the cohomology $H^n(T(P_.))$. Setting $R^nT(M) = H^n(T(P_.))$ defines a functor R^nT in M, which we call the *right derived functor* of the left exact functor T. In the present case we have $R^nT = \mathrm{Ext}_A^n(-, N)$, but we can in general define the right derived functor of a left exact contravariant functor in the same way.

The right derived functor is uniquely determined by the following three properties: (1) $R^0T = T$, (2) if M is projective then $R^nT(M) = 0$ for all $n > 0$, and (3) a short exact sequence $0 \to M' \longrightarrow M \longrightarrow M'' \to 0$ gives rise to a 'natural' long exact sequence

$$0 \to T(M'') \longrightarrow T(M) \longrightarrow T(M')$$
$$\longrightarrow R^1T(M'') \longrightarrow R^1T(M) \longrightarrow R^1T(M')$$
$$\longrightarrow R^2T(M'') \longrightarrow \cdots.$$

(For the meaning of 'natural' see a textbook on homological algebra.) For a left exact covariant functor, we have to replace 'projective' by 'injective' in the above. For right exact functors we can define a left derived functor by taking a projective resolution in the covariant case and an injective resolution in the contravariant case.

Injective hull

Let L be an A-module and $M \subset L$ a submodule; we say that L is an *essential extension* of M if $N \cap M \neq 0$ for every non-zero submodule $N \subset L$, or equivalently if

$$0 \neq x \in L \Rightarrow \text{there exists } a \in A \text{ such that } 0 \neq ax \in M.$$

Theorem B4. An A-module M is injective if and only if it has no essential extensions except M itself.

We leave the proof to the reader. Now suppose that M is a given A-module, and choose an injective module I with $M \subset I$. If we let E be a maximal element among all essential extensions of M in I then by the above theorem E is injective. An injective module E such that $M \subset E$ is an essential extension is called an *injective hull* of M, and written $E(M)$ or $E_A(M)$; this notion plays an important role in §18. If E and E' are injective hulls of M then it is easy to see that there is an isomorphism $\varphi: E \xrightarrow{\sim} E'$ which fixes the elements of M, although φ itself is not necessarily unique.

Let M be an A-module. Take an injective hull I^0 of M, and set $K^1 = I^0/M$. Take an injective hull I^1 of K^1, and set $K^2 = I^1/K^1$. Proceeding in the same way we obtain an injective resolution $0 \to I^0 \longrightarrow I^1 \longrightarrow I^2 \longrightarrow \cdots$ of M, which is called a *minimal injective resolution* of M.

The following two propositions are both famous and useful; the proofs are easy.

The five lemma. Let

$$\begin{array}{ccccccccc} A & \longrightarrow & B & \longrightarrow & C & \longrightarrow & D & \longrightarrow & E \\ {\scriptstyle f_1}\downarrow & & {\scriptstyle f_2}\downarrow & & {\scriptstyle f_3}\downarrow & & {\scriptstyle f_4}\downarrow & & {\scriptstyle f_5}\downarrow \\ A' & \longrightarrow & B' & \longrightarrow & C' & \longrightarrow & D' & \longrightarrow & E' \end{array}$$

be a commutative diagram of modules with exact rows. Then

(1) f_1 surjective, and f_2 and f_4 injective $\Rightarrow f_3$ is injective;
(2) f_5 injective, and f_2 and f_4 surjective $\Rightarrow f_3$ is surjective.

The snake lemma. Let

$$\begin{array}{ccccccc} & & A & \longrightarrow & B & \longrightarrow & C & \rightarrow 0 \\ & & \downarrow{\scriptstyle\alpha} & & \downarrow{\scriptstyle\beta} & & \downarrow{\scriptstyle\gamma} & \\ 0 & \rightarrow & A' & \longrightarrow & B' & \longrightarrow & C' & \end{array}$$

be a commutative diagram of modules with exact rows. Then there exists an exact sequence of the form

$$\mathrm{Ker}(\alpha) \longrightarrow \mathrm{Ker}(\beta) \longrightarrow \mathrm{Ker}(\gamma) \xrightarrow{d}$$

$$\mathrm{Coker}(\alpha) \longrightarrow \mathrm{Coker}(\beta) \longrightarrow \mathrm{Coker}(\gamma).$$

Tensor product of complexes

Given two complexes of A-modules $K_{\bullet}$ and $L_{\bullet}$ the tensor product $K \otimes_A L$ is defined as follows: firstly, set

$$(K \otimes L)_n = \otimes_{p+q=n} K_p \otimes_A L_q,$$

and define the differential d by setting

$$\mathrm{d}(x \otimes y) = \mathrm{d}x \otimes y + (-1)^p x \otimes \mathrm{d}y$$

for $x \in K_p$ and $y \in L_q$. In other words, $K \otimes L$ is the (single) complex obtained from the double complex $W_{\bullet\bullet}$, where $W_{p,q} = K_p \otimes L_q$.

There is an isomorphism of complexes $K \otimes L \simeq L \otimes K$ obtained by sending $x \otimes y$ into $(-1)^{pq} y \otimes x$ for $x \otimes y \in K_p \otimes L_q$. For a third complex of A-modules M, the associative law holds:

$$(K \otimes L) \otimes M = K \otimes (L \otimes M).$$

Hence the tensor product $K^{(1)} \otimes \cdots \otimes K^{(r)}$ of a finite number of complexes can be defined by induction. This is used in §16.

The information on homological algebra given above should be adequate for the purpose of reading this book. However, a student intending to become a specialist in algebra or geometry will require rather more detailed knowledge, including the theory of spectral sequences. We mention here just three representative references, two books by the originators of homological algebra and category theory:

H. Cartan and S. Eilenberg, *Homological Algebra*, Princeton, 1956,

S. Maclane, *Homology*, Springer, 1963,

together with A. Grothendieck's paper

Sur quelques points d'algèbre homologique, *Tohoku Math. J.* **9** (1957), 119–221.

Appendix C

The exterior algebra

(1) Let M and N be modules over a ring A. An r-multilinear map $\varphi: M^r = M \times \cdots \times M \longrightarrow N$ from the direct product of r copies of M is said to be *alternating* if $\varphi(x_1, \ldots, x_r) = 0$ whenever any of the elements $x_1, \ldots, x_r$ appears more than once. If φ is alternating then for any $x_1, \ldots, x_r \in M$ we have

$$\varphi(x_1, \ldots, x_{i-1}, x_i + x_j, x_{i+1}, \ldots, x_{j-1}, x_i + x_j, x_{j+1}, \ldots) = 0,$$

and expanding out the left-hand side gives

$$\varphi(x_1, \ldots, x_i, \ldots, x_j, \ldots) + \varphi(x_1, \ldots, x_j, \ldots, x_i, \ldots) = 0.$$

In other words, on interchanging two of its entries, φ changes sign.

The rth exterior product of M is defined as the module N_0 having a universal alternating r-fold multilinear map $f_0: M^r \longrightarrow N_0$, that is a map satisfying the property that every alternating r-fold multilinear map $f: M^r \longrightarrow N$ factorises as $f = h \circ f_0$ for a unique A-linear map $h: N_0 \longrightarrow N$. We write $N_0 = \wedge^r M$, and use $x_1 \wedge \cdots \wedge x_r$ to denote $f_0(x_1, \ldots, x_r)$. To prove the existence of the exterior product, let N_0 be the quotient of the r-fold tensor product $M \otimes \cdots \otimes M$ by the submodule generated by elements of the form $x_1 \otimes \cdots \otimes x \otimes \cdots \otimes x \otimes \cdots \otimes x_r$. Then N_0 satisfies the above condition. The fact that the exterior product is uniquely determined up to isomorphism is obvious from the definition.

(2) If M is a free A-module of rank n, with basis $e_1, \ldots, e_n$ then $\wedge^r M$ is zero if $r > n$, and if $r \leqslant n$ is the free A-module of rank $\binom{n}{r}$ with basis $\{e_{i_1} \wedge \cdots \wedge e_{i_r} \mid 1 \leqslant i_1 < \cdots < i_r \leqslant n\}$. (If $r > n$ this is easy; if $r \leqslant n$ then the $\binom{n}{r}$ elements given above obviously generate $\wedge^r M$, and the fact that they are linearly independent can also be proved by reducing to the theory of determinants.)

(3) However, if $I \subset A$ is an ideal, then $\wedge^2 A = 0$, but nevertheless $\wedge^2 I$ is not necessarily 0. For example, let k be a field, x and y indeterminates, and $A = k[x, y]$; if $I = xA + yA$ then $\wedge^2 I \neq 0$. Indeed, we can define $\varphi: I \times I \longrightarrow k = A/I$ by

$$\varphi(f, g) = [\partial(f, g)/\partial(x, y)]_{(x,y)=(0,0)},$$

and it is then easy to check that φ is alternating and bilinear, with $\varphi(x, y) = 1$, so that $\varphi \neq 0$, and we must have $\wedge^2 I \neq 0$.

(4) The operation of taking the exterior product commutes with extensions of scalars. That is, let B be an A-algebra and M an A-module, and set $M \otimes_A B = M_B$. Then $(\wedge^r M) \otimes_A B = \wedge^r M_B$, where of course $\wedge^r$ on the right-hand side refers to the exterior product of B-modules. For the proof, according to Appendix A, Formula 11, we have $M_B \otimes_B \cdots \otimes_B M_B = (M \otimes_A \cdots \otimes_A M) \otimes_A B$, so that if we let f_0 be the composite

$$(M_B)^r \longrightarrow \bigotimes_{i=1}^{r} M_B = \left(\bigotimes_{i=1}^{r} M \right) \otimes_A B \longrightarrow (\bigwedge^{r} M) \otimes_A B,$$

then f_0 is an alternating r-fold B-multilinear map. Write $v: M \longrightarrow M_B$ for the natural map $x \mapsto x \otimes 1$. Let N be a B-module and $\varphi: (M_B)^r \longrightarrow N$ be an alternating r-fold B-multilinear map. Then φ induces an alternating r-fold A-multilinear map $\varphi: M^r \longrightarrow N$, and therefore an A-linear map $\bigwedge^r M \longrightarrow N$, and finally a B-linear map $(\bigwedge^r M) \otimes_A B \longrightarrow N$ which we denote h. Then on $v(M^r)$ the two maps φ and $h \circ f_0$ coincide; but M_B is generated over B by $v(M)$, so that $\varphi = h \circ f_0$. Thus f_0 has the universal property, and we can think of $(\wedge^r M) \otimes_A B$ as $\wedge^r M_B$.

Theorem C1. Let A be an integral domain with field of fractions K, and let $I_1, \ldots, I_r$ be ideals of A. Set $M = I_1 \oplus \cdots \oplus I_r$, and let T be the torsion submodule of $\wedge^r M$; then $(\wedge^r M)/T \simeq I_1 \ldots I_r$. Therefore if $J_1, \ldots, J_r$ are ideals of A such that $I_1 \oplus \cdots \oplus I_r \simeq J_1 \oplus \cdots \oplus J_r$ we have $I_1 \ldots I_r \simeq J_1 \ldots J_r$.

Proof. We have $M_K \simeq K \oplus \cdots \oplus K$ (the direct sum of r copies of K), so that $(\wedge^r M) \otimes K = \wedge^r M_K \simeq K$. The kernel of the natural map $\wedge^r M \longrightarrow (\wedge^r M) \otimes K \simeq K$ is obviously T (since tensoring with K is the same thing as the localisation with respect to the zero ideal of A, see §4). In addition, the image is $I_1 \ldots I_r$. Indeed, viewing each I_i as a submodule of K, we can assume that the map is

$$\bigwedge^{r} (I_1 \oplus \cdots \oplus I_r) \longrightarrow \bigwedge^{r} (K \oplus \cdots \oplus K) = Ke_1 \wedge \cdots \wedge e_r \simeq K.$$

and since for $\xi_i = \sum_{j=1}^{r} a_{ij} e_j \in \sum K e_j$ we have $\xi_1 \wedge \cdots \wedge \xi_r = \det(a_{ij}) \cdot e_1 \wedge \cdots \wedge e_r$, it is clear that the above map has image $I_1 \cdots I_r e_1 \wedge \cdots \wedge e_r$. ■

If A is a Dedekind ring then it is known that $I_1 \oplus \cdots \oplus I_r \simeq A^{r-1} \oplus I_1 \ldots I_r$ (see for example [B7], §4, Prop. 24).

Theorem C2. Let A be a ring and M, N A-modules. Then

$$\bigwedge^{r} (M \oplus N) \simeq \bigoplus_{s+t=r} [(\bigwedge^{s} M) \otimes_A (\bigwedge^{t} N)].$$

Proof. $\bigotimes_{i=1}^{r} (M \oplus N)$ can be written as a direct sum of all possible r-fold tensor products of copies of M and N (with 2^r summands). Write $L_{s,t}$ for the submodule which is the direct sum of all tensor products involving s copies of M and t copies of N (with $s+t=r$). Thus

$$\bigotimes_1^r (M \oplus N) = \bigoplus_{s+t=r} L_{s,t}.$$

For example, when $r=2$ we have $L_{2,0} = M \otimes M$, $L_{1,1} = (M \otimes N) \oplus (N \otimes M)$ and $L_{0,2} = N \otimes N$. Now let Q be the submodule of $\bigotimes_1^r (M \oplus N)$ generated by all elements of the form $\cdots \otimes x \otimes \cdots \otimes x \otimes \cdots$; we have $Q = \oplus [Q \cap L_{s,t}]$. We see at once that $Q \cap L_{s,t}$ is the submodule generated by all elements of the forms $\cdots \otimes y \otimes \cdots \otimes y \otimes \cdots$ (with either $y \in M$ or $y \in N$), and $\alpha \otimes y \otimes \beta \otimes z \otimes \gamma + \alpha \otimes z \otimes \beta \otimes y \otimes \gamma$ (with $y \in M$ and $z \in N$). Thus one sees easily that

$$\bigwedge^r (M \oplus N) = \left(\bigotimes_1^r (M \oplus N) \right) / Q = \bigoplus_{s+t=r} (L_{s,t} / Q \cap L_{s,t}),$$

and

$$L_{s,t} / L_{s,t} \cap Q \simeq (\bigwedge^s M) \otimes (\bigwedge^t N). \quad ■$$

(5) Let A be a commutative ring. We say that a (possibly non-commutative) A-algebra E is a *skew-commutative graded algebra* if it has a direct sum decomposition $E = \bigoplus_{n \geq 0} E_n$ as an A-module, such that

(i) $E_p \cdot E_q \subset E_{p+q}$;

(ii) $xy = (-1)^{pq} yx$ for $x \in E_p$ and $y \in E_q$;

(iii) $x^2 = 0$ for $x \in E_{2n+1}$.

For such an algebra E, a *skew-derivation* is an A-linear map $\mathrm{d}: E \longrightarrow E$ such that

(α) $\mathrm{d}(E_n) \subset E_{n-1}$;

(β) $\mathrm{d}(xy) = (\mathrm{d}x)y + (-1)^p x(\mathrm{d}y)$ for $x \in E_p$, $y \in E_q$.

(6) Let A be a ring and M an A-module. We show how to define an A-bilinear map $\Psi: (\bigwedge^p M) \times (\bigwedge^q M) \longrightarrow \bigwedge^{p+q} M$. If we define $\varphi: M^p \times M^q \longrightarrow \bigwedge^{p+q} M$ by $\varphi(x_1, \ldots, x_p, y_1, \ldots, y_q) = x_1 \wedge \ldots \wedge x_p \wedge y_1 \wedge \cdots \wedge y_q$ then for fixed $y_1, \ldots, y_q$ this is an alternating p-multilinear map from M^p to $\bigwedge^{p+q} M$, so that there is a map $\Phi: (\bigwedge^p M) \times M^q \longrightarrow \bigwedge^{p+q} M$ such that $\Phi(\xi, y_1, \ldots, y_q)$ is linear in ξ and satisfies $\Phi(\xi, y_1, \ldots, y_q) = x_1 \wedge \cdots \wedge x_p \wedge y_1 \wedge \cdots \wedge y_q$ if $\xi = x_1 \wedge \cdots \wedge x_p$. Now for fixed ξ, Φ is alternating and multilinear in $y_1, \ldots, y_q$, defining a bilinear map $\Psi: (\bigwedge^p M) \times (\bigwedge^q M) \longrightarrow \bigwedge^{p+q} M$ such that $\Psi(\xi, y_1 \wedge \cdots \wedge y_q) = \Phi(\xi, y_1, \ldots, y_q)$. For $\xi \in \bigwedge^p M$ and $\eta \in \bigwedge^q M$ we write $\xi \wedge \eta$ for $\Psi(\xi, \eta)$; if $\xi = \sum_\alpha x_1^{(\alpha)} \wedge \cdots \wedge x_p^{(\alpha)}$ and $\eta = \sum_\beta y_1^{(\beta)} \wedge \cdots \wedge y_q^{(\beta)}$ then $\xi \wedge \eta = \sum_{\alpha,\beta} x_1^{(\alpha)} \wedge \cdots \wedge x_p^{(\alpha)} \wedge y_1^{(\beta)} \wedge \cdots \wedge y_q^{(\beta)}$. (It might be tempting to make the definition directly in terms of this formula, but the expression of ξ and η in the above form is non-unique, so that this requires an awkward proof.) The multiplication $\wedge$ satisfies the associative law $\xi \wedge (\eta \wedge \zeta) = (\xi \wedge \eta) \wedge \zeta$, so that we can use it to define a product on $\bigoplus_{n=0}^{\infty} \wedge^n M$ (we set $\wedge^0 M = A$), which becomes an A-algebra, and it is easy to see that this satisfies the conditions for a skew-commutative algebra. We write $\wedge M$ for this algebra, and call it the *exterior algebra* of M.

Given any linear map $\alpha: M \longrightarrow A$, there exists a unique skew-derivation d of $\wedge M$ such that d coincides with α on $\wedge^1 M = M$. The uniqueness is clear from the fact that $\wedge M$ is generated as an A-algebra by M: we must have $d(x_1 \wedge \cdots \wedge x_p) = \sum_{r=1}^{p} (-1)^{r-1} \alpha(x_r) x_1 \wedge \cdots \wedge \hat{x}_r \wedge \cdots \wedge x_p$. Conversely, the existence follows easily from the fact that the right-hand side defines an alternating p-fold multilinear map of $x_1, \ldots, x_p$.

In particular, let M be a free A-module of rank n with basis $e_1, \ldots, e_n$, so that $M = Ae_1 \oplus \cdots \oplus Ae_n$. Then taking arbitrary elements $c_1, \ldots, c_n \in A$, we can define $\alpha: M \longrightarrow A$ by $\alpha(e_i) = c_i$, and the skew-derivation of $\wedge M$ takes the form

$$d(e_{i_1} \wedge \cdots \wedge e_{i_p}) = \sum_{r=1}^{p} (-1)^{r-1} c_r e_{i_1} \wedge \cdots \wedge \hat{e}_{i_r} \wedge \cdots \wedge e_{i_p}.$$

This can be identified with the differential operator of the Koszul complex $K_{\bullet c, 1 \ldots n}$ discussed in §16. Thus the Koszul complex can be thought of as the exterior algebra $\wedge (Ae_1 \oplus \cdots \oplus Ae_n)$ with the skew-derivation defined by $d(e_i) = c_i$.

Exercise to Appendix C

C.1. Let $(A, \mathfrak{m}, k)$ be a local ring and M be a finitely generated A-module. Prove that $\min\{r \mid \bigwedge^r M \neq 0\}$ is equal to the minimal number of generators of M.

Solutions and hints for the exercises

(Please be sure to try each exercise on your own before looking at the solution)

§1.

1.1. If $ab = 1 - x$ with $x^n = 0$ then $ab(1 + x \ldots + x^{n-1}) = 1$.

1.2. Set $e_i = (0, \ldots, 1, \ldots, 0) \in A_1 \times \cdots \times A_n$ (with 1 in the ith place); then since $e_i e_j = 0$ for $i \neq j$, any prime ideal $\mathfrak{p}$ of $A_1 \times \cdots \times A_n$ must contain all but one of the e_i.

1.3. (*a*) Use the fact that $\text{rad}(A) = \{x \in A \mid 1 + ax \text{ is a unit of } A, \forall a \in A\}$. Counter-example: $A = \mathbb{Z}$, $B = \mathbb{Z}/(4)$; then $\text{rad}(A) = (0)$, $\text{rad}(B) = 2B$.
(*b*) Let $\mathfrak{m}_1, \ldots, \mathfrak{m}_r$ be the maximal ideals of A and $I = \text{Ker} f$. Suppose $I \subset \mathfrak{m}_i$ for $1 \leqslant i \leqslant s$ and $I \not\subset \mathfrak{m}_i$ for $s < i \leqslant r$; then the maximal ideals of B are $f(\mathfrak{m}_i)$ for $1 \leqslant i \leqslant s$, and $f(\mathfrak{m}_i) = B$ for $i > s$. Now $\text{rad}(A) = \mathfrak{m}_1 \ldots \mathfrak{m}_r$, hence $f(\text{rad}(A)) = f(\mathfrak{m}_1) \ldots f(\mathfrak{m}_r) = f(\mathfrak{m}_1) \ldots f(\mathfrak{m}_s) = \text{rad}(B)$.

1.5. The first half is easy; for the second, use Zorn's lemma.

1.6. We can assume that there are no inclusions among $P_1, \ldots, P_r$. When $r = 2$, take $x \in I - P_1$, $y \in I - P_2$; then one of x, y, $x + y$ is not in P_1 or P_2. When $r > 2$, we can take $x \in I - (P_1 \cup \cdots \cup P_{r-1})$ by induction. Also, since P_r is prime, $P_r \not\supset IP_1 \ldots P_{r-1}$, so take $y \in IP_1 \ldots P_{r-1} - P_r$; then either x or $x + y$ satisfies the condition.

§2.

2.1. By NAK there is an $e \in I$ such that $(1 - e)I = 0$. One sees easily that then $I = Ie = Ae$ and $e^2 = e$.

2.2. If $x \in \text{ann}(M/IM)$ then $xM \subset IM$, so that by Theorem 1 there exists $y \in I$ such that $(x^n + y)M = 0$.

2.3. $(M + N)/N \simeq M/(M \cap N)$ shows that M is finite, and similarly for N.

2.4. (*a*) If $M \simeq A^n$ and P is a maximal ideal of A with $k = A/P$ then $M/PM \simeq k^n$; for a field the result is well-known.
(*b*) The first part is easy by the theory of determinants; the second half comes from the fact that A^n has n linearly independent elements, but any $n + 1$ elements are linearly dependent.
(*c*) Use Theorem 3, (iii).

2.5. (*a*) If F and F' are free modules and $\alpha: F \longrightarrow L$, $\beta: F' \longrightarrow N$ are surjections then there is a mapγ making

$$\begin{array}{ccccccccc} 0 & \rightarrow & L & \longrightarrow & M & \longrightarrow & N & \rightarrow & 0 \\ & & \alpha\uparrow & & \gamma\uparrow & & \beta\uparrow & & \\ 0 & \rightarrow & F & \longrightarrow & F\otimes F' & \longrightarrow & F' & \rightarrow & 0 \end{array}$$

commute. The assertion follows from this and the snake lemma (Appendix B).

(*b*) can be proved similarly.

§3.

3.1. Use the fact that A is isomorphic to a submodule of $(A/I_1)\oplus\cdots\oplus(A/I_n)$.

3.2. Use the previous question.

3.4. If $II^{-1}=A$ then there exist $x_i\in I$ and $y_i\in I^{-1}$ such that $\sum_1^n x_iy_i=1$; then it follows easily that $I=\sum x_iA$.

3.5. If J is a fractional ideal generated by $a_1/b_1,\ldots,a_n/b_n$, with a_i and b_i coprime, then J^{-1} is the principal fractional ideal generated by u/v, where $u=\text{l.c.m.}(b_1,\ldots,b_n)$ and $v=\text{h.c.f.}(a_1,\ldots,a_n)$.

3.6. $\operatorname{Ker}(\varphi^n)=I_n$ for $n=1, 2,\ldots$ is an ascending chain of ideals of A.

3.7. Choose an ideal I of A which is not finitely generated, and set $M=A/I$; then by Theorem 2.6, M cannot be of finite presentation.

§4.

4.5. Write $V(I_\lambda)$ for the complement of U_λ, where I_λ is an ideal of A. Then $\bigcap_\lambda V(I_\lambda)=V(\sum I_\lambda)=\varnothing$, so $1\in\sum I_\lambda$, and therefore a finite sum of I_λ also contains 1.

4.6. If $\operatorname{Spec} A=V(I_1)\cup V(I_2)$ with $V(I_1)\cap V(I_2)=\varnothing$ then $I_1+I_2=A$ and $I_1I_2\subset\operatorname{nil}(A)$. So $1=e_1+e_2$ with $e_i\in I_i$ for $i=1, 2$ and $(e_1e_2)^n=0$. So $1=(e_1+e_2)^{2n}=e_1^nx_1+e_2^nx_2$ with $x_i\in A$. So $e=e_1^nx_1$ satisfies $e(1-e)=0$.

4.10. For $\mathfrak{p}\in\operatorname{Spec} A$, if $V(\mathfrak{p})=V(\mathfrak{a})\cup V(\mathfrak{b})$ then $\mathfrak{p}\in V(\mathfrak{p})$ gives $\mathfrak{p}\supset\mathfrak{a}$ or $\mathfrak{p}\supset\mathfrak{b}$, and hence either $V(\mathfrak{p})=V(\mathfrak{a})$ or $V(\mathfrak{p})=V(\mathfrak{b})$. Conversely, if $V(I)$ is irreducible, then for $x, y\in A$ with $xy\in\sqrt{I}$, from $V=V(I+Ax)\cup V(I+Ay)$ we have, say, $V=V(I+Ax)$, and $x\in\sqrt{I}$; this proves $\sqrt{I}\in\operatorname{Spec} A$.

4.11 If there is a closed subset which cannot be so expressed, let V be a minimal one. Then V must be reducible, but if $V=V_1\cup V_2$ with $V_i\neq V$ then, by minimality, each of V_1 and V_2 is a union of a finite number of irreducible closed set, hence also V, a contradiction.

§5.

5.1. Set $k[X_1,\ldots,X_n]/\mathfrak{p}=k[x_1,\ldots,x_n]$; then by Theorem 6, coht $\mathfrak{p}=$ $\operatorname{tr.deg}_k k(x)$. Suppose this is r, and that $x_1,\ldots,x_r$ is a transcendence basis of $k(x)$ over k, and set $K=k(X_1,\ldots,X_r)$; then $k[X_1,\ldots,X_n]_\mathfrak{p}$ is the localisation of $K[X_{r+1},\ldots,X_n]$ at a prime ideal P, with $\operatorname{ht}\mathfrak{p}=\operatorname{ht}P$. This reduces us to proving that if $r=0$ then $\operatorname{ht}\mathfrak{p}=n$. In this case $x_1,\ldots,x_n$ are algebraic over k, and letting $\mathfrak{p}_i$ be the kernel of $k[X_1,\ldots,X_n]\longrightarrow k[x_1,\ldots,x_i, X_{i+1},\ldots,X_n]$ we get a strictly increasing chain $0\subset\mathfrak{p}_1\subset\mathfrak{p}_2$

$\subset \cdots \subset \mathfrak{p}_n = \mathfrak{p}$, giving $\operatorname{ht} \mathfrak{p} \geqslant n$, but by the corollary to Theorem 6, $\operatorname{ht} \mathfrak{p} \leqslant n$.

5.2. If A is a zero-dimensional Noetherian ring then all prime ideals are minimal, and by Ex. 4.12, there are only finitely many of these. Let these be $\mathfrak{p}_1, \ldots, \mathfrak{p}_r$; then since $\mathfrak{p}_1 \ldots \mathfrak{p}_r = \operatorname{nil}(A)$ there is an n such that $(\mathfrak{p}_1 \ldots \mathfrak{p}_r)^n = 0$. For any ideal I and any i, the module $I/I\mathfrak{p}_i$ is a finite-dimensional vector space over $A/\mathfrak{p}_i$, so that $l(I/I\mathfrak{p}_i) < \infty$. It follows easily that $l(A) < \infty$, so that A is Artinian.

§6.

6.1. $\operatorname{Ass} M = \{(0), (3)\}$. ($\supset$ is obvious, $\subset$ from Theorem 3.)

6.2. No. Let M be as in the previous question, $M_1 = \{(a, \bar{a}) | a \in \mathbb{Z}\}$ and $M_2 = \{(a, 0) | a \in \mathbb{Z}\}$; then $M = M_1 + M_2$, but each $M_i \simeq \mathbb{Z}$.

6.3. Since $xA/x^nA \simeq A/x^{n-1}A$, there is an exact sequence $0 \to A/x^{n-1}A \longrightarrow A/x^nA \longrightarrow A/xA \to 0$.

6.4. Use a primary decomposition of I.

§7.

7.2. For $b \in B$ write $b = y/x$ with $x, y \in A$. Then $y = bx \in xB \cap A = xA$ (by Theorem 7.5, (ii)), so $b \in A$.

7.3. Write $N \subset M$ for the A-submodule generated by $\{m_\lambda\}$; then $B \otimes (M/N) = 0$, so $M/N = 0$.

7.4. Set $M = \prod_\lambda M_\lambda$. It is enough to show that $I \otimes M \longrightarrow IM$ is injective for an ideal $I = \sum_1^n a_i A$ of (Theorem 6). Define $f: A^n \longrightarrow A$ by $f(x_1, \ldots, x_n) = \sum a_i x_i$, and set $K = \operatorname{Ker} f$. Then $0 \to K \longrightarrow A^n \xrightarrow{f} A$ is exact, hence also $0 \to K \otimes M_\lambda \longrightarrow (M_\lambda)^n \longrightarrow M_\lambda$, and if $\sum_1^n a_i \otimes \xi_i \in I \otimes M$ satisfies $\sum a_i \xi_i = 0$ then writing $\xi_{i\lambda}$ for the λth coordinate of $\xi_i \in M$ we have $\sum a_i \xi_{i\lambda} = 0$ for all λ, and hence $(\xi_{1\lambda}, \ldots, \xi_{n\lambda}) \in K \otimes M_\lambda$. Now since A is Noetherian we can write $K = A\beta_1 + \cdots + A\beta_r$ with $\beta_j = (b_{1j}, \ldots, b_{nj}) \in K \subset A^n$ for $1 \leqslant j \leqslant r$. Thus we can write $\xi_{i\lambda} = \sum_j b_{ij} \eta_{j\lambda}$ with $\eta_{j\lambda} \in M_\lambda$. Since $\sum_i a_i b_{ij} = 0$, setting $\eta_j = (\eta_{j\lambda})_\lambda \in M$, we get $\xi_i = \sum_{j=1}^r b_{ij} \eta_j$, and $\sum_i a_i \otimes \xi_i = \sum_i \sum_j a_i b_{ij} \otimes \eta_j = 0$.

7.5. Tensor the exact sequence $0 \to A \xrightarrow{a} A$ with N to get the exact sequence $0 \to N \xrightarrow{a} N$.

7.8. Tensor product does not commute with infinite direct products. If $\{p_i\}$ as an infinite set of prime numbers then $\bigcap_i p_i \mathbb{Z} = (0)$, but $\bigcap_i (p_i \mathbb{Z} \otimes \mathbb{Q}) = \bigcap_i \mathbb{Q} = \mathbb{Q}$.

7.9. If I is an ideal of A then $IB \cap A = I$, so that given a chain $I_1 \subset I_2 \subset \cdots$ of ideals of A, $I_1 B \subset I_2 B \subset \cdots$ eventually terminates, hence so does the given chain.

§8.

8.1. $(I + J)^{2n} \subset I^n + J^n$, so that given $x_1, x_2, \ldots$ such that $x_{n+1} - x_n \in (I + J)^{2n}$ we can write $x_{n+1} - x_n = u_n + v_n$ with $u_n \in I^n$ and $v_n \in J^n$. Thus $\{x_n\}$

converges to $x_1+\sum_1^\infty u_i+\sum_1^\infty v_i$. Also, $I,\ J\subset\text{rad}(A)$, so that $I+J\subset\text{rad}(A)$, and therefore $\bigcap_n(I+J)^n=(0)$.

8.2. If $\{x_n\}$ satisfies $x_{n+1}-x_n\in J^n$ then there is a limit x for the I-adic topology. For any i, taking m large enough we get $x_m-x\in I^i$, so that $x_n-x=x_n-x_m+x_m-x\in J^n+I^i$, and since by Theorem 10, (i) we have $\bigcap_i(J^n+I^i)=J^n$, we get $x_n-x\in J^n$, so that x is also a J-adic limit.

8.3. Let $\mathfrak{a}\hat{A}=\alpha\hat{A}$ and $\alpha=\sum a_i\xi_i$ with $a_i\in\mathfrak{a}$ and $\xi_i\in\hat{A}$. Let I be an ideal of definition of A. Take $x_i\in A$ such that $x_i-\xi_i\in I\hat{A}$ and set $a=\sum a_ix_i$. Then $a\in\mathfrak{a}$ and $\mathfrak{a}\hat{A}\subset a\hat{A}+I\mathfrak{a}\hat{A}$, so that by NAK, $\mathfrak{a}\hat{A}=a\hat{A}$, so $\mathfrak{a}=a\hat{A}\cap A=aA$.

8.4. $e=(e-X)(e+eX+eX^2+\cdots)$.

8.7. $A/\mathfrak{m}^n$ is Artinian, so that there exists $t(n)$ such that $\mathfrak{a}_{t(n)}+\mathfrak{m}^n=\mathfrak{a}_j+\mathfrak{m}^n$ for $j>t(n)$. We can assume that $t(n)<t(n+1)<\cdots$. Supposing that $\mathfrak{a}_{t(r)}\not\subset\mathfrak{m}^r$ for some r, then we take $a_r\in\mathfrak{a}_{t(r)}-\mathfrak{m}^r$, then $a_{r+1}\in\mathfrak{a}_{t(r+1)}$ such that $a_{r+1}-a_r\in\mathfrak{m}^r$, and proceed in the same way taking $a_i\in\mathfrak{a}_{t(i)}$ such that $a_i-a_{i-1}\in\mathfrak{m}^{i-1}$ for $i\geqslant r$. Then $\lim a_i$ belongs to $\bigcap_\nu\mathfrak{a}_\nu$, but not to $\mathfrak{m}^r$, which is a contradiction.

8.8. This can be done by following the proof of Theorem 5, and replacing the use of homogeneous polynomials by multihomogeneous polynomials.

8.9. We can assume that A is a Noetherian local ring with maximal ideal P. There is an $x\neq0$ such that $xP=0$, and then since $\bigcap_nP^n=(0)$ there exists c such that $x\notin P^c$. Then if $I\subset P^c$, $I:x=P$.

8.10. Let $\mathfrak{m}=(X,Y)\subset k[X,Y]$ and set $A=k[X,Y]_{\mathfrak{m}}$; let $\varphi(X)=\sum_1^\infty a_iX^i\in k[\![X]\!]$ be transcendental over $k(X)$ and set $\mathfrak{a}_\nu=(X^{\nu+1},\ Y-\sum_1^\nu a_iX^i)$.

§9.

9.1. $B_\mathfrak{p}$ is integral over $A_\mathfrak{p}$, so that any maximal ideal of $B_\mathfrak{p}$ lies over $\mathfrak{p}A_\mathfrak{p}$ and therefore coincides with $PB_\mathfrak{p}$. Hence $B_\mathfrak{p}$ is a local ring, and the elements of $B-P$ are units of $B_\mathfrak{p}$.

9.2. $\leqslant$ from the going-up theorem, and $\geqslant$ from Theorem 3, (ii).

9.3. Replacing A and B by $A_\mathfrak{p}$ and $B_\mathfrak{p}$ we can assume $\mathfrak{p}$ is maximal; then set $k=A/\mathfrak{p}$, so that $B/\mathfrak{p}B$ is a finite k-module, hence an Artinian ring.

9.4. If $ax^n\in A$ for all n then $A[x]$ is a submodule of the finite A-module $a^{-1}A$; if A is Noetherian then $A[x]$ is also a finite A-module.

9.6. Suppose $f=gh$ with $g,\ h\in K[X]$ monic. Roots of g are roots of f, hence integral over A, and expressing the coefficients of g in terms of the roots, we have that the coefficients of g are integral over A; since A is integrally closed, $g\in A[X]$, and similarly for h.

9.8. By Theorem 3, (ii).

9.10. $L[X]$ is a free module over $K[X]$, hence flat; and if L is algebraic over K then $L[X]$ is integral over $K[X]$. The first part follows from the previous two questions, together with Theorem 5. If f,g have a common factor $\alpha(X)$ in $L[X]$ then set $P=(\alpha(X))$, so that ht $P=1$ (it can easily be seen that a non-zero principal prime ideal in a Noetherian integral domain has height

1). Hence ht $\mathfrak{p} \leqslant 1$. But $f, g \in \mathfrak{p}$, so that ht $\mathfrak{p} = 1$. There is an irreducible divisor h of f in $\mathfrak{p}$, and $\mathfrak{p} = (h)$, so $h|g$.

§10.

10.2. Use the previous question and Theorem 7.7.

10.3. In the proof of Theorem 4 we can choose $\mathfrak{p}$ to contain $(\mathfrak{m}_A, y)A[y]$.

10.4. Let $0 \subset \mathfrak{p}_1 \subset \mathfrak{p}_2$ be a strictly increasing chain of prime ideals of R and let $0 \neq b \in \mathfrak{p}_1, a \in \mathfrak{p}_2 - \mathfrak{p}_1$; thus $ba^{-n} \in R$ for all $n > 0$. Take $f = \sum_1^\infty u_i X^i$ to be a root of $f^2 + af + X = 0$. Then $u_1 = -a^{-1}$, and for all i we have $u_i \in a^{-2i+1}R$, so that $bf(X) \in R[\![X]\!]$ but $f(X) \notin R[\![X]\!]$.

10.5. The first part comes from Theorem 1. For the second part, by §9, Lemma 1 the integral closure of R in K is not the whole of K, and therefore coincides with R, so R is integrally closed; on the other hand, for $x \in K - R$ we have $R[x] = K$, hence $x^{-1} \in R[x]$, so that x^{-1} is integral over R and hence in R. Thus R is a valuation ring. If dim $R > 1$ then there is a prime ideal $\mathfrak{p}$ of R distinct from (0) and from the maximal ideal, and thus $R_\mathfrak{p}$ is intermediate between R and K.

10.8. Let $v: L^* \longrightarrow G$ be the additive valuation corresponding to S, and choose $x_1, \ldots, x_e \in L$ such that $v(x_1), \ldots, v(x_e)$ represent the different cosets of G' in G, and $y_1, \ldots, y_f \in S$ such that their images in k are linearly independent over k'. It follows easily from the previous question that the ef elements $x_i y_j$ are linearly independent over K.

10.9. If $S \subset S_1$ then the residue field k_1 of S_1 contains a valuation ring $A \neq k_1$ such that S is the composite of S_1 and A. We have $k \subset A \subset k_1$, but by the previous question k_1 is an algebraic extension field of k, hence integral over A. But A is integrally closed, therefore $A = k_1$, a contradiction.

§11.

11.1. Let B be a valuation ring of $\bar{K}$ dominating A, and G its value group. Then for $\alpha \in \mathfrak{m}_B$ we have $\sqrt{\alpha} \in \mathfrak{m}_B$, so that G has no minimal element. Also it is easy to see that some multiple of $v(\alpha)$ belongs to $v(K^*)$, so that G is Archimedean.

11.2. If B is a valuation ring of L dominating A and G its value group, set $H = v(K^*)$ and $e = [G:H]$. Then $x \in G \Rightarrow ex \in H$. Hence G is isomorphic to a subgroup of H, and $G \simeq \mathbb{Z}$.

11.5. Just use Forster's theorem (5.7).

11.6. By Ex. 9.7, $A = \mathbb{Z}[\sqrt{10}] = \mathbb{Z}[X]/(X^2 - 10)$. Then $A/3A \simeq \mathbb{Z}[X]/(3, X^2 - 1) = (\mathbb{Z}/3\mathbb{Z})[X]/(X-1)(X+1)$, so that $P = (3, \sqrt{10} - 1)$ is a prime ideal of A. This is not principal, since if $P = (\alpha)$ with $\alpha = a + b\sqrt{10}$, then one and sees easily that the norm $N(\alpha) = a^2 - 10b^2$ would have to be ± 3, but this is impossible since the congruence $a^2 \equiv \pm 3 \bmod 5$ has no solution.

11.7. Let $P_1, \ldots, P_r$ be the maximal ideals of A; choose an element $\alpha \in P_1$ such that $\alpha \notin P_1^2 \cup P_2 \cup \cdots \cup P_r$. Then $\alpha A = P_1$, and similarly each of the prime ideals is principal. Thus by Theorem 6 any ideal is principal. (Of course this also follows from Theorem 5.8.)

§12.

12.1. Suppose that L is normal over K, and let $G = \mathrm{Aut}_K(L)$. Let S_1 and S be valuation rings of L dominating R, and let $S_1, S_2, \ldots, S_r$ be the conjugates of S_1 by elements of G, and $A = S_1 \cap \cdots \cap S_r \cap S$. If $S \neq S_i$ for any i then by Ex. 10.9, there are no inclusions among $S_1, \ldots, S_r$, S, and we can apply Theorem 2. Letting $\mathfrak{n}$, $\mathfrak{n}_i$ be the maximal ideals of S, S_i, and setting $\mathfrak{n} \cap A = \mathfrak{p}$, $\mathfrak{n}_i \cap A = \mathfrak{p}_i$ we have $\mathfrak{p}_1 \ldots \mathfrak{p}_r \not\subset \mathfrak{p}$, so that we can choose $x \in \mathfrak{p}_1 \cap \cdots \cap \mathfrak{p}_r$ with $x \notin \mathfrak{p}$. Then $x \notin \mathfrak{n}$, but since $x \in (\mathfrak{n}_1)^{\sigma^{-1}}$ for all $\sigma \in G$, all the conjugates of x over K belong to $\mathfrak{n}_1$, and the coefficients of the minimal polynomial of x over K belong to $\mathfrak{n}_1 \cap K = \mathrm{rad}(R)$. Thus it follows easily that $x \in \mathfrak{n}$, a contradiction.

12.2. By Ex. 10.3, $\bar{R}$ is the intersection of all valuation rings of L dominating R. Ex. 10.9 can easily be extended to the infinite case, so that the second part follows from the first. For the first part, reduce to the finite case and use the previous question and Theorem 2.

12.4. Let $\mathscr{P}$ be the set of height 1 prime ideals of A, and for $\mathfrak{p} \in \mathscr{P}$ set $I_\mathfrak{p} = a_\mathfrak{p} A_\mathfrak{p}$. Then $x \in \tilde{I} \Leftrightarrow xI^{-1} \subset A_\mathfrak{p}$ for all $\mathfrak{p} \in \mathscr{P} \Leftrightarrow x \in a_\mathfrak{p} A$ for all $\mathfrak{p} \in \mathscr{P}$. Hence $\tilde{I}$ is the intersection of $I_\mathfrak{p} \cap A$ taken over the finitely many $\mathfrak{p} \in \mathscr{P}$ such that $I_\mathfrak{p} \neq A_\mathfrak{p}$.

§13.

13.3. Let $P \in \mathrm{Ass}(A)$ be such that $\mathrm{ht}\, P \geqslant 1$, and let $\mathfrak{p}_1, \ldots, \mathfrak{p}_n$ be the prime divisors of (a). If $P \not\subset \mathfrak{p}_i$ for all i, then there exists $x \in P$ such that $(a):x = (a)$. This is a contradiction, since x is a zero-divisor, but if $xy = 0$ then $y \in \bigcap_n a^n A$. Hence $P + (a) \subset \mathfrak{p}_i$ for some i, and then $\mathrm{ht}\, \mathfrak{p}_i \leqslant 2$.

13.4. (i) is easy. (ii) The homogeneous elements of P are nilpotent mod Q^*, hence so are all elements of P. Now we show that if $f \notin P$, $g \notin Q^*$ then $fg \notin Q^*$. Let $f = f_1 + \cdots + f_r$, $g = g_1 + \cdots + g_s$ with f_i and g_j homogeneous, and $\deg f_1 < \deg f_2 < \cdots$, $\deg g_1 < \deg g_2 < \cdots$; we work by induction on $r + s$. If $r = s = 1$ there is nothing to prove. Also, since we can assume that $g_1 \notin Q^*$ we have $g_1 \notin Q$. If $f_1 \notin P$ then $f_1 g_1 \notin Q^*$. Next, suppose $f_1 \in P$. If $f_1 g \in Q^*$ then $fg \in Q^*$, since $(f_2 + \cdots + f_r) g \in Q^*$. If $f_1 g \in Q^*$ then $f_1^t g \notin Q^*$ and $f_1^{t+1} g \in Q^*$ for some $t \geqslant 1$ (since $f_1^n \in Q^*$ for $n \gg 0$). Replacing g by $f_1^t g$ reduces to the case $f_1 g \in Q^*$, so that $f f_1^t g \notin Q^*$.

13.6. First half: let S be the multiplicative set made up of homogeneous elements of R not in P; then $R_S / P^* R_S$ can be viewed as the localisation of R/P^* with respect to all non-zero homogeneous elements, and by the previous question this is $\simeq K[X, X^{-1}]$, which is a one-dimensional ring. Second half: proof by induction on $\mathrm{ht}\, P = n$; take a prime ideal $Q \subset P$ with $\mathrm{ht}\, Q = n - 1$. If $Q \neq P^*$ then Q is inhomogeneous, $Q^* \subset P^*$ and $\mathrm{ht}(P/Q^*) \geqslant 2$, so by the first half, $P^* \neq Q^*$, hence $\mathrm{ht}\, P^* \geqslant \mathrm{ht}\, Q^* + 1 = n - 1$.

§14.

14.7. Let $\mathfrak{p} \in \mathrm{Spec}\, A$, $f, g \in \mathfrak{m} - \mathfrak{p}$, and $r = \dim A/\mathfrak{p}$. If $r = 1$ then $\mathfrak{p} A_f$ is a maximal ideal. Suppose that $r > 1$. We can choose $x_2, \ldots, x_r \in \mathfrak{m}$ such that

$\mathrm{ht}(\mathfrak{p}, f, x_2, \ldots, x_i)/\mathfrak{p} = \mathrm{ht}(\mathfrak{p}, g, x_2, \ldots, x_i)/\mathfrak{p} = i$ for $2 \leqslant i \leqslant r$. Then any minimal prime divisor P of $(\mathfrak{p}, x_2, \ldots, x_r)$ satisfies $\dim A/P = 1, f \notin P$ and $g \notin P$, so $g \notin PA_f \in \text{m-Spec}(A_f)$.

§15.

15.1. Set $Z = X/Y$ so that $X = YZ$, $B = k[Y, Z] \supset A = k[YZ, Y]$, and $\mathfrak{p}B = YB$. So $B/\mathfrak{p}B \simeq k[Z]$ and $\dim B_P/\mathfrak{p}B_P = 1$. Now let $\mathfrak{p}' = (X - \alpha Y)A$ for $0 \neq \alpha \in k$; then any height 1 prime ideal of B containing $X - \alpha Y = Y(Z - \alpha)$ must be YB or $(Z - \alpha)B$, but since $YB \cap A \neq \mathfrak{p}'$ and $(Z - \alpha)B \not\subset P$ there does not exist any prime ideal of B contained in P and lying over $\mathfrak{p}'$.

15.2. No. Set $f = XY - 1$. Then fB is a prime ideal of B, and $fB \cap A = (0)$. Since $fB + XB = B$ there does not exist any prime ideal of B containing fB and lying over XA. Note that the fibres of $A \longrightarrow B$ are all one-dimensional.

§16.

16.1. $\leqslant$ is easy; for $\geqslant$ consider a system of parameters of M'.

16.2. $\mathrm{Hom}_A(A/\mathfrak{a}, A/\mathfrak{b}) = 0$ by Theorem 9.

16.3. For $P \in \mathrm{Ass}(A/I)$, grade $P \geqslant k$ is clear from $P \supset I$. If grade $P > k$ then by Ex. 16.2, $I : P = I$, which is a contradiction.

16.5. Suppose that $(A, \mathfrak{m})$ is local; then if $\mathfrak{m} \in \mathrm{Ass}(A)$ we have depth $A = 0$, but if $\mathrm{ht}\, P > 0$ and $P \notin \mathrm{Ass}(A)$ then depth $A_P > 0$, so that $M = A$ gives a counter-example. For example, $A = k[X, Y, Z]_{(X,Y,Z)}/(X, Y, Z)^2 \cap (Z)$ and $P = (x, z)A$ satisfy these conditions.

16.7. Let $\underline{x} = (x_1, \ldots, x_n)$ be a maximal M-sequence in $\mathfrak{m}$, and set $M' = M/\sum x_i M$; then there exists $0 \neq \xi \in M'$ such that $\mathfrak{m}\xi = 0$. Thus $\mathfrak{m}B\xi = 0$, but $\mathfrak{n}^\nu \subset \mathfrak{m}B$ so $\mathfrak{n}^\nu \xi = 0$ and $\mathfrak{n} \in \mathrm{Ass}_B(M')$, therefore $\underline{x}$ is also a maximal M-sequence in $\mathfrak{n}$.

16.10. (i) For $r = 1$ the proof is similar to Theorem 14.3. If $r > 1$, applying the case $r = 1$ gives that $(a_1, \ldots, a_{r-1})$ is prime, and we can then use induction. (ii) For any $Q \in \mathrm{Ass}(A)$ we have $QA_Q \in \mathrm{Ass}(A_Q)$; if we had $P \subset Q$ then by (i), A_Q is an integral domain, a contradiction. Hence $P \not\subset Q$. Therefore using Ex. 16.8, we see that P can be generated by an A-sequence. For a counter-example let $A = k[x, y, z] = k[X, Y, Z]/(X(1 - YZ))$; $P = (x, y, z) = (y, z) = (y - y^2 z, z)$ is a prime ideal of height 2, but $y - y^2 z$ is a zero-divisor in A.

§17.

17.1. (*b*) Let k be a field; then $A = k[X, Y]/(XY, Y^2)$ is a one-dimensional ring which is not CM.

17.2. x^3, y^3 is an A-sequence, hence also an R-sequence, so that R is CM. The ring $k[x^4, x^3y, xy^3, y^4]$ is not CM.

17.3. By localisation we need only consider the case of an integral domain, and it then follows from Theorem 11.5, (i).

17.4. We can assume that A is a local ring. Since A/J is CM we have $\dim A/J = \operatorname{depth} A/J = r$, and if we set k for the residue class field of A then $\operatorname{Ext}^i_A(k, A/J) = 0$ for $i < r$. Using the exact sequence $0 \to J^\nu/J^{\nu+1} \longrightarrow A/J^{\nu+1} \longrightarrow A/J^\nu \to 0$ and the fact that $J^\nu/J^{\nu+1}$ is isomorphic to a direct sum of a number of copies of A/J we get by induction that $\operatorname{Ext}^i_A(k, A/J^\nu) = 0$ for $i < r$.

17.5. (i) Let $x_1, \ldots, x_r$ be a maximal A-sequence in P, and extend to a maximal A-sequence in $\mathfrak{m}, x_1, \ldots, x_r, y_1, \ldots, y_s$. There exists $Q \in \operatorname{Ass}_A(A/(x_1, \ldots, x_r))$ containing P, so that by Theorem 2, $\dim A/P \geqslant \dim A/Q \geqslant \operatorname{depth} A/(x_1, \ldots, x_r) = s$.

(ii) $\dim A - \operatorname{ht} P \geqslant \dim(A/P) \geqslant \operatorname{depth} A - \operatorname{depth}(P, A) \geqslant \operatorname{depth} A - \operatorname{depth} A_P$.

§18.

18.1. Using Ex. 16.1, we see that A is CM $\Leftrightarrow B$ is CM. Assuming CM, we need only use condition (5′) of Theorem 1 as a criterion.

18.3. Given a prime ideal P of $A[X]$, by localising A at $P \cap A$ and factoring out by a system of parameters we reduce to proving that if $(A, \mathfrak{m}, k)$ is a zero-dimensional Gorenstein local ring, and P a prime ideal of $A[X]$ such that $P \cap A = \mathfrak{m}$ then $B = A[X]_P$ is Gorenstein. Then P is generated by $\mathfrak{m}$ together with a monic polynomial $f(X)$, and the image of f in $k[X]$ is irreducible. Since f is B-regular, if we set $C = B/(f)$ then the maximal ideal of C is $\mathfrak{m}C$, and $C \simeq A[X]/(f)$; this is a free A-module of finite rank, so that $\operatorname{Hom}_C(C/\mathfrak{m}C, C) = \operatorname{Hom}_A(k, A) \otimes_A C$ (by Ex. 7.7) $\simeq C/\mathfrak{m}C$. So C is Gorenstein, therefore B also.

18.4. $R/(x^3, y^3) = k[x^3, y^3, x^2y, xy^2]/(x^3, y^3) \simeq k[U, V]/(U^2, V^2, UV)$. In this ring (0) is not irreducible, so that R is not Gorenstein.

18.5. For $0 \neq a \in A$, the ideal aA is $\simeq A/I$ with $I \neq A$, so there exists a non-zero map $\varphi: aA \longrightarrow k$; viewed as a map $aA \longrightarrow E$, this extends to $A \longrightarrow E$, so that $0 \neq \operatorname{Im} \varphi \subset aE$.

18.6. We can consider M as a submodule of $E = E_A(k)$. By faithfulness, $A \subset \operatorname{Hom}_A(M, M) \subset \operatorname{Hom}_A(M, E)$. But $0 \to \operatorname{Hom}_A(E/M, E) \longrightarrow \operatorname{Hom}_A(E, E) = A \longrightarrow \operatorname{Hom}_A(M, E) \to 0$ is exact, hence $E/M = 0$.

§19.

19.3. If $0 \to P_r \longrightarrow \cdots \longrightarrow P_0 \longrightarrow M \to 0$ is an exact sequence and each P_i is finite and projective, and if $P_0 \oplus A^n \simeq A^m$, then $\cdots \longrightarrow P_2 \longrightarrow P_1 \oplus A^n \longrightarrow P_0 \oplus A^n \longrightarrow M \to 0$ is again exact, with $P_0 \oplus A^n$ free. Proceeding in the same way, adding a free module to P_i at each stage, we get an FFR $0 \to L_{r+1} \longrightarrow L_r \longrightarrow \cdots \longrightarrow L_0 \longrightarrow M \to 0$.

19.4 For every maximal ideal $\mathfrak{m}$ of A, since the A-module $A/\mathfrak{m}$ has an FFR, also the $A_\mathfrak{m}$-module $A_\mathfrak{m}/\mathfrak{m}A_\mathfrak{m}$ has an FFR, so that the projective dimension over $A_\mathfrak{m}$ is finite. Thus $A_\mathfrak{m}$ is a regular local ring.

§20.

20.4. Use Theorem 5 and Ex. 8.3.

20.5. This follows from Theorem 1. If an ideal I is locally principal it is finitely generated, and principal by Theorem 5.8.

§21.

21.1. For $P \in \operatorname{Spec} R$ with $P \supset I$ set $P/I = \mathfrak{p}$; then $A_\mathfrak{p}$ is c.i.$\Leftrightarrow I_P$ is generated by an R_P-sequence$\Leftrightarrow (I/I^2)_\mathfrak{p}$ is free over $A_\mathfrak{p}$. Now I/I^2 is a finite A-module, so that by Theorem 4.10, $\{\mathfrak{p} \in \operatorname{Spec} A \mid (I/I^2)_\mathfrak{p}$ is free$\}$ is an open subset of Spec A.

21.2. We can assume that A is complete. Then $A = R/I$ with R regular and $\dim R = \dim A + 1$. Now $\operatorname{ht} I = 1$ and A is a CM ring, so that all the prime divisors of I have height 1. Since R is a UFD, I is principal.

21.3. $A = k[x, y, z] = k + kx + ky + kz + kx^2$, with $x^2 = y^2 = z^2$ and $xy = yz = zx = 0$. Therefore $0:\mathfrak{m} = kx^2$, and A is a zero-dimensional Gorenstein ring. Set $I = (X^2 - Y^2, Y^2 - Z^2, XY, YZ, ZX)$ and $M = (X, Y, Z)$; then $I/MI \longrightarrow M^2/M^3$ has five-dimensional image, so that at least five elements are needed to generate I.

§22.

22.2. Algebraic independence comes from Theorem 16.2, (i). To prove flatness, setting $I = \sum x_i C$ and using Theorem 3, we reduce to proving that $\operatorname{Tor}_1^C(k, A) = 0$. Since $\underline{x}$ is a C-sequence, the Koszul complex $L_\bullet = K_\bullet(\underline{x}, C)$ constructed from C and $\underline{x}$ is a free resolution of the C-module $k = C/I$, and $\operatorname{Tor}_1^C(k, A) = H_1(L_\bullet \otimes_C A)$. However, $L_\bullet \otimes_C A$ is just the Koszul complex constructed from A and $\underline{x}$, and since $\underline{x}$ is an A-sequence, $H_1(L_\bullet \otimes A) = 0$.

22.3. We need only show that $\operatorname{Tor}_1^A(k, M) = 0$. By Lemma 2 of §18, $\operatorname{Tor}_1^A(k, M) = \operatorname{Tor}_1^{A/xA}(k, M/xM)$, but by assumption the right-hand side is 0.

§24.

24.1. Use $0 \to I^i/I^{i+1} \longrightarrow A/I^{i+1} \longrightarrow A/I^i \to 0$ to deduce that $\operatorname{Ass}(A/I^i) = \operatorname{Ass}(A/I)$ for all i.

24.2. By Theorem 5, it is enough to show that for a prime ideal $\mathfrak{p}$ of A, $\operatorname{CM}(A/\mathfrak{p})$ contains a non-empty open. Let P be the inverse image of $\mathfrak{p}$ in R, so that $A/\mathfrak{p} = R/P$. If $x_1, \ldots, x_n \in P$ are chosen to form a system of parameters of R_P then since R_P is CM, they form an R_P-sequence. Thus passing to a smaller neighbourhood of P, we can assume (i) P is the unique minimal prime divisor of $(\underline{x}) = (x_1, \ldots, x_n)R$, and (ii) $\underline{x}$ is an R-sequence. Now replacing R by $R/(\underline{x})$, we can assume that P is nilpotent; moreover, we can take P^i/P^{i+1} to be free R/P-modules. Now using the previous question it follows easily that R is CM implies R/P is CM.

24.3. After a preliminary reduction as in the previous question, use the proof of Theorem 6.

§25.

25.2. If $xy=0$ we prove that $y\in\bigcap x^nA$; suppose that $y\in x^nA$, and set $y=x^nz$; then $0=D(x^{n+1}z)=(n+1)x^nz+x^{n+1}Dz$, and $y\in x^{n+1}A$.

25.4. $0\to I\longrightarrow A\otimes_k A\longrightarrow A\to 0$ is a split exact sequence, so that $0\to I\otimes_k k'\longrightarrow A\otimes_k A\otimes_k k'=A'\otimes_{k'}A'\longrightarrow A'\to 0$ is also exact, and hence $\Omega_{A'/k}=(I\otimes k')/(I\otimes k')^2=(I/I^2)\otimes_k k'=\Omega_{A/k}\otimes_k k'$. For A_S, use the fact that A_S is 0-etale over A and Theorem 1.

25.5. See Theorem 27.3.

§26.

26.1. (i) Let $\alpha\in K\cap K'$; then $1,\alpha\in K$ are linearly dependent over K', hence also over k, so $\alpha\in k$. (ii) Assume that $\alpha_1,\ldots,\alpha_n\in K$ are linearly independent over k', and that $\sum\alpha_i\xi_i=0$ with $\xi_i\in k'(K')$; we show that $\xi_i=0$. Clearing denominators, we can assume that $\xi_i\in k'[K']$. Choosing a basis $\{\omega_j\}$ of K' over k we can write $\xi_i=\sum c_{ij}\omega_j$ with $c_{ij}\in k'$. Then since $\sum_{i,j}c_{ij}\alpha_i\omega_j=0$ we get $\sum_i c_{ij}\alpha_i=0$, therefore $c_{ij}=0$ for all i,j.

26.2. It is enough to show that $K((T))$ and $L^p((T^p))$ are linearly disjoint over $K^p((T^p))$. Assume that $\omega_1(T),\ldots,\omega_r(T)\in K((T))$ are linearly independent over $K^p((T^p))$, and that $\sum\varphi_i\omega_i=0$ with $\varphi_i\in L^p((T^p))$; we show that $\varphi_i=0$ for all i. Clearing denominators, we can assume that $\omega_i\in K[\![T]\!]$ and $\varphi_i\in L^p[\![T^p]\!]$. Letting $\{\xi_\lambda\}$ be a basis of L over K we can in a unique way write $\varphi_i=\sum\xi_\lambda^p\varphi_{i\lambda}(T^p)$ with $\varphi_{i\lambda}(T^p)\in K^p[\![T^p]\!]$. Here $\sum\xi_\lambda^p\varphi_{i\lambda}$ is in general an infinite sum, but only a finite number of terms appear in the sum for the coefficient of some monomial in the T's, so that the sum is meaningful. Then $\sum_\lambda\xi_\lambda^p(\sum_i\varphi_{i\lambda}(T^p)\omega_i(T))=0$, so that $\sum_i\varphi_{i\lambda}(T^p)\omega_i=0$ for all λ, so $\varphi_{i\lambda}=0$ for all i,λ.

§28.

28.1. Let N be a B-module satisfying $\mathfrak{m}^\nu N=0$, and $D:B\longrightarrow N$ a derivation over A; then D induces a derivation $\bar{D}:B_0=B/\mathfrak{m}B\longrightarrow N/\mathfrak{m}N$. If B_0 is 0-unramified over k then $\bar{D}=0$, so that $D(B)\subset\mathfrak{m}N$. Proceeding as before gives $D(B)\subset\mathfrak{m}^2N,\ldots$ so that $D=0$. The statement about etale is just putting together those for smooth and unramified.

28.2. Since some $D\in\mathrm{Der}(k)$ with $Da\neq 0$ can be extended to a derivation of A, $a\notin A^p$. If k' were a coefficient field containing k then we would have to have $a\in k'^p\subset A^p$. Also, A is 0-smooth over k because $k[X]$ is.

§29.

29.1. Suppose that $C=R[\![t]\!]$ with R a DVR; if we let u be a uniformising element of R then pR is a power of uR, so that pC has the single prime divisor uC. However, in fact, in our case $C/pC=(B/pB)[X]/(X(X+y))$, so that pC has the two prime divisors (p,x) and $(p,x+y)$.

29.2. R is pR-etale over $\mathbb{Z}_{p\mathbb{Z}}$ (see Ex. 28.1).

§30.

30.1. φ is the composite of $E_r: A \longrightarrow A[\![t]\!]$ and of the map $A[\![t]\!] \longrightarrow A$ obtained by substituting $-x$ for t. Since $\varphi(x) = 0$ we have $xA \subset \operatorname{Ker} \varphi$. For any $a \in A$ we can write $\varphi(a) = a + xb$, so that $\varphi(\varphi(a)) = \varphi(a)$, and therefore $C \cap xA = (0)$ and $A = C + xA$, therefore $A = C[\![x]\!]$. Now $E_t(x) = x + t$, and it is easy to see that this is a non-zero-divisor of $A[\![t]\!]$, so that x is a non-zero-divisor of A. If $c_r x^r + c_{r+1} x^{r+1} + \cdots = 0$ with $c_i \in C$ then dividing by x^r we get $c_r \in C \cap xA = (0)$, so that x is analytically independent over C.

30.2. No. If A is 0-smooth over k' then so is the field of fractions L of A, so L/k' is separable, hence also k/k', and this is not the case.

30.3. (1)$\Rightarrow$(3) is easy using Theorem 28.7. If $[k:k^p] = \infty$ then there are counter-examples to (3)$\Rightarrow$(1) (see [G1], (22.7.7).

References

BOOKS

[AB] M. Auslander and M. Bridger, *Stable module theory.* Memoirs of the AMS **94** (1969)

[AM] M.F. Atiyah and I.G. Macdonald, *Introduction to commutative algebra.* Addison-Wesley, 1969

[An 1] M. André, *Méthodes simpliciales en algèbre homologique et algèbre commutative.* LNM **32**, Springer, 1967

[An 2] M. André, *Homologie des algèbres commutatives.* Springer, 1974

[B 1-9] N. Bourbaki, *Algèbre commutative*, Chap. 1-Chap. 9. Hermann, 1961-83

[Conf] *Conference on commutative algebra, Kansas 1972*, LNM **311**, Springer, 1973

[Diff] R. Berger, R. Kiehl, E. Kunz, and H.-J. Nastold, *Differentialrechung in der analytischen Geometrie.* LNM **38**, Springer, 1967

[EG] E. G. Evans and P. Griffith, *Syzygies.* LMS Lecture Note Series **106**, Cambridge University Press, 1985

[F] R. Fossum, *The divisor class group of a Krull domain.* Springer Ergebnisse **74**, 1973

[Ful] W. Fulton, *Intersection Theory.* Ergebnisse d. Math. 3. Folge Vol. 2, Springer, 1984

[G1] A. Grothendieck, *Éléments de géométrie algébrique*, IV_1. Publ. Math. IHES **20**, 1964

[G 2] A. Grothendieck, *Éléments de géométrie algébrique*, IV_2. Publ. Math. IHES **24**, 1965

[G 2] A. Grothendieck, *Éléments de géométrie algébrique*, IV_3. Publ. Math. IHES **28**, 1966

[G 4] A. Grothendieck, *Éléments de géométrie algébrique*, IV_4. Publ. Math. IHES **32**, 1967

[G 5] A. Grothendieck *et al.*, *Cohomologie locale des faisceaux cohérents et théorèmes de Lefschetz locaux et globaux* (SGA 2). North-Holland, 1968

[G 6] A. Grothendieck, *Local cohomology.* LNM **41**, Springer, 1967

[GL] T. H. Gulliksen and G. Levin, *Homology of local rings.* Queen's papers in pure and appl. math., **20**, Queen's Univ. Kingston, 1969

[GS] S. Greco and P. Salmon, *Topics in m-adic topologies.* Springer Ergenbnisse **58**, 1971

[H] M. Hochster, *Topics in the homological theory of modules over commutative rings.* Regional conference series **24**, AMS 1975

[Ha] R. Hartshorne, *Algebraic geometry.* Springer, 1977

[HK] J. Herzog and E. Kunz, *Der kanonische Modul eines Cohen-Macaulay Rings.* LNM **238**, Springer, 1971

[HSV] M. Hermann, R. Schmidt and W. Vogel, *Theorie der normalen Flachheit.* Teubner Texte zur Math., Leipzig, 1977

[J] N. Jacobson, *Lectures in abstract algebra, Vol. 3. Theory of fields and Galois theory.* Van Nostrand, 1964

[K] I. Kaplansky, *Commutative rings.* Allyn and Bacon, 1970

[Kunz 1] E. Kunz, *Introduction to commutative algebra and algebraic geometry.* Birkhäuser, 1985

[Kunz 2] E. Kunz, *Kähler Differentials.* Vieweg Advanced Lectures in Math., Vieweg & Sohn, Braunschweig, 1986

[KPR] H. Kurke, G. Pfister and M. Roczen, *Henselsche Ringe und algebraische Geometrie.* VEB Deutscher Verlag der Wissen., Berlin, 1975

[KPPRM] H. Kurke, G. Pfister, D. Popescu, M. Roczen and T. Mostowski, *Die Approximationseigenschaft lokaler Ringe.* LNM **634**, Springer, 1978

[Kr] W. Krull, *Idealtheorie.* Springer Ergebnisee, 1935

[L] T.Y. Lam, *Serre's conjecture.* LNM **635**, Springer, 1978

[M] H. Matsumura, *Commutative algebra.* Benjamin, 1970 (second edn. 1980)

[McA] S. McAdam, *Asymptotic prime divisors.* LNM **1023**, Springer, 1983

[Mac] F.S. Macaulay, *Algebraic theory of modular systems.* Cambridge Tracts **19**, Cambridge University Press, 1916

[N 1] M. Nagata, *Local rings.* Interscience, 1962

[N 2] M. Nagata, *Lectures on the fourteenth problem of Hilbert.* Tata Inst., 1965

[N 3] M. Nagata, *Polynomial rings and affine spaces.* Regional conference series **37**, AMS 1978

[Nor 1] D.G. Northcott, *Ideal theory.* Cambridge Tracts **42**, Cambridge University Press, 1952

[Nor 2] D.G. Northcott, *An introduction to homological algebra.* Cambridge University Press, 1960

[Nor 3] D.G. Northcott, *Lessons on rings, modules and multiplicities.* Cambridge University Press, 1968

[Nor 4] D.G. Northcott, *Finite free resolutions.* Cambridge Tracts **71**, Cambridge University Press, 1976

[R] M. Raynaud, *Anneaux locaux Henséliens.* LNM **169**, Springer, 1970

[Rob] P. Roberts, *Homological invariants of modules over commutative rings.* Presses de l'Univ. de Montreal, 1980

[Rot] J. Rotman, *Notes on homological algebra.* Van Nostrand Math Studies **26**, 1968

[S 1] P. Samuel, *Algèbre locale.* Mém. Sci. Math. **123**, Gauthier–Villars, 1953

[S 2] P. Samuel, *Lectures on unique factorisation domains.* Tata Inst., 1964

[S 3] P. Samuel, *Méthodes d'algèbre abstraite en géométrie algébrique.* Springer Ergebnisse, 1955

[Sa] J. Sally, *Numbers of generators of ideals in local rings.* Lecture Notes in Pure and Appl. Math. **35**, Marcel Dekker, 1978

[Sch] P. Schenzel, *Dualisierende Komplexe in der lokalen Algebra und Buchsbaum-Ringe.* LNM **907**, Springer 1982

[Sh] I.R. Shafarevich, *Basic algebraic geometry.* Springer, 1974

[Se] J.-P. Serre, *Algèbre locale, multiplicités.* LNM **11**, Springer, 1965

[St] R.P. Stanley, *Combinatorics and commutative algebra.* Birkhäuser, 1983

[Su] S. Suzuki, *Differentials of commutative rings.* Queen's papers in pure and appl. math., **29**, Queen's Univ., Kingston, 1971

[SV] J. Stückrad and W. Vogel, *Buchsbaum Rings and Applications.* VEB Deutscher Verlag d. Wissenschaften, Berlin 1986

[Va] W. Vasconcelos, *Divisor theory in module categories.* North-Holland Math. Studies **14**, North-Holland, 1974

[ZS] O. Zariski and P. Samuel, *Commutative algebra, Vols. I and II.* Van Nostrand 1958, 1960 (new edn. Springer, 1975)

RESEARCH PAPERS

S. Abhyankar

[1] On the valuations centered in a local domain. *Amer. J. Math.* **78** (1956), 321–48.

[2] Local rings of high embedding dimension. *Amer. J. Math.* **89** (1967), 1073–7.

S. Abhyankar, W. Heinzer and P. Eakin

[1] On the uniqueness of the coefficient ring in a polynomial ring. *J. Alg.* **23** (1972), 310–42.

Y. Akizuki

[1] Einige Bemerkungen über primäre Integritätsbereiche mit Teilerkettensatz. *Proc. Phys.-Math. Soc. Japan* **17** (1935), 327–36.

[2] Teilerkettensatz und Vielfachensatz. *Proc. Phys.-Math. Soc. Japan* **17** (1935), 337–45.

[3] Zur Idealtheorie der einartigen Ringbereiche mit dem Teilerkettensatz. *Japan. J. Math.* **14** (1938), 85–102; II. *Japan J. Math.* **14** (1938), 177–87; III. *Japan J. Math.* **15** (1939), 1–11.

M. André

[1] Localisation de la lissité formelle. *manuscripta math.* **13** (1974), 297–307.

Y. Aoyama,

[1] Some basic results on canonical modules. *J. Math. Kyoto Univ.* **23** (1983), 85–94.

M. Auslander

[1] Modules over unramified regular local rings. *Illinois J. Math.* **5** (1961), 631–47.

[2] On the purity of the branch locus. *Amer. J. Math.* **84** (1962), 116–25.

M. Auslander and D. Buchsbaum

[1] Homological dimension in local rings. *Trans. AMS* **85** (1957), 390–405.

[2] Codimension and multiplicity. *Ann. Math.* **68** (1958), 625–57.

[3] Unique factorization in regular local rings. *Proc. Natl. Acad. Sci. USA* **45** (1959), 733–4.

L. Avramov

[1] Flat morphisms of complete intersections. *Soviet Math. Dokl.* **16** (1975), 1413–17.

[2] Complete intersections and symmetric algebras. *J. Alg.* **73** (1981), 248–63.

[3] Homology of local flat extensions and complete intersection defects. *Math. Ann.* **228** (1977), 27–37.

H. Bass

[1] On the ubiquity of Gorenstein rings. *Math. Z.* **82** (1963), 8–28.

M. Boratyński

[1] A note on set-theoretical complete intersection ideals. *J. Alg.* **54** (1978), 1–5

[2] Generating ideals up to radical and system of parameters of graded rings. *J. Alg.* **78** (1982), 20–4.

A. Brezuleanu and C. Ionescu

[1] On the localization theorems and completion of *P-rings*. *Rev. Roum. Math. Pures Appl.* **29** (1984), 371–80.

A. Brezuleanu and N. Radu

[1] Geometric regularity and formal smoothness. *Rend. Accad. Nat. Lincei* **51** (1971), 326–7.

[2] On the localisation of formal smoothness. *Rev. Roum. Math. Pures Appl.* **20** (1975), 189–200.

[3] Excellent rings and good separation of the module of differentials. *Rev. Roum Math. Pures Appl.* **23** (1978), 1455–70.

A. Brezuleanu and C. Rotthaus

[1] Eine Bemerkung über Ringe mit geometrisch normalen formalen Fasern. *Archiv. d. Math.* **39** (1982), 19–27.

M. Brodmann

[1] Rees rings and form rings of almost complete intersections. *Nagoya Math. J.* **88** (1982), 1–16.

[2] Asymptotic stability of Ass(M/I^nM). *Proc. AMS* **74** (1979), 16–18.

M. Brodmann and C. Rotthaus

[1] Über den regulären Ort in ausgezeichneten Ringen. *Math. Z.* **175** (1980), 81–5.

[2] Local domains with bad sets of formal prime divisors. *J. Alg.* **75** (1982), 386–94.

W. Bruns

[1] 'Jede' endliche freie Auflösung ist freie Auflösung eines von drei Elementen erzeugten Ideals. *J. Alg.* **39** (1976), 429–39.

[2] Zur Konstruktion basischer Elemente. *Math. Z.* **172** (1980), 63–75.

[3] The Eisenbud–Evans generalized principal ideal theorem and determinantal ideals. *Proc. AMS* **83** (1981), 19–24.

D. Buchsbaum and D. Eisenbud

[1] What makes a complex exact? *J. Alg.* **25** (1973), 259–68.

[2] Some structure theorems for finite free resolutions. *Adv. in Math.* **12** (1974), 84–139.

[3] What annihilates a module? *J. Alg.* **47** (1977), 231–43.

L. Burch

[1] On ideals of finite homological dimension in local rings. *Proc. Camb. Phil. Soc.* **64** (1968), 941–52.

A. Caruth

[1] A short proof of the principal ideal theorem. *Quart. J. Math. Oxford* **31** (1980), 401.

S. Chase

[1] Direct product of modules. *Trans. AMS* **97** (1960), 457–73.

C. Chevalley

[1] On the theory of local rings. *Ann. Math.* **44** (1943), 690–708.

[2] Intersection of algebraic and algebroid varieties. *Trans. AMS* **57** (1945), 1–85.

L. Chiantini

[1] On form ideals and Artin–Rees condition. *manuscripta math.* **36** (1981), 125–45.

G. Chiriacescu

[1] On the theory of Japanese rings. *Rev. Roum. Math. Pures Appl.* **27** (1982), 945–8.

I.S. Cohen

[1] On the structure and ideal theory of complete local rings. *Trans. AMS* **59** (1946), 54–106.

[2] Commutative rings with restricted minimum condition. *Duke Math. J.* **17** (1950), 27–42.

[3] Lengths of prime ideal chains. *Amer. J. Math.* **76** (1954), 654–68.

I.S. Cohen and A. Seidenberg

[1] Prime ideals and integral dependence. *Bull. AMS* **52** (1946), 252–61.

C. Cowsik and M.V. Nori

[1] Affine curves in characteristic p are set-theoretic complete intersection. *Inv. math.* **45** (1978), 111–14.

E. Davis

[1] Ideals of the principal class, *R*-sequences and a certain monoidal transformation. *Pacific J. Math.* **20** (1967), 197–205.

C. DeConcini, D. Eisenbud and C. Procesi

[1] Young diagrams and determinantal varieties. *Inv. math.* **56** (1980), 126–65.

[2] Hodge algebras. *Astérisque* **91** (1982), 1–87.

C. DeConcini and C. Procesi

[1] A characteristic-free approach to invariant theory. *Adv. in Math.* **21** (1976), 330–54.

J. Dieudonné

[1] On regular sequences. *Nagoya Math. J.* **27** (1966), 355–6.

S. Dutta

[1] Symbolic powers, intersection multiplicity, and asymptotic behaviour of Tor. *J. London Math. Soc.* **28** (1983), 261–81.

[2] Frobenius and multiplicities. *J. Alg.* **85** (1983), 424–48.

S. Dutta, M. Hochster and J. E. McLaughlin

[1] Modules of finite projective dimension with negative intersection multiplicities. *Inv. math.* **79** (1985), 253–91.

J.A. Eagon

[1] Examples of Cohen–Macaulay rings which are not Gorenstein. *Math. Z.* **109** (1969), 109–11.

J.A. Eagon and M. Hochster

[1] *R*-sequences and indeterminates. *Quart. J. Math. Oxford* **25** (1974), 61–71.

J.A. Eagon and D.G. Northcott

[1] Ideals defined by matrices and a certain complex associated with them. *Proc. Roy. Soc. London* **269** (1962), 188–204.

[2] Generically acylic complexes and generically perfect ideals. *Proc. Roy. Soc. London* **299** (1967), 147–72.

P. Eakin

[1] The converse to a well known theorem on Noetherian rings. *Math. Ann.* **177** (1968), 278–82.

B. Eckmann and A. Schopf

[1] Über injektiven Moduln. *Archiv d. Math.* **4** (1953), 75–8.

D. Eisenbud

[1] Some directions of recent progress in commutative algebra. *Proc. Symposia in Pure Math.* **29** (1975), 111–28.

[2] Introduction to algebras with straightening laws. In *Ring Theory and Algebra III*. pp. 243–68. Dekker, 1980.

D. Eisenbud and D. Buchsbaum

[1] Algebra structures for finite free resolutions, and some structure theorems for ideals of codimension 3. *Amer. J. Math.* **99** (1977), 447–85.

[2] Gorenstein ideals of height 3. In Seminar Eisenbud/Singh/Vogel, *Teubner Texte zur Math. Bd.* **48**, 30–48. Teubner, Leipzig, 1980.

D. Eisenbud and G. Evans

[1] Every algebraic set in n-space is the intersection of n hypersurfaces. *Inv. math.* **19** (1973), 107–12.

[2] Generating modules efficiently: Theorems from algebraic K-theory. *J. Alg.* **27** (1973), 278–305.

[3] A generalized principal ideal theorem. *Nagoya Math. J.* **62** (1976), 41–53.

G. Evans, and P. Griffith

[1] The syzygy problem. *Ann. Math.* **114** (1981), 323–33.

[2] Order ideals of minimal generators. *Proc. AMS* **86** (1982), 375–8.

G. Faltings
[1] Ein einfacher Beweis, dass geometrische Regularität formale Glattheit impliziert. *Arch. d. Math.* **30** (1978), 284–5.
[2] Über Annulatoren lokaler Kohomologiegruppen. *Arch. d. Math.* **30** (1978), 473–6.
[3] Zur Existenz dualisierender Komplexe. *Math. Z.* **162** (1978), 75–86.
[4] Über Macaulayfizierung. *Math. Ann.* **238** (1978), 175–92.

D. Ferrand
[1] Suite régulière et intersèction complète. *C.R. Acad. Paris* **264** (1967), 427–8.
[2] Trivialisation des modules projectifs. La méthode de Kronecker. *J. Pure Appl. Alg.* **24** (1982), 261–4.

D. Ferrand and M. Raynaud
[1] Fibres formelles d'un anneau local noethérien. *Ann. Sci. Ecole Norm. Sup.* **3** (1970), 293–311.

M. Fiorentini
[1] On relative regular sequences. *J. Alg.* **18** (1971), 384–9.

H. Fitting
[1] Die Determinantenideale eines Moduls. *Jahresberichte Deut. Math. Verein.* **46** (1936), 195–228.

H. Flenner
[1] Die Sätze von Bertini für lokale Ringe. *Math. Ann.* **229** (1977), 97–111.

E. Formanek
[1] Faithful Noetherian modules, *Proc. AMS* **41** (1973), 381–3.

O. Forster
[1] Über die Anzahl der Erzeugenden eines Ideals in einem Noetherschen Ring. *Math. Z.* **84** (1964), 80–7.

R. Fossum
[1] Duality over Gorenstein rings. *Math. Scand.* **26** (1970), 165–76.
[2] The structure of indecomposable injective modules. *Math. Scand.* **36** (1975), 291–312.
[3] Commutative extensions by canonical modules are Gorenstein rings. *Proc. AMS* **40** (1973), 395–400.

R. Fossum and H.-B. Foxby
[1] The category of graded modules. *Math. Scand.* **35** (1974), 288–300.

R. Fossum, H.-B. Foxby, P. Griffith and I. Reiten
[1] Minimal injective resolutions with applications to dualizing modules and Gorenstein modules. *Publ. Math. IHES* **45** (1976), 193–215.

H.-B. Foxby
[1] On the μ^i in a minimal injective resolution. *Math. Scand.* **29** (1971), 175–186, II: *Math. Scand.* **41** (1977), 19–44.
[2] Injective modules under flat base change. *Proc. AMS* **50** (1975), 23–7.
[3] Isomorphisms between complexes with applications to the homological theory of modules. *Math. Scand.* **40** (1977), 5–19.
[4] Bounded complexes of flat modules. *J. Pure Appl. Alg.* **5** (1979), 149–72.
[5] A homological theory of complexes of modules. Preprint series No. 19a, 19b., Copenhagen Univ. Math. Inst., 1981.

H.-B. Foxby and A. Thorup
[1] Minimal injective resolutions under flat base change. *Proc. AMS* **67** (1977), 27–31.

O. Goldman
[1] Hibert rings and the Hilbert Nullstellensatz. *Math. Z.* **54** (1951), 136–40.

S. Goto
[1] On Buchsbaum rings. *J. Alg.* **67** (1980), 272–9.
[2] On Buchsbaum rings obtained by gluing, *Nagoya Math. J.* **83** (1981), 123–35.
[3] Buchsbaum rings of multiplicity 2. *J. Alg.* **74** (1982), 494–508.
[4] On the associated graded rings of parameter ideals in Buchsbaum rings. *J. Alg.* **85** (1983), 490–534.
[5] A note on quasi-Buchsbaum rings. *Proc. AMS* **90** (1984), 511–16.

S. Goto and Y. Shimoda
[1] On the Rees algebra of Cohen–Macaulay local rings. In R. Draper (ed.), *Commutative algebra, analytical methods*, pp. 201–231. Dekker, 1982.

S. Goto and N. Suzuki
[1] Index of reducibility of parameter ideals in a local rings. *J. Alg.* **87** (1984), 53–88.

S. Goto, N. Suzuki and K. Watanabe
[1] On affine semigroup rings. *Japan. J. Math.* **2** (1976), 1–12.

S. Goto and K. Watanabe
[1] On graded rings, I, II. I *J. Math. Soc. Japan* **30** (1978), 179–213; II ($\mathbb{Z}^n$-graded rings) *Tokyo J. Math.* **1** (1978), 237–61.
[2] The structure of one-dimensional *F*-pure rings. *J. Alg.* **49** (1977), 415–31.

S. Greco
[1] Henselization of a ring with respect to an ideal. *Trans. AMS* **144** (1969), 43–65.
[2] Two theorems on excellent rings. *Nagoya Math. J.* **60** (1976), 139–46.
[3] A note on universally catenary rings. *Nagoya Math. J.* **87** (1982), 95–100.

S. Greco and M. Marinari
[1] Nagata's criterion and openness of loci for Gorenstein and complete intersections. *Math. Z.* **160** (1978), 207–16.

S. Greco and N. Sankaran
[1] On the separable and algebraic closedness of a Hensel couple in its completion. *J. Alg.* **39** (1976), 335–48.

S. Greco and C. Traverso
[1] On seminormal schemes. *Compositio Math.* **40** (1980), 325–65.

L. Gruson and M. Raynaud
[1] Critères de platitude et de projectivité. *Inv. math.* **13** (1971), 1–89.

R. Hartshorne
[1] Complete intersections and connectedness. *Amer. J. Math.* **84** (1962), 497–508.
[2] A property of *A*-sequences. *Bull. Soc. Math. France* **94** (1966), 61–6.

L. Harper
[1] On differentiably simple algebras. *Trans. AMS* **100** (1961), 63–72.

H. Hasse and F.K. Schmidt
[1] Noch eine Begründung der Theorie der höheren Differential-quotienten in einem algebraischen Funktionenkörper einer Unbestimmten. *J. reine u. angew. Math.* **177** (1937), 215–37.

W. Heinzer
[1] Some remarks on complete integral closure. *J. Australian Math.* **9** (1969), 310–14.
[2] Quotient overrings of an integral domain. *Mathematika* **17** (1970), 139–48.

W. Heinzer and J. Ohm
[1] Locally Noetherian commutative rings. *Trans. AMS* **158** (1971), 273–84.
[2] Noetherian intersections of integral domains. *Trans. AMS* **167** (1972), 291–308.

M. Herrmann and S. Ikeda

.1] Remarks on lifting of Cohen–Macaulay property. *Nagoya Math. J.* **92** (1983), 121–32.

[2] On the Gorenstein property of Rees algebras. *manuscripta math.* **59** (1987), 471–90.

M. Herrmann and U. Orbanz

[1] Faserdimension von Aufblasungen lokaler Ringe und Äquimultiplizität, *J. Math. Kyoto Univ.* **20** (1980), 651–9.

[2] On equimultiplicity. *Math. Proc. Camb. Phil. Soc.* **91** (1982), 207–13.

[3] Two notes on flatness. *manuscripta math.* **40** (1982), 109–33.

M. Herrmann and R. Schmidt

[1] Zur Transitivität der normalen Flachheit. *Inv. math.* **28** (1975), 129–36.

B. Herzog

[1] On a relation between the Hilbert functions belonging to a local homomorphism. *J. London Math. Soc.* **25** (1982), 458–66.

J. Herzog

[1] Generators and relations of Abelian semigroups and semigroup rings, *manuscripta math.* **3** (1970), 175–93.

[2] Ein Cohen–Macaulay-Kriterium mit Anwendungen auf den Konormalenmodul und den Differentialmodul. *Math. Z.* **163** (1978), 149–62.

[3] Deformationen von Cohen-Macaulay Algebren. *J. reine angew. Math.* **318** (1980), 83–105.

J. Herzog and E. Kunz

[1] Die Werthalbgruppe eines lokalen Rings der Dimension 1. *S.-B. Heidelberger Akad. Wiss.* (1971), 27–67.

J. Herzog, A. Simis and W. Vasconcelos

[1] Approximation complexes of blowing-up rings. *J. Alg.* **74** (1982), 466–93; II. *J. Alg.* **82** (1983), 53–83.

T. Hibi

[1] Every affine graded ring has a Hodge algebra structure. *Rend. Sem. Mat. Univers. Politecn. Torino*, **44–2** (1986), 277–86.

[2] Distributive lattices, affine semigroup rings and algebras with straightening laws. In *Commutative Algebra and Combinatorics*, Advanced Studies in Pure Math. **11**, 1987, pp. 93–109. North-Holland, Amsterdam.

D. Hilbert

[1] Über die Theorie der algebraischen Formen. *Math. Ann.* **36** (1890), 471–534.

[2] Über die vollen Invariantensysteme. *Math. Ann.* **42** (1893), 313–73.

H. Hironaka

[1] Resolution of singularities of an algebraic variety over a field of characteristic zero. *Ann. Math.* **79** (1964), 109–326.

M. Hochster

[1] Rings of invariants of tori, Cohen–Macaulay rings generated by monomials, and polytopes. *Ann. Math.* **96** (1972), 318–37.

[2] Non-openness of loci in Noetherian rings. *Duke Math. J.* **40** (1973), 215–19.

[3] Contracted ideals from integral extensions of regular rings. *Nagoya Math. J.* **51** (1973), 25–43.

[4] Constraints on systems of parameters. In *Ring Theory*, Proc. Oklahoma Conf., pp. 121–61, Dekker, 1974.

[5] Cohen–Macaulay rings, combinatorics, and simplicial complexes. In *Ring Theory II*, Proc. Ohio Conf., pp. 171–223. Dekker, 1977.

[6] The Zariski–Lipman conjecture for homogeneous complete intersections. *Proc. AMS* **49** (1975), 261–2.

[7] Cyclic purity versus purity in excellent Noetherian rings. *Trans. AMS* **231** (1977), 464–88.
[8] Some applications of the Frobenius in characteristic 0. *Bull. AMS* **84** (1978), 886–912.
[9] Canonical elements in local cohomology modules and the direct summand conjecture. *J. Alg.* **84** (1983), 503–53.

M. Hochster and J. Eagon
[1] Cohen–Macaulay rings, invariant theory, and generic perfection of determinantal loci. *Amer. J. Math.* **93** (1971), 1020–58.

M. Hochster and C. Huneke
[1] Tightly closed ideals. *Bull. AMS* **18** (1988), 45–8.

M. Hochster and L.J. Ratliff
[1] Five theorems on Macaulay rings. *Pacific J. Math.* **44** (1973), 147–72.

M. Hochster and J.L. Roberts
[1] Rings of invariants of reductive groups acting on regular rings are Cohen–Macaulay. *Adv. in Math.* **13** (1974), 115–75.

C. Hopkins
[1] Rings with minimum condition for left ideals. *Ann. Math.* **40** (1939), 712–30.

C. Huneke
[1] The theory of *d*-sequences and powers of ideals. *Adv. in Math.* **46** (1982), 249–79.
[2] On the symmetric and Rees algebra of an ideal generated by a *d*-sequence. *J. Alg.* **62** (1980), 268–75.
[3] A remark concerning multiplicities. *Proc. AMS* **85** (1982), 331–2.
[4] Linkage and the Koszul homology of ideals. *Amer. J. Math.* **104** (1982), 1043–62.
[5] Numerical invariants of liaison classes. *Inv. Math.* **75** (1984), 301–25.

C. Huneke and B. Ulrich
[1] Divisor class groups and deformations. *Amer. J. Math.* **107** (1985), 1265–303.
[2] The structure of linkage. *Ann. Math.* **126** (1987), 277–334.

S. Ikeda
[1] The Cohen–Macaulayness of the Rees algebras of local rings. *Nagoya Math. J.* **89** (1983), 47–63.
[2] On the Gorensteinness of Rees algebras over local rings. *Nagoya Math. J.* **102** (1986).

F. Ischebeck
[1] Eine Dualität zwischen den Funktoren Ext und Tor. *J. Alg.* **11** (1969), 510–31.

B. Iversen
[1] Amplitude inequalities for complexes. *Ann. sci. Éc. Norm. Sup.* **10** (1977), 547–58.

P. Jothilingam
[1] A note on grade. *Nagoya Math. J.* **59** (1975), 149–52.

V. Kac and K. Watanabe
[1] Finite linear groups whose ring of invariants is a complete intersection. *Bull. AMS* **6** (1982), 221–3.

I. Kaplansky
[1] Modules over Dedekind rings and valuation rings. *Trans. AMS* **72** (1952), 327–40.
[2] Projective modules. *Ann. Math.* **68** (1958), 372–7.

M. Kersken

[1] Cousinkomplex und Nennersysteme, *Math. Z.* **182** (1983), 389–402.

T. Kimura and H. Niitsuma

[1] On Kunz's conjecture. *J. Math. Soc. Japan* **34** (1982), 371–8.

D. Kirby

[1] Coprimary decomposition of Artinian modules. *J. London Math. Soc.* **6** (1973), 571–6.

K.H. Kiyek

[1] Anwendung von Ideal-Transformationen. *manuscripta math.* **34** (1981), 327–53.

W. Krull

[1] Primidealketten in allgemeinen Ringbereichen. *S.-B. Heidelberg Akad. Wiss.* **7** (1928).

[2] Ein Satz über primäre Integritätsbereiche. *Math. Ann.* **103** (1930), 450–65.

[3] Allgemeine Bewertungstheorie. *J. reine angew. Math.* **167** (1931), 160–96.

[4] Über die Zerlegung der Hauptideale in allgemeinen Ringen. *Math. Ann.* **105** (1931), 1–14.

[5] Beiträge zur Arithmetik kommutativer Integritätsbereiche. I. *Math. Z.* **41** (1936), 545–77; II. *Math. Z.* **41** (1936), 665–79; III. *Math. Z.* **42** (1937), 745–66; IV. *Math. Z.* **42** (1937), 767–73; V. *Math. Z.* **43** (1938), 768–82; VI. *Math. Z.* **45** (1939), 1–19; VII. *Math. Z.* **45** (1940), 319–34; VIII. *Math. Z.* **48** (1942/43), 533–52.

[6] Dimensionstheorie in Stellenringen. *J. reine angew. Math.* **179** (1938), 204–26.

[7] Jacobsonsche Ringe, Hilbertscher Nullstellensatz, Dimensionstheorie. *Math. Z.* **54** (1951), 354–87.

E. Kunz

[1] Characterizations of regular local rings of characteristic *p*. *Amer J. Math.* **91** (1969), 772–84.

[2] On Noetherian rings of characteristic *p*. *Amer. J. Math.* **98** (1976), 999–1013

[3] Almost complete intersections are not Gorenstein rings. *J. Alg.* **28** (1974), 111–15

A. Kustin and M. Miller

[1] Deformation and linkage of Gorenstein algebras. *Trans. AMS* **284** (1984), 501–23.

[2] Tight double linkage of Gorenstein algebras. *J. Alg.* **95** (1985), 384–97.

A. Kustin, M. Miller and B. Ulrich

[1] Linkage theory for algebras with pure resolutions. *J. Alg.* **102** (1986), 199–228.

K. Langman and C. Rotthaus

[1] Über den regulärem Ort bei japanischen lokalen Ringen. *manuscripta math.* **27** (1979), 19–30.

A. Lascoux,

[1] Syzygies des variétés déterminantales. *Adv. in Math.* **30** (1978), 202–37.

T. Larfeldt and C. Lech

[1] Analytic ramifications and flat couples of local rings. *Acta math.* **146** (1981), 201–8.

D. Lazard

[1] Autour de la platitude. *Bull. Soc. Math. France* **97** (1969), 81–128.

C. Lech

[1] Inequalities related to certain couples of local rings. *Acta math.* **112** (1964), 69–89.

G. Levin and W. Vasconcelos

[1] Homological dimensions and Macaulay rings. *Pacific J. Math.* **25** (1968), 315–23.

S. Lichtenbaum

[1] On the vanishing of Tor in regular local rings. *Illinois J. Math.* **10** (1966), 220–6.

J. Lipman

[1] Free derivation modules on algebraic varieties. *Amer. J. Math.* **87** (1965), 874–98.

[2] On the Jacobian ideal of the module of differentials. *Proc. AMS* **21** (1969), 422–6.

I.G. MacDonald

[1] Secondary representation of modules over a commutative ring. *Symposia Mathematica* **XI** (1973), 23–43.

I.G. MacDonald and R.Y. Sharp

[1] An elementary proof of the non-vanishing of certain local cohomology modules. *Quart. J. Math. Oxford* **23** (1972), 197–204.

R. MacRae

[1] On the homological dimension of certain ideals. *Proc. AMS* **14** (1963), 746–50.

[2] On an application of the Fitting invariants. *J. Alg.* **2** (1965), 153–69.

S. McAdam

[1] Finite coverings by ideals. In *Ring Theory*, Proc. Oklahoma Conf, pp. 163–71, Dekker, 1974.

[2] Going down. *Duke J. Math.* **39** (1972), 633–6.

[3] Saturated chains in Noetherian rings. *Indiana Univ. Math. J.* **23** (1974), 719–28.

[4] Two applications of asymptotic prime divisors. *Proc. AMS* **84** (1982), 179–80.

S. McAdam and P. Eakin

[1] The asymptotic Ass. *J. Alg.* **61** (1979), 71–81.

J. Marot

[1] Sur les anneaux universellement japonais. *C. R. Acad. Paris*, **277** (1973), 1029–31; *C.R. Acad. Paris* **278** (1974), 1169–72.

C. Massaza and P. Valabrega

[1] Sul apertura di luoghi in uno schema localmente noetheriano. *Boll. UMI* **14-A** (1977), 564–74.

J. Matijevic

[1] Maximal ideal transforms of Noetherian rings, *Proc. AMS* **54** (1976), 49–52.

J. Matijevic and P. Roberts

[1] A conjecture of Nagata on graded Cohen–Macaulay rings. *J. Math. Kyoto Univ.* **14** (1974), 125–8.

E. Matlis

[1] Injective modules over Noetherian rings. *Pacific J. Math.* **8** (1958), 511–28.

[2] Reflexive domains. *J. Alg.* **8** (1968), 1–33.

[3] The Koszul complex and duality. *Comm. Alg.* **1** (1974), 87–144.

[4] The higher properties of *R*-sequences. *J. Alg.* **50** (1978), 77–112.

H. Matsumura

[1] Formal power series rings over polynomial rings, I. In *Number Theory, Algebraic Geometry and Commutative Algebra in honor of Y. Akizuki*, pp. 511–20. Kinokuniya, Tokyo, 1973.

[2] Noetherian rings with many derivations. In *Contributions to Algebra (dedicated to E. Kolchin)*, pp. 279–94. Academic Press, New York, 1977.

[3] Quasi-coefficient rings. *Nagoya Math. J.* **68** (1977), 123–30.

[4] Integrable derivations. *Nagoya Math. J.* **87** (1982), 227–45.

N. Mohan Kumar

[1] Complete intersections. *J. Math. Kyoto Univ.* **17** (1977), 533–8.

[2] On two conjectures about polynomial rings. *Inv. math.* **46** (1978), 225–36.

Y. Mori

[1] On the integral closure of an integral domain, I, II. I *Mem. Coll. Sci. Univ. Kyoto* **27** (1953), 249–56; II *Bull. Kyoto Gakugei Univ.* **7** (1955), 19–30.

P. Murthy

[1] A note on factorial rings. *Arch. d. Math.* **15** (1964), 418–21.

K.R. Nagarajan

[1] Groups acting on Noetherian rings. *Nieuw Archief voor Wiskunde* **16** (1968), 25–29.

M. Nagata

[1] On the theory of Henselian rings, I, II, III. I *Nagoya Math. J.* **5** (1953), 45–57; II *Nagoya Math. J.* **7** (1954), 1–19; III *Mem. Coll. Sci. Univ. Kyoto* **32** (1959/60), 93–101.

[2] Basic theorems on general commutative rings. *Mem. Coll. Sci. Univ. Kyoto* **29** (1955), 59–77.

[3] On the derived normal rings of Noetherian integral domains. *Mem. Coll. Sci. Univ. Kyoto* **29** (1955), 293–303.

[4] The theory of multiplicity in general local rings. In *Proc. Intern. Symp. Tokyo-Nikko 1955*, Science Council of Japan, 1956, pp. 191–226.

[5] On the chain problem of prime ideals. *Nagoya Math. J.* **10** (1956), 51–64.

[6] A. Jacobian criterion of simple points. *Illinois J. Math.* **1** (1957), 427–32.

[7] On the purity of branch loci in regular local rings. *Illinois J. Math.* **3** (1959), 328–33.

[8] On the closedness of singular loci. *Publ. Math. IHES* **2** (1959), 29–36.

[9] A type of subrings of a Noetherian ring. *J. Math. Kyoto Univ.* **8** (1968), 465–7.

[10] Flatness of an extension of a commutative ring. *J. Math Kyoto Univ.* **9** (1969), 439–48.

Y. Nakai

[1] Theory of differentials in commutative rings. *J. Math. Soc. Japan* **13** (1961), 68–84.

[2] High order derivations I. *Osaka J. Math.* **7** (1970), 1–27.

[3] On a ring with plenty of high order derivations. *J. Math. Kyoto Univ.* **13** (1972), 159–64.

[4] On locally finite iterative higher derivations. *Osaka J. Math.* **15** (1978), 655–62.

Y. Nakai, K. Kosaki and Y. Ishibashi

[1] High order derivations II. *J. Sci. Hiroshima Univ. Ser. A-I* **34** (1970), 17–27.

E. Noether

[1] Idealtheorie in Ringbererreichen. *Math. Ann.* **83** (1921), 24–66.

[2] Abstrakter Aufbau der Idealtheorie in algebraischen Zahl- und Funktionenkörpern. *Math. Ann.* **96** (1926), 26–61.

J. Nishimura

[1] Note on Krull domain. *J. Math. Kyoto Univ.* **15** (1975), 397–400.

[2] Note on integral closures of a Noetherian integrai domain. *J. Math. Kyoto Univ.* **16** (1976), 117–122.

[3] On ideal-adic completion of Noetherian rings. *J. Math. Kyoto Univ.* **2** (1981), 153–69.

M. Nomura

[1] Formal power series rings over polynomial rings, II. In *Number Theory, Algebraic Geometry and Commutative Algebra in honor of Y. Akizuki*, pp. 521–8. Kinokuniya, Tokyo, 1973

D.G. Northcott

[1] On unmixed ideals in regular local rings. *Proc. London Math. Soc.* **3** (1953), 20–8.

[2] A note on classical ideal theory. *Proc. Camb. Phil. Soc.* **51** (1955), 766–7.

[3] On homogeneous ideals. *Proc. Glasgow Math. Assoc.* **2** (1955), 105–11.

[4] On irreducible ideals in local rings. *J. London Math. Soc.* **32** (1957), 82–8.

[5] Syzygies and specializations. *Proc. London Math. Soc.* **15** (1965), 1–25.

[6] Hilbert functions and the Koszul complex. *Bull. London Math. Soc.* **2** (1970), 69–72.

[7] Grade sensitivity and generic perfection. *Proc. London Math. Soc.* **20** (1970), 597–618.

[8] Injective envelopes and inverse polynomials. *J. London Math. Soc.* **8** (1974), 290–296.

[9] Generalized Koszul complexes and Artinian modules. *Quart. J. Math. Oxford* **23** (1972), 289–297.

D.G. Northcott and D. Rees

[1] Reductions of ideals in local rings. *Proc. Camb. Phil. Soc.* **50** (1954), 145–58.

[2] A note on reductions of ideals with an application to the generalized Hilbert function. *Proc. Camb. Phil. Soc.* **50** (1954), 353–9.

[3] Principal systems. *Quart. J. Math. Oxford* **8** (1957), 119–27.

[4] Extensions and simplifications of the theory of regular local rings. *J. London Math. Soc.* **32** (1957), 367–74.

L. O'Carroll

[1] Generalized fractions, determinantal maps, and top cohomology modules. *J. Pure Appl. Alg.* **32** (1984), 59–70.

[2] On the generalized fractions of Sharp and Zakeri. *J. London Math. Soc.* **28** (1983), 417–27.

T. Ogoma

[1] Non-catenary pseudo-geometric normal rings. *Japan. J. Math.* **6** (1980), 147–63.

[2] Cohen–Macaulay factorial domain is not necessarily Gorenstein. *Mem. Fac. Sci. Kochi Univ.* (*Math.*) **3** (1982), 65–74.

[3] Some examples of rings with curious formal fibres. *Mem. Fac. Sci. Kochi Univ. Math.* **1** (1980), 17–22.

[4] Existence of dualizing complexes. *J. Math. Kyoto Univ.* **24** (1984), 27–48.

J. Ohm

[1] Semi-valuations and groups of divisibility. *Can. J. Math.* **21** (1969), 576–91.

U. Orbanz

[1] Höhere Derivationen und Regularität. *J. reine angew. Math.* **262/263** (1973), 194–204.

C. Peskine

[1] Une généralisation du 'main theorem' de Zariski. *Bull. Sci. Math.* **90** (1966), 119–27.

C. Peskine and L. Szpiro

[1] Dimension projective finie et cohomologie locale. *Publ. Math. IHES* **42** (1973), 47–119.

[2] Liaison des variétés algébriques. I. *Inv. math.* **26** (1974), 271–302.

[3] Syzygies et multiplicité. *C. R. Acad. Sci. Paris*, Ser. A-B. **27B** (1974), 1421–4.

G. Pfister

[1] Schlechte Henselsche Ringe. *Bull. Acad. Polon. Sci.* **25** (1977), 1083–8.

G. Pfister and D. Popescu

[1] Die strenge Approximationseigenschaft lokaler Ringe. *Inv. math.* **30** (1975), 145–74.

[2] Die Approximationseigenschaft von Primidealen. *Bull. Acad. Polon. Sci.* **27** (1979), 771–8.

D. Popescu

[1] General Néron desingularization. *Nagoya Math. J.* **100** (1985), 97–126.

[2] General Néron desingularization and approximation. *Nagoya Math. J.* **104** (1986), 85–115.

J. Querré

[1] Sur un théorème de Mori–Nagata. *C.R. Acad. Paris* **285** (1977), 323–4.

[2] Sur une propriété des anneaux de Krull. *Bull. Sci. Math. France*, **95** (1971), 341–54.

D. Quillen

[1] Projective modules over polynomial rings. *Inv. math.* **36** (1976), 167–71.

S. Rabinowitch

[1] Zum Hilbertschen Nullstellensatz. *Math. Ann.* **102** (1929), 520.

N. Radu

[1] Un critère differentiel de lissité formelle. *C.R. Acad. Paris* **272** (1971), 1166–8.

[2] Une caracterisation des algèbres noetheriennes régulières sur un corps de caracteristique zero. *C.R. Acad. Paris* **270** (1970), 851–3.

[3] Sur les algèbres dont le module des différentielles est plat. *Rev. Roum. Math. Pures et Appl.* **21** (1976), 933–9.

[4] Sur un critère de lissité formelle. *Bull. Math. Soc. Sci. Roum.* **21** (1977), 133–5.

[5] Sur la structure des algèbres locales noethériennes formellement lisses. *Analele Univ. Bucuresti, Mathematica*, **29** (1980), 81–4.

L.J. Ratliff

[1] On quasi-unmixed local domains, the altitude formula, and the chain condition for prime ideals. I *Amer. J. Math.* **91** (1969), 508–28; II *Amer. J. Math.* **92** (1970), 99–144.

[2] Characterizations of catenary rings. *Amer. J. Math.* **93** (1971), 1070–108.

[3] Catenary rings and the altitude formula. *Amer. J. Math.* **94** (1972), 458–66.

D. Rees

[1] Valuations associated with a local ring. I *Proc. London Math. Soc.* **5** (1955), 107–28; II *J. London Math. Soc.* **31** (1956), 228–35.

[2] Valuations associated with ideals. I *Proc. London Math. Soc.* **6** (1956), 161–74: II *J. London Math. Soc.* **31** (1956), 221–8.

[3] Two classical theorems of ideal theory. *Proc. Camb. Phil. Soc.* **52** (1956), 155–7.

[4] A theorem of homological algebra, *Proc. Camb. Phil. Soc.* **52** (1956), 605–10.

[5] The grade of an ideal or module. *Proc. Camb. Phil. Soc.* **53** (1957), 28–42.

[6] A note on form rings and ideals. *Mathematika* **4** (1957), 51–60.

[7] A note on general ideals. *J. London Math. Soc.* **32** (1957), 181–6.

[8] On a problem of Zariski. *Illinois J. Math.* **2** (1958), 145–9.

[9] A note on analytically unramified local rings. *J. London Math. Soc.* **36** (1961), 24–8.

[10] Rings associated with ideals and analytic spread. *Proc. Camb. Phil. Soc.* **89** (1981), 423–32.

D. Rees and R.Y. Sharp

[1] On a theorem of B. Teissier on multiplicities of ideals in local rings. *J. London Math. Soc.* **18** (1978), 449–63.

G.A. Reisner

[1] Cohen–Macaulay quotients of polynomial rings. *Adv. in Math.* **21** (1976), 30–49.

G. Restuccia and H. Matsumura

[1] Integrable derivations in rings of unequal characteristic. *Nagoya Math. J.* **93** (1984), 173–8.

L. Robbiano and G. Valla

[1] On normal flatness and normal torsion-freeness. *J. Alg.* **43** (1976), 552–60.

P. Roberts

[1] Two applications of dualizing complexes over local rings. *Ann. Sci. Éc. Norm. Sup.* **9** (1976), 103–6.

[2] The vanishing of intersection multiplicities of perfect complexes. *Bull. AMS* **13** (1985), 127–30.

[3] Le théorème d'intersection. *C. R. Acad. Sci. Paris*, **304** (1987), 177–80.

C. Rotthaus

[1] Nicht ausgezeichnete, universell japanische Ringe. *Math. Z.* **152** (1977), 107–25.

[2] Universell japanische Ringe mit nicht offenem regulärem Ort. *Nagoya Math. J.* **74** (1979), 123–35.

[3] Komplettierung semilokaler quasiausgezeichneter Ringe. *Nagoya Math. J.* **76** (1979), 173–80.

[4] Zur Komplettierung ausgezeichneter Ringe. *Math. Ann.* **253** (1980), 213–26.

[5] Potenzreihenerweiterung und formale Fasern in lokalen Ringen mit Approximationseigenschaft. *manuscripta math.* **42** (1983), 53–65.

J. Sally

[1] On the number of generators of powers of an ideal. *Proc. AMS* **53** (1975), 24–6.

[2] Bounds for numbers of generators of Cohen–Macaulay ideals. *Pac. J. Math.* **63** (1976), 517–20.

[3] On the associated graded ring of a local Cohen–Macaulay ring. *J. Math. Kyoto Univ.* **17** (1977), 19–21.

[4] Cohen–Macaulay local rings of maximal embedding dimension. *J. Alg.* **56** (1979), 168–83.

[5] Reductions, local cohomology and Hilbert functions of local rings. In *Commutative Algebra, Durham 1981*, pp. 231–41; Cambridge University Press 1982.

P. Samuel

[1] La notion de multiplicité en algèbre et en géométrie algébrique. *J. math. pure et appl.* **30** (1951), 159–274.

[2] On unique factorization domains. *Illinois J. Math.* **5** (1961), 1–17.

[3] Sur les anneaux factoriels. *Bull. Soc. Math. France* **89** (1961), 155–73.

G. Scheja and U. Storch

[1] Differentielle Eigenschaften der Lokalisierungen analytischer Algebren. *Math. Ann.* **197** (1972), 137–70.

[2] Über Spurfunktionen bei vollständigen Durchschnitten. *J. reine angew. Math.* **278/279** (1975), 174–90.

P. Schenzel

[1] Applications of dualizing complexes to Buchsbaum rings. *Adv. in Math.* **44** (1982), 61–77.

[2] Multiplizitäten in verallgemeinerten Cohen–Macaulay-Moduln. *Math. Nachr.* **88** (1979), 295–306.

P. Schenzel, Trung, Ngo viet and Cuong, Nguyen tu

[1] Verallgemeinerte Cohen–Macaulay-Moduln. *Math. Nachr.* **85** (1978), 57–73.

A. Seidenberg

[1] Derivations and integral closure. *Pacific J.* **16** (1966), 167–73.

J.-P. Serre

[1] Sur la dimension homologique des anneaux et des modules noethériens. *Proc. Intern. Symp., Tokyo-Nikko, 1955*, Science Council of Japan, 1956, 175–89.

H. Seydi

[1] Anneaux henséliens et conditions de chaînes. I *Bull. Soc. Math. France* **98** (1970), 9–31; II *C.R. Acad. Paris* **270** (1970), 696–8; III *Bull. Soc. Math. France* **98** (1970), 329–36.

[2] Sur deux théorèmes d'algèbre commutative. *C.R. Acad. Paris* **271** (1970), 1105–8.

[3] Sur deux théorémes d'algébre commutative. *C.R. Acad. Paris* **271** (1970), 1169–72.

[4] Un théorème de descente effective universelle er une application. *C.R. Acad. Paris* **270** (1970), 801–3.

R.Y. Sharp

[1] The Cousin complex for a module over a commutative Noetherian ring. *Math. Z.* **112** (1969), 340–56.

[2] Gorenstein modules. *Math. Z.* **115** (1970), 117–39.

[3] Local cohomology theory in commutative algebra. *Quart. J. Math. Oxford* **21** (1970), 425–34.

[4] Dualizing complexes for commutative Noetherian rings. *Math. Proc. Camb. Phil. Soc.* **78** (1975), 369–86.

[5] Acceptable rings and homomorphic images of Gorenstein rings. *J. Alg.* **44** (1977), 246–61.

[6] A commutative Noetherian ring which possesses a dualizing complex is acceptable. *Math. Proc. Camb. Phil. Soc.* **82** (1977), 197–213.

[7] On the attached prime ideals of certain. Artinian local cohomology modules. *Proc. Edinburgh Math. Soc.* **24** (2) (1981), 9–14.

[8] Secondary representations for injective modules over commutative Noetherian rings. *Proc. Edinburg Math. Soc.* **20** (2) (1976), 143–51

R.Y. Sharp and P. Vamos

[1] The dimension of the tensor product of a finite number of field extensions. *J. Pure. Appl. Alg.* **10** (1977), 249–52.

R.Y. Sharp and H. Zakeri

[1] Modules of generalized fractions. *Mathematika* **29** (1982), 32–41.

[2] Modules of generalized fractions and balanced big Cohen–Macaulay modules. In R.Y. Sharp (ed.), *Commutative Algebra, Durham 1981* pp. 61–82; Cambridge University Press, 1982.

Y. Shimoda

[1] A note on Rees algebras of two-dimensional local domains. *J. Math. Kyoto Univ.* **19** (1979), 327–33.

B. Singh,

[1] On two numerical criteria for normal flatness in codimension one. *Math. Nachr.* **83** (1978), 241–6.

[2] Maximally differential prime ideals in a complete local ring. *J. Alg.* **82** (1983), 331–9.

R. Stanley

[1] Relative invariants of finite groups generated by pseudoreflections. *J. Alg.* **49** (1977), 134–48.

[2] Hilbert functions of graded algebras. *Adv. in Math.* **28** (1978), 57–83.

U. Storch

[1] Bemerkung zu einem Satz von M. Kneser. *Arch. d. Math.* **23** (1972), 403–4.

J. Stückrad

[1] Über die kohomologische Charakterisierung von Buchsbaum-Moduln. *Math. Nachr.* **95** (1980), 265–72.

J. Stückrad and W. Vogel

[1] Toward a theory of Buchsbaum singularities. *Amer. J. Math.* **100** (1978), 727–46.

N. Suzuki

[1] On the generalized local cohomology and its duality. *J. Math. Kyoto Univ.* **18** (1978), 71–85.

[2] On the Koszul complex generated by a system of parameters for a Buchsbaum module. *Bull. Dept. General Education of Shizuoka College of Pharmacy* **8** (1979), 27–35.

[3] Canonical duality for Buchsbaum modules – an application of Goto's lemma on Buchsbaum modules. *Bull. Dept. General Education of Shizuoka College of Pharmacy* **13** (1984), 47–60.

[4] On a basic theorem for quasi-Buchsbaum modules. *Bull. Dept. General Education of Shizuoka College of Pharmacy* **11** (1982), 33–40.

S. Suzuki

[1] Some results on Hausdorff m-adic modules and m-adic differentials. *J. Math. Kyoto Univ.* **2** (1963), 157–82.

[2] On torsion of the module of differentials of a locality which is a complete intersection. *J. Math. Kyoto Univ.* **4** (1966), 471–5.

[3] On Negger's numbers of discrete valuation rings. *J. Math. Kyoto Univ.* **11** (1971), 373–5.

R. Swan

[1] The number of generators of a module. *Math. Z.* **102** (1967), 318–22.

[2] On seminormality. *J. Alg.* **67** (1980), 210–29.

H. Tanimoto

[1] Some characterisations of smoothness. *J. Math. Kyoto Univ.* **23** (198), 695–706.

[2] Smoothness of Noetherian rings. *Nagoya Math. J.* **95** (1984), 163–79.

J. Tate

[1] Homology of Noetherian rings and local rings. *Illinois J. Math.* **1** (1957), 246–61.

B. Teissier

[1] Cycles evanescents, sections plane et conditions de Whitney, *Astérisque* **7/8** (1973), 285–362.

[2] Sur une égalité à la Minkowski pour les multiplicités. *Ann. Math.* **106** (1977), 38–44.

[3] On a Minkowski-type inequality for multiplicities – II. In *C.P. Ramanujam – a tribute*, pp. 347–61. Tata Inst. Studies in Math. **8**, Springer, 1978.

Trung, Ngo viet

[1] On the symbolic powers of determinantal ideals. *J. Alg.* **58** (1979), 361–9.

[2] A class of imperfect prime ideals having the equality of ordinary and symbolic powers. *J. Math. Kyoto Univ.* **21** (1981), 239–50.

[3] Absolutely superficial sequences. *Math. Proc. Camb. Phil. Soc.* **93** (1983), 35–47.

B. Ulrich

[1] Gorenstein rings as specializations of unique factorization domains. *J. Alg.* **86** (1984), 129–40.

[2] Gorenstein rings and modules with high number of generators. *Math. Z.* **188** (1984), 23–32.
[3] Liaison and deformation. *J. Pure Appl. Alg.* **39** (1986), 165–75.
[4] Rings of invariants and linkage of determinantal ideals. *Math. Ann.* **274** (1986), 1–17.

P. Valabrega
[1] On the excellent property for power series rings over polynomial rings. *J. Math. Kyoto Univ.* **15** (1975), 387–95.
[2] A few theorems on completion of excellent rings. *Nagoya Math. J.* **61** (1976), 127–33.
[3] Formal fibres and openness of loci. *J. Math. Kyoto Univ.* **18** (1978), 199–208.

G. Valla
[1] Certain graded algebras are always Cohen–Macaulay. *J. Alg.* **42** (1976), 537–48.
[2] On the symmetric and Rees algebras of an ideal. *manuscripta math.* **30** (1980), 239–55.

P. Valabrega and G. Valla
[1] Form rings and regular sequences. *Nagoya Math. J.* **72** (1978), 93–101.

U. Vetter,
[1] Generische determinantielle Singularitäten: homologische Eigenschaften des Derivationenmoduls. *manuscripta math.* **45** (1984), 161–91.

W. Vasconcelos
[1] Ideals defined by *R*-sequences. *J. Alg.* **6** (1967), 309–16.
[2] On finitely generated flat modules. *Trans. AMS* **138** (1969), 505–12.

W. Vogel
[1] Über eine Vermutung von D.A. Buchsbaum. *J. Alg.* **25** (1973), 106–12.
[2] A non-zero-divisor characterization of Buchsbaum modules. *Michigan Math. J.* **28** (1981), 147–55.

J. Watanabe
[1] Some remarks on Cohen–Macaulay rings with many zero-divisors. *J. Alg.* **39** (1976), 1–14.
[2] A note on Gorenstein rings of embedding codimension three. *Nagoya Math. J.* **50** (1973), 227–32.
[3] *m*-full ideals. *Nagoya Math. J.* **106** (1987), 101–11.
[4] The Dilworth number of artinian rings and finite posets with rank function. In *Commutative Algebra and Combinatorics*, Advanced Studies in Pure Math. **11** (1987), pp. 303–12. North-Holland, Amsterdam.

K. Watanabe
[1] Some examples of one dimensional Gorenstein domains. *Nagoya Math. J.* **49** (1973), 101–9.
[2] Certain invariant subrings are Gorenstein, I, II. I *Osaka J. Math.* **11** (1974), 1–8; II *Osaka J. Math.* **11** (1974), 379–88.
[3] Invariant subrings which are complete intersections, I. Invariant subrings of finite Abelian groups. *Nagoya Math. J.* **77** (1980), 89–98.

K. Watanabe, T. Ishikawa, S. Tachibana and K. Otsuka
[1] On tensor products of Gorenstein rings. *J. Math. Kyoto Univ.* **15** (1975), 387–95.

K. Watanabe and D. Rotillon
[1] Invariant subrings of $\mathbb{C}[X, Y, Z]$ which are complete intersections. *manuscripta math.* **39** (1982), 339–57.

Y. Yoshino

[1] Brauer–Thrall type theorem for maximal Cohen–Macaulay modules. *J. Math. Soc. Japan* **39** (1987), 719–39.

Yuan, Shuen

[1] Differentiably simple rings of prime characteristic. *Duke, Math. J.* **31** (1964), 623–30.

[2] Inseparable Galois theory of exponent one. *Trans. AMS* **149** (1970), 163–70.

[3] On logarithmic derivations. *Bull. Soc. math. France*, **96** (1968), 41–52.

O. Zariski

[1] Analytical irreducibility of normal varieties. *Ann. Math.* **49** (1948), 352–61.

[2] Sur la normalité analytique des variétés normales. *Ann. Inst. Fourier* **2** (1950), 161–4.

[3] On the purity of the branch locus of algebraic functions. *Proc. Natl. Acad. Sci. U.S.* **44** (1958), 791–6.

dex

Bold numbers **nn** indicate a definition